Benchmark Papers in Systematic and Evolutionary Biology

Series Editor: Howell V. Daly
University of California, Berkeley

PUBLISHED VOLUMES AND VOLUMES IN PREPARATION

MULTIVARIATE STATISTICAL METHODS: Among-Groups Covariation
William R. Atchley and Edwin H. Bryant
MULTIVARIATE STATISTICAL METHODS: Within-Groups Covariation
Edwin H. Bryant and William R. Atchley
NUMERICAL TAXONOMY
M. W. Moss
ZOOGEOGRAPHY
E. I. Schlinger
CONCEPTS OF SPECIES
C. N. Slobodchicoff

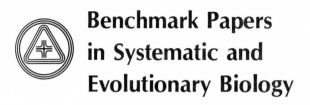

Benchmark Papers in Systematic and Evolutionary Biology / 1

A BENCHMARK® Books Series

MULTIVARIATE STATISTICAL METHODS
Among-Groups Covariation

Edited by

WILLIAM R. ATCHLEY
Texas Tech University

and EDWIN H. BRYANT
University of Houston

Dowden, Hutchinson & Ross, Inc.
Stroudsburg, Pennsylvania

Distributed by
HALSTED PRESS *A Division of John Wiley & Sons, Inc.*

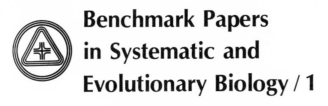

Benchmark Papers in Systematic and Evolutionary Biology, Volume 1
Library of Congress Catalog Card Number: 75-9893
ISBN: 0-470-03595-1

77 76 75 1 2 3 4 5
Manufactured in the United States of America.

LIBRARY OF CONGRESS CATALOGING IN PUBLICATION DATA
Main entry under title:

Multivariate statistical methods, among-groups covariation

(Benchmark papers in systematic and evolutionary biology ; 1)
1. Biometry--Addresses, essays, lectures. 2. Multivariate analysis--Addresses, essays, lectures. 3. Variation (Biology)--Mathematical models--Addresses, essays, lectures. I. Atchley, William R. II. Bryant, Edwin H.
QH323.5.M838 519.5'3 75-9893
ISBN 0-470-03595-1

Exclusive Distributor: **Halsted Press**
A Division of John Wiley & Sons, Inc.

ACKNOWLEDGMENTS AND PERMISSIONS

ACKNOWLEDGMENTS

CROP SCIENCE SOCIETY OF AMERICA—*Crop Science*
The Races of Maize: II. Use of Multivariate Analysis of Variance to Measure Morphological Similarity

GROWTH PUBLISHING COMPANY, INC.—*Growth*
A Note on Geographical Variation in European *Rana*

NATIONAL INDIAN SCIENCE ACADEMY—*Proceedings of the National Institute of Science of India*
On the Generalized Distance in Statistics

THE SOCIETY FOR THE STUDY OF EVOLUTION—*Evolution*
Changes in Microgeographic Variation Patterns of *Pemphigus populitransversus* over a Six-Year Span
Multivariate Geographical Variation in the Wolf *Canis lupus* L.

THE SOCIETY OF SYSTEMATIC ZOOLOGY—*Systematic Zoology*
Adaptive Hierarchical Clustering Schemes
Distance Analysis in Biology
An Empirical Comparison of Three Ordination Techniques in Numerical Taxonomy
Karyology and Morphometrics of Peters' Tent-Making Bat, *Uroderma bilobatum* Peters (Chiroptera, Phyllostomatidae)
On the Cophenetic Correlation Coefficient

THE UNIVERSITY OF KANSAS—*University of Kansas Science Bulletin*
A Statistical Method for Evaluating Systematic Relationships

PERMISSIONS

The following papers have been reprinted with permission of the authors and copyright holders.

ACADEMIC PRESS, INC.—*Multivariate Analysis,* Volume II
A Comparison of Some Methods of Simultaneous Inference in MANOVA

THE BIOMETRIC SOCIETY—*Biometrics*
A Comparison of Some Methods of Cluster Analysis
Growth-Invariant Discriminant Functions and Generalized Distances

BLACKWELL SCIENTIFIC PUBLICATIONS LTD., OXFORD (FOR THE BRITISH ECOLOGICAL SOCIETY)—*Journal of Ecology*
Multivariate Methods in Plant Ecology: I. Association-Analysis in Plant Communities
Multivariate Methods in Plant Ecology: V. Similarity Analysis and Information-Analysis

Acknowledgments and Permission

THE BRITISH PSYCHOLOGICAL SOCIETY and O. POREBSKI—*The British Journal of Mathematical and Statistical Psychology*
On the Interrelated Nature of the Multivariate Statistics Used in Discriminatory Analysis

DUKE UNIVERSITY PRESS (FOR THE ECOLOGICAL SOCIETY OF AMERICA)—*Ecology*
A Multivariate Statistical Approach to the Hutchinsonian Niche: Bivalve Molluscs of Central Canada

THE INSTITUTE OF MATHEMATICAL STATISTICS—*Annals of Mathematical Statistics*
The Generalization of Student's Ratio
Some Contributions to Anova in One or More Dimensions

STICHTING INTERNATIONAL BUREAU FOR PLANT TAXONOMY AND NOMENCLATURE, UTRECHT—*Taxon*
The Comparison of Dendrograms by Objective Methods
Perspectives on the Application of Multivariate Statistics to Taxonomy

N. JARDINE and R. SIBSON—*Computer Journal*
The Construction of Hierarchic and Non-hierarchic Classifications

PRESIDENT AND FELLOWS OF HARVARD COLLEGE—*Harvard Educational Review*
The Generalized Discriminant Function: Mathematical Formulation and Computational Routine

PSYCHOMETRIC SOCIETY—*Psychometrika*
Multidimensional Scaling by Optimizing Goodness of Fit to a Nonmetric Hypothesis

ROYAL STATISTICAL SOCIETY—*The Journal of the Royal Statistical Society*
The Utilization of Multiple Measurements in Problems of Biological Classification

UNIVERSITY COLLEGE, LONDON and UNIVERSITY OF ADELAIDE, AUSTRALIA—*Annals of Eugenics*
The Use of Multiple Measurements in Taxonomic Problems

SERIES EDITOR'S PREFACE

The volumes of the Benchmark Series in Systematics and Evolutionary Biology will make accessible to students and scientists alike a broad selection of topics from these interrelated fields. While other series will be primarily ecological, genetic, or behavioral, this series will concern the classification of organisms, techniques useful in systematics, and those processes or aspects of biological relationships that are essentially evolutionary in nature. Within the past decade or two, entirely new areas of research have emerged as techniques were developed and used and concepts altered by discoveries. Many exciting developments have sprung from the juncture of systematic and evolutionary biology with other disciplines: biogeography and the geological drift of continents; molecular evolution and immunological cross-reactions; genetics of natural populations and the biochemical electrophoresis of proteins; evolutionary strategies and mathematical models. In these examples and others, we are witnessing the convergence of investigations on the central questions of evolution by scientists of disparate backgrounds. It is not an easy task for the undergraduate or researcher to acquire information drawn from so many sources. This Benchmark series will not only provide reproductions of selected original papers, but will assist in interpreting and indexing current, interdisciplinary contributions.

Other volumes will be mainly historical, because systematists retain a strong interest in the early attempts to cope with classifying organisms and to understand their lineages. The development of the theory of evolution, nature of species, and biological nomenclatures are among these retrospective topics planned for the series. Facsimile reproductions and translations will make classic papers readily available, some for the first time.

The first two volumes in this series were jointly prepared by William R. Atchley and Edwin H. Bryant. With the advent of electronic computers, methods of multivariate statistics are now widely employed in the solution of biological problems. New applications in systematics, ecology, and genetics are being recognized as more biologists become familiar with both the techniques and the machines. Although young, Atchley and Bryant are nevertheless seasoned investigators and carry with them the excellent training of the University of Kansas, which has been a center for biometrical instruction and research. They have divided the subject and senior authorship as follows: The first volume, by Atchley and Bryant, deals with covariation of measurements among groups; the second volume, by Bryant and Atchley, concerns covariation within groups. Consequently, systematists who wish to compare the geographic variation of populations, phenetic similarity of taxa, or devise identification procedures will find Part I especially useful. Where the subgroups are not yet known or the interrelationships of characters are of major concern, the sections of Part II on principal-component, factor, and regression analysis will be of interest. Both volumes should be at hand because the methods can be used in a variety of contexts.

HOWELL V. DALY

PREFACE

The purpose of this volume of readings in multivariate statistics is to accumulate some of the more important classical papers on the analysis of among-groups covariation and integrate them with recent applications from the field of biology. We have primarily concerned ourselves with those techniques that have greatest applicability to the disciplines of systematics and ecology. The battery of techniques examined in this volume includes multivariate analysis of variance, discriminant analysis, cluster analysis, and nonmetric scaling. A companion volume, published separately in the series, is concerned with those multivariate procedures relating to within-groups covariation.

In compiling this small set of what we construe to be benchmark papers in multivariate statistics, we were forced to select from among a large number of excellent articles. In the end, we selected primarily those papers relating to statistical theory which seemed to us to represent either the most important historical developments or in some cases the most lucid summaries on a particular facet of that method of analysis. We also tried to include papers dealing with biological applications which discuss clearly and in some detail the methodology in question and its application to a specific biological problem.

The greatest obstacle in this endeavor, however, other than the practical one of being able to include only a limited number of papers, was to surmount the problem that a large segment of the biologists in systematics and ecology (including the authors) have limited preparation in mathematics. Thus, if we included only the classical papers in mathematical statistics, the volumes would have very limited utility to the average biologist. However, we feel that the same would be true had we included only papers on the applications of multivariate methods in biology. It is our opinion that too many biologists analyze quantitative data by imitation of previous work and without any real understanding of the underlying assumptions of the procedures being used. This is an unfortunate drawback of the wide availability of computers and easily used multivariate statistics programs.

In general, biologists have begun only relatively recently to use multivariate statistical procedures to analyze complicated sets of data.

This is no doubt because larger electronic computers have been widely available only about the last 15 years or so. As a result, the papers that deal with applications of multivariate techniques in systematics and ecology are generally of much more recent vintage than the theoretical papers.

The statistical applications in biology included in this volume tend to be of two types. In some instances we have included studies which might seem to be rather "elementary" in that they work through the analysis in a stepwise fashion using a calculator and only a small number of variables. These papers were included because of their great heuristic value to biologists having little familiarity with multivariate statistics. The remaining papers dealing with applications, however, involve much more complicated sets of biological data and, we hope, convey evidence as to the widespread usefulness of multivariate techniques for the analysis of complex biological phenomena. Finally, we selected papers because of the statistical applications and no attempt was made to evaluate the biological validity of the conclusions that were reached.

By way of organization, each topic is briefly introduced, a set of theoretical papers given, and then papers with biological applications are presented. In each case we have tried to justify the inclusion of a paper, and in some cases why other papers were not included. Finally, a short bibliography of additional references on theory and applications has been included.

We have not provided a formal theoretical introduction to each major statistical technique but rather have generally used a historical approach in introducing the subject matter. Since this is a volume of "classical" papers, we felt that this approach was justified. For those readers who might be unfamiliar with these methods, we have provided a brief description of the primary aims of each technique in the form of a "taxonomic key" as well as reprinting a recent survey paper by F. J. Rohlf on multivariate analysis. Further, in each part there is generally at least one reprinted paper that contains a lucid theoretical introduction to that particular type of analysis.

The chapter on multivariate analysis of variance suffers considerably due to the absence of three very important papers by J. Wishart, S. S. Wilks, and K. R. Gabriel. An unfortunate policy by the Trustees of Biometrika prevented us from including these papers, and we ask the readers to overlook these unfortunate omissions.

In conclusion, we would like to express our deep appreciation to our colleague, F. James Rohlf, who read a draft of this volume and offered many valuable suggestions. Any omissions or errors, however, remain our own.

WILLIAM R. ATCHLEY
EDWIN H. BRYANT

CONTENTS

PART II: DISCRIMINANT ANALYSIS

PART III: CLUSTER ANALYSIS

CONTENTS BY AUTHOR

INTRODUCTION

Most biologists are familiar with statistical procedures involving a single variable. By quantifying the uncertainty of inductive inference, these methods permit one to evaluate various hypotheses about statistical populations by sample observations. For example, based upon observed differences in bill width among samples of certain finches, a biologist may conclude that the finch populations are heterogeneous with respect to bill width. This conclusion may then lead him to speculate further on possible adaptive variation in feeding habits of these birds. More likely, he would wish to consider several characters before making any general conclusions on such adaptive divergence. Because organisms are integrated units where characters are intercorrelated to varying degrees, the actual information content per additional character may be much less than one might initially think, however. By not taking character interdependency into account, numerous univariate analyses will misleadingly overestimate the dimensionality of divergence. Hence, these characters should not be analyzed separately, but rather the correct approach must necessarily be multivariate.

In general, multivariate procedures are concerned with data in which several variables have been assessed for each object under study. In systematics these objects may be specimens of one or more species collected from one or more localities and the variables would be the characters recorded on these specimens. In ecology, the objects may be localities or collection sites and the "variables" may be the species residing at each of these sites. Whatever kinds of variables or objects are under study, the basic data matrix displays the observed values of the variables on the objects. A typical data matrix will then consist of N rows representing the objects in the sample(s) and p columns, representing the variables.

An important distinction in multivariate analysis is whether the N objects represent samples from a single population or from several populations. For example, a data matrix in systematics may contain measurements of several variables on specimens of one species from a single locality, or on specimens from several localities of the same or different species. In the first case one might be interested in size and shape relationships among individuals within a population, while in the second, one may wish to describe the pattern of adaptive divergence among populations. This book concerns analyses of among-group covariation; a companion volume (Bryant and Atchley, 1975) concerns those analyses appropriate to within-group covariation. To clarify the relationships and distinctions among the various multivariate procedures, we have provided a brief "key" to the analyses covered in both books. In addition, we have reprinted a short article by F. J. Rohlf (Paper 1), which gives an excellent nonmathematical introduction to multivariate methods in taxonomy. The reader may also consult Reyment (1969) for a similar discussion of multivariate methods and Sokal (1965) for a general review of statistical methods in systematics. We also provide a list of general texts in multivariate methods for the reader who wishes a more rigorous introduction to these methods.

One difficulty for biologists attempting to study multivariate methods is that one needs to have a good working knowledge of matrix algebra to fully appreciate the various derivations. We therefore encourage readers with limited background in this area to consult a basic text on matrix algebra, such as Searle (1966), Horst (1963), or Graybill (1969), before attempting to read many of the more theoretical papers that we have included in this volume. Another paper that may be helpful is one by J. C. Gower (1967), which discusses the geometric implications of several multivariate procedures.

The immense computations required for most multivariate methods necessitates computer assistance. Today, however, most computer centers carry subroutines for standard multivariate procedures, and so, few biologists will be faced with writing their own programs. A biologist can then devote his efforts to posing meaningful multivariate questions and choosing the appropriate methodology for answering them. We hope that the papers included in this volume increase the understanding necessary for making such an appropriate choice.

KEY TO MULTIVARIATE ANALYSES

1.	The objects were sampled from more than one population	2
	The objects were sampled from a single population	3
2.	The main purpose of the analysis is to determine if the samples could have been drawn from a single statistical population; i.e., are the mean vectors of the populations equal?	*multivariate analysis of variance*
	The main purpose of the analysis is to find linear combinations of the variables that maximize differences among preexisting populations; i.e., one wishes to sort the objects into their appropriate populations with minimal error	*discriminant analysis*
	The main purpose of the analysis is to sort a previously unpartitioned heterogeneous collection of objects into a series of sets; i.e., one wishes to identify sets and allocate objects to these sets simultaneously	*cluster analysis*
	The main purpose of the analysis is to arrange the objects graphically in few dimensions, while retaining maximal fidelity to the original interobject relationships	*nonmetric scaling*
3.	The variables can be logically divided into two (or more) sets and one wishes to establish maximal linear relationships among these sets	*multiple regression and correlation*

	The variables logically belong to a homogeneous set	**4**
4.	The main purpose of the analysis is to describe parsimoniously the total variance in a sample in a few dimensions; i.e., one wishes to reduce the dimensionality of the original data while minimizing loss of information. These few dimensions are the linear combinations of the original variables that successively account for the major independent patterns of variation in the sample	*principal components*
	The main purpose of the analysis is to resolve the intercorrelations among variables into their putative underlying causes; i.e. one wishes to reproduce only the intercorrelations among variables rather than their total variances	*factor analysis*

BIBLIOGRAPHY

Anderson, T. W. 1958. *An Introduction to Multivariate Statistical Analysis.* John Wiley & Sons, Inc., New York. 374 pp.

Bartlett, M. S. 1965. Multivariate statistics, in *Theoretical and Mathematical Biology* (T. H. Waterman and H. J. Morowitz, eds.). Xerox College Publishing, Lexington, Mass. pp. 201–224.

Blackith, R. E., and R. A. Reyment. 1971. *Multivariate Morphometrics.* Academic Press, New York. 412 pp.

Bryant, E. H., and W. R. Atchley. 1975. *Multivariate Statistical Methods: Within-Groups Covariation.* Dowden, Hutchinson & Ross, Inc., Stroudsburg, Pa.

Cooley, W. W., and P. R. Lohnes. 1962. *Multivariate Procedures for the Behavioral Sciences.* John Wiley & Sons, Inc., New York. 211 pp.

———, and P. R. Lohnes. 1971. *Multivariate Data Analysis.* John Wiley & Sons, Inc., New York. 364 pp.

Cramer, E. M., and R. D. Bock. 1966. Multivariate analysis. *Rev. Educ. Res. 36:*604–614.

Dempster, A. P. 1969. *Elements of Continuous Multivariate Analysis.* Addison-Wesley Publishing Company, Inc., Menlo Park, Calif. 388 pp.

———. 1971. An overview of multivariate data analysis. *J. Multivariate Anal. 1:*316–346.

DuBois, P. H. 1957. *Multivariate Correlation Analysis.* Harper & Row, Inc., New York. 202 pp.

Gower, J. C. 1967. Multivariate analysis and multidimensional geometry. *The Statistician 17:*13–28.

Graybill, F. A. 1969. *Introduction to Matrices with Applications in Statistics.* Wadsworth Publishing Co., Inc., Belmont, Calif. 372 pp.

Horst, P. 1963. *Matrix Algebra for Social Scientists.* Holt, Rinehart and Winston, Inc., New York. 517 pp.

Kendall, M. G. 1969. *A Course in Multivariate Analysis.* Hofman, New York, 185 pp.

Kshirsagar, A. M. 1972. *Multivariate Analysis.* Marcel Dekker, Inc., New York. 534 pp.

Lee, P. J. 1970. Multivariate analysis for the fisheries biology. *Fish. Res. Board Canada Tech. Rept. 244.* 182 pp. (mimeo).

Morrison, D. F. 1967. *Multivariate Statistical Methods.* McGraw-Hill Book Company, New York. 338 pp.

Overall, J. E., and C. J. Klett. 1972. *Applied Multivariate Analysis.* McGraw-Hill Book Company, New York. 500 pp.

Rao, C. R. 1952. *Advanced Statistical Methods in Biometric Research.* John Wiley & Sons, Inc., New York. 390 pp.

———. 1960. Multivariate analysis: an indispensable statistical aid in applied research. *Sankhya 22:*317–338.

———. 1965. *Linear Statistical Inference and Its Applications.* John Wiley & Sons, Inc., New York. 522 pp.

———. 1972. Recent trends of research work in multivariate analysis. *Biometrics 28:*3–22.

Reyment, R. S. 1969. Biometrical techniques in systematics, in *Systematic Biology.* National Academy of Science, Washington, D. C. pp. 542–587.

Seal, H. 1964. *Multivariate Statistical Analysis for Biologists.* Meuthen & Company Ltd., London. 209 pp.

Searle, S. R. 1966. *Matrix Algebra for the Biological Sciences.* John Wiley & Sons, Inc., New York. 296 pp.

Sokal, R. R. 1965. Statistical methods in systematics. *Biol. Rev. 40:337–*391.

Tatsuoka, M. M. 1971. *Multivariate Analysis: Techniques for Education and Psychological Research.* John Wiley & Sons, Inc., New York. 310 pp.

Van de Geer, J. P. 1971. *Introduction to Multivariate Analysis for the Social Sciences.* W. H. Freeman and Company, San Francisco. 293 pp.

1

Reprinted from *Taxon,* **20**(1), 85–90 (1971)

PERSPECTIVES ON THE APPLICATION OF MULTIVARIATE STATISTICS TO TAXONOMY *

F. James Rohlf *

Summary

A brief outline is given of the principal types of multivariate statistical techniques which have found use in taxonomy. Techniques such as correlation, principal components, canonical correlation, and factor analyses are described for problems dealing with analysis of covariation within a single sample. Techniques such as canonical variate, cluster, multidimensional scaling, and network analyses are described for dealing with analyses of among sample variation. The purpose of this account is to give an intuitive understanding of what the various techniques have to offer to research in taxonomy.

Introduction

Since taxonomy is concerned with the classification of organisms based upon relationships (both cladistic and phenetic) inferred from characteristics of the *whole* organism, statistical analysis in this field must take into consideration the simultaneous covariation of many characters of the organism as possible. Thus taxonomy differs in an important way from fields such as physiology or biochemistry where investigations often are concerned with the effect of a certain combination of treatments upon a single variable of particular interest. In taxonomy there is often no special interest in the particular characters used. They are a means to an end, needed in order to compare samples of organisms taken from different localities or from what are believed to be different taxa. For these reasons, the techniques of multivariate statistics are of particular importance in taxonomy.

In the account given below I have outlined a variety of techniques which have found use in taxonomy. The account is purposely nonmathematical. Its intention is to give one a general intuitive feeling for the types of questions which can be answered using presently available multivariate techniques and to introduce some of the jargon of the field so that one can communicate the type of analysis desired to someone who can arrange for the actual computations to be performed (since most of the analysis require an enormous amount of arithmetic, the actual numerical computations will almost always have to be done on a highspeed digital computer).

Several texts are available dealing with the applications of multivariate statistics, e.g., Morrison (1967), Seal (1964), Cooley and Lohnes (1962), and Rao (1952). While these texts all have brief introductions to matrix algebra, the books Searle (1966) and Graybill (1969) should be consulted for a more complete understanding.

The account given below is divided into two main sections. The first one discusses techniques which analyze patterns of covariation found within a single sample and the second is concerned with analysis of variation between samples.

---

* Contribution number 29 from the Program in Ecology and Evolution, State University of New York at Stony Brook.

** Department of Ecology and Evolution, State University of New York at Stony Brook, Stony Brook, New York 11790.

There are a number of ways in which the patterns of variation and covariation within a single sample can be described. If 3 or fewer characters have been used, then frequency distributions and scatter diagrams can give one a useful intuitive appreciation of the variation found. If many characters have been used, then one can still plot scatter diagrams for various combinations of the characters taken 2 or 3 at a time, but it is usually difficult to fully appreciate complex patterns of covariation. A conventional statistical description of the sample would involve the computation of the mean for each variable and the variance-covariance matrix (a symmetrical table containing the variances of each character down the main diagonal and covariances in the off diagonal cells). If the sample represents a random sample from a multivariate normal distribution then such statistics contain sufficient information for estimating various properties of the population from which the sample was drawn. If the population was not normally distributed, then other statistics must be computed. In univariate statistics, one can compute higher moments such as g_1 and g_2 to measure skewness and kurtosis (Sokal and Rohlf, 1969). The analogous matrices in multivariate statistics are difficult to interpret. For this reason most workers resort to graphical techniques in such situations.

It is difficult to test for goodness of fit of an observed sample to a multivariate normal distribution. One can test whether each character taken separately fits a univariate normal distribution. If even one character does not fit a normal distribution, then the entire suite of characters does not fit a multivariate normal distribution. However, even if they all fit it does not guarantee that the entire suite of characters is consistant with a multivariate normal, since there can be a variety of complex interactions among the characters. If one has very large samples, the p-dimensional space can be partitioned into a series of regions and compare the frequency of observations in each region with that which would be expected based on a multivariate normal distribution (using the sample means and covariances). With samples of the size usually employed in taxonomy, this is not practical unless only a very few characters are used. The only alternative is to perform some sort of multidimensional scaling analysis (ordination) which will enable one to reduce the dimensionality of the system which needs to be considered. That is, to construct a few axes which contain most of the information about the covariation among the observations found in the original characters. If 3 or fewer axes are sufficient, then one can examine scatter diagrams constructed by projecting the specimens onto these axes and then plotting them against one another. Techniques such as non-metric multidimensional scaling (Kruskal, 1964 and Rohlf, 1970) and principal components analysis (Seal, 1964; Rohlf, 1970; Jolicoeur and Mosiman, 1960) have been used in taxonomy.

If one is satisfied that the data are consistant with the multivariate normal distribution, then principal components analysis can serve as a particularly compact means to describe the variation found in ones sample. The first principal component indicates the direction in hyperspace in which the observations differ most (the relative magnitude of the first eigenvalue indicates the extent to which the observations vary in this direction. This direction often corresponds to variation in overall size of the specimens (but it can sometimes represent the directions in which polymorphs vary if the sample is heterogeneous). The other principal components are often more difficult to interpret but they usually correspond to various shape differences between specimens. These differences are usually expressed as contrasts (high positive coefficients for some characters and high negative coefficients for other characters). The particular contrasts which result from the analysis are a consequence of the structure of the correlations between the characters. The information given by a principal components analysis can also be used to construct equal frequency ellipses which enclose regions expected to enclose $(1-\alpha)$ 100% of the observations. An example of this construction for the 2-dimensional case is given in Sokal and Rohlf (1969).

If a major purpose of the analysis is the investigation of patterns of intercorrelations among the characters, it is often useful to perform a factor analysis with rotation to simple structure (Harmon, 1967). This type of analysis expresses basically the same information but displays the correlation structure present in a much simpler form. Here each axis (or factor) corresponds to a group of characters as indicated by high (in absolute value) correlations between each factor and a set of characters. Characters not belonging to a set should have correlations near zero. Examples of the use of factor analysis in systematics are: Rohlf and Sokal (1958), Gould and Garwood (1969). Some other examples are listed in Seal (1964).

When the suite of characters can be logically divided into two sets and the relationships (if any) between the two sets is of interest, one can employ canonical correlation analysis. This technique obtains that a linear combination of the characters from each set of characters is such that these two linear functions have the highest possible correlation. This largest correlation is called the canonical correlation and measures the extent to which relationships in one set of characters can be predicted by a knowledge of the other set of characters. For example, one could use this type of analysis to locate those features in the adult stage which can be predicted based upon a knowledge of the larval stage. Morrison (1967) gives an outline of the necessary computations.

Description of variation among samples

There are several approaches to the study of variation among samples. The "proper" approach depends upon the statistical model and the purposes of the analysis (i.e., the questions being asked).

The question most commonly asked is: "Are the samples homogeneous?" If each sample can be assumed to have been drawn from a multivariate normal distribution then we can use a generalization of Bartlet's test to test for homogeneity of the variance-covariance matrices (Seal, 1964; Reyment, 1969). If they are homogeneous then we can use the techniques of multivariate analysis of variance to test whether the means of the samples are significantly heterogeneous. Of course, we must remember that if the samples were drawn from different geographic regions of a species or from different species, then we *know* that the true means (and probably also the variances and covariances) are different in different statistical populations. What we are testing is whether we have sufficient evidence to demonstrate that such differences exist and to set confidence intervals on the magnitude of the differences. If the test of significance yields a significant result, then it usually will be of interest to isolate those characters whose differences between various samples were most important in contributing to the overall significance test (just as we would turn to either *a priori* or to multiple comparison tests in a similar situation in univariate anova). However, it is difficult to know how to fully break down the overall multivariate test in the most meaningful way. If one has designed the sampling so that one can test a variety of *a priori* hypotheses, then one is relatively well off. One can then partition the overall test into a series of tests reflecting differences due to time of year vs. locality vs. sex vs. food plant, etc. If, however, one simply has those samples which are available one must use some kind of *a posteriori* test. Several multivariate multiple comparisons tests have been devised. Gabriel and Sokal (1969) described a test which can be used for this purpose, but it has the disadvantage that it produces "too many" answers. The voluminous output of this procedure reflects the fact that there are a very large number of ways in which multivariate samples can differ from one another. The results of this type of test are usually expressed in terms of so-called maximum nonsignificant subsets. These sets have the property that the addition of any other sample or variable to the set would cause it to be significantly heterogeneous.

The description of the patterns of variation among samples in terms of sets (which may be partially overlapping) is usually not very convenient. Other techniques (which have less statistical rigor) have been developed to more conveniently express statistical relationships among the variables over the samples.

There are three main classes of techniques which are used to reveal the relationships among the samples in the p-dimensional space: multidimensional scaling, cluster analysis, and network analysis. These techniques all come under the heading of multivariate data analysis since their main purpose is to give insight into ones data and to place less stress on tests of significance. In biology these techniques are associated with the field of numerical taxonomy where they have been found very useful in elucidating taxonomic relationships.

Multidimensional scaling is used when one wishes to express relationships among the sample means in terms of their coordinates on a few specially constructed coordinate axes. The goal is to preserve as much of the information about interpoint distances as possible while reducing the number of variables to be considered from p down to k (where k is 1, 2, 3, or perhaps 4 at the most). If the variation within all of the samples is homogeneous (or at least the orientation of the scatter ellipsoids are similar) then one can validly compute a pooled within group variance-covariance matrix and then perform a canonical variates analysis (see Jolicoeur, 1959; Seal, 1964). In this type of analysis relationships among the samples are expressed relative to the average covariation found within the samples.

This type of analysis is also sometimes called a generalized discriminant analysis since in the special case where there are only two samples the canonical variable is the discriminant function. When there are more than two samples, the canonical variables constitute a set of linear combinations of the variables which best discriminate between the groups. They can be used to form a probabilistic identification scheme (Cooley and Lohnes, 1962).

If the within group variation is not homogeneous (particularly if the orientation of the scatter ellipsoids differ) then it is difficult to make use of the within sample information and one must base ones analyses on the among sample variation. For example, one can perform a principal components analysis on the among sample correlation matrix to obtain vectors which indicate the major trends of variation among groups. One can then project the standardized sample means onto these axes in order to be able to prepare a scatter diagram depicting the among group variation relative to the total amount of variation found among the samples (since the correlation matrix and standardized data were used).

When the samples correspond to higher taxa then one expects the within sample covariation to be heterogeneous. This is one reason why there is seldom any attempt to take within sample covariation into account in numerical taxonomy. Often the taxa being sampled are sufficiently distinct that only a few specimens are used to represent each taxon. This is a valid shortcut whenever the among sample variation is much larger than the within sample variation as would be expected, for example, when the samples correspond to different species sampled throughout a family. In such cases there seems little point in worrying about tests of significance — the species are obviously different from one another. What is uncertain is their relative degrees of overall similarity and the way in which this can be most simply expressed. Another alternative (which sometimes is capable of expressing the relationships in fewer dimensions) is non-metric multidimensional scaling (see Kruskal, 1964; Rohlf, 1970). If a sufficient amount of the among group variation can be expressed in k-dimensions ($k << p$) then one can visually look for patterns in the differences among the samples (results are mostly intuitive, few tests of significance are possible here, but one often gains considerable insight into ones data).

Cluster analysis sorts the samples into a series of sets. These sets may be mutually exclusive, hierarchic, or partially overlapping in various ways. Hierarchical clustering schemes have been used most commonly in taxonomy. Typically these techniques begin with a matrix of distances between sample means (computed in various ways) and a search for other points which are relatively close together and separated by gaps from other such groups. The distance coefficient can be computed in such a way as to take the within sample covariation into consideration if this is desired. The generalized distance D is one way in which this can be done (Rohlf, 1970; Seal, 1964; Rao, 1952). For data in which the relationships among the samples are hierarchic cluster analyses works rather efficiently. They tend to be somewhat unsatisfactory on data in which the distribution of points in the p-dimensional space form very elongated clusters or where there are many points which are intermediate between clusters. These techniques also do not reveal the fact that some clusters may be in between other clusters (see Rohlf, 1970 for a general discussion).

Network analyses express relationships in terms of a graph (in the sense of graph theory, Ore, 1963) which consists of vertices (corresponding to the samples) and edges (which are connections between vertices). The existence of an edge implies that the two vertices so connected share some relation between them (e.g., they are nearest neighbors in the p-dimensional space). The shortest simply connected network has been found to be useful in numerical taxonomy since it indicates in a convenient fashion the closest neighbor of each point. Kruskal (1956) and Prim (1957) give algorithms for constructing such networks. Jardine and Sibson (1968) have suggested the use of networks which are more than simply connected and thus more capable of summarizing multivariate relationships (and hence more complex to understand). An example of the use of a shortest connection network in taxonomy is given in Rohlf (1970).

Comprehension of multivariate relationships is difficult. This difficulty is not helped by the fact that classical multivariate statistical techniques tend to result in a single number which is used for tests of significance. Such statistics are often difficult to interpret in terms of the particular samples and variables under investigation. For this reason more emphasis has been placed in the last few years upon a variety of graphical techniques which allow one to visualize many parameters of the sample simultaneously. The account given above is an attempt to give one a brief intuitive introduction to the types of techniques which are apt to be found useful in taxonomy. A number of workers are attempting to develop new mathematical tools which will allow a simple but efficient graphical summarization of multivariate relationships. Such developments, if successful, could have a large impact upon taxonomic methodology.

References

COOLEY, W. W. and P. R. LOHNES 1962 – Multivariate procedures for the behavioral sciences. Wiley: New York 211 pp.

GABRIEL, K. R. and R. R. SOKAL 1969 – A new statistical approach to geographic variation analysis. Systematic Zool. 18: 259–278.

GRAYBILL, F. A. 1969 – Introduction to matrices with applications in Statistics. Wadsworth: Belmont, Calif. 372 pp.

GOULD, S. J. and R. A. GARWOOD 1969 – Levels of integration in mammalian dentitions: An analysis of correlations in *Nesophontes micrus* (Insectivora) and *Oryzomys couesi* (Rodentia). Evolution, 23: 276–300.

HARMON, H. H. 1967 – Modern factor analysis. Chicago, 470 pp.

JARDINE, N. and R. SIBSON 1968 – The construction of hierarchic and non-hierarchic classifications. Computer Jour. 11: 177–184.

JOLICOEUR, P. 1959 – Multivariate geographical variation in the wolf, *Canis Lupus* L. Evolution, 13: 283–299.

JOLICOEUR, P. and J. E. MOSIMANN 1960 – Size and shape variation in the painted turtle. A principal component analysis. Growth, 24: 339–354.

KRUSKAL, J. B. 1956 – On the shortest spanning subtree of a graph and the traveling salesman problem. Proc. Amer. Math. Soc., 7: 48–50.

KRUSKAL, J. B. 1964 – Non-metric multidimensional scaling. Psychometrica, 29: 1–27.

MORRISON, D. F. 1967 – Multivariate statistical methods. McGraw-Hill: New York, 338 pp.

ORE, O. 1963 – Graphs and their uses. Random House: New York, 131 pp.

PRIM, R. C. 1957 – Shortest connection networks and some generalizations. Bell System Technical Jour. 36: 1389–1401.

RAO, C. R. 1952 – Advanced statistical methods in biometrical research. Wiley: New York, 390 pp.

REYMENT, R. A. 1969 – Biometrical techniques in systematics. In Systematic Biology. Publ. 1692 National Academy of Sciences. pp. 541–594.

ROHLF, F. J. 1970 – Adaptive hierarchical clustering schemes. Systematic Zool., 19: 58–82.

ROHLF, F. J. and R. R. SOKAL 1958 – The description of taxonomic relationships by factor analysis. Systematic Zool., 11: 1–16.

SEAL, H. 1964 – Multivariate statistical analysis for biologists. Wiley: New York, 207 pp.

SEARLE, S. R. 1966 – Matrix algebra for the biological sciences. Wiley: New York, 296 pp.

SOKAL, R. R. and F. J. ROHLF 1969 – Biometry. Freeman: San Francisco, 776 pp.

Part I

MULTIVARIATE ANALYSIS OF VARIANCE

Editors' Comments on Papers 2 Through 7

2 **HOTELLING**
The Generalization of Student's Ratio

3 **ROY and GNANADESIKAN**
Some Contributions to ANOVA in One or More Dimensions: II

4 **GABRIEL**
A Comparison of Some Methods of Simultaneous Inference in MANOVA

5 **REYMENT**
A Note on Geographical Variation in European Rana

6 **SOKAL, HERYFORD, and KISHPAUGH**
Changes in Microgeographic Variation Patterns of Pemphigus populitransversus *over a Six-Year Span*

7 **GOODMAN**
The Races of Maize: II. Use of Multivariate Analysis of Variance to Measure Morphological Similarity

Although the beginnings of analysis of variance or ANOVA can be traced back into the 1800s, much of what we presently understand about ANOVA is due to the early efforts of R. A. Fisher, who is responsible for formally introducing the terms *variance* and *analysis of variance* into the statistics literature. It was Fisher who first systematically described the partitioning of the total sums of squares into components associated with recognizable sources of variation and thus introduced this powerful statistical tool for hypothesis testing and statistical inference.

Multivariate analysis of variance or MANOVA probably began with the publication by Wishart (1928) of the joint distribution of variances and covariances in samples derived from p-variate normal distribution. The Wishart distribution, as it has since become known, provided the multivariate analog of the earlier bivariate case outlined by Fisher in 1915 and thus provided the cornerstone for further development of procedures related to multivariate hypothesis testing.

The development of multivariate hypothesis testing to a large extent followed the same chronological pattern as the univariate methods. In univariate methodology, Gosset (writing under the pseudonym "Student") introduced the Student's *t* test for differences between two sample means, which was followed several years later by Fisher's derivation of the multiple group procedures we now refer to as analysis of variance. It is thus appropriate that the first important development in multivariate hypothesis testing following the derivation of the Wishart distribution was the extension by Hotelling (Paper 2) of the Student's *t* statistic to the multivariate T^2 statistic. While the *t* statistic tests the null hypothesis that the sample statistics estimate the same parameter, the T^2 statistic has a null hypothesis that the two centroids coincide in p-dimensional variate space. When $p = 1$, the T^2 statistic reduces to the square of Student's *t*.

In his T^2 paper, Hotelling noted early in the history of multivariate analysis some of the advantages of multivariate techniques such as the T^2 statistic over repetitive univariate procedures to answer the same question, i.e., equality of a series of means obtained from the same observations. This simple statistical relationship is still not appreciated by many biologists.

While its development constituted the first important step in multivariate hypothesis testing, the T^2 statistic, like its univariate counterpart, was obviously limited in its application since only pairs of groups could be considered. In order to extend such a test to K samples, Wilks (1932) employed the Neyman–Pearson likelihood ratio principle to generate a multivariate variance ratio test now known as the Wilks' lambda criterion. In univariate analysis of variance, the F test of significance is a ratio of the among-groups to the within-groups mean square and is used to ascertain if the divergence among groups is significantly greater than the variation within groups. Wilks' procedure is analogous to the univariate procedures in that it involves partitioning the total sums-of-squares and cross-products matrix (T) into within-groups (W) and among-groups (A) components. Since the multivariate analog of the univariate variance is the determinant of a covariance matrix, the Wilks' lambda criterion is thus the ratio of the within-groups determinant to the total.

We deeply regret an unfortunate policy by the Trustees of Biometrika which has prevented us from reprinting these classical papers by Wishart (1928) and Wilks (1932), since they are among the most important contributions in the development of multivariant analysis of variance. However, the excellent paper by Porebski (Paper 12) included in Part II, Discriminant Analysis, provides a lucid account of the Wilks' lambda, Hotelling's T^2, and related tests useful in MANOVA.

The contributions of Hotelling and Wilks, fundamental steps in the generalization of univariate analysis of variance to the multivariate case, still represented techniques of limited utility. The T^2 statistic has utility only for the two-group test, while the Wilks' lambda test is, in fact, a single-classification MANOVA. As such, the Wilks' test is very difficult to apply to more complex experimental designs, such as those having unequal numbers of cells or unequal frequencies within cells.

For a complete generalization of the univariate analysis of variance to multivariate data and designs, a prime requisite was to circumvent the need for different calculator procedures for each type of design, which could only be accomplished through the use of matrix and vector methodology. Three papers stand out at this point in the development of a completely generalized multivariate analysis of variance. Tukey (1949), expanding on the earlier ideas of M. S. Bartlett, outlined vector methods in analysis of variance, which he called dyadic ANOVA. Approximately ten years after the appearance of Tukey's paper on dyadic ANOVA, a two-part article was published by Roy and Gnanadesiken (1959), which provided a complete generalization of ANOVA to MANOVA based on matrix operations. These authors extended MANOVA design to very complex situations, e.g., complete and incomplete designs, and proportionate and disproportionate subclass numbers. We have included the second article of this series by Roy and Gnanadesiken as Paper 3, because of its importance in the development of MANOVA methodology.

In univariate analysis of variance, the analysis is usually not thought to be complete until some type of mean separation or multiple comparisons procedure has been carried out to delimit homogeneous subgroups of samples. This is most commonly done by *a posteriori* testing procedures such as Student–Newman–Keuls, simultaneous test procedures, or related methods. The same need exists in multivariate analysis of variance, but unfortunately it is only very recently that mean vector separation procedures have been proposed. Among the more sophisticated attempts to resolve this problem are those of K. R. Gabriel, who has developed methods to delimit homogeneous subsets in multivariate data using simultaneous test procedures.

An excellent introduction to Gabriel's work on simultaneous test procedures in MANOVA is his 1968 paper. However, rather than include the 1968 work, we selected a more generalized paper on multiple comparisons by Gabriel (Paper 4) for this volume. This particular paper summarizes a variety of procedures and methods useful in MANOVA.

The great majority of applications of multivariate analysis of variance in systematics and ecology have been single-classification MANOVA carried out in conjunction with other multivariate procedures, such as discriminant analysis. As examples where MANOVA was the primary statistical tool, we have included two papers from the field of systematics and one from agricultural genetics. Reyment (Paper 5) has used Hotelling's T^2 and the Wilks' lambda criterion to study geographic variation among populations of *Rana temporaria*. Reyment's paper was included because it provides lucid examples of the calculations involved in multivariate hypothesis testing and the use of MANOVA in biology. As an example of MANOVA and mean vector separation procedures, we have included a paper by Sokal et al. (Paper 6), who used multivariate analysis of variance followed by simultaneous test procedures, as outlined by Gabriel, to delimit homogeneous subsets of samples in an aphid *Pemphigus populitransversus*. The third example of MANOVA techniques included in this volume is from the fields of agriculture and genetics. This interesting paper by Goodman (Paper 7) extends MANOVA procedures to a randomized block design and thereby makes the technique more useful to studies in agriculture and ecology.

Other papers on multivariate analysis of variance of interest to biologists include Smith et al., (1962), Cramer and Bock (1966) and Rempe and Weber (1972). In addition, Morrison (1967) includes a good introduction to multivariate analysis of variance in his text.

BIBLIOGRAPHY

Baker, R. J., W. R. Atchley, and V. R. McDaniel. 1972. Karyology and morphometrics of Peter's tent-making bat, *Uroderma bilobatum* Peters (Chiroptera, Phyllostomatidae). *Syst. Zool. 21:*414–429.

Bartlett, M. S. 1947. Multivariate analysis. *J. Roy. Stat. Soc. Suppl. 9:*176–197.

Bhapkar, V. P. 1966. Some non-parametric tests for the multivariate several sample location problem, in *Multivariate Analysis* (P. R. Krishnaiah, ed.). Academic Press, New York. pp. 29–41.

Bock, R. D. 1963. Programming univariate and multivariate analysis of variance. *Technometrics 5:*95–117.

Cramer, E. M., and R. D. Bock. 1966. Multivariate analysis. *Rev. Educ. Res. 36:*604–614.

Fisher, R. A. 1915. Frequency distribution of the values of the correlation coefficient in samples from an indefinitely large population. *Biometrika 10:*507–521.

Gabriel, K. R. 1968. Simultaneous test procedures in multivariate analysis of variance. *Biometrika 55:*489–504.

Hotelling, H. 1951. A generalized *T* test and measure of multivariate dispersion, in *Proc. 2nd Berkeley Symp. Multivariate Stat. Probability* (J. Neyman, ed.). University of California Press, Berkeley, Calif. pp. 23–41.

Jensen, D. R. 1972. Some simultaneous multivariate procedures using Hotelling's T^2 statistic. *Biometrics 28:*39–54.

Krishnaiah, P. R. 1965. On a multivariate generalization of the simultaneous analysis of variance test. *Ann. Inst. Stat. Math. 17:*167–173.

———. 1965. On the simultaneous ANOVA and MANOVA tests. *Ann. Inst. Stat. Math. 17:*35–53.

Morrison, D. F. 1967. *Multivariate Statistical Methods.* McGraw-Hill Book Company, New York. 338 pp.

Porebski, O. R. 1969. On the interrelated nature of the multivariate statistics used in discriminatory analysis. *British J. Math. Stat. Psychol. 19:*197–214.

Potthoff, R. F., and S. N. Roy. 1964. A generalized multivariate analysis useful especially for growth curve problems. *Biometrika 51:*313–326.

Rempe, V., and E. E. Weber, 1972. An illustration of the principal ideas of MANOVA. *Biometrics 28:*235–238.

Roy, S. N., and R. Gnanadesikan. 1959. Some contributions to ANOVA in one or more dimensions. I. *Ann. Math. Stat. 30:*304–317.

Roy, S. N., and R. C. Bose. 1953. Simultaneous confidence interval estimation. *Ann. Math. Stat. 24:*513–536.

Schatzoff, M. 1966. Exact distributions of Wilks' likelihood ratio criterion. *Biometrika 53:*347–358.

Smith, H., R. Gnanadesikan, and J. B. Hughes. 1962. Multivariate analysis of variance (MANOVA). *Biometrics 18:*22–41.

Srivastava, J. N. 1966. Some generalizations of multivariate analysis of variance, in *Multivariate Analysis* (P. R. Krishnaiah, ed.). Academic Press, New York. pp. 129–145.

Tukey, J. W. 1949. Dyadic ANOVA, an analysis of variance for vectors. *Human Biol. 21:*65–110.

Wilks, S. S. 1932. Certain generalizations in the analysis of variance *Biometrika. 24:*471–494.

Wilks, S. S. 1946. Sample criteria for testing equality of means, equality of variances and equality of covariances in a multivariate distribution. *Ann. Math. Stat. 17:*257–281.

Wishart, J. 1928. The generalized product moment distribution in samples from a normal multivariate population. *Biometrika 20:*32–58.

2

Reprinted from *Ann. Math. Stat.*, 2(3), 360–378 (1931)

THE GENERALIZATION OF STUDENT'S RATIO*

By

Harold Hotelling

The accuracy of an estimate of a normally distributed quantity is judged by reference to its variance, or rather, to an estimate of the variance based on the available sample. In 1908 "Student" examined the ratio of the mean to the standard deviation of a sample.[1] The distribution at which he arrived was obtained in a more rigorous manner in 1925 by R. A. Fisher,[2] who at the same time showed how to extend the application of the distribution beyond the problem of the significance of means, which had been its original object, and applied it to examine regression coefficients and other quantities obtained by least squares, testing not only the deviation of a statistic from a hypothetical value but also the difference between two statistics.

Let ξ be any linear function of normally and independently distributed observations of equal variance, and let s be the estimate of the standard error of ξ derived by the method of maximum likelihood. If we let t be the ratio to s of the deviation of ξ from its mathematical expectation, Fisher's result is that the probability that t lies between t_1 and t_2 is

---

*Presented at the meeting of the American Mathematical Society at Berkeley, April 11, 1931.

[1] Biometrika, vol. 6 (1908), p. 1.

[2] Applications of Student's Distribution, Metron, vol. 5 (1925), p. 90.

$$(1) \qquad \frac{1}{\sqrt{\pi n}} \qquad \frac{\Gamma\left(\frac{n+1}{2}\right)}{\Gamma\left(\frac{n}{2}\right)} \int_{t_1}^{t_2} \frac{dt}{(1+t^2/n)^{\frac{n+1}{2}}}$$

where n is the number of degrees of freedom involved in the estimate s .

It is easy to see how this result may be extended to cases in which the variances of the observations are not equal but have known ratios and in which, instead of independence among the observations, we have a known system of intercorrelations. Indeed, we have only to replace the observations by a set of linear functions of them which are independently distributed with equal variance. By way of further extension beyond the cases discussed by Fisher, it may be remarked that the estimate of variance s^2 may be based on a body of data not involved in the calculation of ξ . Thus the accuracy of a physical measurement may be estimated by means of the dispersion among similar measurements on a different quantity.

A generalization of quite a different order is needed to test the simultaneous deviations of several quantities. This problem was raised by Karl Pearson in connection with the determination whether two groups of individuals do or do not belong to the same race, measurements of a number of organs or characters having been obtained for all the individuals. Several "coefficients of racial likeness" have been suggested by Pearson and by V. Romanovsky with a view to such biological uses. Romanovsky has made a careful study[1] of the sampling distributions, assuming in each case that the variates are independently and normally

[1]V. Romanovsky, On the criteria that two given samples belong to the same normal population (on the different coefficients of racial likeness), Metron, vol. 7 (1928), no. 3, pp. 3-46; K. Pearson, On the coefficient of racial likeness, Biometrika, vol. 18 (1926), pp. 105-118.

distributed. One of Romanovsky's most important results is the exact sampling distribution of L, a constant multiple of the sum of the squares of the values of t for the different variates. This distribution function is given by a somewhat complex infinite series. For large samples and numerous variates it slowly approximates to the normal form; for 500 individuals, Romanovsky considers that an adequate approach to normality requires that no fewer than 62 characters be measured in each individual. When it is remembered that all these characters must be entirely independent, and that it is usually hard to find as many as three independent characters, the difficulties in application will be apparent. To avoid these troubles, Romanovsky proposes a new coefficient of racial likeness, H, the average of the ratios of variances in the two samples for the several characters. He obtains the exact distribution of H, again as an infinite series, though it approaches normality more rapidly than the distribution of L. But H does not satisfy the need for a comparison between magnitudes of characters, since it concerns only their variabilities.

Joint comparisons of correlated variates, and variates of unknown correlations and standard deviations, are required not only for biologic purposes, but in a great variety of subjects. The eclipse and comparison star plates used in testing the Einstein deflection of light show deviations in right ascension and in declination; an exact calculation of probability combining the two least-square solutions is desirable. The comparison of the prices of a list of commodities at two times, with a view to discovering whether the changes are more than can reasonably be ascribed to ordinary fluctuation, is a problem dealt with only very crudely by means of index numbers, and is one of many examples of the need for such a coefficient as is now proposed. We shall generalize Student's distribution to take account of such cases.

We consider p variates $x_1, x_2, \ldots, x_p$, each of which is measured for N individuals, and denote by $X_{i\alpha}$ the value of x_i for the α th individual. Taking first the problem of the significance of the deviations from a hypothetical set of mean values $m_1, m_2, \ldots, m_p$, we calculate the means $\bar{x}_1, \bar{x}_2, \ldots, \bar{x}_p$, of the samples, and put

$$\xi_i = (\bar{x}_i - m_i)\sqrt{N}.$$

Then the mean values of the ξ_i will all be zero, and the variances and covariances will be the same as for the corresponding x_i, since the individuals are supposed chosen independently from an infinite population.[1] In order to estimate them with the help of the deviations

$$x_{i\alpha} = X_{i\alpha} - \bar{x}_i$$

from the respective means, we call $n = N - 1$ the number of degrees of freedom and take as the estimates of the variances and covariances,

$$a_{ji} = a_{ij} = \frac{1}{n}\sum_{\alpha=1}^{N} x_{i\alpha} x_{j\alpha}$$

We next put:

$$a = \begin{vmatrix} a_{11} & a_{12} & \cdots & a_{1p} \\ a_{21} & a_{22} & \cdots & a_{2p} \\ \cdot & \cdot & \cdots & \cdot \\ a_{p1} & a_{p2} & \cdots & a_{pp} \end{vmatrix}$$

[1] "Mean Value" is used in the sense of mathematical expectation; the variance of a quantity whose mean value is zero is defined as the expectation of its squares; the covariance of two such quantities is the expectation of their product. Thus the correlation of the two in a hypothetical infinite population is the ratio of their covariance to the geometric mean of the variances.

$$(3) \qquad A_{ij} = A_{ji} = \frac{\text{cofactor of } a_{ij} \text{ in } a}{a}$$

The measure of simultaneous deviations which we shall employ is

$$(4) \qquad T^2 = \sum_{i=1}^{p} \sum_{j=1}^{p} A_{ij} \xi_i \xi_j .$$

For a single variate it is natural to take $A_{11} = 1/a_{11}$; then T reduces to t , the ordinary "critical ratio" of a deviation in a mean to its estimated standard error, a ratio which has "Student's distribution," (1). For examining the deviations from zero of two variates x and y ,

$$T = \frac{N}{1-r^2}\left\{\frac{\bar{x}^2}{s_1^2} - \frac{2r\bar{x}\bar{y}}{s_1 s_2} + \frac{\bar{y}^2}{s_2^2}\right\},$$

where

$$s_1^2 = \frac{\Sigma(X-\bar{x})^2}{N-1}, \qquad s_2^2 = \frac{\Sigma(Y-\bar{y})^2}{N-1},$$

$$r = \frac{\Sigma(X-\bar{x})(Y-\bar{y})}{\sqrt{\Sigma(X-\bar{x})^2 \Sigma(Y-\bar{y})^2}}$$

For comparing the means of two samples, one of N_1 and the other of N_2 individuals, we distinguish symbols pertaining to the second sample by primes, and write

$$(5) \qquad \xi_i = \frac{\bar{x}_i - \bar{x}'_i}{\sqrt{1/N_1 + 1/N_2}}$$

$$n = N_1 + N_2 - 2,$$

$$(6) \qquad a_{ij} = \frac{1}{n}\left[\Sigma(X_{i\alpha} - \bar{x}_i)(X_{j\alpha} - \bar{x}_j) + \Sigma(X'_{i\alpha} - \bar{x}'_i)(X'_{j\alpha} - \bar{x}'_j)\right]$$

$$= \frac{1}{n}\left[\Sigma X_{i\alpha} X_{j\alpha} - N_1 \bar{x}_i \bar{x}_j + \Sigma X'_{i\alpha} X'_{j\alpha} - N_2 \bar{x}'_i \bar{x}'_j\right]$$

and take as our "coefficients of racial likeness" the value (4) of T^2, in which the ξ_i are calculated from (5) and the A_{ij} from (6) and (3).

Other situations to which the measure T^2 of simultaneous deviations can be applied include comparisons of regression coefficients and slopes of lines of secular trend, comparisons which for single variates have been explained by R. A. Fisher.[1] In each case we deal for each variate with a linear function ξ_i of the observed values, such that the sum of the squares of the coefficients is unity, so that the variance is the same as for a single observation, and such that the expectation of ξ_i is, on the hypothesis to be tested, zero. Deviations $x_{i\alpha}$ of the observations from means, or from trend lines or other such estimates, are used to provide the estimated variances and covariances a_{ij} by (2). The number of degrees of freedom n is the difference between the number N of individuals and the number q of independent linear relations which must be satisfied by the quan-

[1]Metron, loc. cit., and Statistical Methods for Research Workers, Oliver and Boyd, third edition (1928).

tities $x_{i1}, x_{i2}, \ldots, x_{iN}$ on account of their method of derivation. For all the variates, these relations and n must be the same.

The general procedure is to set up what may be called normal values $\bar{x}_{i\alpha}$ for the respective $X_{i\alpha}$, putting

$$(7) \qquad x_{i\alpha} = X_{i\alpha} - \bar{x}_{i\alpha}.$$

The underlying assumption is that $X_{i\alpha}$ is composed of two parts, of which one, $\varepsilon_{i\alpha}$, is normally and independently distributed about zero with variance σ_i^2 which is the same for all the observations on x_i. The other component is determined by the time, place, or other circumstances of the α'th observation in some regular manner, the same for all the variates. Denoting this part by $\eta_{i\alpha}$, we have

$$X_{i\alpha} = \eta_{i\alpha} + \varepsilon_{i\alpha}.$$

Specifically, we take $\eta_{i\alpha}$ to be a linear function, with known coefficients $g_{\alpha s}$, of q unknown parameters $\zeta_{i1}, \ldots, \zeta_{iq}$ where $q < N$:

$$(8) \qquad \eta_{i\alpha} = \sum_{s=1}^{q} g_{\alpha s} \zeta_{is}.$$

Thus in dealing with a secular trend representable by a polynomial in the time, we may take the g's as powers of the time-variable, the ζ's as the coefficients. For differences of means, the g's are 0's and 1's, and the ζ's the true means.

We estimate the ζ's by minimizing

$$(9) \qquad 2V_i = \sum_{\alpha=1}^{N} \varepsilon_{i\alpha}^2 = \sum_{\alpha=1}^{N} (X_{i\alpha} - \eta_{i\alpha})^2.$$

Substituting from (8), differentiating with respect to ζ_{is}, and replacing $\eta_{i\alpha}$ by $\bar{x}_{i\alpha}$ for the minimizing value, we obtain:

$$(10) \qquad \sum_{\alpha=1}^{N} g_{\alpha s}(X_{i\alpha} - \bar{x}_{i\alpha}) = 0, \qquad (s = 1, 2, \cdots, q)$$

or by (7),

$$(11) \qquad \sum_{\alpha=1}^{N} g_{\alpha s}\, x_{i\alpha} = 0 \qquad (s = 1, 2, \cdots, q)$$

Denoting also the minimizing values of ζ_{is} by z_{is}, we have made from (8),

$$\bar{x}_{i\alpha} = \sum_{s=1}^{q} g_{\alpha s}\, z_{is}$$

Subtracting (8),

$$(12) \qquad \bar{x}_{i\alpha} - \eta_{i\alpha} = \sum_{s=1}^{q} g_{\alpha s}(z_{is} - \zeta_{is})$$

From (9),

$$2V = \sum_{\alpha=1}^{N} \left[(X_{i\alpha} - \bar{x}_{i\alpha}) + (\bar{x}_{i\alpha} - \eta_{i\alpha}) \right]^2$$

$$(13) \qquad = \sum_{\alpha=1}^{N} (X_{i\alpha} - \bar{x}_{i\alpha})^2 + 2\sum_{\alpha=1}^{N} (X_{i\alpha} - \bar{x}_{i\alpha})(\bar{x}_{i\alpha} - \eta_{i\alpha})$$

$$+ \sum_{\alpha=1}^{N} (\bar{x}_{i\alpha} - \eta_{i\alpha})^2$$

The middle term, by (12), equals

$$2\sum_{\alpha=1}^{N}\sum_{s=1}^{q} g_{\alpha s}(X_{i\alpha} - \bar{x}_{i\alpha})(z_{is} - \zeta_{is}),$$

this, by (10), is zero. Hence, by (7) and (13),

$$U_i = V_i + W_i ,$$

where

$$2V_i = \sum_{\alpha=1}^{N} x_{i\alpha}^2$$

$$2W_i = \sum_{\alpha=1}^{N} (\bar{x}_{i\alpha} - \eta_{i\alpha})^2$$

If the q equations (10) be solved for $\bar{x}_{i1}$, $\bar{x}_{i2}$, ..., $\bar{x}_{iN}$, the values of these quantities will be found to be homogeneous linear functions of the observations $X_{i\alpha}$. By (7), therefore, the quantities

$$\bar{x}_{i1}, \bar{x}_{i2}, \ldots\ldots, \bar{x}_{iN}$$

are homogeneous linear functions of the $X_{i\alpha}$. But they are not linearly independent functions, since they are connected by the q relations (11). Hence V is a quadratic form of rank

$$n = N - q \,.\,.$$

Since V_i, by (9), is of rank N, W is of rank q.

This shows that Np new quantities $x'_{i\alpha}$, given by equations of the form

$$x'_{i\alpha} = \sum_{\beta=1}^{N} c_{\alpha\beta} x_{i\beta} = \sum_{\beta=1}^{N} c_{\alpha\beta} X_{i\beta}, (\alpha = 1, 2, \cdots, n)$$

(14)

$$x'_{i\alpha} = \sum_{\beta=1}^{N} c_{\alpha\beta} (\bar{x}_{i\beta} - \eta_{i\beta}) = \sum_{\beta=1}^{N} (c_{\alpha\beta} X_{i\beta} - c_{\alpha\beta}\, \eta_{i\beta}), (\alpha = n+1, \cdots, N)$$

can be found such that

$$(15)\qquad 2V_i = \sum_{\alpha=1}^{N} x_{i\alpha}^2 = \sum_{\alpha=1}^{N} x_{i\alpha}'^2 ,$$

$$2W_i = \sum_{\alpha=n+1}^{N} x_{i\alpha}'^2 ,$$

and therefore

$$(16)\qquad 2U_i = \sum_{\alpha=1}^{N} x_{i\alpha}'^2 .$$

Substituting (14) in (15) and equating like coefficients,

$$(17)\qquad \sum_{\alpha=1}^{n} c_{\alpha\beta}\, c_{\alpha\gamma} = \delta_{\beta\gamma}$$

where $\delta_{\beta\gamma}$ is the Kronecker delta, equal to 1 if $\beta = \gamma$, to 0 if $\beta \neq \gamma$

The coefficients $c_{\alpha\beta}$ depend only on the $g_{\alpha\beta}$, which have been assumed to be the same for all the p variates. Thus (14) may be written

$$x_{j\alpha}' = \sum_{\gamma=1}^{N} c_{\alpha\gamma}\, x_{j\gamma} .$$

Multiplying by (14), summing with respect to α from 1 to n, and using (17),

$$(18)\qquad \sum_{\alpha=1}^{n} x_{i\alpha}' x_{j\alpha}' = \sum_{\alpha=1}^{n} \sum_{\beta=1}^{N} \sum_{\gamma=1}^{N} c_{\alpha\beta}\, c_{\alpha\gamma}\, x_{i\beta}\, x_{j\gamma}$$

$$= \sum_{\beta=1}^{N} \sum_{\gamma=1}^{N} \delta_{\beta\gamma}\, x_{i\beta}\, x_{j\gamma} = \sum_{\beta=1}^{N} x_{i\beta}\, x_{j\beta}$$

Just as in (2), we define a_{ij} in this generalized case by

(19) $$a_{ij} = \frac{1}{n}\sum_{\alpha=1}^{N} x_{i\alpha} x_{j\alpha} .$$

Then by (18),

(20) $$a_{ij} = \frac{1}{n}\sum_{\alpha=1}^{N} x'_{i\alpha} x'_{j\alpha} .$$

Of the last equation, (6) is a special case.

The random parts $\varepsilon_{i\alpha}$ of the observations on x_i have by hypothesis the distribution

$$\frac{1}{(\sigma_i\sqrt{2\pi})^N} e^{-U_i/2\sigma_i^2} d\varepsilon_{i1}\, d\varepsilon_{i2} \cdots d\varepsilon_{iN} ,$$

where V_i is given by (9). From what has been shown, it is clear that this may be transformed into

$$\frac{1}{(\sigma_i\sqrt{2\pi})^N} e^{-(x'^2_{i1} + x'^2_{i2} + \cdots + x'^2_{iN})/2\sigma_i^2} dx'_{i1} \cdots dx'_{iN} ,$$

showing that $x'_{i1}, \ldots, x'_{iN}$ are normally and independently distributed with equal variance σ_i^2.

The statistic ξ_i must be independent of the quantities $x'_{i1}, x'_{i2}, \ldots, x'_{in}$ entering into (20), its mean value must be zero, and its variance must be σ_i^2. These conditions are satisfied in the cases which have been mentioned, and are satisfied in general if ξ_i is a linear homogeneous function of $x'_{i,n+1}, \ldots, x'_{iN}$ with the sum of the squares of the coefficients equal to unity.

The measure of simultaneous discrepancy is

$$T^2 = \sum_{i=1}^{p}\sum_{j=1}^{p} A_{ij}\, \xi_i\, \xi_j ,$$

A_{ij} being defined by (3) on the basis of (19). It is evident that

$$(21)\quad T^2 = -\frac{\begin{vmatrix} 0 & \xi_1 & \xi_2 & \cdots & \xi_p \\ \xi_1 & a_{11} & a_{12} & \cdots & a_{1p} \\ \xi_2 & a_{21} & a_{22} & \cdots & a_{2p} \\ \cdots & a \cdot & \cdots & \cdots & \cdots \\ \xi_p & a_{p1} & a_{p2} & \cdots & a_{pp} \end{vmatrix}}{\begin{vmatrix} a_{11} & a_{12} & \cdots & a_{1p} \\ a_{21} & a_{22} & \cdots & a_{2p} \\ \cdots & \cdots & \cdots & \cdots \\ a_{p1} & a_{p2} & \cdots & a_{pp} \end{vmatrix}}$$

as appears when the numerator is expanded by the first row, and the resulting determinants by their first columns.

A most important property of T is that it is an absolute invariant under all homogeneous linear transformations of the variates $x_i, \ldots, x_p$. This may be seen most simply by tensor analysis; for ξ_i is covariant of the first order and A_{ij} is contravariant of the second order.

The invariance of T shows that in seeking its sampling distribution we may, without loss of generality, assume that the variates $x_1, \ldots, x_p$ have, in the normal population, zero correlations and equal variances for they may always by a linear transformation be replaced by such variates.

Let us now take

$$\xi_i,\ x'_{i1},\ x'_{i2},\ \ldots,\ x'_{in}$$

as rectangular coordinates of a point P_i in space V_{n+1} of $n+1$ dimensions. Since these quantities are normally and independently distributed with equal variance about zero, the probability density for P_i has spherical symmetry about the origin. Indefinite repetition of the sampling would result in a globular cluster of representative points for each variate. Actually the sample in hand fixes the points P_1, P_2, . . ., P_p, which may be regarded as taken independently.

We shall now show that T is a function of the angle θ between the ξ-axis and the flat space V_p containing the points P_1, P_2, . . ., P_p and the origin 0. We shall denote by A the point on the ξ-axis of coordinates 1, 0, 0, . . ., 0, and by V_n the flat space containing the remaining axes. Since in V_{n+1} one equation specifies V_n and $n+1-p$ equations V_p, the intersection of V_n and V_p is specified by all these $n+2-p$ equations, and is therefore of $p-1$ dimensions. Call it V_{p-1}

If P_1, P_2, . . ., P_p be moved about in V_p, θ will not change, and neither will T, since T is invariant under linear transformations, equivalent to such motions of the P_i. Hence T always has the value which it takes if all the lines OP_1, OP_2, . . ., OP_p are perpendicular, with the last $p-1$ of these lines lying in V_{p-1}. In this case the angle AOP_1 equals θ. Applying to the coordinates of A and of P_1 the formula for the cosine of an angle at the origin of lines to (x_1, x_2, . . .) and (y_1, y_2, . . .), namely,

$$\cos\theta = \frac{\Sigma xy}{\sqrt{\Sigma x^2 \Sigma y^2}} \tag{22}$$

We obtain

$$\cos\theta = \frac{\xi}{\sqrt{\xi_1^2 + x_{11}'^2 + \cdots + x_{1n}'^2}}.$$

Since $x_{11}'^2 + \cdots + x_{1n}'^2 = na_{11}$,

it follows that

$$(23) \qquad n \cot^2 \theta = \xi_1^2 / a_{11}.$$

The fact that $P_2, P_3, \ldots, P_p$ lie in V_{p-1}, and therefore in V_n, shows that in this case

$$\xi_2 = \xi_3 = \cdots = \xi_p = 0.$$

Because $OP_1', OP_2', \ldots, OP_p'$ are mutually perpendicular, (20) and (22) show that $a_{ij} = 0$ whenever $i \neq j$. Hence, by (21) and (23),

$$(24) \qquad T = \xi_1 / a_{11} = \sqrt{n} \cot \theta.$$

By this result the problem of the sampling distribution of T is reduced to that of the angle θ between a line OA in V_{n+1} and the flat space V_p containing p other lines drawn independently through the origin. The distribution will be unaffected if we suppose V_p fixed and OA drawn at random, with spherical symmetry for the points A.[1] Let us then, abandoning the coordinates hitherto used, take new axes of rectangular coordinates $y_1, y_2, \ldots, y_{n+1}$, of which the first p lie in V_p. A unit hypersphere about 0 is defined in terms of the general-

[1]This geometrical interpretation of T shows its affinity with the multiple correlation coefficient, whose interpretation as the cosine of an angle of a random line with a V_p enabled R. A. Fisher to obtain its exact distribution (Phil. Trans., vol. 213B, 1924, p. 91; and Proc. Roy. Soc., vol. 121A, 1928, p. 654). The omitted steps in Fisher's argument may be supplied with the help of generalized polar coordinates as in the text. Other examples of the use of these coordinates in statistics have been given by the author in The Distribution of Correlation Ratios Calculated from Random Data, Proc. Nat. Acad. Sci., vol. 11 (1925), p. 657, and in The Physical State of Protoplasm, Koninklijke Akademie van Wetenschappen te Amsterdam, verhandlingen, vol. 25 (1928), no. 5, pp. 28-31.

ized latitude-longitude parameters $\phi_1, \ldots, \phi_n$ if we put

$$
\begin{aligned}
y_1 &= \sin\phi_1 \sin\phi_2 \sin\phi_3 \cdots \sin\phi_{p-1} \cos\phi_p \\
y_2 &= \cos\phi_1 \sin\phi_2 \sin\phi_3 \cdots \sin\phi_{p-1} \cos\phi_p \\
y_3 &= \cos\phi_2 \sin\phi_3 \cdots \sin\phi_{p-1} \cos\phi_p \\
y_4 & \cos\phi_3 \cdots \sin\phi_{p-1} \cos\phi_p \\
&\cdots\cdots\cdots\cdots\cdots\cdots \\
y_p &= \cos\phi_{p-1} \cos\phi_p \\
y_{p+1} &= \sin\phi_p \cos\phi_{p+1} \\
&\cdots\cdots\cdots\cdots\cdots\cdots \\
y_n &= \sin\phi_p \sin\phi_{p+1} \cdots \cos\phi_n \\
y_{n+1} &= \sin\phi_p \sin\phi_{p+1} \cdots \sin\phi_n,
\end{aligned}
$$

for the sum of the squares is unity.. Since

$$y_{p+1}^2 + \cdots + y_{n+1}^2 = \sin^2\phi_p$$

we have

$$\phi_p = \theta .$$

The element of probability is proportional to the element of generalized area, which is given by

$$\sqrt{D}\, d\phi_1\, d\phi_2 \cdots d\phi_n ,$$

where D is an n-rowed determinant in which the element in the ith row and jth column is

$$\sum_{k=1}^{n+1} \frac{\partial y_k}{\partial \phi_i} \cdot \frac{\partial y_k}{\partial \phi_j}$$

For $i \neq j$, this is zero. Of the diagonal elements, the first p-1 contain the factor $\cos^2 \phi_p$; the pth is unity; and the remaining $n-p$ elements contain the factor $\sin^2 \phi_p$. Since ϕ is not otherwise involved, the element of area is the product of

$$\cos^{p-1} \phi_p \sin^{n-p} \phi_p \, d\phi_p$$

by factors independent of ϕ_p. The distribution function of θ is obtained by replacing ϕ_p by θ and integrating with respect to the other parameters. Since θ lies between 0 and $\pi/2$, we divide by the integral between these limits and obtain for the frequency element,

$$\frac{2\Gamma\left(\frac{n+1}{2}\right)}{\Gamma\left(\frac{p}{2}\right)\Gamma\left(\frac{n-p+1}{2}\right)} \cos^{p-1}\theta \sin^{n-p}\theta \, d\theta.$$

Substituting from (24) we have as the distribution of T:

$$(25) \qquad \frac{2\Gamma\left(\frac{n+1}{2}\right)}{\Gamma(p/2)\Gamma\left(\frac{n-p+1}{2}\right) n^{p/2}} \quad \frac{T^{p-1}\, dT}{\left(1+\frac{T^2}{n}\right)^{\frac{n+1}{2}}}$$

For $p = 1$ this reduces to the form of Student's distribution given by Fisher and tabulated in the issue of Metron cited; however, as T may be negative as well as positive in this case, Fisher omits the factor 2.

For $p = 2$ the distribution becomes

$$\frac{n-1}{n} \quad \frac{T\, dT}{\left(1+\frac{T^2}{n}\right)^{\frac{n+1}{2}}}.$$

From this it is easy to calculate as the probability that a given value of T will be exceeded by chance,

$$(26) \qquad P = \frac{1}{\left(1+\frac{T^2}{n}\right)^{\frac{n-1}{2}}}$$

a very convenient expression.

The probability integral for higher values of p may be calculated in various ways, the most direct being successive integration by parts, giving a series of terms analogous to (26) to which, if p is odd, is added an integral which may be evaluated with the help of the tables of Student's distribution. If p is large, this process is laborious; but other methods are available.

The probability integral is reduced to the incomplete beta function if we put

$$x = (1 + T^2/n)^{-1},$$

for then the integral of (25) from T to infinity becomes

$$P = I_x\left(\frac{n-p+1}{2}, \frac{p}{2}\right),$$

the notation being.

$$B_x(p,q) = \int_0^x x^{p-1}(1-x)^{q-1}\,dx,$$

$$B(p,q) = \int_0^1 x^{p-1}(1-x)^{q-1}\,dx,$$

$$I_x(p,q) = \frac{B_x(p,q)}{B(p,q)}.$$

Many methods of calculation have been discussed by H. E. Soper[1] and by V. Romanovsky.[2] An extensive table of the incomplete beta function being prepared under the supervision of Professor Karl Pearson has not yet been published.

Perhaps the most generally useful method now available is

[1]Tracts for Computers, no. 7 (1921).

[2]On certain expansions in series of polynomials of incomplete B-functions (in English), Recueil Math. de la Soc. de Moscou, vol. 33 (1926), pp. 207-229.

to make the substitution

$$z = \frac{1}{2} \log_e (n-p+1) T^2 - \frac{1}{2} \log_e np,$$

$$n_1 = p$$

$$n_2 = n-p+1,$$

reducing (25) to a form considered by Fisher. Table VI in his book, Statistical Methods for Research Workers, gives the values of z which will be exceeded by chance in 5 per cent and in 1 per cent of cases. If the value of z obtained from the data is greater than that in Fisher's table, the indication is that the deviations measured are real.

If the variances and covariances are known a priori, they are to be used instead of the a_{ij} ; the resulting expression T has the well known distribution of χ, with p degrees of freedom. For very large samples the estimates of the covariances from the sample are sufficiently accurate to permit the use of the χ distribution for T. This is well shown by (25), in which, as n increases, the factor involving T approaches

$$T^{p-1} e^{-T^2/2} dT,$$

which is proportional to the frequency element for χ when χ is put for T.

As Pearson pointed out, the labor of calculating χ, which we replace by T, is prohibitive when forty or fifty characters are measured on each individual. With two, three, or four characters, however, the labor is very moderate, and the results far more accurate than any attainable with the Pearson coefficient. The great advantage of using T is the simplicity of its distribution, with its complete independence of any correlations among the variates which may exist in the population.

To means of a single variate it is customary to attach a

"probable error," with the assumption that the difference between the true and calculated values is almost certainly less than a certain multiple of the probable error. A more precise way to follow out this assumption would be to adopt some definite level of probability, say $P = .05$, of a greater discrepancy, and to determine from a table of Student's distribution the corresponding value of t, which will depend on n; adding and subtracting the product of this value of t by the estimated standard error would give upper and lower limits between which the true values may with the given degree of confidence be said to lie. With T an exactly analogous procedure may be followed, resulting in the determination of an ellipse or ellipsoid centered at the point ξ_1, ξ_2, . . ., ξ_p. Confidence corresponding to the adopted probability P may then be placed in the proposition that the set of true values is represented by a point within this boundary.

3

Reprinted from *Ann. Math. Stat.*, **30**(2), 318–340 (1959)

SOME CONTRIBUTIONS TO ANOVA IN ONE OR MORE DIMENSIONS: II

BY S. N. ROY AND R. GNANADESIKAN[1]

University of North Carolina

0. Introduction and notation. This paper presents certain natural extensions, to the multi-dimensional or multivariate situation, of the results contained in the first paper [10] by the authors. We shall use the same notation as before and, in addition, we shall use the following notation: $c(A)$ will denote all the characteristic roots of the matrix A, and if A is at least positive semi-definite, then $c_{\min}(A)$ and $c_{\max}(A)$ will denote, respectively, the smallest and the largest of these roots; $D_a(p \times p)$ will denote a diagonal matrix whose elements are a_1, $a_2, \cdots, a_p$; $\tilde{T}(p \times p)$ will denote a triangular matrix whose non-zero elements are along and below the diagonal; $|A|$ will denote the determinant of a square matrix A; and, $A(p \times p) \cdot \times B(q \times q)$ will denote the Kronecker product or right direct product [5] of the matrices A and B. Also min (p, q) will denote the lesser of the two real numbers p and q.

1. Resume of problems and results under the multivariate Model I of ANOVA.

1.1 *The multivariate Model I*: Let $X(p \times n) = p[\mathbf{x}_1 \, \mathbf{x}_2 \cdots \mathbf{x}_n]$ be a set of n observable stochastic p-vectors such that

$$X'(n \times p) = A(n \times m)\xi(m \times p) + \epsilon(n \times p), \qquad m < n,$$

(1.1.1)
$$= n\underset{r \quad (m-r)}{[A_I \quad A_D]}\begin{bmatrix}\xi_I\\ \xi_D\end{bmatrix}\begin{matrix} r \\ (m-r)\end{matrix} + \epsilon(n \times p) \quad \text{(say)},$$
$$p$$

where, as in the univariate situation, A is the *design* matrix with rank $(A) = r \leqq m \leqq n$, and A_I is a basis of A with a consequent partitioning of ξ into $\xi_I(r \times p)$ and $\xi_D(\overline{m - r} \times p)$, and where

(i) $\xi(m \times p)$ is a set of unknown parameters;

(ii) $\epsilon(n \times p)$, whose elements are physically of the nature of errors, is a random sample of size n from the non-singular p-variate normal $N[\mathbf{O}(p \times 1), \Sigma(p \times p)]$. Furthermore, we assume here that $p \leqq (n - r)$.

Under this model it is seen that $\mathbf{x}_i(p \times 1)$, for $i = 1, 2, \cdots, n$, are n independent stochasic p-vectors such that $\mathbf{x}_i$ is $N[E(\mathbf{x}_i), \Sigma]$, where the unknown dispersion matrix $\Sigma(p \times p)$ is the same for all the n vectors, and $E(\mathbf{x}_i)$, for $i = 1, 2, \cdots, n$, is given by

(1.1.2)
$$E[X'](n \times p) = A(n \times m)\xi(m \times p) = [A_I \; A_D]\begin{bmatrix}\xi_I\\ \xi_D\end{bmatrix}.$$

Received July 24, 1957.

[1] This research was sponsored by the United States Air Force through the Office of Scientific Research of the Air Research and Development Command.

Here again, as in the univariate case, the assumption of normality in (ii) of the above model is not necessary for purposes of linear estimation. In fact, since the linear estimation part of the present problem can be easily handled and may not be of much additional interest, we skip it and proceed directly to the solutions of the problems of testing of linear hypotheses and the associated confidence bounds.

1.2 *Testing of linear hypotheses.* The hypothesis that we seek to test, under the model of section 1.1, is

$$H_0\colon C(s \times m)\xi(m \times p)M(p \times u) = 0(s \times u),$$

$$\text{(1.2.1)} \qquad \text{or,} \quad s\begin{bmatrix} \underset{r}{C_1} & \underset{(m-r)}{C_2} \end{bmatrix} \begin{bmatrix} \xi_I(r \times p) \\ \xi_D((m-r) \times p) \end{bmatrix} M(p \times u) = 0(s \times u),$$

against

$$H\colon C\xi M = \eta(s \times u) \neq 0(s \times u),$$

where C and M are matrices given by the hypothesis, and hence called the *hypothesis* matrices, such that rank $(C) = s \leqq r \leqq m < n$ and rank $(M) = u \leqq p$ and $\eta(s \times u)$ is an arbitrary unspecified nonnull matrix. Notice that s may be greater than or equal to or less than u. One can, of course, verify that (1.2.1) is by no means the most general type of linear hypothesis imaginable, although it includes a wide variety of linear hypotheses in which we might be interested. The main results follow. [12, pp. 84a–84i]

(1.2.2) All the following results are invariant under the choice of a basis A_I of A (with a consequent determination of ξ_I and C_1).

(1.2.3) Whether we use the likelihood ratio criterion or the one used by the authors, [12], we have a similar notion of *testability* for this situation as for the univariate case, and the *testability* condition is the same as (1.3.3) of [10].

(1.2.4) The test itself is given by the following rule:

Reject H_0 against H if $c_{\max}(S_1S^{-1}) \geqq c_\alpha(u, s, n-r)$ and accept (do not reject) H_0 against H otherwise, where $c_{\max}(S_1S^{-1})$ denotes the largest characteristic root (necessarily positive except on a set of probability measure zero) of S_1S^{-1}, $c_\alpha(u, s, n-r)$ to be called c_α, for shortness, is a constant which depends on the level of significance α and the degrees of freedom u, s and $(n-r)$ and which is being tabulated from the relation

$$\text{(1.2.5)} \qquad P[c_{\max}(S_1S^{-1}) \geqq c_\alpha \mid H_0] = \alpha,$$

the distribution involved being long available [11, 12]. Here, S_1 is an $u \times u$ symmetric and at least positive semi-definite matrix of rank, almost everywhere, (i.e., except on a set of probability measure zero), $\min(u, s)$, being given by

$$\text{(1.2.6)} \quad sS_1(u \times u) = M'XA_I(A_I'A_I)^{-1}C_1'[C_1(A_I'A_I)^{-1}C_1']^{-1}C_1(A_I'A_I)^{-1}A_I'X'M,$$

and S in an $u \times u$ symmetric and, almost everywhere, positive definite matrix of rank u (necessarily), given by

$$\text{(1.2.7)} \qquad (n-r)S(u \times u) = M'X[I(n) - A_I(A_I'A_I)^{-1}A_I']X'M.$$

We shall call the matrix on the right side of (1.2.6) the *matrix due to the hypothesis* (1.2.1) and the matrix on the right side of (1.2.7) the *matrix due to error*.

The reduction, to a canonical form, of the relevant distribution problem is one in which the characteristic roots of S_1S^{-1} are the same as those of $[((n - r) / s)Y_1Y_1'(YY')^{-1}]$, where $Y_1(u \times s)$ and $Y(u \times (n - r))$ have, in general, i.e., under H, the distribution

$$(1.2.8)\qquad (2\pi)^{-[u(n-r+s)]/2} \exp\left[\frac{-1}{2}\left\{\operatorname{tr}(Y_1Y_1' + YY') + \sum_{i=1}^{\min(u,s)} \gamma_i - 2\sum_{i=1}^{\min(u,s)} (Y_1)_{ii}\sqrt{\gamma_i}\right\}\right] dY_1\, dY,$$

where γ_i's $(i = 1, 2, \cdots, \min(u, s))$ are the possibly non-zero characteristic roots of the $u \times u$ matrix $\eta'[C_1(A_I'A_I)^{-1}C_1']^{-1}\eta(M'\Sigma M)^{-1}$. It is to be noted that the u characteristic roots of this matrix are all non-negative and t of them are positive while the rest, $(u - t)$ in number, are zero, where $t(\leqq \min(u, s))$ is the rank of η. All the roots are zero if, and only if, $\eta = 0$, i.e., under H_0, and in this case we have, for Y_1 and Y, the distribution

$$(1.2.9)\qquad (2\pi)(^{-[u(n-r+s)]/2} \exp[-\tfrac{1}{2}\operatorname{tr}(Y_1Y_1' + YY')]dY_1\, dY.$$

The distribution of $c_{\max}(S_1S^{-1})$, i.e., of $c_{\max}[(n - r / s)Y_1Y_1'(YY')^{-1}]$, on the null hypothesis H_0, was obtained earlier [11, 12] starting from (1.2.9), and this forms the basis of the tables, now being prepared, giving $c_\alpha(u, s, n - r)$ when α, u, s and $(n - r)$ are prescribed. It may be noted, from (1.2.8) and (1.2.9), that Y_1 and Y are independently distributed.

We can introduce here, just as in the univariate case, the notion of two or more different hypotheses like (1.2.1) being testable in a *quasi-independent* manner and can derive a set of necessary and sufficient conditions for this. In fact, when the hypotheses differ only in their C matrices and have the same M matrix, so that, we have, for instance

$$(1.2.10)\qquad H_{0i}\colon\ C_i(s_i \times m)\xi(m \times p)M(p \times u) = 0(s_i \times u), \qquad \text{for } i = 1, 2, \cdots, k,$$

against respective alternatives H_i, like H defined under (1.2.1), where rank $(C_i) = \operatorname{rank}[\underset{r}{C_{i1}}\ \underset{(m-r)}{C_{i2}}]s_i = s_i$, and $\sum_{i=1}^{k} s_i \leqq r \leqq m < n$, then, the necessary and sufficient condition for being able to test the k hypotheses (1.2.10) in a *quasi-independent* manner is that

$$(1.2.11)\qquad C_{i1}(A_I'A_I)^{-1}C_{j1}' = 0(s_i \times s_j), \qquad (i \neq j = 1, 2, \cdots, k),$$

which is the same as condition (1.3.7) of [10] for the univariate case.

1.3. *The associated confidence bounds.* Going back to (1.2.1), we observe that $\eta(\neq 0)$ represents a deviation from H_0. The main results follow [9].

With a joint confidence coefficient $\geqq (1 - \alpha)$, for a preassigned α, we have the following simultaneous confidence bounds:

$$(1.3.1)\qquad \begin{aligned} c_{\max}^{\frac{1}{2}}(sS_1) - [sc_\alpha]^{\frac{1}{2}}c_{\max}^{\frac{1}{2}}(S) &\leq c_{\max}^{\frac{1}{2}}[\eta'(C_1(A_I'A_I)^{-1}C_1')^{-1}\eta] \\ &\leq c_{\max}^{\frac{1}{2}}(sS_1) + [sc_\alpha]^{\frac{1}{2}}c_{\max}^{\frac{1}{2}}(S); \end{aligned}$$

and similar confidence bounds in terms of the same c_α but of truncated η, S_1 and S obtained by going back to (1.2.1) for η, to (1.2.6) for S_1 and to (1.2.7) for S, and then (i) cutting out any row of η' and the corresponding row of M', or any two rows of η' and the corresponding two rows of M' and so on till we get down to just any row of η' and the corresponding row of M', and also (ii) cutting out any column of η' and the corresponding row of C_1, any two columns of η' and the two corresponding rows of C_1 and so on; and finally (iii) combining any case of truncation under (i) with any case of truncation under (ii). Thus, with a joint probability $\geq (1 - \alpha)$, we have $(2^u - 1) \times (2^s - 1)$ statements of which (1.3.1) is the first one.

As in the univariate case [10, Section 1.4], it is to be observed that, taken together, the $(2^u - 1) \times (2^s - 1)$ confidence statements enable us to put confidence bounds on parametric functions which are not only measures of deviation from the total null hypotheses (1.2.1) but also on all the component parts of it.

It may be noted that if, in the H_0 of (1.2.1), $M(p \times u)$ is not present, as will be seen in the hypotheses of Section 2.4, then all the above results go through if we replace u by p and $M(p \times u)$ by the identity matrix $I(p)$.

2. Multivariate Variance Components.

2.1 *The multivariate Model II*: Let $X(p \times n) = [\mathbf{x}_1\mathbf{x}_2 \cdots \mathbf{x}_n]$ be a set of n observable stochastic p-vectors such that

$$(2.1.1)\qquad \begin{aligned} X'(n \times p) &= A(n \times m)\xi(m \times p) + \epsilon(n \times p), \qquad m < n, \\ &= {}_n[A_1\, A_2 \cdots A_k] \begin{bmatrix} \xi_1 \\ \xi_2 \\ \vdots \\ \xi_k \end{bmatrix}_p \begin{matrix} m_1 \\ m_2 \\ \vdots \\ m_k \end{matrix} + \epsilon(n \times p), \quad \sum_{i=1}^k m_i = m, \quad \text{(say)}, \\ &\qquad\ \ m_1\, m_2 \cdots m_k \end{aligned}$$

where A is the *design* matrix of rank $r \leq m < n$, and where

(i) $\xi_i(m_i \times p)$ is a random sample of size m_i from the p-variate non-singular normal population $N[\mathbf{\mu}_i(p \times 1), \Sigma_i(p \times p)]$ for $i = 1, 2, \cdots, k$, and $\epsilon(n \times p)$ and ξ_i's (for $i = 1, 2, \cdots, k$) are mutually independent;

(ii) $\epsilon(n \times p)$, whose elements are physically of the nature of errors, is a random sample of size n from the p-variate non-singular $N[\mathbf{O}(p \times 1), \Sigma(p \times p)]$. Furthermore, we assume that $p \leq (n - r)$.

Writing

$$X'(n \times p) = {}_n[\mathbf{x}_1\, \mathbf{x}_2 \cdots \mathbf{x}_p] \quad \text{and} \quad \mathbf{x}(pn \times 1) = \begin{bmatrix} \mathbf{x}_1 \\ \mathbf{x}_2 \\ \vdots \\ \mathbf{x}_p \end{bmatrix},$$

we see that, under the above model, the elements of $X'(n \times p)$, i.e., of $\mathbf{x}(pn \times 1)$, have a pn-variate non-singular normal distribution $N[E(\mathbf{x}), \Sigma^*(pn \times pn)]$, where

$$E(\mathbf{x})(pn \times 1) = A^*(pn \times pm) \begin{bmatrix} \mu_{11}\cdot\mathbf{1} \\ \vdots \\ \mu_{k1}\cdot\mathbf{1} \\ \hline \vdots \\ \hline \mu_{1p}\cdot\mathbf{1} \\ \vdots \\ \mu_{kp}\cdot\mathbf{1} \end{bmatrix} \begin{matrix} m_1, \\ \vdots \\ m_k \\ \vdots \\ m_1 \\ \vdots \\ m_k \end{matrix} \tag{2.1.2}$$

and where

$$\boldsymbol{\mu}_i(p \times 1) = \begin{bmatrix} \mu_{i1} \\ \mu_{i2} \\ \vdots \\ \mu_{ip} \end{bmatrix} \quad \text{and} \quad A^*(pn \times pm) = A(n \times m)\cdot \times I(p)$$
$$= \begin{matrix} n \\ n \\ \cdot \\ n \end{matrix} \begin{bmatrix} A & 0 & \cdots & 0 \\ 0 & A & \cdots & 0 \\ \cdot & \cdot & \cdots & \cdot \\ 0 & 0 & \cdots & A \end{bmatrix}, \\ \quad\ \ m \quad m \quad \cdots \quad m$$

and

$$\begin{aligned} \Sigma^*(pn \times pn) &= E(\mathbf{xx}') - E(\mathbf{x})E(\mathbf{x}') \\ &= A_1A_1'\cdot\times\Sigma_1 + A_2A_2'\cdot\times\Sigma_2 + \cdots + A_kA_k'\cdot\times\Sigma_k \\ &\quad + I(n)\cdot\times\Sigma \end{aligned} \tag{2.1.3}$$

if we recall and use the Kronecker product notation $A\cdot\times B$.

We shall, in this paper, consider, in detail, only the relatively more restricted model wherein

$$\Sigma_i(p \times p) = \sigma_i^2\Sigma(p \times p), \tag{2.1.4}$$

since, as will be shown in section 2.3, the more general set-up of the model defined above does not lend itself to an easy mathematical treatment. We shall call the model defined at the beginning of this section, taken together with the restriction (2.1.4), as the restricted multivariate Model II of ANOVA. Federer [3] points out that models, where dispersion matrices are proportional, have been tentatively proposed for a certain type of genetical problem so that our restricted model might still be meaningful in certain physical situations.[2]

[2] Since this paper was written up and submitted in July, 1957 further investigation showed that even without this (rather severe and unrealistic) restriction it was still possible to go ahead with (i) point estimation, (ii) testing of hypothesis and (iii) confidence interval estimation, but in terms of a different set of statistics leading up to results less sharp than those aimed at here. The mathematical tools needed are those given here plus some further tools. Thus, from a physical standpoint, this paper might be regarded as an indispensable first step toward handling the more realistic situation that does not involve the very restrictive assumption of proportionality. The justification of the present paper from a physical standpoint, in terms of a possible genetical application, is thus today entirely redundant.

Our objectives will be: (i) to estimate any *estimable* linear function of the elements of $\mathbf{u}_1, \cdots, \mathbf{u}_k$ and to test *testable* linear hypotheses on $\mathbf{u}_1, \cdots, \mathbf{u}_k$; (ii) to obtain estimates of, and test hypotheses on, the multivariate variance components, viz., the characteristic roots $c(\Sigma_1), c(\Sigma_2), \cdots, c(\Sigma_k)$ and $c(\Sigma)$; and (iii) to obtain confidence bounds (simultaneous and/or separate) on $c(\Sigma_1), \cdots, c(\Sigma_k)$ and $c(\Sigma)$. Under the restricted Model II, of course, (ii) is equivalent to obtaining estimates of, and testing hypotheses on, $\sigma_1^2, \cdots, \sigma_k^2$ and $c(\Sigma)$, while (iii) is equivalent to obtaining confidence bounds on $\sigma_1^2, \sigma_2^2, \cdots, \sigma_k^2$ and $c(\Sigma)$.

2.2 *Linear estimation and testing of linear hypotheses.* Recall that for the restricted k-way classification the design matrix $A(n \times m)$ is such that, for all $i = 1, 2, \cdots, k$, the submatrix $A_i(n \times m_i)$ has one and only one non-zero element, equal to unity, in each row, such that rank $(A) = (m - k + 1)$. When we select n individuals under this design and measure each on not one but p variates we have a multivariate restricted k-way classification analogous to the univariate restricted k-way classification discussed in [10]. Multivariate analogues of the usual complete and incomplete *connected* designs are included under this general case. Using a result given in [12] we can establish the following lemma [4, pp. 96–97]:

LEMMA 1: *For the multivariate restricted k-way classification, the necessary and sufficient condition for the estimability of* $\sum_{i=1}^{k} \mathbf{l}_i'(1 \times p)\mathbf{u}_i(p \times 1)$ *is that*

$$\mathbf{l}_1'(1 \times p) = \mathbf{l}_2'(1 \times p) = \cdots = \mathbf{l}_k'(1 \times p),$$

so that, a linear function $\sum_{i=1}^{k} \mathbf{l}_i'(1 \times p)\mathbf{u}_i(p \times 1)$ *of all the elements of* $\mathbf{u}_1, \cdots, \mathbf{u}_k$, *which is estimable, and hypotheses on which are testable, is of the form* $\mathbf{l}'(1 \times p)[\mathbf{u}_1 + \cdots + \mathbf{u}_k]$, *and hence, neither linear functions of the elements of each* $\mathbf{u}_i$ *nor the elements of each* $\mathbf{u}_i$ *are separately* estimable *and linear hypotheses on these separate functions or elements are not* testable.

2.3. Estimation of the multivariate variance components. Analogous to the univariate χ^2-distribution, we shall introduce, for the multivariate situation, the pseudo-Wishart distribution a definition of which follows.

Suppose $X(p \times n)$ has the distribution

$$(2.3.1) \qquad (2\pi)^{-pn/2}\, |\Sigma|^{-n/2} \exp\left[\frac{-1}{2} \operatorname{tr} \Sigma^{-1}(X - \zeta)(X' - \zeta')\right] dX$$

where the elements of X, x_{ij} are such that $-\infty < x_{ij} < \infty$ $(i = 1, \cdots, p)$ $(j = 1, \cdots, n)$ so that, $E(X) = \zeta(p \times n)$ and the symmetric positive definite matrix $\Sigma(p \times p)$ is interpreted as, $n\Sigma(p \times p) = E[(X - \zeta)(X' - \zeta')]$. Then we shall call the distribution of the symmetric at least positive semi-definite matrix, $S(p \times p) = (1/n)XX'$, the pseudo-Wishart distribution with degrees of freedom n, and the distribution is central or non-central according as $\zeta = 0$ or $\zeta \neq 0$, i.e., according as $\zeta\zeta'$ is the null matrix $0(p \times p)$ or not. Conversely, we shall say that any symmetric at least positive semi-definite matrix, $S(p \times p)$, has the pseudo-Wishart distribution (in general, non-central) with degrees of freedom n, if we can write $S(p \times p) = (1/n)X(p \times n)X'(n \times p)$, where $X(p \times n)$ has the distribution (2.3.1). Further, if $E(X)E(X')$ is the null matrix $0(p \times p)$ then, and then only, will the distribution be said to be central. In particular, if in the above definition, $p \leqq n$ and rank $(S) = p$ then a pseudo-Wishart distribution for S is equivalent to the ordinary Wishart distribution.

Starting from (2.3.1), it can be shown that the distribution of the ith diagonal element, for $i = 1, 2, \cdots, p$, of XX', where of course $(1/n)XX'$ has a pseudo-Wishart distribution with degrees of freedom n, is distributed as $\sigma_{ii}\chi^2_{(n)}$ where σ_{ii} denotes the ith diagonal element of $\Sigma(p \times p)$ and where $\chi^2_{(n)}$ stands for the χ^2 variate, with degrees of freedom n, being central or non-central according as the pseudo-Wishart distribution of $(1/n)XX'$ is central or non-central.

We shall next proceed to problems of estimating and testing hypotheses on the multivariate variance components. For these purposes, by analogy with the univariate case [10, section 2.3], we shall seek $(k + 1)$ matrices, $S_i(p \times p) = (1/n_i)X(p \times n)Q_i(n \times n)X'(n \times p)$ (for $i = 0, 1, \cdots, k$), where $Q_i(n \times n)$ is symmetric and at least positive semi-definite of rank $n_i \leqq n$, $\sum_{i=0}^{k} n_i \leqq n$, such that

(2.3.2) $(1/\lambda_i)S_i$, of rank $\leqq p$, has a central pseudo-Wishart distribution with degrees of freedom n_i $(i = 0, 1, \cdots, k)$, where λ_i is a positive constant;

(2.3.3) $\dfrac{1}{\lambda_0} S_0, \quad \dfrac{1}{\lambda_1} S_1, \cdots, \dfrac{1}{\lambda_k} S_k$ are mutually independent.

LEMMA 2: *If $X(p \times n)$ has the distribution* (2.3.1), *then a set of necessary and sufficient conditions for $S_0, S_1, \cdots, S_k$ to satisfy the above restrictions is given by*

(a) $Q_i^2(n \times n) = \lambda_i Q_i$ *which is equivalent to the statement that* $Q_i = \lambda_i L_i' L_i$ *where* $L_i(n_i \times n)L_i'(n \times n_i) = I(n_i)(i = 0, 1, \cdots, k)$;

(b) $E(X)Q_iE(X') = 0(p \times p)$;

(c) $Q_iQ_j = 0(n \times n)$ *which, taken together with* (a), *is equivalent to the statement that* $L_i(n_i \times n)L_j'(n \times n_j) = 0(n_i \times n_j)$ $(i \neq j = 0, 1, \cdots, k)$.

PROOF: *Necessity of* (a), (b), *and* (c). Suppose that the matrices $S_0, S_1, \cdots, S_k$ satisfy (2.3.2) and (2.3.3). Then, since

$$(1 / \lambda_i)S_i = (1/n_i\lambda_i)XQ_iX' = XP_iX'$$

(where $P_i(n \times n) = (1/\lambda_i)Q_i(n \times n)$) has a central pseudo-Wishart distribution with degrees of freedom n_i, therefore, by our previous discussion, the jth $(j = 1, 2, \cdots, p)$ diagonal element of XP_iX' is *necessarily* distributed as a constant times $\chi^2_{(n_i)}$ where $\chi^2_{(n_j)}$ denotes a (central) χ^2 variate with n_j degrees of freedom. Now, if

$$X(p \times n) = \underset{n}{\begin{bmatrix} \mathbf{x}_1' \\ \mathbf{x}_2' \\ \vdots \\ \mathbf{x}_p' \end{bmatrix}}$$

(say) has the distribution (2.3.1) then $\mathbf{x}_j'(1 \times n)$ has the distribution

$$\frac{1}{(2\pi\sigma_{jj})^{n/2}} \exp\left[-\frac{1}{2\sigma_{jj}}(\mathbf{x}_j' - \zeta_j')(\mathbf{x}_j - \zeta_j)\right] d\mathbf{x}_j,$$

where

$$\zeta(p \times n) = E(X(p \times n)) = \underset{n}{\begin{bmatrix} \zeta_1' \\ \vdots \\ \zeta_p' \end{bmatrix}}$$

and σ_{jj} is the jth diagonal element of $\Sigma(p \times p)$. Using the result of [2, 6], we have that, if $\mathbf{x}_j(n \times 1)$ has the above distribution, then, in order that the jth diagonal element of XP_iX', i.e., $\mathbf{x}_j'P_i\mathbf{x}_j$, where $P_i(n \times n)$ is symmetric at least positive semi-definite of rank $n_i \leqq n$, may be distributed as $\sigma_{jj}\chi^2_{(n_i)}$, we must *necessarily* have

$$P_i^2 = P_i, \text{ i.e. } \quad Q_i^2 = \lambda_i Q_i .$$

Hence the necessity of (a).

Next, since $Q_i(n \times n)$ is symmetric positive semi-definite of rank n_i and λ_i is a positive constant, therefore, by a well-known result [12, pp. A-16 and A-17], there exists a transformation

$$\frac{1}{\lambda_i} Q_i = \begin{matrix} n_i \\ n - n_i \end{matrix}\underset{n_i}{\begin{bmatrix} \tilde{T}_1 \\ T_2 \end{bmatrix}} [\tilde{T}_1' \;\; T_2'].$$

If now

$$Y_i(p \times n_i) = X(p \times n) \begin{bmatrix} \tilde{T}_1 \\ T_2 \end{bmatrix},$$

then

$$\frac{1}{n_i\lambda_i} XQ_i X' = \frac{1}{n_i} Y_i Y_i' .$$

Thus, if $(1/n_i\lambda_i)XQ_iX' = (1/n_i)Y_iY_i'$ has a central pseudo-Wishart distribution, then, by definition, $E(Y_i)E(Y_i') = 0(p \times p)$, i.e., $E(X)Q_iE(X') = 0(p \times p)$, which proves the necessity of (b).

Finally, if $(1/\lambda_0)S_0, \cdots, (1/\lambda_k)S_k$ are distributed mutually independently in pseudo-Wishart forms with respective degrees of freedom $n_0, \cdots, n_k$, then, necessarily, their lth $(l = 1, 2, \cdots, p)$ diagonal elements, viz.,

$$\mathbf{x}_l'P_0\mathbf{x}_l, \cdots, \mathbf{x}_l'P_k\mathbf{x}_l,$$

where $P_i = (1/\lambda_i)Q_i\,(i = 0, 1, \cdots, k)$, are distributed as constant multiples of mutually independent χ^2 variates with respective degrees of freedom n_0,

$n_1, \cdots$ and n_k. Hence, from [2, 6], we necessarily have

$$P_iP_j = 0(n \times n) \quad \text{or} \quad Q_iQ_j = 0(n \times n)$$

for $i \neq j = 0, 1, \cdots, k$, which proves the necessity of (c).

Sufficiency of (a), (b), *and* (c). We shall now assume that $S_0, S_1, \cdots, S_k$ satisfy (a), (b) and (c), so that

$$\frac{1}{\lambda_i} Q_i(n \times n) = L_i'(n \times n_i)L_i(n_i \times n)$$

where $L_iL_i' = I(n_i)$ for $i = 0, 1, \cdots, k$ and $L_iL_j' = 0(n_i \times n_j)$ for $i \neq j = 0, 1, \cdots, k$, and, also, $E(X)Q_iE(X') = 0(p \times p)$ for $i = 0, 1, \cdots, k$. When $L_0, L_1, \cdots, L_k$ satisfy these conditions, it is well known that we can find a completion $L^*((n - \sum_{i=0}^k n_i) \times n)$ of the matrix

$$\begin{bmatrix} L_0 \\ L_1 \\ \vdots \\ L_k \end{bmatrix}$$

such that the completed matrix

$$L(n \times n) = \begin{bmatrix} L_0 \\ L_1 \\ \vdots \\ L_k \\ L^* \end{bmatrix}$$

is orthogonal. Let us now make the transformation

$$Y(p \times n) = p[\underset{n_0}{Y_0}\, \underset{n_1}{Y_1} \cdots \underset{n_k}{Y_k} \quad \underset{\left(n - \sum_{i=0}^{k} n_i\right)}{Y^*}] = X(p \times n)[L_0'\ L_1' \cdots L_k'\ L^{*\prime}],$$

so that, the Jacobian of the transformation is unity. Notice that

$$YY' = \sum_{i=0}^{k} Y_i Y_i' + Y^* Y^{*\prime} = \sum_{i=0}^{k} \frac{1}{\lambda_i} XQ_i X' + XL^{*\prime}L^*X'.$$

Starting from the distribution (2.3.1) of $X(p \times n)$, we therefore have for the joint distribution of $Y_0, \cdots, Y_k$ and Y^*

$$(2\pi)^{-pn/2} |\Sigma|^{-n/2} \exp\left[\frac{-1}{2} \operatorname{tr} \Sigma^{-1}\left\{\sum_{i=0}^{k} (Y_i - \eta_i)(Y_i' - \eta_i') \right.\right.$$
$$\left.\left. + (Y^* - \eta^*)(Y^{*\prime} - \eta^{*\prime})\right\}\right] \times \prod_{i=0}^{k} dY_i \cdot dY^*,$$

where

$$p[\eta_0 \ \eta_1 \ \cdots \ \eta_k \quad \eta^*] = \zeta(p \times n)[L_0' \ L_1' \ \cdots \ L_k' \ L^{*\prime}],$$
$$n_0 \ n_1 \ \cdots \ n_k \left(n - \sum_{i=0}^{k} n_i\right)$$

so that, $E(Y_i) = \eta_i \,(i = 0, 1, \cdots, k)$, $E(Y^*) = \eta^*$. The elements of each Y matrix, of course, vary between $-\infty$ and ∞. Integrating out over Y^*, we have for the joint distribution of $Y_0, Y_1, \cdots, Y_k$,

$$(2.3.4) \qquad (2\pi)^{\frac{-p(n_0+\cdots+n_k)}{2}} |\Sigma|^{-\frac{n_0+\cdots+n_k}{2}} \cdot \exp\left[\frac{-1}{2} \operatorname{tr} \Sigma^{-1}\left\{\sum_{i=0}^{k} (Y_i - \eta_i)(Y_i' - \eta_i')\right\}\right] dY_0 \cdots dY_k,$$

where, of course, $E(Y_i) = \eta_i$ and $E[(Y_i - \eta_i)(Y_i' - \eta_i')] = n_i\Sigma(p \times p)$ for $i = 0, 1, \cdots, k$. From (2.3.4), it follows, by definition, that

$$(2.3.5) \qquad \frac{1}{n_i} Y_i Y_i' = \frac{1}{n_i} XL_i' L_i X' = \frac{1}{n_i \lambda_i} XQ_i X' = \frac{1}{\lambda_i} S_i,$$

for $i = 0, 1, \cdots, k$, has a pseudo-Wishart distribution with n_i degrees of free dom. Also, if $E(X)Q_iE(X') = 0(p \times p)$, then, since λ_i is a positive constant' $(1/\lambda_i)E(X)Q_iE(X') = E(X)L_i'L_iE(X') = E(Y_i)E(Y_i') = \eta_i\eta_i' = 0(p \times p)$· Hence, again by definition, the pseudo-Wishart distribution of $(1/\lambda_i)S_i$ $(i = 0, 1, \cdots, k)$ is central. Finally, from (2.3.4), we observe that $Y_0, Y_1, \cdots,$ Y_k are mutually independent, and, hence it follows from (2.3.5) that

$$(1/\lambda_0)S_0, \cdots, (1/\lambda_k)S_k$$

are mutually independent.

Hence the sufficiency of (a), (b) and (c).

LEMMA 3: *If* $X(p \times n)$ *has the distribution*

$$(2.3.6) \quad (2\pi)^{-pn/2} |\Sigma|^{-n/2} |B|^{-p/2} \exp[-\tfrac{1}{2} \operatorname{tr} \Sigma^{-1}(X - \zeta)B^{-1}(X' - \zeta')] \, dX, \qquad -\infty < x_{ij} < \infty,$$

where $B(n \times n)$ *and* $\Sigma(p \times p)$ *are symmetric positive definite, then, a set of necessary and sufficient conditions for* $S_0, S_1, \cdots, S_k$ *(defined immediately before* (2.3.2)*) to satisfy the conditions* (2.3.2) and (2.3.3) *is given by*

(α) $\quad Q_iBQ_i = \lambda_iQ_i, \qquad i = 0, 1, \cdots, k;$

(β) $\quad E(X)Q_iE(X') = 0(p \times p), \qquad i = 0, 1, \cdots, k;$ and

(γ) $\quad Q_iBQ_j = 0(n \times n), \qquad i \neq j = 0, 1, \cdots, k.$

PROOF: Since $B(n \times n)$ is symmetric positive definite, therefore, there exists the transformation

$$B(n \times n) = \tilde{T}(n \times n)\tilde{T}'(n \times n),$$

so that,

$$B^{-1} = (\tilde{T}')^{-1}\tilde{T}^{-1}.$$

Writing $Y(p \times n) = X(p \times n)\,(\tilde{T}')^{-1}$, or, $X = Y\tilde{T}'$, and $\theta(p \times n) = \zeta(p \times n)\,(\tilde{T}')^{-1}$, we have the Jacobian of the transformation to be $|\tilde{T}'|^p = |B|^{p/2}$. Then we notice that $(1/n_i\lambda_i)\,XQ_iX' = (1/n_i\lambda_i)Y\tilde{T}'Q_i\tilde{T}Y'$, and, from (2.3.6), the distribution of $Y(p \times n)$ is

$$(2\pi)^{-pn/2}\,|\Sigma|^{-n/2}\exp\,[-\tfrac{1}{2}\operatorname{tr}\Sigma^{-1}(Y - \theta)(Y' - \theta')]\,dY, \qquad -\infty < y_{ij} < \infty,$$

which is of the same form as (2.3.1). Now applying Lemma 2 to the matrices $(1/n_i\lambda_i)Y\tilde{T}'Q_i\tilde{T}Y'$ (notice that rank $(\tilde{T}'Q_i\tilde{T})$ = rank $(Q_i) = n_i$), we obtain that a set of necessary and sufficient conditions for these matrices to satisfy (2.3.2) and (2.3.3) is

(α) $\quad \tilde{T}'Q_i\tilde{T}\tilde{T}'Q_i\tilde{T} = \lambda_i\tilde{T}'Q_i\tilde{T}$, or, $Q_iBQ_i = \lambda_iQ_i \qquad (i = 0, 1, \cdots, k)$;

(β) $\quad E(Y)\tilde{T}'Q_i\tilde{T}E(Y') = 0(p \times p)$, or $E(X)Q_iE(X') = 0(p \times p)$, $(i = 0, 1, \cdots, k)$; and, finally,

(γ) $\quad \tilde{T}'Q_i\tilde{T}\tilde{T}'Q_j\tilde{T} = 0(n \times n)$, or, $Q_iBQ_j = 0(n \times n)$,

$$(i \neq j = 0, 1, \cdots, k).$$

Hence the lemma is proved.

Next, under the general multivariate Model II, we have noted that

$$\mathbf{x}(pn \times 1) = \begin{bmatrix} \mathbf{x}_1 \\ \vdots \\ \mathbf{x}_p \end{bmatrix}$$

has the distribution

$$(2\pi)^{-pn/2}\,|\Sigma^*|^{-1/2}$$

$$\exp\left[-\frac{1}{2}\left\{[(\mathbf{x}_1' \cdots \mathbf{x}_p') - E(\mathbf{x}_1' \cdots \mathbf{x}_p')]\,\Sigma^{*-1}\left[\begin{bmatrix} \mathbf{x}_1 \\ \vdots \\ \mathbf{x}_p \end{bmatrix} - E\begin{bmatrix} \mathbf{x}_1 \\ \vdots \\ \mathbf{x}_p \end{bmatrix}\right]\right\}\right] d\mathbf{x}, \tag{2.3.7}$$

where

$$E(\mathbf{x}) = E\begin{bmatrix} \mathbf{x}_1 \\ \vdots \\ \mathbf{x}_p \end{bmatrix}$$

and Σ^* are defined respectively in (2.1.2) and (2.1.3). In order that this distribution of $X(p \times n)$ be, essentially, of the same form as (2.3.6), we should be able to express the exponent in (2.3.7), except for a constant factor $(-1)/2$, in the form

$$\operatorname{tr} M_2^{-1}\,(X - \zeta)\,M_1^{-1}\,(X' - \zeta'),$$

where $M_1(n \times n)$ and $M_2(p \times p)$ are symmetric positive definite matrices and where $E(X) = \zeta$.

LEMMA 4: *A necessary and sufficient condition for this is that*

$$\Sigma^*(pn \times pn) = M_1(n \times n) \cdot \times M_2(p \times p). \tag{2.3.8}$$

PROOF: *Sufficiency of the condition.* If $\Sigma^* = M_1 \cdot \times M_2$ then it is known [5] that $\Sigma^{*-1} = M_1^{-1} \cdot x\, M_2^{-1}$. Now, let

$$M_1^{-1}(n \times n) = \begin{bmatrix} m_{11}^{(1)} & m_{12}^{(1)} & \cdots & m_{1n}^{(1)} \\ m_{21}^{(1)} & m_{22}^{(1)} & \cdots & m_{1n}^{(1)} \\ \vdots & \vdots & \cdots & \vdots \\ m_{n1}^{(1)} & m_{n2}^{(1)} & \cdots & m_{nn}^{(1)} \end{bmatrix}, \qquad m_{ij}^{(1)} = m_{ij}^{(1)},$$

and

$$M_2^{-1}\,(p \times p) = \begin{bmatrix} m_{11}^{(2)} & m_{12}^{(2)} & \cdots & m_{1p}^{(2)} \\ m_{21}^{(2)} & m_{22}^{(2)} & \cdots & m_{2p}^{(2)} \\ \vdots & \vdots & \cdots & \vdots \\ m_{p1}^{(2)} & m_{p2}^{(2)} & \cdots & m_{pp}^{(2)} \end{bmatrix}, \qquad m_{ij}^{(2)} = m_{ji}^{(2)}.$$

Then,

$$[\mathbf{x}_1' - E(\mathbf{x}_1'), \quad \cdots, \quad \mathbf{x}_p' - E(\mathbf{x}_p')]\, \Sigma^{*-1} \begin{bmatrix} \mathbf{x}_1 & - & E(\mathbf{x}_1) \\ & \vdots & \\ \mathbf{x}_p & - & E(\mathbf{x}_p) \end{bmatrix}$$

$$= \sum_{i,j=1}^{p} [\mathbf{x}_i' - E(\mathbf{x}_i')]\, M_1^{-1}\,(n \times n) \cdot m_{ij}^{(2)}\, [\mathbf{x}_j - E(\mathbf{x}_j)]$$

$$= \operatorname{tr} M_2^{-1} \begin{bmatrix} \mathbf{x}_1' & - & E(\mathbf{x}_1') \\ & \vdots & \\ \mathbf{x}_p' & - & E(\mathbf{x}_p') \end{bmatrix} \qquad M_1^{-1}\, [\mathbf{x}_1 - E(\mathbf{x}_1), \cdots, \mathbf{x}_p - E(\mathbf{x}_p)]$$

$$= \operatorname{tr} M_2^{-1}\, [X - E(X)]\, M_1^{-1}\, [X' - E(X')].$$

Hence the sufficiency of the condition (2.3.8).

Necessity of the condition. Supposing now that

$$[\mathbf{x}_1' - E(\mathbf{x}_1'), \quad \cdots, \quad \mathbf{x}_p' - E(\mathbf{x}_p')\,]\Sigma^{*-1} \begin{bmatrix} \mathbf{x}_1 & - & E(\mathbf{x}_1) \\ & \vdots & \\ \mathbf{x}_p & - & E(\mathbf{x}_p) \end{bmatrix}$$

$$= \operatorname{tr} M_2^{-1}\, [X - E(X)]\, M_1^{-1}\, [X' - E(X')],$$

and writing M_1^{-1} and M_2^{-1} as before, we can argue backwards in the proof of the sufficiency of the condition and obtain that $\Sigma^{*-1} = M_1^{-1} \cdot \times M_2^{-1}$, so that, $\Sigma^* = M_1 \cdot \times M_2$.

Hence the lemma is proved.

Further, we can establish that for $\Sigma^*(pn \times pn)$, defined in (2.1.3), under the general Multivariate Model II with a perfectly general design matrix, to be expressible as $M_1 \cdot\times M_2$ we have the necessary and sufficient conditions

$$\Sigma_i(p \times p) = \sigma_i^2 \, \Sigma(p \times p), \qquad (i = 1, 2, \cdots, k),$$

which yield the restricted Model II. That these conditions on $\Sigma_i(p \times p)$ are sufficient is easily verified. That they are necessary can be demonstrated as follows where, for simplicity of argument, we assume that $p = 2$.

Suppose $\Sigma^*(pn \times pn) = \Sigma^*(2n \times 2n)$, since $p = 2$, here,

$$= M_1(n \times n) \cdot\times M_2(2 \times 2),$$

where M_1 and M_2 are symmetric positive definite matrices. Then, from (2.1.3), we have

$$A_1A_1' \cdot\times \Sigma_1 + \cdots + A_kA_k' \cdot\times \Sigma_k + I(n) \cdot\times \Sigma = M_1 \cdot\times M_2 .$$

From this we obtain the equations

$$\begin{aligned} A_1A_1' [\sigma_{11}^{(1)} - c_1\sigma_{22}^{(1)}] + A_2A_2' [\sigma_{11}^{(2)} - c_1\sigma_{22}^{(2)}] + \cdots \\ + A_kA_k' [\sigma_{11}^{(k)} - c_1\sigma_{22}^{(k)}] + I(n) [\sigma_{11} - c_1\sigma_{22}] = 0(n \times n), \end{aligned} \tag{2.3.9}$$

and

$$\begin{aligned} A_1A_1' [\sigma_{11}^{(1)} - c_2\sigma_{12}^{(1)}] + A_2A_2' [\sigma_{11}^{(2)} - c_2\sigma_{12}^{(2)}] + \cdots \\ + A_kA_k' [\sigma_{11}^{(k)} - c_2\sigma_{12}^{(k)}] + I(n) [\sigma_{11} - c_2\sigma_{12}] = 0(n \times n), \end{aligned}$$

where $c_1 = (M_2)_{11} / (M_2)_{22}$, $c_2 = (M_2)_{11} / (M_2)_{12}$, $(M_2)_{ij}$ is the ijth $(i, j = 1, 2)$ element of $M_2(2 \times 2)$, and where

$$\Sigma_i(2 \times 2) = \begin{bmatrix} \sigma_{11}^{(i)} & \sigma_{12}^{(i)} \\ \sigma_{12}^{(i)} & \sigma_{22}^{(i)} \end{bmatrix}$$

for $i = 1, 2, \cdots, k$ and

$$\Sigma(2 \times 2) = \begin{bmatrix} \sigma_{11} & \sigma_{12} \\ \sigma_{12} & \sigma_{22} \end{bmatrix}.$$

For the equations (2.3.9) to hold we must have either,

$$\sigma_{11}^{(1)} / \sigma_{22}^{(1)} = \sigma_{11}^{(2)} / \sigma_{22}^{(2)} = \cdots = \sigma_{11}^{(k)} / \sigma_{22}^{(k)} = \sigma_{11} / \sigma_{22} = c_1 ,$$

and

$$\sigma_{11}^{(1)} / \sigma_{12}^{(1)} = \sigma_{11}^{(2)} / \sigma_{12}^{(2)} = \cdots = \sigma_{11}^{(k)} / \sigma_{12}^{(k)} = \sigma_{11} / \sigma_{12} = c_2 ,$$

or,

$$A_iA_i' = a_iI(n), \qquad (i = 1, 2, \cdots, k)$$

where a_i is a scalar constant. These latter conditions on the submatrices of the design matrix are too restrictive and unrealistic, so that, for a perfectly general design matrix, the former conditions hold necessarily if $\Sigma^* = M_1 \cdot\times M_2$, and they are verified to be equivalent to the conditions, $\Sigma_i = \sigma_i^2\Sigma$, for $i = 1, 2, \cdots, k$, where σ_i^2 are certain positive constants. The proof of the necessity of the conditions for general p follows exactly along the same lines.

We have thus set up, for reasons of easier mathematical treatment, the restricted multivariate Model II mentioned in section 2.1, and under this restricted set-up we have, for $X(p \times n)$, the distribution

$$(2.3.10) \qquad (2\pi)^{(-pn/2)} |\Sigma|^{(-n/2)} \left|\sum_{i=1}^{k} \sigma_i^2 A_i A_i' + I(n)\right|^{(-p/2)}$$

$$\cdot \exp\left[-\frac{1}{2} \operatorname{tr} \Sigma^{-1}(X - E(X)) \left(\sum_{i=1}^{k} \sigma_i^2 A_i A_i' + I(n)\right)^{-1}(X' - E(X'))\right] dX,$$

since $|\Sigma^*| = |M_1(n \times n) \cdot\times M_2(p \times p)| = |M_1|^p \, |M_2|^n$ by [5].

Next, suppose that Q_i $(n \times n)$ $(i = 0, 1, \cdots, k)$ is a symmetric at least positive semi-definite matrix of rank n_i $(\leqq n)$ such that $E(X)QE(X') = 0(n \times n)$; then, we have, under the multivariate Model II,

$$\Lambda_i\,(p \times p) = \frac{1}{n_i} E(XQ_iX')$$

$$(2.3.11) \qquad = \frac{1}{n_i} E \begin{bmatrix} \mathbf{x}_1' Q_i \mathbf{x}_1, \cdots, \mathbf{x}_1' Q_i \mathbf{x}_p \\ \mathbf{x}_2' Q_i \mathbf{x}_1, \cdots, \mathbf{x}_2' Q_i \mathbf{x}_p \\ \vdots \quad \cdots \quad \vdots \\ \mathbf{x}_p' Q_i \mathbf{x}_1, \cdots, \mathbf{x}_p' Q_i \mathbf{x}_p \end{bmatrix}$$

$$= \frac{1}{n_i}\left[\sum_{j=1}^{k} \operatorname{tr}\,(A_j A_j' Q_i)\, \Sigma_j + (\operatorname{tr} Q_i)\, \Sigma\right],$$

by using Lemma 3 of [10] and simplifying,

$$= \frac{1}{n_i}\left[\sum_{j=i}^{k} \sigma_j^2 \operatorname{tr}\,(A_j A_j' Q_i) + \operatorname{tr} Q_i\right] \Sigma(p \times p),$$

for the restricted multivariate Model II.

Also, if $(1/\lambda_i)S_i\ (p \times p) = (1 / n_i\lambda_i)\ XQ_iX'$ has a central pseudo-Wishart distribution with degrees of freedom n_i (= rank of Q_i) then, under the restricted Multivariate Model II where $X(p \times n)$ has the distribution (2.3.10), we have

$$E(S_i) = \lambda_i\Sigma = \Lambda_i\ (p \times p),$$

so that, from (2.3.11),

$$(2.3.12) \qquad \lambda_i = \frac{1}{n_i}\left[\sum_{j=1}^{k} \sigma_j^2 \operatorname{tr}\,(A_j A_j' Q_i) + \operatorname{tr} Q_i\right].$$

Again, under the restricted multivariate Model II, if we apply the conditions (α), (β), and (γ) of Lemma 3 (remembering that $B(n \times n)$ of (2.3.6) is now replaced by $\sum_{i=1}^{k} \sigma_i^2 A_i A_i' + I(n)$) to the $(k + 1)$ matrices $(1/\lambda_0) S_0, \cdots (1/\lambda_k) S_k$, and then require, as in the univariate case discussed in [10, Section 2.3], that these matrices satisfy the conditions (α) and (γ) for all $\sigma_1^2, \cdots, \sigma_k^2$, we have, after some simplification,

$$\left.\begin{array}{l} Q_i A_l A_l' Q_i = \left[\frac{1}{n_i} \operatorname{tr} A_l A_l' Q_i\right] Q_i, \\ \qquad\qquad\qquad\qquad l = 1, 2, \cdots, k; \\ Q_i^2 = \left[\frac{1}{n_i} \operatorname{tr} Q_i\right] Q_i \end{array}\right\} (i = 0, 1, \cdots, k), \tag{2.3.13}$$

and

$$Q_i A_l A_l' Q_j = 0(n \times n), \quad l = 1, 2, \cdots, k; \quad Q_i Q_j = 0(n \times n) \quad \text{for } i \neq j = 0, 1, \cdots, k. \tag{2.3.14}$$

It is seen that these conditions, (2.3.13) and (2.3.14), are exactly the same as those obtained for the univariate problem [Cf. (2.3.4) and (2.3.5) of [10]]. Thus, for a given design matrix $A(n \times m)$, the same $Q_0, Q_1, \cdots, Q_k$ which satisfy (2.3.4) and (2.3.5) of [10] for the univariate case, also satisfy (2.3.13) and (2.3.14) under the restricted multivariate Model II set-up. Also, for given $Q_0, Q_1, \cdots, Q_k$, the same design matrix $A(n \times m)$, which satisfies (2.3.4) and (2.3.5) of [10] under the univariate Model II set-up, also satisfies (2.3.13) and (2.3.14) under the restricted multivariate Model II set-up.

We shall next present a tie-up, for the multivariate restricted k-way classification, between the analysis under Model I and the analysis under Model II.

2.4. *Tie-up between the analyses under the multivariate Models I and II for the restricted k-way classification.* We recall from section 1.2 that, under the multivariate Model I, we can obtain k matrices due to the k *testable* hypotheses of equality of the row vectors of $\xi_i(m_i \times p)$ $(i = 1, 2, \cdots, k)$, which can, by analogy with (1.2.1), be written as

$$\begin{aligned} H_{0i} : C_i((m_i - 1) \times m)\xi(m \times p) \\ = [\underset{r}{C_{i1}} \quad \underset{m-r}{C_{i2}}]\xi, \qquad & \text{where } r = rank(A) = m - k + 1 \text{ and } m = \sum_{i=1}^{k} m_i, \end{aligned} \tag{2.4.1}$$

$$= (m_i - 1)\left[\begin{array}{c|c|c|cccc|c|c|c} & & & 1 & 0 & \cdots & 0 \; -1 & & & \\ 0 & 0 & \cdots & 0 & 1 & \cdots & 0 \; -1 & 0 & \cdots & 0 \\ & & & \cdot & \cdot & \cdots & \cdot \quad \cdot & & & \\ & & & 0 & 0 & \cdots & 1 \; -1 & & & \\ \end{array}\right] \cdot \xi$$

$$\qquad\qquad m_1 \quad m_2 \quad \cdots \qquad m_i \qquad \cdots \qquad m_k$$

$$= 0((m_i - 1) \times p),$$

so that rank $(C_i) = (m_i - 1)$ $(i = 1, 2, \cdots, k)$. As in section 1.2, we can obtain *k matrices due to the k hypotheses* $H_{01}, H_{02}, \cdots, H_{0k}$, viz.,

$$XA_I(A_I'A_I)^{-1}C_{i1}' [C_{i1}(A_I'A_I)^{-1}C_{i1}']^{-1} C_{i1}(A_I'A_I)^{-1}A_I'X' \tag{2.4.2}$$

for $i = 1, 2, \cdots, k$, and these matrices are symmetric and at least positive semi-definite of rank, almost everywhere, $\min(p, (m_i - 1))$. We have, also, the *matrix due to error*

$$X[I(n) - A_I(A_I'A_I)^{-1}A_I']X' \tag{2.4.3}$$

which is symmetric positive definite (almost everywhere) since we assume in the model that $p \leqq (n - r)$, where for the restricted k-way classification $r = (m - k + 1)$.

Now, under the multivariate Model II, in the notation of section 2.3, suppose we take $n_0 S_0 = XQ_0X'$ as the *matrix due to error* given by (2.4.3) with $n_0 = (n - r) = (n - m + k - 1)$, and $n_i S_i = XQ_iX'$, for $i = 1, 2, \cdots, k$, as the *matrices due to hypothesis* given by (2.4.2) with $n_i = \text{rank } (Q_i) = (m_i - 1)$. Notice that $\sum_{k}^{i=0} n_i = (n - 1) < n$. It is seen that these Q_i, $i = 0, 1, \cdots, k$, are the same as those for the univariate case [Cf. section 2.4 of [10]]. We may verify that all these Q_i's are such that $E(X)Q_iE(X') = 0(p \times p)$ under the multivariate Model II and hence we can obtain that

$$\begin{aligned} \Lambda_i(p \times p) &= \frac{1}{n_i} E(XQ_i X') \\ &= \nu_i \Sigma_i + \Sigma, \end{aligned} \tag{2.4.4}$$

for the general multivariate Model II, where

$$\begin{aligned} \nu_i &= \frac{2}{(m_i - 1)} \left\{ \begin{matrix} \text{sum of the elements along and below the diagonal of} \\ [C_{i1}(A_I' A_I)^{-1}C_{i1}']^{-1} \end{matrix} \right\} \\ &= (\nu_i \sigma_i^2 + 1)\Sigma, \end{aligned}$$

for the restricted multivariate Model II, $(i = 1, 2, \cdots, k)$, and

$$\Lambda_0 = \frac{1}{n_0} E(XQ_0 X') = \Sigma.$$

Therefore, under the restricted multivariate Model II, we have, for the above $Q_i(n \times n)$, that

$$\lambda_i = \nu_i\sigma_i^2 + 1, \qquad i = 1, 2, \cdots, k \text{ and } \lambda_0 = 1. \tag{2.4.5}$$

If now, under the restricted multivariate Model II, we apply the conditions (α), (β), and (γ) of Lemma 3 to the set of matrices $(1/\lambda_0)S_0$, $(1/\lambda_1)S_1, \cdots, (1/\lambda_k)S_k$, taken as above, we notice that they all satisfy (β) so that their distributions are all central (if they are pseudo-Wishart at all). Furthermore, $(1/\lambda_0)S_0 = S_0$ (by 2.4.5)), where n_0S_0, the *matrix due to error*, can be seen to have the central pseudo-Wishart distribution (in fact, the ordinary Wishart distribution, since S_0 is positive definite here) with degrees of freedom n_0, and to be also distributed independently of $(1/\lambda_i)S_i$ for $i = 1, 2, \cdots, k$. Also, by applying (α) and (γ) of Lemma 3 to $(1/\lambda_i)S_i$ $(i = 1, 2, \cdots, k)$, we observe that they are distributed mutually independently in central pseudo-Wishart forms with respective degrees of freedom $n_1, n_2, \cdots, n_k$, if and only if,

$$C_{i1}(A_I' A_I)^{-1}C_{i1}' = \frac{1}{\nu_i}[I(m_i - 1) + J((m_i - 1) \times (m_i - 1))], \quad (i = 1, \cdots, k), \tag{2.4.6}$$

and

$$C_{i1}(A_I'A_I)^{-1}C_{j1}' = 0((m_i - 1) \times (m_j - 1)), \quad (i \neq j = 1, 2, \cdots, k), \tag{2.4.7}$$

where we recall that $I(p)$ denotes the identity matrix of order p and $J(p \times q)$ denotes a matrix of p rows and q columns all of whose elements are equal to unity. The conditions (2.4.6) and (2.4.7), which are independent of the unknown variance components, are the same as (2.4.5) and (2.4.6) of [10] for the univariate case.

Recalling the remarks toward the end of section 2.3, we observe that these conditions, (2.4.6) and (2.4.7), are both satisfied by the multivariate analogues of the usual univariate complete block designs.

Finally, it may be seen from (2.4.4) that we can take $(1/\nu_i)$ $(S_i - S_0)$ as an unbiased estimate of Σ_i $(p \times p)$, for $i = 1, \cdots, k$, and S_0 as an unbiased estimate of $\Sigma(p \times p)$. We may, therefore, use $c[(1/\nu_i)$ $(S_i - S_0)]$ as estimates of $c(\Sigma_i)$ and $c(S_0)$ as estimates of $c(\Sigma)$.

2.5 *Tests of hypotheses on the multivariate variance components.* The usual null hypotheses may be stated as

$$H_{0i}: \quad \Sigma_i(p \times p) = 0(p \times p), \text{ or, equivalently, } c(\Sigma_i) = 0, \quad \text{for } i = 1, 2, \cdots, k. \tag{2.5.1}$$

It is easily seen, from (2.4.4), that for the restricted k-way classification H_{0i} is equivalent to $\Lambda_i(p \times p) = \Lambda_0(p \times p)$ $(i = 1, 2, \cdots, k)$, or, for the restricted multivariate Model II, to the hypotheses $\lambda_i = \lambda_0$ $(i = 1, 2, \cdots, k)$. The alternative to this last form is taken to be H_{1i}: $\lambda_i > \lambda_0$. Assuming that the restricted k-way classification has matrices like (2.4.2) and (2.4.3) which satisfy both (2.4.6) and (2.4.7), we have, by definition of the pseudo-Wishart distribution, that $XQ_0X' = Y_0(p \times n_0)Y_0'(n_0 \times p)$ and $XQ_iX' = Y_i(p \times n_i)Y'(n_i \times p)$ $(i = 1, 2, \cdots, k)$, where Y_0 and Y_i have the joint distribution

$$\text{const.} \exp\left[-\tfrac{1}{2} \operatorname{tr} \Sigma^{-1}\left\{\frac{1}{\lambda_0} Y_0 Y_0' + \frac{1}{\lambda_i} Y_i Y_i'\right\}\right] dY_0\, dY_i,$$

and where $E(Y_0Y_0') = n_0\lambda_0\Sigma$ and $E(Y_iY_i') = n_i\lambda_i\Sigma$.

Consider $\mathbf{a}'$ $(1 \times p)Y_0(p \times n_0)$ and $\mathbf{a}'$ $(1 \times p)Y_i(p \times n_i)$ for all nonnull

$\mathbf{a}(p \times 1)$. Then, $(1/n_0)E(\mathbf{a}'Y_0Y_0'\mathbf{a}) = \lambda_0\mathbf{a}'\Sigma\mathbf{a}$ and $(1/n_i)E(\mathbf{a}'Y_iY_i'\mathbf{a}) = \lambda_i\mathbf{a}'\Sigma\mathbf{a}$. For testing $\lambda_i = \lambda_0$ against $\lambda_i > \lambda_0$ we may take as critical region

$$w_{i\mathbf{a}} : F_{\mathbf{a}}(n_i, n_0) = \frac{\mathbf{a}'Y_i Y_i' \mathbf{a} / n_i}{\mathbf{a}'Y_0 Y_0' \mathbf{a} / n_0} \geqq F_\alpha(n_i, n_0),$$

where $F_\alpha(n_i, n_0)$ is the upper 100α % point of the central F-distribution with n_i and n_0 degrees of freedom. Taking, $w_i = \bigcap_{\mathbf{a}} w_{i\mathbf{a}}$, as a critical region for the hypothesis (2.5.1) we obtain that

$$w_i: \quad c_{\max}(S_i S_0^{-1}) \geqq c_\alpha^*(p, n_i, n_0)$$

which is seen, from (1.2.4), to be the critical region of the test, at a level α^*, for the hypothesis (2.4.1) under the multivariate Model I discussed under Section 1 of this paper.

It must be noted that the above arguments for deriving a test were made solely to obtain, for the customary null hypotheses under the restricted multivariate Model II, if possible, a critical region which is the same as the one for the customary null hypotheses under the multivariate Model I. The use of the union-intersection principle to obtain w_i from $w_{i\mathbf{a}}$ is rather artificial since we do not have H_{0i} itself as an intersection of hypotheses $H_{0i\mathbf{a}}$.

2.6. *Confidence statements.* We shall first assume that we are dealing with restricted k-way classifications that have matrices like (2.4.2) and (2.4.3) satisfying both (2.4.6) and (2.4.7). As observed before, the multivariate analogues of the usual univariate complete block designs satisfy these requirements. Under the restricted multivariate Model II, we shall then obtain simultaneous confidence bounds on $\sigma_1^2, \cdots, \sigma_k^2$ and $c(\Sigma)$. Next, we shall relax the condition (2.4.7), i.e., we shall not require that the pseudo-Wishart distributions of the matrices like (2.4.2) in our analysis be independent. Under this relaxation, we shall obtain an alternate set of confidence bounds for the individual σ_i^2's $(i = 1, 2, \cdots, k)$.

If $(1/\lambda_i)S_i(p \times p) = (1/n_i\lambda_i)XQ_iX'$, for $i = 0, 1, \cdots, k$, have independent central pseudo-Wishart distributions with respective degrees of freedom $n_0, n_1, \cdots$ and n_k, then, by definition, we have $(1/n_i\lambda_i)XQ_iX' =$

$$(1/n_i)Y_i(p \times n_i)Y_i'(n_i \times p),$$

for $i = 0, 1, \cdots, k$, where the joint distribution of $Y_0, Y_1, \cdots, Y_k$ is

$$(2\pi)^{-[p(n-1)]/2}|\Sigma|^{-(n-1)/2}\exp\left[-\tfrac{1}{2}\operatorname{tr}\Sigma^{-1}\left\{\sum_{i=0}^{k} Y_i Y_i'\right\}\right]dY_0 \cdots dY_k, \tag{2.6.1}$$

$$-\infty < \text{all elements of } Y_i < \infty,$$

and where $E(Y_iY_i') = n_i\Sigma(p \times p)$ and $\sum_{k}^{i=0} n_i = (n - 1)$. It is well known that, for the symmetric positive definite matrix Σ, there exists an orthogonal matrix, $\Gamma(p \times p)$, such that $\Sigma(p \times p) = \Gamma' D_\gamma \Gamma$, where the p (non-zero) elements of the diagonal matrix D_γ are the p (positive) characteristic roots of Σ. Now making the transformation

$$D_{1/\sqrt{\gamma}}\ \Gamma Y_i(p \times n_i) = Z_i\ (p \times n_i), \qquad (i = 0, 1, \cdots, k), \tag{2.6.2}$$

we can verify that the Jacobian is $|\Sigma|^{(n-1)/2}$, so that, the joint distribution of $Z_0, Z_1, \cdots, Z_k$ is

$$(2\pi)^{-[p(n-1)]/2} \exp\left[-\tfrac{1}{2}\ \mathrm{tr}\left\{\sum_{i=0}^{k} Z_i Z_i'\right\}\right] dZ_0 \cdots dZ_k, \tag{2.6.3}$$

$$-\infty < \text{all elements of } Z_i < \infty.$$

From (2.6.3), it can be seen, by analogy with the methods used in [8, 9, 12], that we can obtain constants, $\mu_{i1}(p, n_i, \alpha_i) = \mu_{i1}$ (say) and $\mu_{i2}(p, n_i, \alpha_i) = \mu_{i2}$ (say), for $i = 0, 1, \cdots, k$, such that the statement

$$\mu_{i1} \leqq c_{\min}(Z_i Z_i') \leqq c_{\max}(Z_i Z_i') \leqq \mu_{i2}, \tag{2.6.4}$$

has probability $(1 - \alpha_i)$ and the probability of statements like (2.6.4) holding simultaneously for $i = 0, 1, \cdots, k$ is $(1 - \alpha) = \prod_{i=0}^{k} (1 - \alpha_i)$. We note that $Z_0 Z_0' = (n_0/\lambda_0) D_{1/\sqrt{\gamma}}\ \Gamma S_0 \Gamma' D_{1/\sqrt{\gamma}}$, where $n_0 S_0\ (p \times p)$ is the *matrix due to error* given by (2.4.3) and is symmetric positive definite. Therefore, starting from (2.6.4), with $i = 0$, and reasoning exactly as in section 1 of [9], we obtain the confidence statement

$$\frac{n_0}{\mu_{01}}\, c_{\max}(S_0) \geqq c_{\max}(\Sigma) \geqq c_{\min}(\Sigma) \geqq \frac{n_0}{\mu_{02}}\, c_{\min}(S_0) \tag{2.6.5}$$

with confidence coefficient $\geqq (1 - \alpha_0)$.

Next, for any $i = 1, 2, \cdots, k$, we note that (2.6.4) is equivalent to

$$\mu_{i1} \leqq \frac{n_i}{\lambda_i}\, c_{\min}(D_{1/\gamma}\ \Gamma S_i\, \Gamma') \leqq \frac{n_i}{\lambda_i}\, c_{\max}(D_{1/\gamma} \Gamma S_i \Gamma') \leqq \mu_{i2}. \tag{2.6.6}$$

However, it is known that the non-zero characteristic roots of $A(p \times q)$ $B\ (q \times p)$ are the same as those of $B(q \times p) A(p \times q)$ and that $c_{\min}(A_1)$ $c_{\min}\ (A_2) \leqq$ all $c(A_1 A_2) \leqq c_{\max}(A_1)\ c_{\max}\ (A_2)$ where $A_1\ (p \times p)$ is symmetric positive definite and $A_2(p \times p)$ is symmetric at least positive semi-definite. [Cf. pp. A-5 and A-7 of [12] for proofs.] Using these two results we have

$$c_{\min}(\Gamma S_i\, \Gamma' D_{1/\gamma}) \leqq \frac{c_{\min}(S_i)}{c_{\min}(\Sigma)} \quad \text{and} \quad c_{\max}(\Gamma S_i\, \Gamma' D_{1/\gamma}) \geqq \frac{c_{\max}(S_i)}{c_{\max}(\Sigma)},$$

so that, (2.6.6) implies the statement

$$\frac{n_i}{\mu_{i2}} \frac{c_{\max}(S_i)}{c_{\max}(\Sigma)} \leqq \lambda_i \leqq \frac{n_i}{\mu_{i1}} \frac{c_{\min}(S_i)}{c_{\min}(\Sigma)}, \tag{2.6.7}$$

which, therefore, has a probability $\geqq (1 - \alpha_i)$.

Taking the statement (2.6.5) together with all statements like (2.6.7) for

$i = 1, 2, \cdots, k$, we obtain the simultaneous statements

$$
\begin{aligned}
&\frac{n_0}{\mu_{02}} c_{\min}(S_0) \leqq c_{\min}(\Sigma) \leqq c_{\max}(\Sigma) \leqq \frac{n_0}{\mu_{01}} c_{\max}(S_0) \\
&\frac{n_1}{\mu_{12}} \frac{c_{\max}(S_1)}{c_{\max}(\Sigma)} \leqq \lambda_1 \leqq \frac{n_1}{\mu_{11}} \frac{c_{\min}(S_1)}{c_{\min}(\Sigma)} \\
&\cdots \qquad \cdots \qquad \cdots \\
&\frac{n_k}{\mu_{k2}} \frac{c_{\max}(S_k)}{c_{\max}(\Sigma)} \leqq \lambda_k \leqq \frac{n_k}{\mu_{k1}} \frac{c_{\min}(S_k)}{c_{\min}(\Sigma)}
\end{aligned}
\tag{2.6.8}
$$

with a joint probability $\geqq (1 - \alpha) = \prod_{j=0}^{k} (1 - \alpha_j)$.

Recalling now, from (2.4.5), that $\lambda_i = \nu_i \sigma_i^2 + 1$ $(i = 1, 2, \cdots, k)$, and using the leading statement in (2.6.8), we obtain the further statements implied by (2.6.8)

$$
\begin{aligned}
&\frac{n_0}{\mu_{02}} c_{\min}(S_0) \leqq c_{\min}(\Sigma) \leqq c_{\max}(\Sigma) \leqq \frac{n_0}{\mu_{01}} c_{\max}(S_0) \\
&\frac{1}{\nu_1}\left[\frac{n_1}{n_0} \frac{\mu_{01}}{\mu_{12}} \frac{c_{\max}(S_1)}{c_{\max}(S_0)} - 1\right] \leqq \sigma_1^2 \leqq \frac{1}{\nu_1}\left[\frac{n_1}{n_0} \frac{\mu_{02}}{\mu_{11}} \frac{c_{\min}(S_1)}{c_{\min}(S_0)} - 1\right] \\
&\cdots \qquad \cdots \qquad \cdots \\
&\frac{1}{\nu_k}\left[\frac{n_k}{n_0} \frac{\mu_{01}}{\mu_{k2}} \frac{c_{\max}(S_k)}{c_{\max}(S_0)} - 1\right] \leqq \sigma_k^2 \leqq \frac{1}{\nu_k}\left[\frac{n_k}{n_0} \frac{\mu_{02}}{\mu_{k1}} \frac{c_{\min}(S_k)}{c_{\min}(S_0)} - 1\right]
\end{aligned}
\tag{2.6.9}
$$

which, therefore, are a set of simultaneous confidence bounds on $\sigma_1^2, \sigma_2^2, \cdots, \sigma_k^2$ and all $c(\Sigma)$ with a joint confidence coefficient $\geqq (1 - \alpha)$, for a preassigned α. These simultaneous confidence bounds on the set of $c(\Sigma)$ and $\sigma_1^2, \cdots, \sigma_k^2$ are obtained on the assumption of independence between $(1/\lambda_i)S_i$ (for $i = 1, 2, \cdots, k$). If this assumption were relaxed we would still be able to obtain individual confidence bounds on $\sigma_1^2, \cdots, \sigma_k^2$ and the set of all $c(\Sigma)$, although the simultaneous confidence bounds in this situation would be far more difficult to obtain.

We shall next obtain the alternate set of separate confidence bounds for $\sigma_1^2, \sigma_2^2, \cdots$ and σ_k^2.

If $Y_0(p \times n_0)$ and $Y_i(p \times n_i)$, where $p \leqq n_0$ but may be $\gtreqless n_i$, are such that rank $(Y_0Y_0') = p$ and rank $(Y_iY_i') = \min(p, n_i)$, and further if Y_0 and Y_i have the joint distribution

$$
(2\pi)^{-[p(n_0+n_i)]/2} \, |\Sigma|^{-(n_0+n_i)/2} \exp[-\tfrac{1}{2} \operatorname{tr}\Sigma^{-1}\{Y_0Y_0' + Y_iY_i'\}] dY_0 \, dY_i, \tag{2.6.10}
$$

where $\Sigma(p \times p)$ is symmetric positive definite and $E(Y_0Y_0') = n_0\Sigma$, $E(Y_iY_i') = n_i\Sigma$, then, Rao [7], in continuation of the work of Bartlett and Wald, has shown that, for large m_i, $-m_i \log_e \Lambda_i$ has the central χ^2-distribution with pn_i degrees of freedom, where

$$
\Lambda_i = \frac{|Y_0 Y_0'|}{|Y_0 Y_0' + Y_i Y_i'|} \quad \text{and} \quad m_i = \left[n_0 + n_i - \frac{p + n_i + 1}{2}\right]. \tag{2.6.11}
$$

Hence, we can find $\chi^2_{1\alpha_i}$ and $\chi^2_{2\alpha_i}$ such that the statement

(2.6.12) $$\chi^2_{1\alpha_i} \leqq -m_i \log_e \Lambda_i \leqq \chi^2_{2\alpha_i},$$

or, equivalently

$$\mu_{1\alpha_i} \leqq \Lambda_i \leqq \mu_{2\alpha_i}, \qquad [\mu_{1\alpha_i} = \exp(-\tfrac{1}{2}\chi^2_{2\alpha_i}), \qquad \mu_{2\alpha_i} = \exp(-\tfrac{1}{2}\chi^2_{1\alpha_i})],$$

has a probability $(1 - \alpha_i)$ for a preassigned α_i.

Under the restricted multivariate Model II, for a restricted k-way classification, if we take the matrices $(1/\lambda_0)S_0$, $(1/\lambda_1)S_1, \cdots, (1/\lambda_k)S_k$ as in Section 2.4, then we have seen that $(1/\lambda_0)S_0$ is distributed in the central Wishart form with n_0 degrees of freedom and it is distributed independently of $(1/\lambda_1)S_1, \cdots, (1/\lambda_k)S_k$. We have, also, by (2.4.5), that $\lambda_i = \nu_i\sigma_i^2 + 1$ $(i = 1, \cdots, k)$ and $\lambda_0 = 1$. Further, if $(1/\lambda_1)S_1, \cdots, (1/\lambda_k)S_k$ satisfy the condition (2.4.6), so that they have central pseudo-Wishart distributions with degrees of freedom $n_1, \cdots, n_k$ (even though these may not be independent), then, by definition, we can write $n_0S_0 = Z_0(p \times n_0)Z_0'(n_0 \times p)$ and

$$(n_i/\lambda_i)S_i = Z_i(p \times n_i)Z_i'(n_i \times p).$$

The joint distribution of Z_0 and Z_i is then of the same form as (2.6.10), and, by analogy with the statements (2.6.10) – (2.6.12), we can find, for large m_i, constants (depending on a central χ^2-distribution with pn_i degrees of freedom), $\mu_{1\alpha_i}$ and $\mu_{2\alpha_i}$, such that the statement

(2.6.13) $$\mu_{1\alpha_i} \leqq \frac{|n_0 S_0|}{\left|n_0 S_0 + \dfrac{n_i}{\lambda_i} S_i\right|} \leqq \mu_{2\alpha_i},$$

or equivalently,

(2.6.14) $$\frac{1}{\mu_{1\alpha_i}} \geqq |\zeta_i S_i S_0^{-1} + I(p)| \geqq \frac{1}{\mu_{2\alpha_i}},$$

where $\zeta_i = (n_i/n_0\lambda_i) > 0$, will have a probability $(1 - \alpha_i)$. If rank $(S_i) = \min(p, n_i) = s_i$ (say), then (2.6.14) is seen to be equivalent to

(2.6.15) $$\frac{1}{\mu_{1\alpha_i}} \geqq (\zeta_i)^{s_i} \operatorname{tr}_{s_i}(S_i S_0^{-1}) + (\zeta_i)^{s_i-1} \operatorname{tr}_{s_i-1}(S_i S_0^{-1}) + \cdots$$
$$\cdots + \zeta_i \operatorname{tr}(S_i S_0^{-1}) + 1 \geqq \frac{1}{\mu_{2\alpha_i}},$$

where $\operatorname{tr}_s(A)$ denotes the sum of all sth order principal minors of A. Using certain matrix factorization theorems given in [12, pp. A-15–A-17], we can prove that $\operatorname{tr}_s(S_iS_0^{-1}) = \operatorname{tr}_s$ [a symmetric at least positive semi-definite matrix] > 0 for $s \leqq s_i$. Hence, all the coefficients of powers of ζ_i in the middle part of (2.6.15) are real and positive. Next, since $|\zeta_iS_iS_0^{-1} + I(p)| > 1$, in order that the bounds in (2.6.15) may be non-trivial, we should have $1/\mu_{2\alpha_i} > 1$.

Considering now the equality signs in (2.6.15), we obtain the equations

$$(2.6.16)\qquad \begin{aligned} &(\zeta_i)^{s_i}\, \mathrm{tr}_{s_i}(S_i S_0^{-1}) + \cdots + \zeta_i \,\mathrm{tr}\,(S_i S_0^{-1}) - \left(\frac{1}{\mu_{2\alpha_i}} - 1\right) = 0 \\ &(\zeta_i)^{s_i}\, \mathrm{tr}_{s_i}(S_i S_0^{-1}) + \cdots + \zeta_i \,\mathrm{tr}\,(S_i S_0^{-1}) - \left(\frac{1}{\mu_{1\alpha_i}} - 1\right) = 0. \end{aligned}$$

From well-known results in the theory of equations, it now follows that the equations (2.6.16) each have one and only one positive real root. Let these positive real roots be denoted by $\theta_{2\alpha_i}$ and $\theta_{1\alpha_i}$. Then it is seen that (2.6.14) or (2.6.15) is equivalent to

$$(2.6.17)\qquad \theta_{1\alpha_i} \geqq \zeta_i \geqq \theta_{2\alpha_i}$$

with a probability $(1 - \alpha_i)$. Recalling that $\zeta_i = (n_i / n_0\lambda_i)$ and $\lambda_i = \nu_i\sigma_i^2 + 1$, we see that (2.6.17) is equivalent to the confidence interval statement

$$(2.6.18)\qquad \frac{1}{\nu_i}\left[\frac{n_i}{n_0\,\theta_{1\alpha_i}} - 1\right] \leqq \sigma_i^2 \leqq \frac{1}{\nu_i}\left[\frac{n_i}{n_0\,\theta_{2\alpha_i}} - 1\right]$$

with a confidence coefficient $(1 - \alpha_i)$, for a preassigned α_i. We thus have, for $i = 1, 2, \cdots, k$, separate confidence interval statements for each of $\sigma_1^2, \sigma_2^2, \cdots$ and σ_k^2, but, due to the complexity of the distribution problem involved, it would be far more difficult to obtain simultaneous confidence bounds on $\sigma_1^2, \sigma_2^2, \cdots$, and σ_k^2 by this method. Nor would the difficulty be appreciably reduced, under this approach, even if we assumed that $(1/\lambda_i)S_i$'s (for $i = 1, 2, \cdots, k$) were independent as we did under the first approach.

2.7 *Concluding remarks*: After the work presented in this paper and in [10] had been completed, it was brought to the attention of the authors that Bose [1] has, for the univariate case, given a general treatment, using slightly different methods, of a mixed model with one set of random components. A very recent paper by Zelen [13] also has some results, for the univariate case on a mixed model with one set of random components as applied to Incomplete Block Designs, which are contained in [10].

REFERENCES

[1] R. C. Bose, "Versuche in Unvollständigen Blöcken," *Gastvarlesung*, 3 bis 11. März 1955, Universität Frankfurt/M., Naturwissenschaftlische Fakultät.

[2] O. Carpenter, "Note on the extension of Craig's theorem to non-central variates," *Ann. Math. Stat.*, Vol. 21 (1950), pp. 455–457.

[3] W. T. Federer, "Testing proportionality of covariance matrices," *Ann. Math. Stat.*, Vol. 22 (1951), pp. 102–106.

[4] R. Gnanadesikan, "Contributions to multivariate analysis including univariate and multivariate variance components analysis and factor analysis," Institute of Statistics, University of North Carolina, *Mimeo. Series, No.* 158 (1956).

[5] C. C. MacDuffee, *The Theory of Matrices*, J. Springer, Berlin, 1933, pp. 81–88.

[6] J. Ogawa, "On the independence of quadratic forms in a non-central normal system," *Osaka Math. Jour.*, Vol. 2 (1950), pp. 151–159.

[7] C. R. RAO, "Tests of significance in multivariate analysis," *Biometrika* 35 (1948), pp. 58–79.

[8] S. N. ROY AND R. C. BOSE, "Simultaneous confidence interval estimation," *Ann. Math. Stat.*, Vol. 24 (1953), pp. 513–536.

[9] S. N. ROY AND R. GNANADESIKAN, "Further contributions to multivariate confidence bounds," Institute of Statistics, University of North Carolina, *Mimeo Series No.* 155 (1956).

[10] S. N. ROY AND R. GNANADESIKAN, "Some contributions to ANOVA in one or more dimensions: I," *Ann. Math. Stat.*, Vol. 30 (1959), pp. 304–317.

[11] S. N. ROY, "The individual sampling distributions of the maximum, the minimum and any intermediate one of the p-statistics on the null hypothesis," *Sankhya*, Vol. 7 (1945), pp. 133–158.

[12] S. N. ROY, "A report on some aspects of multivariate analysis," Institute of Statistics, University of North Carolina, *Mimeo Series No.* 121 (1954).

[13] M. ZELEN, "The analysis of incomplete block designs," *J. Amer. Stat. Assn.*, Vol. 52 (1957), pp. 204–217.

4

Reprinted from *Multivariate Analysis,* Vol. II, P. R. Kirshnaiah, ed., Academic Press, Inc., New York, 1969, pp. 67–86

A Comparison of Some Methods of Simultaneous Inference in MANOVA[1]

K. R. GABRIEL

HEBREW UNIVERSITY
JERUSALEM, ISRAEL

INTRODUCTION AND SUMMARY

A variety of techniques are available for multiple comparisons on variables and effects in MANOVA. This paper attempts a common listing of these test procedures, step-down methods, and simultaneous confidence bounds in terms of the one-way setup. Comparisons of the properties of different procedures are made mainly in terms of their power to detect the ultimate detailed divergences from the hypotheses. It is brought out that the choice of a suitable technique depends greatly on the family of hypotheses for which inferences are sought. An experimenter interested in contrastwise inferences on all linear combinations of variables does best with the maximum characteristic root statistics, but one interested merely in comparing all pairs of samples on all single variables cannot do better than with a battery of t tests.

Most comparisons can readily be extended to other MANOVA setups.

1. DATA AND NOTATION

Consider a one-way p-variate Normal MANOVA setup comparing c classes by means of samples of sizes $n_1, n_2, \ldots, n_c$, respectively. Denote the αth observation of sample k with respect to variable i by

$$y_{k\alpha}^{(i)}, \qquad i = 1, \ldots, p, \quad k = 1, \ldots, c, \quad \alpha = 1, \ldots, n_k$$

and the corresponding p-variable vector observation by

$$\mathbf{y}'_{k\alpha} = (y_{k\alpha}^{(1)}, \ldots, y_{k\alpha}^{(p)}), \qquad k = 1, \ldots, c, \quad \alpha = 1, \ldots, n_k.$$

[1] This research was supported by the National Center for Health Statistics grant NCHS-IS-1.

Next, write the kth sample mean vector as

$$\bar{\mathbf{y}}_k = n_k^{-1} \sum_{\alpha=1}^{n_k} \mathbf{y}_{k\alpha}, \qquad k = 1, \ldots, c$$

and the overall mean as

$$\bar{\bar{\mathbf{y}}} = n^{-1} \sum_{k=1}^{c} \sum_{\alpha=1}^{n_k} \mathbf{y}_{k\alpha},$$

where

$$n = \sum_{k=1}^{c} n_k.$$

Correspondingly, define the sums of squares and products between samples and within samples as, respectively,

$$H = \sum_{k=1}^{c} n_k(\bar{\mathbf{y}}_k - \bar{\bar{\mathbf{y}}})(\bar{\mathbf{y}}_k - \bar{\bar{\mathbf{y}}})'$$

and

$$S = \sum_{k=1}^{c} \sum_{\alpha=1}^{n_k} (\mathbf{y}_{k\alpha} - \bar{\mathbf{y}}_k)(\mathbf{y}_{k\alpha} - \bar{\mathbf{y}}_k)'.$$

Further, consider any subset V of p_V $(1 \leq p_V \leq p)$ of the variables and any subgroup G of c_G $(2 \leq c_G \leq c)$ of the classes. Let a superscript V on a matrix (vector) denote the $p_V \times p_V$ principal minor ($p_V \times 1$ subvector) of the rows and columns (rows) belonging to V. Thus, if, for example, $V = \{p\}$ $(p_V = 1)$ the subset consisting of the last variable only, then $\mathbf{y}_{k\alpha}^V = y_{k\alpha}^{(p)}$, $\bar{\mathbf{y}}_k{}^V = \bar{y}_k^{(p)}$, $\bar{\bar{\mathbf{y}}}^V = \bar{\bar{\mathbf{y}}}^{(p)}$, where the scalar means are defined analogously to the vector means, and also $H^V = \sum_k n_k(\bar{y}_k^{(p)} - \bar{\bar{y}}^{(p)})^2$ and $S^V = \sum_k \sum_\alpha (y_{k\alpha}^{(p)} - \bar{y}_k^{(p)})^2$. Further, let a subscript G on a matrix (vector) denote that the summation or averaging of the matrix (vector) is restricted to the samples from the classes of subgroup G. Thus $\bar{\mathbf{y}}_G = n_G^{-1} \sum_{k \in G} \sum_\alpha \mathbf{y}_{k\alpha}$ where $n_G = \sum_{k \in G} n_k$ and $H_G = \sum_{k \in G} n_k (\bar{\mathbf{y}}_k - \bar{\mathbf{y}}_G)(\bar{\mathbf{y}}_k - \bar{\mathbf{y}}_G)'$. An example of the joint use of superscripts and subscripts for $V = \{1\}$ and $G = \{3, 4\}$, is $H_G{}^V = n_3(\bar{y}_3^{(1)} - \bar{y}_G^{(1)})^2 + n_4(\bar{y}_4^{(1)} - \bar{y}_G^{(1)})^2$.

For a vector space $\tilde{V}$ spanned by the variables in V and for a linear set of contrasts G^* generated by the differences in group G, the definitions of $\mathbf{y}_{k\alpha}^{\tilde{V}}, \bar{\mathbf{y}}_k^{\tilde{V}}, \bar{\mathbf{y}}_{G^*}^{\tilde{V}}$, $H_{G^*}^{\tilde{V}}$, and $S^{\tilde{V}}$ are the same as those for V and G. For other vector spaces $\tilde{V}$ of variables and linear sets G^* of contrasts, the definitions are analogous.

In the more general nonparametric model allowing for shift alternatives, the data are ranked separately for each variable. Analogously to the within matrix S, one defines T based on the within sample rankings. Thus, if $R_{k\alpha}^{(1)}$

is the rank of observation α within sample k on variable 1, one uses vectors $\mathbf{r}'_{k\alpha} = (n_k + 1)^{-1}(R_{k\alpha}^{(1)}, \ldots, R_{k\alpha}^{(p)})$ to compute T exactly as one computes S from the vectors $\mathbf{y}'_{k\alpha}$, above. Pairwise sample comparisons are carried out by jointly ranking the observations of both samples, and comparing mean ranks. Thus, for samples k and l the ith variable comparison uses

$$\frac{\bar{R}_k^{(i)} - \bar{R}_l^{(i)}}{n_k + n_l + 1},$$

where $\bar{R}_k^{(i)}$ is the mean of the n_k kth sample i-variable observations ranks in the joint ranking with sample l. Note that $\binom{c}{2} + 1$ separate rankings are involved, one for each pair of samples and one within samples. Computationally, this may be prohibitive without use of a computer. A FORTRAN program carrying out all these rankings and obtaining the summary L statistics [6] for overall testing and simultaneous inference is available from the author. This program also allows for permutation of the data to provide randomization tests.

2. MODEL AND HYPOTHESES

Assume the Normal model $\mathbf{y}_{k\alpha} \sim \mathcal{N}(\boldsymbol{\mu}_k, \Sigma)$ independently for $1, \ldots, n_k$ $k = 1, \ldots, c$ with any $(c \times p)$ expectation matrix $M = (\boldsymbol{\mu}_1, \ldots, \boldsymbol{\mu}_c)'$ and any $(p \times p)$ positive definite variance matrix Σ. The overall MANOVA hypothesis is that of expectation equality of the c classes on all p variables,

$$\omega_0: \quad \boldsymbol{\mu}_1 = \cdots = \boldsymbol{\mu}_c.$$

More specific hypotheses in MANOVA are those of equality within a subgroup G of classes with respect to a subset V of the variables,

$$\omega_G{}^V: \quad \boldsymbol{\mu}_k{}^V = \bar{\boldsymbol{\mu}}_G{}^V, \qquad \text{say}, \qquad \forall\, k \in G.$$

In particular, $\omega_0 = \omega_{G_0}^{V_0}$, where V_0 is the set of all p variables and G_0 the group of all classes.

Consider the *family* of hypotheses of within subgroup expectation equality on subset variables. This may be written

$$\Omega = \{\omega_G{}^V \mid V \subset V_0, G \subset G_0\}.$$

There are $(2^c - c - 1) \times (2^p - 1)$ hypotheses in Ω, including ω_0 which is called the *total* or *intersection* hypothesis because it implies all other hypotheses of Ω. Further implication relations exist in Ω, in that $\omega_G{}^V \subset \omega_F{}^W$ whenever $W \subset V$ and $F \subset G$, as is readily seen from the above definition of the hypotheses.

The family Ω of hypotheses may be extended in two ways. First, one may consider comparisons not only on single variables but also on linear combinations of variables as well. If $\tilde{V}$ is a vector space of such combinations, then $\omega_G^{\tilde{V}}$ is the hypothesis that all classes of group G have equal expectations on every variable of space $\tilde{V}$. Second, one may consider contrasts between class expectations and not only differences. If G^* is a linear set of such contrasts, $\omega_{G^*}^V$ becomes the hypothesis of nullity of all contrasts in G^* on each of the variables of V. The extension to $\omega_{G^*}^{\tilde{V}}$ is obvious. Thus one obtains four families of hypotheses, as follows:

variable sets, class differences	$\Omega = \{\omega_G{}^V \mid V \subset V_0, G \subset G_0\}$,
variable spaces, class differences	$\tilde{\Omega} = \{\omega_G{}^{\tilde{V}} \mid \tilde{V} \subset \tilde{V}_0, G \subset G_0\}$,
variable sets, class contrasts	$\Omega^* = \{\omega_{G^*}^V \mid V \subset V_0, G^* \subset G_0{}^*\}$,
variable spaces, class contrasts	$\tilde{\Omega}^* = \{\omega_{G^*}^{\tilde{V}} \mid \tilde{V} \subset \tilde{V}_0, G^* \subset G_0{}^*\}$,

where $\tilde{V}_0$ is the space spanned by V_0, and $G_0{}^*$ the linear set of contrasts in G_0. Indeed, if $\tilde{V}$ is spanned by V and G^* by G, one obtains

$$\omega_G{}^V = \omega_G{}^{\tilde{V}} = \omega_{G^*}^V = \omega_{G^*}^{\tilde{V}},$$

so that the four families are seen to have the same intersection hypothesis. Yet it is important to distinguish the four families, for $\tilde{\Omega}$ contains Ω as well as hypotheses on spaces of variables not spanned by any set of the original variables, e.g., the sum of the variables, and Ω^* contains Ω as well as hypotheses on linear sets of contrasts other than those generated by a set of class differences. Similarly, $\tilde{\Omega}^*$ contains both $\tilde{\Omega}$ and Ω^* and additional hypotheses. More detailed simultaneous inferences are possible on $\tilde{\Omega}^*$ than on either $\tilde{\Omega}$ or Ω^* and least detailed on Ω. A question that arises here, but not in classical hypothesis testing, is what detail, or which atomic hypotheses, are relevant to the required inferences. The appropriate family will then be chosen accordingly ([14], Chapter 1).

The hypotheses $\omega_{G^*}^{\tilde{V}}$ of any one of these families may be classified by the dimension $p_{\tilde{V}}$ of the variables space and the degrees of freedom $c_{G^*} - 1$ of the linear set of contrasts. In particular, for V and G these become the number of variables p_V in set V and the number of classes c_G in group G. One may define

$$\text{rank } \omega_{G^*}^{\tilde{V}} = \min(p_{\tilde{V}}, c_{G^*} - 1),$$

so that rank one hypotheses are univariate, $p_{\tilde{V}} = 1$ and/or unicontrast, $c_{G^*} - 1 = 1$. Among rank one hypotheses those with $p_{\tilde{V}} = c_{G^*} - 1 = 1$ concern the nullity of a single contrast on a single linear variable combination; within Ω, they concern the equality of two classes on a single variable. These hypotheses are referred to as *atomic* in that such an *atom* implies no other hypothesis of the family, but every nonatomic hypothesis $\omega_{G^*}^{\tilde{V}}$ is equivalent to the intersection of all its atoms within the family, that is,

$$\omega_{G^*}^{\tilde{V}} = \bigcap_{\tilde{W}} \bigcap_{F^*} \{\omega_{F^*}^{\tilde{W}} \mid \tilde{W} \subset \tilde{V},\ F^* \subset G^*,\ p_{\tilde{W}} = c_{F^*} - 1 = 1\}.$$

As an alternative model, one may consider more generally independent vectors $\mathbf{y}_{k\alpha}$, $\alpha = 1, \ldots, n_k$, $k = 1, \ldots, c$, where

$$F(\mathbf{y}_{k\alpha}) = F(\mathbf{y} + \boldsymbol{\mu}_k),$$

with unspecified p-variate distribution function F and any $(c \times p)$ location matrix $M = (\boldsymbol{\mu}_1, \ldots, \boldsymbol{\mu}_c)'$. Hypotheses ω_0 and $\omega_G{}^V$ of Ω may be considered here analogously to those of the above Normal model. Indeed, the Normal model and its hypotheses in Ω might be considered special cases of these latter ones. The extension to other families of hypotheses is not available at present.

3. AVAILABLE PROCEDURES FOR SIMULTANEOUS INFERENCE

Two types of simultaneous inference procedures are available. *Simultaneous Test Procedures* (STP's) use the original variables without any ordering and allow inferences with respect to them. *Step-down* procedures, on the other hand, analyze successive deviations of the variables from their regressions on all previous variables. The order of variables entering into step-down analysis is arbitrary, and the analysis of regression residuals cannot be readily interpreted in terms of the original variables, except that a hypothesis on *all* variables corresponds to one on *all* residuals. To stress the different types of variables analyzed by step-down methods, the families of hypotheses corresponding to Ω and Ω^*, but relating to regression residuals, are denoted $\Omega(R)$ and $\Omega^*(R)$, respectively.

An STP [4, 5] uses a statistic Z_i for each hypothesis ω_i of a given family and rejects ω_i if $Z_i > \zeta$, where ζ is a critical value common to all ω_i of the family. This ζ is chosen so as to ensure a probability of at least α—the *level* of the STP—that no hypothesis of the family be wrongly rejected. A listing of STP's in MANOVA is given in Table I with some detail about the statistics used and the family of hypotheses on which it provides inferences. The table also lists some MANOVA step-down procedures.

Some of these test procedures were originally proposed as *simultaneous confidence bounds* (SCB's) and others can be inverted to provide such bounds. Table II, therefore, lists SCB's corresponding to the STP's of Table I, showing bounds on deviations from the atomic hypotheses of Ω as well as bounds

TABLE I

Procedure	Statistic $Z_G{}^V$ for samples group G on variables set V	Critical value ζ	
		Distribution	Parameters
STP's based on the original (unordered) variables			
Maximum root	$c_{\max}[(S^{\tilde{V}})^{-1}H^{\tilde{V}}_{G*}]$	$c^{e}_{\max}$	$p, c-1, n-c$
Increasing root functions[a]	$\psi[(S^{\tilde{V}})^{-1}H^{\tilde{V}}_{G*}]$	ψ	$p, c-1, n-c$
Maximum T^2 [b]	$\max_{(k,l)\in G} c_{\max}[(S^{\tilde{V}})^{-1}H^{\tilde{V}}_{(k,l)}]$	χ^2 ratio[f,g]	$p, n-c-p+1$
Maximum F [c]	$\max_{i\in V} [H^{(i)}_{G*}/S^{(i)}]$	χ^2 ratio	$c-1, n-c$
Maximum t^2 [d]	$\max_{i\in V} \max_{(k,l)\in G} H^{(i)}_{(k,l)}/S^{(i)}$	χ^2 ratio	$1, n-c$
Root ratio R	$c_{\max}[H^{V}_{G*}]/c_{\max}[S^{V}]$	—	as $c_{\max}$
B-K bounds	As max root	—	as $c_{\max}$
Symmetric gauge functions	σ : various (incl. above two)	—	as $c_{\max}$, also others
Nonparametric	$\frac{1}{2}\sum_{G} n_k n_l \sum_{V} U^{(i)}_{(k,l)} U^{(j)}_{(k,l)} T^{ij}_{G}$	χ^2 or permutation distribution	$p(c-1)$
Maximum W	Wilcoxon-Mann-Whitney maximum	Tables	
Methods based on successive (ordered) deviations from regressions on previous variables			
Step-down	$H^{(j)}_{G*}/S^{(j)}$ cond. on $1, \ldots, j-1$	χ^2 ratio	$c-1, n-c-j+1$
Finite-Intersection (any $c-1$ contrasts)	$\max[t^{2(j)} \mid \text{contrasts}]$ cond. on $1, \ldots, j-1$	Mult. F [h]	$c-1, 1, n-c-j+1$
Multiv. SANOVA ($c-1$ indpt. contrasts)	As above	Max *St.* χ^2 [i]	$c-1, 1, n-c-j+1$

[a] Includes as special cases the likelihood ratio statistic, Hotelling-Lawley's trace and Pillai's trace. The distribution of ψ is not known in general.

[b] $T^2 = (n-c)c_{\max}$ when $c_G = 2$.

[c] $F = (n-c)c_{\max}/(c_G - 1)$ when $p_V = 1$.

[d] $t^2 = (n-c)c_{\max}$ when $c_G - 1 = p_V = 1$.

[e] Distribution tabulated as $\theta = c_{\max}/(1 + c_{\max})$ with parameters $\min(p, c-1)$, $(|p-c+1|-1)/2$, $(n-c-p+1)/2$.

[f] F ratio times ratio of numerator to denominator degrees of freedom.

[g] A method of getting approximate percentage points for $T^2_{\max}$ is given by Siotani [21] (in Japanese). See also Siotani [22]. Bonferroni inequalities were used by Kaskey *et al.* [7].

Test Procedures for Simultaneous Inference in One-Way MANOVA

Point	Overall rejection prob.	Family	Coherence	Consonance	Original ref. (see also Table II)	Fortran program
α	α	$\tilde{\Omega}^{*}$ [j]	Yes	$\tilde{\Omega}^{*}$	S. N. Roy[k]	Yes
α	α	$\tilde{\Omega}^{*}$	Yes	—	Gabriel [5]	Yes
$\alpha/\binom{c}{2}$	$<\alpha$	$\tilde{\Omega}$	Yes	$\tilde{\Omega}$	Roy and Bose [19]	
α/p	$<\alpha$	Ω^{*}	Yes	Ω^{*}		
$\alpha/p\binom{c}{2}$	$<\alpha$	Ω	Yes	Ω		
—	$<\alpha$	$\tilde{\Omega}^{*}$ [j]	No	—		Yes
—	α	$\tilde{\Omega}^{*}$	Yes	$\tilde{\Omega}^{*}$		
—	$<\alpha$	$\tilde{\Omega}^{*}$	Some	?		
α	$\approx\alpha$	Ω	Yes	—	Gabriel and Sen [6]	Yes
$\alpha/p\binom{c}{2}$	$<\alpha$	Ω	Yes	Ω		
$1-(1-\alpha)^{1/p}$	α	$\Omega^{*}(R)$	Yes	$\Omega^{*}(R)$	J. Roy [17]	
$1-(1-\alpha)^{1/p}$	α	$\Omega(R)$	Yes	$\Omega(R)$	Krishnaiah [12]	
$1-(1-\alpha)^{1/p}$	α	$\Omega(R)$	Yes	$\Omega(R)$	Krishnaiah [11]	

[h] Multivariate F, distribution tabulated by Krishnaiah and Armitage [13].

[i] Studentized largest chi-square, distribution tabulated by Armitage and Krishnaiah [1].

[j] Krishnaiah also proposed the use of this type of procedure for a more restricted family of orthogonal hypotheses [10]. However, little appears to be known about the requisite distributions.

[k] The method is implicit in Roy's writing and teaching though he does not seem to have formulated it explicitly.

TABLE II

Confidence Bounds for Simultaneous Inference in One-Way MANOVA—*Confidence at Least* $1 - \alpha$

Procedure	General: Bounds on $c_{\max}^{1/2}[\sum_{k \in G} n_k(\boldsymbol{\mu}_k^V - \bar{\boldsymbol{\mu}}_G^V)(\boldsymbol{\mu}_k - \bar{\boldsymbol{\mu}}_G^V)']$ for group G and subset V	Atomic: Bounds on $(\mu_k^{(i)} - \mu_l^i)/(n_k^{-1} + n_l^{-1})$	Original ref. (see also Table I)
Methods based on the original (unordered) variables			
Maximum root	Not available		Roy and Bose [19]
Increasing root functions	Not available		Gabriel [5]
Maximum T^2	Not available		
Maximum F	Not available		
Maximum t^2	Not available		
Roots ratio R	$c_{\max}^{1/2}[H_G{}^V] \pm \zeta^{1/2} c_{\max}^{1/2}[S^V]$	$(\bar{y}_k^i - \bar{y}_l^i)/(n_k^{-1} + n_l^{-1}) \pm \zeta^{1/2}(S^i)^{1/2}$	Roy and Gnanadesikan [20]
B-K bounds	$c_{\max}^{1/2}[S^V]\{c_{\max}^{1/2}[(S^V)^{-1}H_G{}^V] - \zeta^{1/2}\}$ lower bound, upper bound as by Roy-Gnanadesikan		Bhapkar [2], Khatri [8]
Symmetric gauge functions	Various (incl. above two)		Mudholkar [15]
Nonparametric L	Not available	See Gabriel and Sen [6]	Gabriel and Sen [6]
Maximum W	Not available	As above	
Methods based on (successive) ordered deviations from regressions on previous variables			
No bounds available on original variables except one (see Roy [17], Krishnaiah [11, 12]).			

relevant to nonatomic hypotheses, where available. Original references are cited in either Table I or II, according to whether the method was originally proposed for tests or for bounds, respectively.

Tables I and II show a number of unique STP's as well as two classes, each of which include STP's based on some of the well known MANOVA test criteria. The class of increasing root function STP's [5] includes all procedures based on functions ψ of $S^{-1}H$ such that (i) ψ is monotone increasing in all roots[2] of $S^{-1}H$, (ii) ψ is equal for any two such matrices with equal nonzero roots, irrespective of the number of zero roots either matrix may have, (iii) $\psi = c_{\max}$ if $S^{-1}H$ has only one nonzero root. Wilks's likelihood ratio as well as Hotelling-Lawley's and Pillai's traces are statistics belonging to this class. The class of σ-STP's uses symmetric gauge function of the square roots of the characteristic roots of $S^{-1}H$. For a discussion of symmetric gauge functions see Mudholkar [15].

All SCB's of Table II provide bounds for the atomic comparisons, i.e., univariate comparisons of pairs of sample means. Roy and Gnanadesikan's R procedure as well as Bhapkar-Khatri's and, more generally, Mudholkar's procedures, further provide bounds for multivariate comparisons of several (3 or more) means. Corresponding to hypothesis $\omega_{G^*}^{\bar{V}}$, bounds are set on the noncentrality parameter $c_{\max}^{1/2}[\chi_{G^*}^{\bar{V}}]$, where $\chi_{G^*}^{\bar{V}}$ is related to expectations as $H_{G^*}^{\bar{V}}$ is related to sample means, that is,

$$((\chi_G^V))_{ij} = \sum_{k \in G} n_k(\mu_k^{(i)} - \bar{\mu}_G^{(i)})(\mu_k^{(j)} - \bar{\mu}_G^{(j)}),$$

where $i \in V$ and $j \in V$.

The functions $c_{\max}^{1/2}[\chi_{G^*}^{\bar{V}}]$ are unlikely to have physically meaningful interpretations in many practical applications. The utility of the bounds on these functions therefore seems doubtful, and, indeed, no useful application seems to be known to date. However, $c_{\max}^{1/2}[\chi_G^V]$ is positive definite in all expectation differences, that is, it is zero if, and only if, the expectations of all classes are equal. Thus, the main use of these SCB's is in determining whether $\omega_{G^*}^{\bar{V}}$ is true according to whether the bounds for $c_{\max}^{1/2}[\chi_{G^*}^{\bar{V}}]$ include zero. As such, the SCB's function as STP's, and are thus presented in Table I. It will be noted that the STP derived from B-K SCB's is equivalent to the $c_{\max}$ STP.

4. PROPERTIES AND COMPARISONS OF PROCEDURES

4.1. Coherence and Consonance

A method of simultaneous inference should preserve, as far as possible, the implication relations between the hypotheses of the family $\tilde{\Omega}^*$, or other

[2] This definition differs slightly from that of Gabriel [5], essentially only in that the present definition excludes $c_{\max}$ from the ψ class.

family under consideration. A basic requisite is that any such method be *coherent* in the sense that, when $\tilde{W} \subset \tilde{V}$ and $F^* \subset G^*$, i.e., when $\omega_{G^*}^{\tilde{V}} \subset \omega_{F^*}^{\tilde{W}}$, acceptance of $\omega_{G^*}^{\tilde{V}}$ is always accompanied by acceptance of $\omega_{F^*}^{\tilde{W}}$. Equivalently, $\omega_{G^*}^{\tilde{V}}$ is always rejected when $\omega_{F^*}^{\tilde{W}}$ is ([4], Section 3). A further desirable quality is *consonance* ([4]), Section 3). This means that any $\omega_{G^*}^{\tilde{V}}$ is rejected only if at least one hypothesis $\omega_{F^*}^{\tilde{W}}$ implied by it is also rejected. Thus a coherent and consonant procedure rejects a hypothesis if, and only if, at least one of the implied atoms is rejected.

An STP is coherent and consonant if its statistics satisfy Roy's union-intersection principle within the family of hypotheses considered ([4], Corollary 2). In other words, the statistic for any hypothesis must be the maximum of the statistics for all atoms it implies ([4], Theorem 3). It is well known that $c_{\max}$ satisfies this requirement for $\tilde{\Omega}^*$, and clearly max T^2 satisfies it for $\tilde{\Omega}$ and max F for Ω^*. Both max t^2 and max W obviously satisfy it for Ω. This establishes the coherence and consonance properties of STP's in Table I. Those for step-down methods similarly follow.

If a statistic is coherent for a wider family of hypotheses, it obviously is coherent also for any subfamily contained in it. Hence, each of the above statistics is coherent for Ω. Coherence of the increasing root function class of STP's was proved for $\tilde{\Omega}^*$ by Gabriel [5], and that of L for Ω by Gabriel and Sen [6]. The only incoherent procedure is that based on the ratio of roots[3] R (examples of rejecting a hypothesis implied by a nonrejected hypothesis are shown in Section 5, Table VI). Presumably, Roy and Gnanadesikan [20] intended not to test hypotheses implied by accepted hypotheses, thus avoiding incoherences by ignoring them. (The most common univariate methods of multiple comparisons proceed in the same way. [4], Section 9.)

For nonatomic hypotheses the SCB's on $c_{\max}^{1/2}[\chi_G{}^V]$ present some further difficulties. First, note that this noncentrality parameter is never smaller for an implying hypothesis than for one implied by it. Thus, if $W \subset V$ and $F \subset G$, so that $\omega_G{}^V \subset \omega_F{}^W$, then $c_{\max}[\chi_G{}^V] \geq c_{\max}[\chi_F{}^W]$. This may be proved as follows:

$$
\begin{aligned}
c_{\max}[\chi_G{}^V] &= c_{\max}\left[\left(\sum_{k\in G} n_k(\mu_k^{(i)} - \bar{\mu}_G^{(i)})(\mu_k^{(j)} - \bar{\mu}_G^{(j)})\right)_{i,j\in V}\right] \\
&= \max_{\boldsymbol{\alpha}}\left\{\sum_{i\in V}\sum_{j\in V}\alpha_i\alpha_j\sum_{k\in G} n_k(\mu_k^{(i)} - \bar{\mu}_G^{(i)})(\mu_k^{(j)} - \bar{\mu}_G^{(j)}) \,\Big|\, \sum_{i\in V}\alpha_i^2 = 1\right\} \\
&= \max_{\boldsymbol{\alpha}}\left[\sum_{k\in G} n_k\left(\sum_{i\in V}\alpha_i\mu_k^{(i)} - \sum_{i\in V}\alpha_i\bar{\mu}_G^{(i)}\right)^2 \,\Big|\, \sum_{i\in V}\alpha_i^2 = 1\right] \\
&\geq \max_{\boldsymbol{\alpha}}\left[\sum_{k\in G} n_k\left(\sum_{i\in W}\alpha_i\mu_k^{(i)} - \sum_{i\in W}\alpha_i\bar{\mu}_G^{(i)}\right)^2 \,\Big|\, \sum_{i\in W}\alpha_i^2 = 1\right] \\
&\geq \max_{\boldsymbol{\alpha}}\left[\sum_{k\in F} n_k\left(\sum_{i\in W}\alpha_i\mu_k^{(i)} - \sum_{i\in W}\alpha_i\bar{\mu}_F^{(i)}\right)^2 \,\Big|\, \sum_{i\in W}\alpha_i^2 = 1\right] \\
&= c_{\max}[\chi_F{}^W].
\end{aligned}
$$

[3] In terms of the definition in [4], the R procedure is not strictly an STP.

The first inequality follows since a maximum over a narrower set of α's must be no larger than over a wider set. The second inequality is clear since $\sum_{i \in W} \alpha_i \bar{\mu}_G^{(i)}$ is the mean of the values $\sum_{i \in W} \alpha_i \mu_k^{(i)}$ with weights n_k, $k \in G$ (see also Gabriel [3]). Now, note that the bounds set on $c_{\max}^{1/2}[\chi_G{}^V]$ may include lower values than those within the bounds for $c_{\max}^{1/2}[\chi_F{}^W]$, thus leading to incoherent results. This is readily seen from the following example: set

$$H_G^{(1,2,3)} = \begin{pmatrix} 9 & 0 & 0 \\ 0 & 6\frac{1}{4} & 0 \\ 0 & 0 & 4 \end{pmatrix} \quad \text{and} \quad S^{(1,2,3)} = \begin{pmatrix} 4 & 0 & 0 \\ 0 & \frac{1}{4} & 0 \\ 0 & 0 & 1 \end{pmatrix},$$

so that

$$H_G^{(2)} = 6\tfrac{1}{4} \quad \text{and} \quad S^{(2)} = \tfrac{1}{4}.$$

The following Roy-Gnanadesikan bounds will be obtained:

$$3 - \sqrt{\zeta}2 \le c_{\max}^{1/2}[\chi_G^{(1,2,3)}] \le 3 + \sqrt{\zeta}2$$

and

$$2\tfrac{1}{2} - \sqrt{\zeta}\tfrac{1}{2} \le c_{\max}^{1/2}[\chi_G^{(2)}] \le 2\tfrac{1}{2} + \sqrt{\zeta}\tfrac{1}{2}.$$

If $\sqrt{\zeta} = \frac{1}{2}$, this means that $c_{\max}^{1/2}[\chi_G^{(2)}]$ cannot be less than $2\frac{1}{4}$ whereas $c_{\max}^{1/2}[\chi_G^{(1,2,3)}]$ can be as little as 2. As the latter parameter was just proved to be no less than the former, these bounds clearly are not coherent. (For a discussion of coherence of confidence statements see Gabriel [4] Section 5). In the same example, one obtains the following B-K bounds

$$2 - \sqrt{\zeta} \le c_{\max}^{1/2}[\chi_G^{(1,3)}] \le 3 + 2\sqrt{\zeta}$$

and

$$3 - 2\sqrt{\zeta} \le c_{\max}^{1/2}[\chi_G^{(1)}] \le 3 + 2\sqrt{\zeta}.$$

Again, with $\sqrt{\zeta} = \frac{1}{2}$, $c_{\max}^{1/2}[\chi_G^{(1)}]$ may not be below 2, whereas $c_{\max}^{1/2}[\chi_G^{(1,3)}]$, which was proved to be no less, may be as low as $1\frac{1}{2}$, another incoherence.

Not only are these bounds set on functions of limited practical interest, but furthermore they are incoherent with one another. One may doubt whether there is much practical utility to them.

4.2. Numerical Comparisons

The statistics of all STP's enumerated in Table I for family Ω, are—except for the nonparametric statistics L and max W—functions of the two matrices S^V and $H_G{}^V$. It is instructive to consider the numerical relations among the various statistics. Table III has been set up for this purpose, allowing comparisons for four distinct cases: multivariate multiple comparisons ($p_V > 1$,

$c_G > 2$), multivariate single comparison ($p_V > 1$, $c_G = 2$), univariate multiple comparisons ($p_V = 1$, $c_G > 2$), and univariate single or atomic comparison, ($p_V = 1$, $c_G = 2$).

TABLE III

Numerical Comparison of Statistics[a]

Multivariate multiple $p_V > 1$, $c_G > 2$
$\psi_G{}^V > c_G{}^V > (R_G{}^V, \max_G T^V_{k,l}, \max_V f_G{}^i) > \max_{G,V} t^i_{k,l}$
Multivariate simple $p_V > 1$, $c_G = 2$
$R^V_{k,l} < \psi^V_{k,l} = c^V_{k,l} = T^V_{k,l} > \max_V f^i_{k,l} = \max_V t^i_{k,l}$
Univariate multiple $p_V = 1$, $c_G > 2$
$R_G{}^i = \psi_G{}^i = c_G{}^i = f_G{}^i > \max_G T^i_{k,l} = \max_G t^i_{k,l}$
Univariate simple $p_V = 1$, $c_G = 2$
$R^i_{k,l} = \psi^i_{k,l} = c^i_{k,l} = T^i_{k,l} = f^i_{k,l} = t^i_{k,l}$

[a] *Notation:* For group G on variable set V, $c_G{}^V = c_{\max}$, $T_G{}^V = T^2/(n-c)$, $f_G{}^V = F(c_G - 1)/(n-c)$ $t_G{}^V = t^2/(n-c)$. Max_G and $\max_V$ are short for $\max_{(k,l) \in G}$ and $\max_{i \in V}$, respectively. All inequalities are strict with probability one.

Proofs: The $R \leq c$ inequality follows from Roy's theorem $c_{\max}[H]/c_{\max}[S] \leq c_{\max}[S^{-1}H]$ [18]. It is readily seen that the inequality is strict with probability one except when H and S are scalars, and then it is obviously an equality.

The $\psi \geq c_{\max}$ inequalities are proved by Gabriel [5]. The inequality with respect to maxima over G and/or V follow from the fact that the left-hand statistics are similar maxima over the wider families G^* and/or $\tilde{V}$, respectively.

4.3. Resolution and Atomic Power

The purpose of simultaneous inference is to point out what the rejection of a hypothesis is due to, that is, to locate the atoms of the hypothesis which must also be rejected. Consonant procedures ensure that there always is some such *resolution* of rejection into rejection of atoms. In that sense, consonant procedures are the most desirable and other procedures may be compared according to the extent of resolution they do provide. If one procedure always rejects all atoms rejected by another and sometimes rejects more, it will be said to be strictly *more resolvent* than the other. If it merely has a higher chance of rejecting atoms, it will be said to have a higher *resolution probability*.

In evaluating a simultaneous inference procedure, it will therefore be necessary to consider not only the total—experimentwise—probability of rejection, i.e., the level of significance under the hypothesis and the power under the alternatives but also, one will need to consider the resolution probabilities, at least that between the total hypothesis and the atoms. The product of the total power and the resolution probability is the *atomic power*, that is, the power of a procedure to reject an atomic hypothesis under its alternative. Indeed, this atomic power may be ultimately the criterion by which a simultaneous inference procedure should be evaluated. The purpose of such a procedure being the establishment of significant detailed departures from the hypothesis, its advantages should be measured in terms of the probability of establishing such departures.

Where the distribution of the test statistics is the same for all atomic hypotheses, the atomic power may be represented by a single function $\beta(\delta)$ of a suitable noncentrality parameter δ. In that case, an evaluation of a procedure may well be made in terms of the total level α and the atomic power function $\beta(\delta)$. Contrast this with the classical testing situation in which one is concerned with the probability of rejecting the total hypothesis ω_0 if it is true—level α_0—and with total power $\beta_0(\Delta)$, where Δ is a noncentrality parameter for ω_0. The levels may well be equal—$\alpha = \alpha_0$—in both types of inference, but total power β_0 is quite a different matter from atomic power β. For a given $\alpha = \alpha_0$ a good test has large β_0 whereas a good method of simultaneous inference has large β. For confidence statements the corresponding comparison is in terms of narrowness of bounds on Δ and on δ, respectively.

Comparisons of the performance of STP's must be made with respect to a given family of hypotheses. Of all STP's of level α which use the same statistics for atomic hypotheses, the consonant STP will be the most resolvent ([4], Corollary 3), and generally one STP will be strictly more resolvent than another if its statistics for nonatomic hypotheses are smaller than or equal to those of the other ([4], Theorem 5). Such comparisons can be made among Normal procedures all of which use the common t statistics for atomic decisions.

For the most general family $\tilde{\Omega}^*$ the c_{max} STP is, therefore, the most resolvent. The other available methods are the incoherent ratio of roots R procedure, the increasing root functions ψ STP's which are less resolvent [5], and the other symmetric gauge function σ STP's which are also less resolvent than the c_{max} STP. (Mudholkar [15] shows the analogous result in terms of width of SCB's.)

For pairwise class differences family $\tilde{\Omega}$ the max T^2 STP is the most resolvent, the c_{max} STP being inferior, and, *a fortiori*, the other ones mentioned above. (The corresponding result for SCB's was shown by Krishnaiah [9].)

For contrastwise univariate family Ω^* the most resolvent STP is that using max F, c_{max} and other ones being inferior. Max T^2 is not available.

Max t^2 provides the most resolvent STP for family Ω. Less resolvent are max T^2 and max F, the $c_{\max}$ STP being less resolvent than either of these and the ψ and σ STP's even less so.

In view of Table III, the R procedure is equal to the $c_{\max}$ STP for all univariate tests, hence its atomic power is the same.

The above comparisons are strictly valid only when all the STP's are of exact level α. The max T^2, max F, and max t^2 STP's (and also that using max W) of Table I merely have the level α as an upper bound by the first Bonferroni inequality. Hence, it is not clear whether they will actually be more resolvent than the $c_{\max}$ STP of exact level α (see, however, Section 5 for a numerical illustration).

No general statements can be made about comparisons with the nonparametric STP's which use different statistics for atomic hypotheses and can be used only for family Ω.

Among step-down methods the Finite Intersection procedure is strictly more resolvent within $\Omega(R)$ than J. Roy's original procedure. (The corresponding confidence bound width comparison was made by Krishnaiah [12].) However J. Roy's procedure extends to the contrast family $\Omega^*(R)$.

This evaluation of simultaneous procedures deals exclusively with atomic hypotheses, stressing all of them equally. If decisions on other hypotheses are also taken into account, or different atoms stressed differently, the evaluation of such procedures becomes much more difficult.

4.4. Robustness

As far as robustness is concerned very little is known. All parametric procedures lean on Normality assumptions, though probably those using max T^2 and max F do so less heavily than that using $c_{\max}$, but more heavily than the max t^2 procedure. The nonparametric procedures are not distribution-free, except asymptotically, though the permutation distribution ensures robustness under the null hypothesis.

4.5. General Linear Setups

Last, generalization to other multivariate setups are immediate for the Normal parametric procedures, with all the comparisons going over as well. The nonparametric procedures, on the other hand, do not carry over to those situations.

5. AN EXAMPLE

As an illustration consider Reeve's anteater data which have a trivariate six class (localities) one-way MANOVA setup [16]. These data were analyzed in an earlier paper [5] by means of the $c_{\max}$ STP and increasing root function

STP's. For a nonparametric STP Table IV shows the L statistic for each of the 399 hypotheses of family Ω italicizing all values significant at overall level 5%. (Compare with a similar table of $c_{\max}$ by Gabriel [5].) Asymptotic theory [6] suggested a critical value from the chi-square distribution with $p_{V_0}(c_{G_0} - 1) = 15$ df. However, 230 permutations of the data showed this distribution to be unsuitable—Table V—and gave an estimated upper 5% critical value of 29.8 (as compared with 25.0 from the chi-square distribution which was exceeded by 32 out of the 230 L values).

Another nonparametric procedure is the max W STP. First, the Wilcoxon-Mann-Whitney statistic is computed for each atom, i.e., each univariate two sample comparison, and the absolute value W of the corresponding Normalized statistic noted. Then max W is obtained for all hypotheses of ω as the maximum of the W's over all atoms of the hypothesis. For a 5% level STP any hypothesis is declared significant if its max W exceeds the upper $0.025/[3 \times \binom{6}{2}] = 0.00055$ point of the Normal distribution. In the present example, however, the samples were too small to allow use of the Normal approximation, so each atomic hypothesis was tested at 0.0011 (two-sided) by means of tables of the Wilcoxon-Mann-Whitney statistic and each non-atomic hypothesis declared significant if any of its atoms was. The results of this procedure appear in Table VI.

A comparison of several of the STP's for family Ω of class differences on variable sets is presented in Table VI which shows, for each hypothesis, which of the 5% level STP's would have rejected it. (Groups of 3 and 4 localities have been omitted from the table for reasons of space.) The relation between the STP's for atomic hypotheses are as expected from the foregoing discussion, even though the latter do not attain the exact 5% level: the order of atomic power is max $t^2 >$ max $F >$ max $T^2 > c_{\max} = R$. The reverse order is found for nonatomic hypotheses, with the $c_{\max}$ and max T^2 STP's showing more significant results than the max F and max t^2 STP's.

Actual calculation of the atomic power function for this set-up—Table VII—bring the comparisons out strikingly. The power of locating pairwise class differences on single variables is appreciably higher with the max t^2 STP than with the $c_{\max}$ STP. The advantage of the max t^2 *for this purpose* is stressed by recalling a similar comparison [5] showing the $c_{\max}$ STP to have considerably more atomic power than STP's with the other common MANOVA statistics, that is, the likelihood ratio determinantal ratio and Hotelling-Lawley's and Pillai's traces.

The R procedure coincides with the $c_{\max}$ STP for univariate hypotheses. For multivariate hypotheses it is clearly inferior in providing fewer rejections, and incoherent ones at that. Thus, for instance, it rejects equality of the group of localities 2, 3, 4, 5, 6 on the single variable Z but not on all three variable XYZ together.

TABLE IV *L Statistics for All Groups of Localities on All Sets of Variables*[a]

Localities	XYZ	XY	XZ	YZ	X	Y	Z
		Variables					
123456	*71.944*	*44.812*	*65.945*	*54.496*	*36.247*	*34.191*	20.709
12345	*58.571*	*35.714*	*53.300*	*42.697*	28.405	26.036	12.926
1234 6	*50.678*	*32.137*	*47.895*	*41.390*	29.755	27.155	9.741
123 56	*60.684*	*36.110*	*57.244*	*47.909*	*31.197*	29.011	17.670
12 456	*46.410*	*32.039*	*42.890*	*34.000*	26.155	23.709	15.714
1 3456	*52.599*	*36.014*	*46.995*	*42.047*	28.700	29.275	18.359
23456	18.832	7.232	15.458	9.943	0.697	1.577	8.428
1234	*38.606*	23.420	*36.456*	*30.216*	22.074	19.001	2.287
123 5	*47.789*	27.190	*44.996*	*36.347*	23.434	20.894	10.096
12 45	*35.348*	28.328	*32.542*	24.149	18.518	15.857	9.644
1 345	*40.368*	26.976	*35.456*	*31.355*	20.994	21.176	11.397
2345	13.602	6.229	10.467	6.023	0.434	1.181	5.354
123 6	*44.160*	26.684	*41.929*	*35.603*	24.889	22.158	7.448
12 4 6	*30.357*	21.184	29.282	23.713	19.671	17.162	7.268
1 34 6	*33.960*	23.994	*31.363*	*30.573*	22.211	22.415	8.745
234 6	4.951	1.131	4.656	4.065	0·422	0.730	3.475
12 56	*35.758*	23.541	*34.792*	28.000	21.266	18.732	13.251
1 3 56	*41.539*	27.471	*38.425*	*35.614*	23.707	24.224	15.451
23 56	12.806	3.445	11.588	8.164	0.294	1.024	6.763
1 456	27.282	23.389	24.074	21.579	18.629	18.794	13.384
2 456	10.486	4.676	7.996	4.559	0.381	0.581	3.594
3456	14.649	6.213	11.667	7.019	0.560	1.215	6.099
123	*32.565*	18.144	*30.888*	24.667	17.207	14.041	0.204
12 4	20.595	12.852	20.122	14.488	12.115	9.309	1.528
12 5	25.173	15.007	24.825	18.386	13.629	10.918	7.391
12 6	24.446	15.934	23.920	18.513	14.966	12.368	5.553
1 34	23.030	15.336	21.030	20.508	14.586	14.316	2.113
1 3 5	29.786	18.610	27.284	25.160	16.001	16.163	8.699
1 3 6	27.641	18.698	25.529	24.940	17.402	17.547	6.584
1 45	17.361	14.736	14.815	12.836	11.048	10.998	8.136
1 4 6	13.855	13.187	12.884	12.925	12.147	12.423	6.293
1 56	16.829	15.049	16.109	15.733	13.797	13.947	11.053
234	1.023	0.509	0.871	0.770	0.240	0.334	0.730
23 5	8.054	2.619	6.995	4.481	0.031	0.665	3.899
23 6	3.667	0.592	3.521	3.086	0.202	0.359	2.556
2 45	7.567	4.060	5.286	2.587	0.244	0.488	2.233
2 4 6	1.819	0.395	1.636	1.500	0.114	0.223	1.163
2 56	5.068	1.093	4.730	3.367	0.140	0.231	2.506
345	10.561	5.270	7.783	4.206	0.353	0.875	3.847
34 6	3.395	0.767	3.283	2.773	0.287	0.544	2.500
3 56	8.823	2.585	7.929	5.394	0.214	0.791	4.565
456	6.519	3.805	4.340	1.664	0.265	0.221	1.286

TABLE IV (*Cont.*) *L Statistics for All Groups of Localities on All Sets of Variables*[a]

	Variables						
Localities	XYZ	XY	XZ	YZ	X	Y	Z
12	15.161	7.779	15.159	9.525	7.410	4.554	0.022
1 3	17.188	10.217	15.594	15.113	9.777	9.486	0.161
23	0.216	0.147	0.135	0.029	0.021	0.001	0.021
1 4	5.234	4.915	4.831	4.808	4.648	4.627	1.374
2 4	0.200	0.158	0.132	0.154	0.058	0.129	0.132
34	0.607	0.204	0.605	0.587	0.162	0.204	0.577
1 5	7.385	6.574	7.248	7.228	6.217	6.188	6.015
2 5	2.626	0.654	2.418	1.633	0.003	0.176	1.355
3 5	5.213	1.819	4.442	2.819	0.008	0.489	2.523
45	4.741	3.248	2.736	0.800	0.184	0.183	0.747
1 6	8.143	8.095	7.655	7.879	7.500	7.759	4.709
2 6	1.142	0.059	1.106	1.108	0.056	0.056	0.822
3 6	2.310	0.386	2.280	1.949	0.126	0.302	1.713
4 6	0.477	0.177	0.398	0.238	0.000	0.038	0.210
56	1.300	0.380	1.206	0.626	0.081	0.000	0.329

[a] 5% significant statistics set in italics. Critical value 29.8.

TABLE V

Sample Permutation Distribution of L — 230 Permutations

L	Frequency of permutations	Expected frequency under $\chi^2(15)$-asymptotic
0 —	5	11.5
7.261—	8	11.5
8.547—	12	23
10.307—	15	23
11.721—	22	23
13.030—	22	23
14.339—	21	23
15.773—	20	23
17.322—	21	23
19.311—	29	23
22.307—	23	11.5
24.996—	14	5.75
27.488—	11	3.45
30.578—	7	2.3
	230	230.00

TABLE VI

Significance Decisions by Various Procedures for Groups of Localities on Sets of Variables[a]

Localities	Variables						
	XYZ	*XY*	*XZ*	*YZ*	*X*	*Y*	*Z*
123456	*RcTFtLW*	*RcTFtLW*	*RcTFtLW*	*RcTFtLW*	*RcTFtLW*	*RcTFtLW*	*RcTFt W*
12345	*RcTFtLW*	*RcTFtLW*	*RcTFtLW*	*RcTFtLW*	*RcTFt W*	*RcTFt W*	*RcTFt W*
1234 6	*RcTFtLW*	*RcTFtLW*	*RcTFtLW*	*cTFtLW*	*RcTFt W*	*RcTFt W*	
123 56	*RcTFtLW*	*RcTFtLW*	*RcTFtLW*	*RcTFtLW*	*RcTFtLW*	*RcTFt W*	*RcTFt W*
12 456	*RcTFtLW*	*RcTFtLW*	*RcTFtLW*	*RcTFtLW*	*RcTFt W*	*RcTFt W*	*RcTFt W*
1 3456	*RcTFtLW*	*RcTFtLW*	*RcTFtLW*	*RcTFtLW*	*RcTFt W*	*RcTFt W*	*RcTFt W*
23456	*cTFt W*	*cT*	*RcTFt W*	*RcTFt W*			*RcTFt W*
12	*cTFt W*	*RcTFt W*	*cTFt W*	*cTFt*	*RcTFt W*	*TFt*	
1 3	*cTFt W*	*RcTFt W*	*cTFt W*	*cTFt W*	*RcTFt W*	*Ft W*	
23							
1 4	*cTFt*	*TFt*	*cTFt*		*Ft*		
2 4							
34							
1 5	*RcTFt W*	*cTFt W*	*RcTFt W*	*RcTFt W*	*t W*	*RcTFt W*	*RcTFt W*
2 5	*cTFt*	*cT*	*cTFt*	*cTFt*			*Ft*
3 5	*cTFt W*	*cT*	*cTFt W*	*cTFt W*			*TFt W*
45	*cT*		*cT*				
1 6	*cTFt W*	*cTFt W*	*cTFt W*	*cTFt W*	*RcTFt W*	*RcTFt W*	
2 6							
3 6							
4 6							
56							

[a] (For brevity groups of 3 or 4 localities are omitted). All procedures test at a level of 5% or less, rejections are indicated as follows: *R*-ratio of roots, *c*-maximum root, *T*-max T^2, *F*-max *F*, *t*-max t^2, *L*-nonparametric *L* and *W* for the maximum Wilcoxon-Mann-Whitney statistic.

TABLE VII

Atomic Powers of Various 5% Level MANOVA *STP's* (6 classes, 3 variables, 42 error df)

Statistic	Noncentrality parameter $\delta = (\mu - \mu_0)/\sigma$					Critical value	θ
	0	1.6	3.2	4.8	6.4		
Maximum root	<0.0001	0.004	0.116	0.592	0.951	$\theta_{0.05} = 0.3323$	0.3323
Max T^2	0.0001	0.012	0.215	0.740	0.981	$F_{(3,40)0.05/15} = 5.55$	0.294
Max *F*	0.0002	0.015	0.240	0.775	0.986	$F_{(5,42)0.05/3} = 3.20$	0.276
Max t^2	0.0006	0.041	0.403	0.892	0.997	$F_{(1,42)0.05/45} = 12.2$	0.225

It should be mentioned that the max t^2 STP which comes out so well in this comparison is not available for exploration either of contrasts or linear combinations of variables. The additional insight into the data brought out by exploring the sum and differences of the original variables in the earlier $c_{\max}$ STP analysis [5] could not have been obtained by the max t^2 STP. (Though the max T^2 STP would also have provided it). Clearly, the additional power on the atoms of Ω is gained by ignoring other atoms, such as those in $\tilde{\Omega}$. This illustrates the importance of proper choice of the family of hypotheses on which inferences are to be made.

The nonparametric L-STP does not come out well on detailed hypotheses. It does not show significance on any atomic hypothesis. Use of this technique in lieu of the parametric STP's results in giving up most power of resolution. (An attempt was made to obtain greater detailed power by modifying $L_G{}^V$ to $L_G{}^V N_{G_0}^{V_0}/N_G{}^V$ (see Gabriel and Sen [6], Section 4), but this resulted in a large number of incoherences.)

The max W STP performs much better. It is closely related in its structure to the max t^2 STP, and indeed the results of these two STP's in Table VI do not differ very much, though the max t^2 STP provides several more significance decisions than the max W STP.

REFERENCES

1. ARMITAGE, J. V., and KRISHNAIAH, P. R. (1964). Tables for the studentized largest chi-squared distribution and their applications. ARL 64-188. Aerospace Res. Labs., Wright-Patterson Air Force Base, Ohio.
2. BHAPKAR, V. P. (1965). Lectures in multivariate analysis. Univ. of North Carolina, Chapel Hill, North Carolina.
3. GABRIEL, K. R. (1964). A procedure for testing the homogeneity of all sets of means in analysis of variance. *Biometrics* **20** 459–477.
4. GABRIEL, K. R. (1969). Simultaneous test procedures—some theory of multiple comparisons. *Ann. Math. Statist.* **40** 224–250.
5. GABRIEL, K. R. (1968). Simultaneous test procedures in multivariate analysis of variance. *Biometrika* **55** 489–504.
6. GABRIEL, K. R., and SEN, P. K. (1968). Simultaneous test procedures for one-way ANOVA and MANOVA based on rank scores. *Sankhyā Ser. A* **30** 303–312.
7. KASKEY, G., KRISHNAIAH, P. R., and AZZARI, A. (1962). Cluster formation and diagnostic significance in psychiatric symptom evaluation. *Proc. Fall Joint Comput. Conf., Philadelphia, 1962.* Spartan Books, Washington, D.C.
8. KHATRI, G. C. (1966). A note on a MANOVA model applied to problems in growth curves. *Ann. Inst. Statist. Math.* **18** 75–86 (also, Mimeo Seri. No. 399, 1964, Inst. Statist. Univ. North Carolina, Chapel Hill, North Carolina).
9. KRISHNAIAH, P. R. (1964). Multiple comparison tests in multivariate case. ARL 64-124. Aerospace Res. Labs., Wright-Patterson Air Force Base, Ohio.
10. KRISHNAIAH, P. R. (1965). On the simultaneous ANOVA and MANOVA tests. *Ann. Inst. Statist. Math.* **17** 35–53.
11. KRISHNAIAH, P. R. (1965). On a multivariate generalization of the simultaneous analysis of variance test. *Ann. Inst. Statist. Math.* **17** 167–173.

12. KRISHNAIAH, P. R. (1965). Multiple comparison tests in multi-response experiments. *Sankhyā Ser. A* **27** 65–71.
13. KRISHNAIAH, P. R., and ARMITAGE, J. V. (1965). Probability integrals of the multivariate F distribution, tables and applications. ARL 65-236. Aerospace Res. Labs., Wright-Patterson Air Force Base, Ohio.
14. MILLER, R. G. (1966). "Simultaneous Statistical Inference." McGraw-Hill, New York.
15. MUDHOLKAR, G. S. (1966). On confidence bounds associated with multivariate analysis of variance and non-independence between two sets of variates. *Ann. Math. Statist.* **37** 1736–1746.
16. REEVE, E. C. R. (1941). A statistical analysis of taxonomic differences within the genus *Tamandua* Gray (Xenartha). *Proc. Zool. Soc. London Ser. A* **111** 279–302.
17. ROY, J. (1958). Step-down procedure in multivariate analysis. *Ann. Math. Statist.* **29** 1177–1187.
18. ROY, S. N. (1954). A useful theorem in matrix theory. *Proc. Amer. Math. Soc.* **5** 635–638.
19. ROY, S. N., and BOSE, R. C. (1953). Simultaneous confidence interval estimation. *Ann. Math. Statist.* **24** 513–536.
20. ROY, S. N., and GNANADESIKAN, R. (1957). Further contributions to multivariate confidence bounds. *Biometrika* **44** 399–410.
21. SIOTANI, M. (1959). On the range in multivariate case (in Japanese). *Proc. Inst. Statist. Math.* **6** 155–165.
22. SIOTANI, M. (1960). Notes on multivariate confidence bounds. *Ann. Inst. Statist. Math.* **11** 167–182.

5

Reprinted from *Growth,* **25**(3), 219–227 (1961)

A NOTE ON GEOGRAPHICAL VARIATION IN EUROPEAN *RANA*

RICHARD A. REYMENT

Department of Geology, University of Stockholm, Stockholm, 6

Received April 10, 1961

INTRODUCTION

European populations of *Rana* have recently been the subject of a monograph by Kauri (1959) in connection with a critical discussion of the "climatic rules" of morphologic differentiation. In particular, the most important "race- and species-forming" criteria were reviewed. His investigations indicate that most of these features are little more than modifications of slight taxonomic importance.

One of the characters that has often been considered to be of great taxonomic importance is the size of the prehallux. Kauri's work shows this to be largely an ontogenetic-conditioned feature. Other dimensions taken up in the monograph under discussion are body length (measured between snout and posterior end of *os coccygis*), length of tibia and, in some cases, length of femur, cranial breadth and cranial length (the former measured across the point of jaw articulation and the latter measured from snout tip to the occipital margin). Details of the mensurational techniques employed are given in Kauri (1959, p. 16); all measurements in mm. Kauri's studies were made with the aid of standard techniques of univariate and bivariate analysis. In the present paper a few notes are presented on conclusions that may be drawn from Kauri's published data by the use of "multivariate" biometric analysis. The calculations are based on mature individuals, as classified by Kauri. The reasoning involved in the following is more fully discussed in a recent paper by the writer (Reyment, 1960), to which reference is also made for details concerning pertinent literature.

SEXUAL DIMORPHISM IN CRANIAL DIMENSIONS

A good suite of cranial measurements for *Rana esculenta* from Vienna were available which permit the following conclusions.

The sample of mature females consists of 35 specimens and that of mature males contains 14 specimens. The two cranial dimensions measured are length of cranium, x_1, and breadth of cranium, x_2. Testing difference of means by a univariate t-test gave for the difference between cranial lengths $t = 0.78$, and for the difference between cranial breadths $t = 1.01$, neither of which are significant, and on the basis of these results significant sexual dimorphism in the two dimensions considered cannot be claimed to exist.

The data were then tested in a T^2-test. The required statistics are:

1. *Covariance matrices*

$$S_1 = \begin{bmatrix} 17.683 & 20.290 \\ 20.290 & 24.407 \end{bmatrix}$$

$$S_2 = \begin{bmatrix} 18.479 & 19.095 \\ 19.095 & 20.755 \end{bmatrix}.$$

2. *Mean vectors*

$$\bar{x}_1 = \begin{bmatrix} 22.860 \\ 24.397 \end{bmatrix} \quad \text{and} \quad \bar{x}_2 = \begin{bmatrix} 21.821 \\ 22.843 \end{bmatrix}.$$

3. *Inverse of S_λ, the pooled covariance matrix*

$$S_\lambda^{-1} = \begin{bmatrix} 5.504 & -4.800 \\ -4.800 & 4.407 \end{bmatrix}.$$

In order to obtain the T^2-statistic it is first required to expand the quadratic form

$$Q = (\bar{x}_1 - \bar{x}_2)S_\lambda^{-1}(\bar{x}_1 - \bar{x}_2)'.$$

The difference vector

$$d = \bar{x}_1 - \bar{x}_2 = \begin{bmatrix} 1.039 \\ 1.554 \end{bmatrix}.$$

From these values we obtain $Q = 1.192$.

$$T^2 = Q \cdot \frac{N_1 N_2}{N_1 + N_2} = 11.92.$$

Using

$$F = 11.92. \frac{(N_1 + N_2 - p - 1)}{p(N_1 + N_2 - 2)} = 5.83$$

as a variance ratio with 2 and 46 degrees of freedom the difference in mean vectors is significant on the 1%-level. There is thus definite sexual dimorphism in the two dimensions studied.

Mean Vectors of Populations

For the purposes of this investigation samples from populations of *Rana esculenta* from Southern Sweden (N = 17), the environs of Vienna (N = 49), Austria, the Balkan area (N = 34), and North Africa (N = 25) were considered. It was desired to ascertain whether significant differences in the mean vectors occur; i.e., does

$$\mu_1 = \mu_2 = \mu_3 = \mu_4,$$

where μ_i denotes the mean vector of the ith population. Owing to the results obtained in Kauri's investigations there are objections to using any one of body length, x_1, tibia, x_2, or prehallux, x_3, on its own in an analysis of variance. A preliminary analysis of variance using x_3 actually shows it to be significantly different in the material studied, but this may be fortuitous. It therefore would be desirable to be able to base studies on the geographical unlikeness of populations of *Rana esculenta* on several characters instead of just the doubtful prehallux dimension. The three variables were therefore investigated in a trivariate dispersion analysis for four samples from four populations.

The following matrices are required for this analysis (matrices of sums and products).

1. *Total dispersion matrix* (N = 125).

$$T = \begin{bmatrix} 18303.70 & 9377.23 & 1092.55 \\ 9377.23 & 5403.04 & 534.35 \\ 1092.55 & 534.55 & 123.72 \end{bmatrix}.$$

2. *Total dispersion matrix within samples.*

$$U = \begin{bmatrix} 16861.24 & 8862.01 & 835.10 \\ 9962.01 & 5094.31 & 461.98 \\ 835.19 & 461.98 & 69.82 \end{bmatrix}.$$

The ratio of the determinants of these matrices provides the statistic Λ:

$$\Lambda = \frac{|U|}{|T|} = 0.278124.$$

The significance of this value is obtained by calculating the statistic

$$V = m \log_e \Lambda = -\left(n - \frac{p+q+1}{2}\right) \log_e \Lambda = 154.20.$$

V is roughly distributed as χ^2 with pq = 9 degrees of freedom. The above value is significant on even the 0.1% level and there is no doubt that the four populations of *Rana esculenta* differ highly significantly with respect to the three characters treated.

Comparative "Variability" of *Rana esculenta* and *Rana temporaria*

A concept that may prove useful in taxonomic work is a generalization of the univariate coefficient of variation. It was recently proposed by the writer (Reyment, 1960). By definition it is

$$V_k = \frac{|S|^{1/2k}}{|\bar{x}|},$$

the ratio of the 1/2k th root of the generalized variance, divided by the absolute value of the mean vector, where k is the number of characters investigated.

In the present study it was of interest to obtain an impression of the relative variability of *R. esculenta,* compared with that of *R. temporaria.* With this end in view the Southern Swedish populations of both species were analyzed, first by univariate analysis of body length, tibia and prehallux and then all three taken together. The results are shown in the following table. (V_L = coefficient of variation for the length dimension, V_T is that of the tibia and V_P that of the prehallux; V_3 is the trivariate coefficient.)

Coefficient	*Rana esculenta*	*Rana temporaria*
V_L	10.09	21.22
V_T	12.10	19.71
V_P	9.41	23.13
V_3	11.31×10^{-2}	13.19×10^{-2}

These figures indicate the higher variability in the sample of *Rana temporaria.*

Diagrammatic Representation of the Trivariate Geographical Variation

It is useful to be able to give an idea of the geographic unlikeness by means of a diagrammatic representation. This was accomplished with the aid of a discriminant function between a sample from the population of *Rana esculenta* in Skåne and the Balkans population. By substituting sample mean vectors in the resultant equation a series of values for the various geographic occurrences is obtained. The three dimensions used in the calculations are those of body length, x_1, tibia, x_2, and prehallux, x_3. The following data were used in the construction of the discriminant function. [$N_1 = 34$, $N_2 = 17$.]

1. *Pooled covariance matrix inverse*

$$S_\lambda^{-1} = \begin{bmatrix} 0.02398 & 0.08309 & -1.01075 \\ 0.08309 & -0.61596 & 3.92600 \\ -1.01075 & 3.92600 & 16.51220 \end{bmatrix}.$$

2. *Sample mean vectors*

$$\bar{x}^{(1)} = \begin{bmatrix} 67.747 \\ 31.394 \\ 4.718 \end{bmatrix} \text{ and } \bar{x}^{(2)} = \begin{bmatrix} 62.797 \\ 33.791 \\ 3.544 \end{bmatrix}.$$

The three discriminant function coefficients obtained from these data are: $\delta_1 = -1.339$, $\delta_2 = 6.248$, $\delta_3 = -30.767$. The discriminant function is

$$z = x_1 - 4.666x_2 + 22.978x_3.$$

A rough idea of the effectivity of the function was obtained by calculating $d_{\bar{z}/s_z} = 24.837$, which indicates a high probability of effectivity.

The values obtained by substituting the mean vectors of the samples of the various populations in the above discriminant function are:

	Rana esculenta								*Rana temporaria*
Population	Skåne	Balkans	Vienna	North Africa	South Spain	Italy	Asia Minor	Syria	Skåne
z	22.67	—13.44	9.15	—17.78	—18.35	—0.98	—9.84	—11.12	—17.80

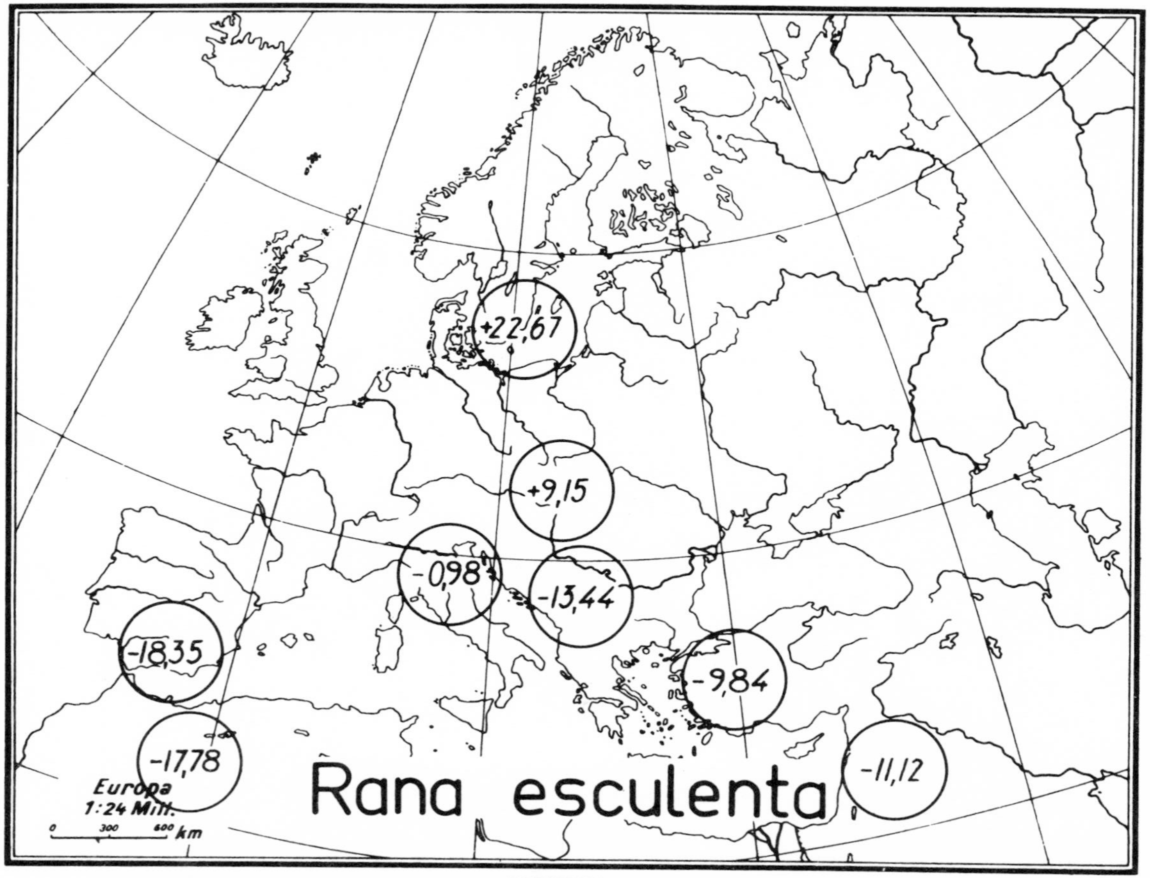

FIGURE 1

Map of Europe and parts of North Africa and the Middle East showing the location of the places from which the study material of *Rana esculenta* was derived. Details of the localities are: Skåne, Southern Sweden; environs of Vienna, Austria; Granada and Algeciras, Spain; Firenze, Italy; Morocco, Lambesi, Ain Zarah, El Eerayet, Biskra, Tuggurt, North Africa; Fort Opus, Solonika, Chania, Nicaria, Arta, Berac, Balkans region; Alexandretta, Asia Minor; Homs, Syria.

The table also contains a value for *Rana temporaria* based on a sample from the population of this species in the county of Skåne, Southern Sweden. The values are shown on the map (p. 224) in relation to the localities from which the study material was derived.

The figure conveys the impression of a gradual southerly trend in morphologic change with North Africa lying beyond the southerly "end point."

Generalized Distances Between Populations

The Mahalanobis' generalized distance is one of the most useful modern aids to taxonomy yet devised. The generalized distance is obtained from the quadratic form

$$D^2 = dS^{-1}d',$$

where d is the vector of differences between mean vectors.

The square root of this value, D, is equivalent to the generalized Euclidean distance in two-, three-, or n-space between the centroids of the respective probability masses. In the following table D-values calculated between the Skåne frogs and the samples from the Balkans, Vienna, and North African populations are shown. The distance between the Scanian sample of *Rana esculenta* and a sample of Southern Swedish *Rana temporaria* is also given.

Some Generalized Distances for Rana Esculenta and Rana Temporaria

R. esculenta			*R. temporaria*
Skåne—Balkans	Skåne—Vienna	Skåne—North Africa	Skåne—S. Sweden
7.33	1.10	4.38	4.72

(It should be noted that the material of *Rana temporaria* is derived from several localities in Southern Sweden, of which the county of Skåne is a part.)

Homogeneity of Covariance Matrices

All of the procedures employed in the foregoing require the assumption of homogeneity of covariance matrices. This may be tested by constructing the generalized Bartlett test of homogeneity

$$V = \frac{\Pi|A|^{n_1/2}}{|A|^{n/2}},$$

where $n_i = N_i - 1$ and $n = \Sigma n_i = N - q$, where q is the number of populations involved. In order to ascertain the significance of the above statistic one calculates $-\log_e W$, where

$$W = V \cdot \frac{n^{pn/2}}{\Pi n_i^{pn_i/2}},$$

and where p is the number of variables.

Using this procedure on the covariance matrices of three samples of *Rana esculenta* and one of *R. temporaria* a value of $V = 0.6936 \times 10^{-77}$ was obtained and $W = 0.111 \times 10^{25}$. The significance of the latter statistic is found by the use of an asymptotic expression of the form

$$Pr(-\varrho \log_e W \leqslant z) = Pr(\chi_f^2 \leqslant z) + \ldots\ldots$$

It is clear that $-\log_e W$ will be very small and there is no need to proceed further with the calculations. The covariance matrices are clearly homogeneous.

Summary

Notes are first presented on sexual dimorphism in two cranial dimensions of a sample of frogs from Vienna and environs. The main part of the study is concerned with establishing trends in morphologic differentiation. A multivariate analysis of dispersion indicates significant differences in mean vectors (body length, tibia, prehallux) for samples from Southern Sweden, Austria, Balkans and North Africa. Use of a discriminant function between Swedish and Balkans frogs is made in order to illustrate the north-southerly gradation in the three characters involved by inserting values taken from the other populations studied. The trend, or "multicline," shown indicates a general change from north to south with perhaps a slight tendency towards a regression in the North African material. An interesting result is provided by the relatively extreme z-value given by the Balkans sample, which is more widely divergent from the most northerly sample than those of Asia Minor and Syria. This is also reflected in the generalized distance values calculated between the sample from Skåne and the samples from the Balkans, Austria and North Africa. These indicate that the statistical distance between Skåne and the Balkans is greater than that between Skåne and North Africa, and that the distance between the sample from Skåne and that from Austria is small.

A test of homogeneity of covariance matrices of the samples drawn from the Skåne, Vienna, and North Africa populations of *Rana esculenta* and the Skåne population of *Rana temporaria* indicates a high degree of homogeneity.

References

Kauri, H. 1959. Die Rassenbildung bei europäischen *Rana*-Arten und die Gültigkeit der Klimaregeln. Doctoral dissertation at the university of Lund. Gleerupska Universitetsbokhandelns Förlag AB, Lund, 171 pp.

Reyment, R. A. 1960. Studies on Nigerian Upper Cretaceous and Lower Tertiary ostracoda. Part 1: Senonian and Maestrichtian ostracoda. *Stockh. Contr. Geol.*, **7**, 238 pp., 23 pls. Almqvist & Wicksell, Stockholm.

6

Reprinted from *Evolution,* **25**(3), 584–590 (1971)

CHANGES IN MICROGEOGRAPHIC VARIATION PATTERNS OF *PEMPHIGUS POPULITRANSVERSUS* OVER A SIX-YEAR SPAN[1,2,3]

Robert R. Sokal, Norman N. Heryford[4] and John R. L. Kishpaugh

Division of Biological Sciences, State University of New York at Stony Brook, Stony Brook, N. Y. 11790

Received January 27, 1971

Eight characters of the alate fundatrigeniae of the gall producing aphid *Pemphigus populitransversus* Riley were measured in samples from eight closely situated localities during the years 1961, 1964, and 1966. It was of interest to find whether the pattern of geographic variation described by Rinkel (1965) from samples collected in 1961 is retained or changed in collections from 1964 and 1966. The magnitude of such changes in a microgeographic situation must be known if changes in a wider space or time frame are to be interpreted correctly. The inferences that can be drawn from certain variation patterns about the population biology of *P. populitransversus* specifically, and other organisms in general, will also be discussed.

The present emphasis on population studies in evolutionary research has led to an active interest in population phenetics and the inferences about population structure and evolutionary mechanisms that may be drawn from such investigations (see Sneath and Sokal, 1972, for a general review). In spite of an increasing number of recent studies in this field (Ehrlich, 1965; Soulé, 1967; Johnston, 1969; Koehn et al., 1971; Selander, 1970) we still are not in a position to make general statements about classes of phenetic variation patterns to be found in nature nor about the types of population structure such patterns would reflect.

Materials and Methods

The biology of *P. populitransversus* has been described in several of the papers in this series and in earlier publications to which references are given in the cited papers. At this time we need mention only that galls of *P. populitransversus* are found on the petioles of leaves of the Eastern Cottonwood, *Populus deltoides* Bartr. ex Marsh. The alate fundatrigeniae, born parthenogenetically and viviparously to the single stem mother in each gall, emerge during the summer months through the emergence slit in the gall. Further life history information is deferred to the Discussion Section, where it will be needed for an interpretation of the findings.

Of the 15 localities from northeastern Douglas County and western Leavenworth County, Kansas sampled by Rinkel (1965) in 1961, only 13 yielded collections in 1964, and by 1966 samples could be obtained from only eight localities. This study is therefore restricted to these eight localities for the three years. All collections were made in early September. The localities are indicated in Figure 1, where for convenience new numbers from 1–8 are assigned to them. Their code numbers in

---

[1] Contribution number 33 from the program in Ecology and Evolution at the State University of New York at Stony Brook.

[2] Part 7 of a study of variation in the aphid genus *Pemphigus.* Parts 1–6, respectively, are the following publications: Sokal (1952), Sokal (1962), Sokal and Rinkel (1963), Rinkel (1965), Sokal and Thomas (1965) and Heryford and Sokal (1971).

[3] This investigation was supported in part by grant GB-20501 from the National Science Foundation. This research was started at the Department of Entomology, The University of Kansas and continued at the State University of New York at Stony Brook.

[4] Captain, U. S. Army Medical Service Corps, Department of Entomology, Kansas State University, Manhattan, Kansas 66502. The opinions stated herein are those of the authors and should not be construed as official or reflecting the view of the Department of the Army.

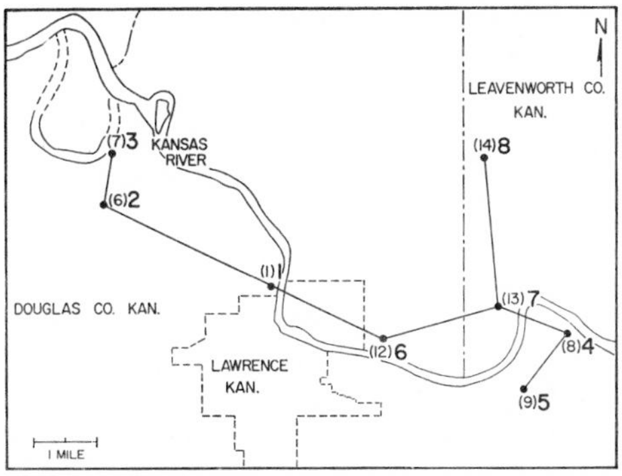

FIG. 1. Map giving locations of numbered localities. Numbers in parentheses are those employed for the same localities by Rinkel (1965). Contiguity relationships following the definition of Gabriel and Sokal (1969) are indicated by a network connecting the localities.

Rinkel's study are given in parentheses. The exact locations of these localities are given in Table 1 of Rinkel (1965).

The eight characters measured are defined by Sokal (1952). They are forewing length, head width, thorax length, thorax width, femur length, tarsus length, antenna III (length), and antenna IV (length).

SIMULTANEOUS TEST PROCEDURES

These methods of multiple comparisons (see Sokal and Rohlf, 1969) were developed in a series of papers starting with Gabriel (1964), and have been proposed for geographic variation analysis by Gabriel and Sokal (1969). Since multiple comparisons tests frequently result in a large number of homogeneous and overlapping subsets, Gabriel and Sokal (1969) stressed the need for more stringent criteria of acceptable homogeneous sets of localities. They proposed that each homogeneous set of localities should consist of a *continuous region* with each locality contiguous to at least one other locality. Two localities A and B are said to be *contiguous* if, and only if, all other localities in the study are outside the circle whose diameter is defined by the line AB. The contiguity relations connecting the eight localities of the present study have been indicated in Figure 1.

UNIVARIATE ANALYSES

The means of the eight characters for each of the three collection years and each of the eight localities are shown in Table 1. For any one character, differences among localities and among the three years within any one locality are evident. This is confirmed when each of the characters is subjected to a two-way analysis of variance with replication. All main effect and interaction mean squares are highly significant (the mean square among years for antenna IV is significant at $P < 0.025$; all others at $P < 0.001$). The significant interaction mean squares substantiate the impression gained from Table 1 that the changes in any one character over the three different years varied with the locality, making a rather complicated, shifting geographic variation pattern from year to year, differing for each of the characters.

MULTIVARIATE ANALYSES

The multivariate analyses of variance were carried out on a two-way design of the eight localities against the three years and also on three one-way designs, separately for each year. Each manova was based on the eight characters in the present study. The overall hypothesis of equality of mean vectors for the eight localities or the three years could be rejected at $P < 0.01$ using Roy's maximum characteristic root test. The simultaneous test criterion employed was also based on Roy's maximum characteristic root technique as recommended by Gabriel (1968). Each of the three years was significantly different from the other two. Having established this fact, most of our attention was focused on multiple comparisons tests of the differences among the eight localities jointly for the three years and separately for each of these years (manovas among localities were also significant each year).

For the joint three-year study the maximally acceptable connected sets (homogeneous continuous regions) are shown in Figure 2. Two major groups are distinguishable, a western one and an eastern

TABLE 1. *Means of all characters for all localities in each of the three years.*

Locality	Year	Forewing length	Head width	Thorax length	Thorax width	Femur length	Tarsus length	Antenna III length	Antenna IV length
1	1961	2.07[1]	.378	.556	.604	.429	.171	.166	.075
	1964	1.99	.386	.562	.608	.451	.152	.172	.072
	1966	2.21	.390	.601	.697	.477	.155	.203	.079
2	1961	2.02	.369	.519	.612	.451	.178	.161	.072
	1964	2.14	.391	.633	.680	.482	.150	.173	.072
	1966	2.50	.395	.629	.724	.500	.158	.199	.077
3	1961	2.20	.384	.546	.662	.462	.180	.168	.076
	1964	2.07	.386	.616	.672	.478	.154	.176	.076
	1966	2.04	.390	.578	.657	.472	.152	.169	.063
4	1961	2.07	.369	.530	.596	.431	.173	.162	.070
	1964	1.76	.370	.546	.618	.436	.140	.156	.068
	1966	2.01	.370	.560	.614	.444	.151	.187	.072
5	1961	1.97	.358	.504	.569	.430	.175	.155	.067
	1964	1.94	.382	.568	.601	.437	.146	.173	.069
	1966	1.85	.376	.539	.603	.457	.149	.169	.065
6	1961	2.20	.386	.575	.663	.466	.178	.156	.077
	1964	1.85	.377	.560	.606	.437	.141	.163	.069
	1966	2.28	.395	.612	.692	.501	.160	.165	.070
7	1961	2.04	.362	.515	.596	.440	.178	.162	.077
	1964	2.14	.394	.635	.693	.496	.153	.182	.075
	1966	2.20	.385	.588	.683	.481	.156	.156	.069
8	1961	1.97	.373	.538	.610	.451	.170	.153	.070
	1964	1.98	.389	.570	.626	.451	.150	.172	.073
	1966	2.32	.394	.608	.674	.490	.158	.164	.071
Average standard error of the means based on 15 galls		.055	.0046	.0121	.0145	.0107	.0025	.0052	.0022

[1] All variables are given in millimeters. Means are based on 15 galls per locality, with 2 alates measured per gall. Average standard errors have 336 *df*.

one, the former consisting of two overlapping sets of two localities, the latter of three overlapping sets comprising the eastern five localities.

However, we believe that the simplicity of Figure 2 is misleading. In Figure 3 *all* the maximally acceptable sets of the eight localities are plotted (regardless of whether they are continuous regions) and many more interrelationships among localities are indicated.

The different patterns among the localities for the separate years 1961, 1964, and 1966, respectively, are shown in Figures 4, 5, and 6. These illustrate maximally acceptable connected sets and should be compared with the pattern for the joint three-year study in Figure 2. The simple intergrading pattern of 1961 becomes more fragmented in 1964 and again simplified in 1966, although the break between localities 1 and 6 found in the joint three-year study is not found for any one of the three years. This is an example of *dissonance*, discussed in connection with manova STP by Gabriel (1968). Again many more relationships are present among the localities when contiguity is ignored.

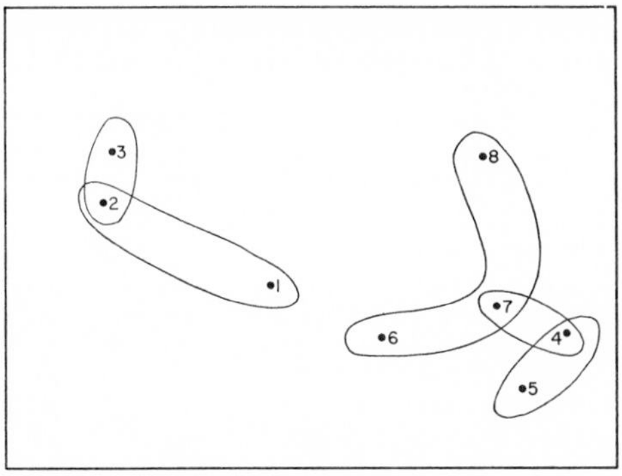

FIG. 2. Maximal acceptable connected sets (at the 1% level) based on manova of joint three-year study. Localities and their connections are indicated in Figure 1.

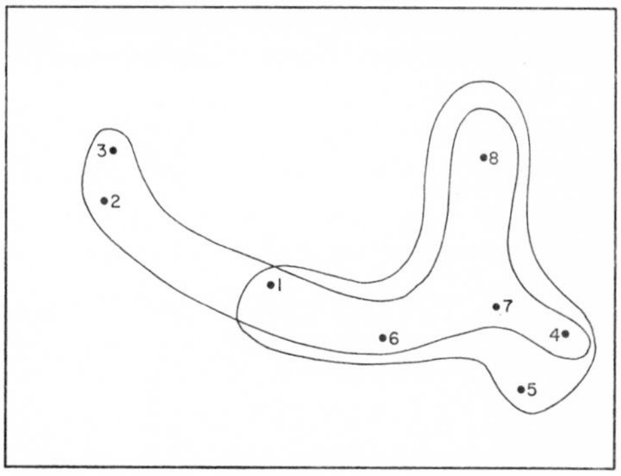

FIG. 4. Maximal acceptable connected sets (at the 1% level) based on manova for the year 1961. Localities and their connections are indicated in Figure 1. Compare with subsequent years (Figs. 5 and 6) and with three-year study (Fig. 2).

DISCUSSION AND CONCLUSIONS

The present study confirms the findings of Rinkel (1965) that significant microgeographic phenetic differentiation of alate fundatrigeniae of *P. populitransversus* occurs in this study area. In *Pemphigus* the genetic composition of the population of stem mothers and alates for a given year depends on the sexuparae (remigrants from the herbaceous secondary hosts) of the previous fall. We do not know from how large a sampling area the remigrants return to the cottonwood, their primary host. Are the large aggregations of *Pemphigus* sexuparae and sexuales found in sheltering cracks in the bark of any one cottonwood, descendants from a single migration, i.e., a "swarm" of sexuparae returning from a single source of secondary hosts under the influence of a particular climatic pattern? Or do numerous sexuparae from many sources arrive at neighboring cottonwoods over a period of days and congregate in the most suitable places for giving birth to the sexuales? The second case would give rise to greater genetic variability of aphids on any one tree and less intertree

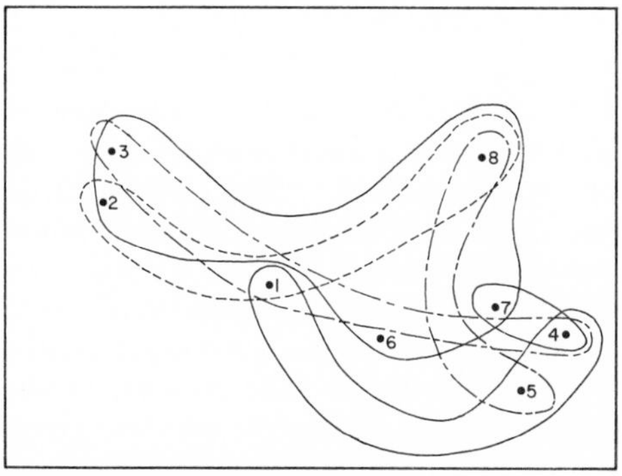

FIG. 3. Maximal acceptable sets (at the 1% level) based on manova of joint three-year study. Localities and their connections are indicated in Figure 1. Note that all subsets in Figure 2 are proper subsets of the subsets in this figure.

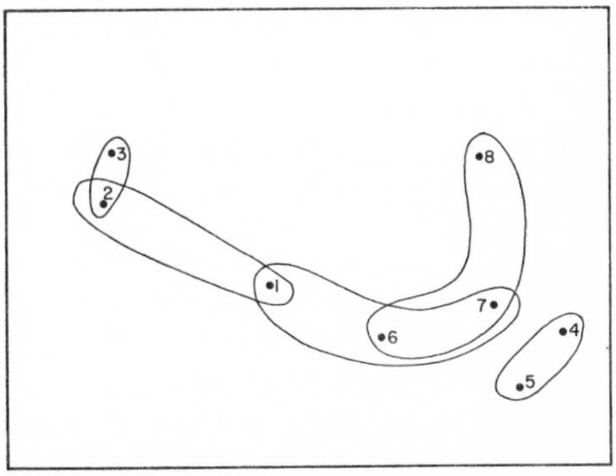

FIG. 5. Maximal acceptable connected sets (at the 1% level) based on manova for the year 1964. Localities and their connections are indicated in Figure 1. Compare with other years (Figs. 4 and 6) and with three-year study (Fig. 2).

differentiation at any one locality, and seems on biological grounds the more likely of the two alternatives. It is supported by the finding of Rinkel (1965) that aphids from three different trees at the same locality showed no significant intertree variation for 17 of 18 characters measured.

Are secondary hosts nearby or far away? The known secondary hosts such as *Erigeron, Solidago, Brassica, Lactuca* and others, are frequently seen close to infected cottonwoods, and it may be inferred that they serve as a major source of remigrating aphids. Very localized sources of remigrants may give rise to local differentiation among the fundatrigeniae, especially if the aphids on the roots are subjected to differential selective processes because of differences in climatic conditions or in intermediary host species from place to place or year to year. Gene flow can only take place during sexual reproduction among the sexuales produced by the remigrant sexuparae during the fall. Therefore, gene flow between local populations necessitates the intermingling of genetically differing alienicolae on secondary hosts (less likely in *Pemphigus*), or the admixture of diverse remigrants on the primary host.

We should distinguish five important potential outcomes of the two-way manova. It could lead to findings of significant heterogeneity among localities (L^+), or lack thereof (L^-). Similarly, a significant differentiation among years could result (Y^+), or the pattern could remain uniform (Y^-) for the three years examined. The above dichotomies lead to the classes, L^+Y^+, L^+Y^-, L^-Y^+, and L^-Y^-. Each combination could additionally exhibit $L \times Y$ interaction. We shall show below that all cases of significant interaction lead to a similar biological interpretation, regardless of the L,Y-combination associated with it.

It does not seem profitable to couch the discussion of the possible outcomes in general terms applicable to all types of organisms. It will be more meaningful to discuss the various possibilities for population-phenetic variation patterns in terms of

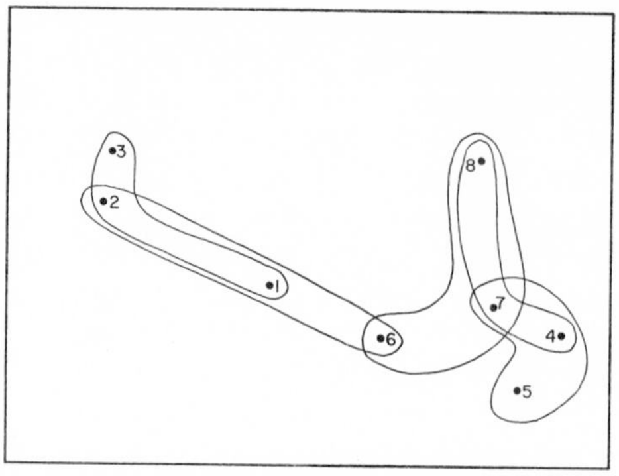

FIG. 6. Maximal acceptable connected sets (at the 1% level) based on manova for the year 1966. Localities and their connections are indicated in Figure 1. Compare with previous years (Figs. 4 and 5) and with three-year study (Fig. 2).

the particular organisms (*Pemphigus, Populus* and the intermediate hosts) interacting here. Yet given an example of this type of reasoning, the reader should not find it difficult to transfer the discussion to other organisms with their own peculiar set of biological properties.

The combination L^+Y^- would reflect a stable pattern of local differentiation. Such a pattern would be exhibited by a species with very localized populations (low vagility), or by a species with moderate localization but intensive locally diverse selective forces creating a population polymorphism by intensive disruptive selection. Such a situation might also be mimicked by strong local differences in environmental effects on the aphid phenotypes, such as differences in the genotypes of individual cottonwoods or their specific edaphic factors. If the several localities depended on sources of sexuparae from different populations of perennial secondary hosts, such a pattern could also arise. Local differences in such a generally small area would require the secondary hosts to be so close to the primary hosts that each cottonwood receives sexuparae from its own source.

The combination L^+Y^+ must also lead to the assumption of localized populations or locally selected or environmentally affected populations, yet we must also explain the differences from year to year. Since the genome of individual trees is fixed and edaphic factors change relatively little, we would have to assume marked annual changes in general environmental factors. Since there is no interaction, the pattern of differentiation among localities would retain its configuration but be translated in character hyperspace for the three years.

The combination L^-Y^+ would imply that a genetically homogeneous population inhabited the area and that genetic differences among trees and differences in edaphic factors have relatively trivial effects. The phenetic changes from year to year might imply general immigration of sexuparae into the local area from some distance, with the source area changing from year to year depending on weather and wind patterns. Or we could suppose that annual differences in climate from the previous fall until maturity of the alates would affect the morphology of the alates in any given year.

The combination L^-Y^- implies a fairly stable interbreeding population (which may in fact be a subset of a yet larger homogeneous population), whose remigrants come from the same gene pool and who are not subject to great environmental fluctuations from year to year at either the primary or secondary host sites. They would also not be much affected by genetic differences among cottonwoods, or edaphic and microclimatic heterogeneity among separate localities.

Finally, a fifth outcome would be the presence of $L \times Y$ interaction. Each year the pattern of locality means would change in phenetic hyperspace. Local environmental differences could, of course, affect the primary host plant (e.g., their sap) or the microenvironment in the galls, either of which could alter the morphology of the alates among localities and among years. Alternatively, chance allocation of remigrants from varying sources, or from different areas subject to varying environmental conditions which impose selection upon the aphids while they are on the secondary hosts, could easily give rise to differences in patterns among years. Significant interaction could be associated with any of the four possible combinations of L and Y mentioned above. In such an instance L^+ or Y^+ would indicate that the forces differentiating localities or years were of such magnitude as to maintain significant differences among the appropriate mean vectors in addition to the heterogeneity implied by the interaction term.

The results of our study, L^+Y^+ with $L \times Y$ interaction (see Figures 4, 5, and 6, and compare with Figure 2), and a consideration of the biology of this species would lead us to suppose that the remigrants came from intermediary hosts nearby, but quite likely from several species of such hosts and also several local stands of each. Although the aphids on a given tree may derive from different clones—witness the considerable genotypic variability of aphids on any one tree, demonstrated by Sokal (1952, 1962), the local differentiation observed in this study probably stems from annual and local differentiation of the source population. This comes from secondary hosts that vary genetically, in species composition and in habitats. Stands of these hosts are found close to the cottonwoods and are often isolated one from the other, so that the apterous parthenogenetic alienicolae cannot migrate among them. Local and temporal environmental factors will enhance local phenetic differentiation, that varies independently from year to year.

The pattern of variation differs over the six-year span of the study, without any consistent trend being apparent either among the eight characters considered singly, or all characters as studied by manova. The presence of such a trend might have reflected a consistent change in the breeding structure of the population. For example, had previous differentiation into distinct homogeneous subsets changed to fewer but larger subsets, increased gene flow among the previous subpopulations might be one of several plausible hypotheses. But barring strong local counter-selection affecting one or more characters, or, more plausibly, strong local environmental factors affecting only one or a few of the characters, such a trend should be uniformly observable over all characteristics, which is not true in the data at hand.

Summary

Samples from eight localities, none more than eight miles apart, were made of populations of the gall-producing aphid *Pemphigus populitransversus*. Eight characters of the alate fundatrigeniae were measured and analyzed by univariate and multivariate analyses of variance, and by the simultaneous test procedure as modified by Gabriel and Sokal (1969). Interpretations are given for the various possible outcomes of the statistical analysis as these might relate to the population structure of various populations. Univariate and multivariate heterogeneity was found among localities and among years, and the geographic patterns found were not consistent among years. The outcome of the statistical analysis makes it likely that the gene pool of each locality is based on remigrants from intermediary hosts nearby, leading to local differentiation based on interclonal selection on the secondary hosts, or due to changes in local microclimate.

Acknowledgment

We are indebted to F. James Rohlf for statistical and computational advice and for comments on a draft of this manuscript.

Literature Cited

Ehrlich, P. R. 1965. The population biology of the butterfly, *Euphydryas editha*. II. The structure of the Jasper Ridge colony. Evolution 19:327–336.

Gabriel, K. R. 1964. A procedure for testing the homogeneity of all sets of means in analysis of variance. Biometrics 20:459–477.

Gabriel, K. R. 1968. Simultaneous test procedures in multivariate analysis of variance. Biometrika 55:489–504.

Gabriel, K. R., and R. R. Sokal. 1969. A new statistical approach to geographic variation analysis. Syst. Zool. 18:259–278.

Heryford, N. N., and R. R. Sokal. 1971. Seasonal morphometric variation in *Pemphigus populitransversus*. J. Kans. Entom. Soc. (*in press*).

Johnston, R. F. 1969. Character variation and adaption in European sparrows. Syst. Zool., 18:206–231.

Koehn, R. K., J. E. Perez, and R. B. Merritt. 1971. Esterase enzyme function and genetical structure of populations of the freshwater fish *Notropis stramineus*. Amer. Natur. 105:51–69.

Rinkel, R. C. 1965. Microgeographic variation and covariation in *Pemphigus populi-transversus*. Univ. Kans. Sci. Bull. 46:167–200.

Selander, R. K. 1970. Behavior and genetic variation in natural populations. Amer. Zool. 10:53–66.

Sneath, P. H. A., and R. R. Sokal. 1972. Numerical taxonomy: The principles and practice of numerical classification. W. H. Freeman and Co., San Francisco, (*in press*).

Sokal, R. R. 1952. Variation in a local population of *Pemphigus*. Evolution 3:296–315.

Sokal, R. R. 1962. Variation and covariation of characters of alate *Pemphigus populi-transversus* in eastern North America. Evolution 16:227–245.

Sokal, R. R., and R. C. Rinkel. 1963. Geographic variation of alate *Pemphigus populi-transversus* in eastern North America. Univ. Kans. Sci. Bull. 44:467–507.

Sokal, R. R., and F. J. Rohlf. 1969. Biometry. W. H. Freeman and Co., San Francisco. 776 p.

Sokal, R. R., and P. A. Thomas. 1965. Geographic variation of *Pemphigus populi-transversus* in eastern North America: Stem mothers and new data on alates. Univ. Kans. Sci. Bull. 46:201–252.

Soulé, M. 1967. Phenetics of natural populations. I. Phenetic relationships of insular populations of the side-blotched lizard. Evolution 21:584–591.

7

Reprinted from *Crop Sci.*, **8**, 693–698 (Nov.–Dec. 1968)

The Races of Maize: II. Use of Multivariate Analysis of Variance to Measure Morphological Similarity[1]

Major M. Goodman[2]

ABSTRACT

The racial means and the residual covariance matrix from the multivariate analysis of variance of an experiment based on a randomized block design involving 15 races of maize (*Zea mays* L.) from southeastern South America were used to calculate generalized distances between the races. Sixteen characters commonly used in taxonomic studies of the races of maize were employed. The effects of transformations designed to eliminate some of the heterogeneity among the within-race, within-row covariance matrices were studied, and the effects of within-plot sampling were investigated. It was shown that the use of transformations had very little effect on the relative distances (and hence that highly significant heterogeneity — as measured by the multivariate extension of Bartlett's test — had little effect on the analysis of the non-transformed data). Similarly, essentially the same relative distances were obtained when only a single plant from each race was used per block (eight blocks were used). In all cases the distances obtained were relatively very similar (i.e., Spearman rank correlation coefficient between 0.91 and 0.99) to the distances obtained from the commonly employed Mahalanobis' generalized distance technique.

***Additional key words:* *Zea mays*, generalized distance, Mahalanobis, taxometrics, classification, numerical taxonomy.**

APPROXIMATELY 25 years ago methods were developed for the study and identification of races of maize, *Zea mays* L. (Anderson and Cutler, 1942; Anderson, 1943). Since then these methods have been modified somewhat and applied widely. By 1963 essentially all of the Latin American races had been identified and described (Grant et al., 1963). Very little is known about the relationships among these races, although there has been much speculation about the role hybridization has played (see Wellhausen et al., 1952; and Brieger et al., 1958; for examples). In cases such as this, where the quantity of pertinent data and the number of taxa are very large, numerical taxonomy has frequently been suggested as an invaluable tool (Sokal and Sneath, 1963).

This particular study concerns the use of multivariate analysis to derive a system of classification for 15 races of maize from the region encompassed by southern Brazil, Bolivia, Paraguay, Uruguay, and northern Argentina. The goal of the study was to evaluate the possible utility of the multivariate analysis of variance in the derivation of a system of classification based upon the phenotypic similarities among these races of maize. This study is a part of a project which has as its ultimate goal the development of a practical system of classification for the races of maize of Latin America. Various other techniques of numerical taxonomy are currently under investigation.

[1] Contribution from the Instituto de Genética, Escola Superior de Agricultura 'Luiz de Queiroz,' Universidade de Sao Paulo, Piracicaba. S.P., Brazil, and the Department of Experimental Statistics, North Carolina Agr. Exp. Sta., Raleigh, N.C. 27607. Paper No. 2613 of the Journal Series of the N. C. State University Agr. Exp. Sta. Received Apr. 13, 1968.

[2] Assistant Professor of Experimental Statistics, North Carolina State University at Raleigh (Formerly NSF Postdoctoral Fellow, Instituto de Genética, E.S.A.L.Q., Piracicaba, S. P., Brazil). A part of the computations were carried out at the Centro de Computacao Electronica, Escola Politécnica, Universidade de Sao Paulo, Sao Paulo, Brazil. This investigation was supported in part by Public Health Service Grant GM 11546. The cooperation of A. Blumenschein, E. Paterniani, M. R. Alleoni, and the Cadeira de Matemática e Estatistica, E.S.A.L.Q., and the assistance of H. Kuniyuki are gratefully acknowledged.

LITERATURE REVIEW

Two early measures of similarity between taxa were Pearson's (1926) Coefficient of Racial Likeness (C.R.L.) and Mahalanobis' (1936) generalized distance. Although Pearson's C.R.L. has fallen into disuse (largely because it is really apropriate only in cases where the characters are essentially non-correlated), the growing popularity of numerical taxonomy has resulted in a renewed interest in Mahalanobis' generalized distance (Olson, 1964; Sokal, 1965; Murty and Arunachalam, 1966), which is usually calculated as (see Rao, 1952)

$$D_{ij} = [\ (\bar{x}_i - \bar{x}_j)' \ \hat{\Sigma}^{-1} \ (\bar{x}_i - \bar{x}_j)\ ]^{1/2} \qquad [1]$$

where $\bar{x}_i$ is the vector of character means for taxon i, and $\hat{\Sigma}$ is the estimate of the common within-taxa covariance matrix.

Hotelling (1954) has shown that D is equivalent to the multivariate generalization of "Student's" t test of the equality of two means. Furthermore, just as the t test can be used to measure significance of the difference between two treatment means from an analysis of variance,

$$t^2 = r\,(\bar{x}_i - \bar{x}_j)^2 / (2\hat{\sigma}^2), \qquad [2]$$

where $\bar{x}_i$ is the mean of treatment i, r is the number of replications, and $\hat{\sigma}^2$ is the appropriate mean square from the analysis of variance (Anderson and Bancroft, 1952; Snedecor, 1956); D can be used as a measure of similarity (or as a test of the equality) of two mean vectors from the corresponding multivariate analysis of variance;

$$T^2 = rD^2/\ (2) \qquad [3]$$

$$D^2 = (\bar{x}_i - \bar{x}_j)'\ \hat{\Sigma}_e^{-1}\ (\bar{x}_i - \bar{x}_j), \qquad [4]$$

where $\hat{\Sigma}_e$ is the mean product matrix from the multivariate analysis of variance corresponding to $\hat{\sigma}^2$ (from the univariate analysis of variance) of equation [2]. The relationship between the T^2 statistic of Hotelling and D^2 of Mahalanobis was not widely recognized for some time (Kendall, 1957), which probably accounts for the fact that in taxonomic work D^2 has been used only in the case of experiments based upon completely randomized designs where the basic experimental unit has been the individual plant (or animal).

In the case of experimental studies in plant taxonomy and especially in the case of cultivated plants, the basic experimental unit is quite frequently the plot or row mean. Individual plant variation within plots may be essentially irrelevant. Furthermore, experimental designs involving various types of blocking are usually used to eliminate at least some of the environmental effects from the comparisons to be made. Thus it is Hotelling's generalization of Mahalanobis' generalized distance that is applicable in such cases.

MATERIALS AND METHODS

With one exception the races of maize included in this study have been described by Brieger et al. (1958) and/or Paterniani (In preparation). The nomenclature adopted is that of the latter, whenever possible. The races and, whenever possible, the population or collection numbers of the samples used are listed in Table 1. Paterniani has described how the populations were synthesized, and the original collections are listed in the reports of the Committee on Preservation of Indigenous Strains of Maize (1954, 1955). "Cateto Sao Simao," which is not described in the publications of Brieger et al. and Paterniani, is a commercial variety of yellow, flint corn which resulted from a mass selection program carried out at the Estacao Experimental de Sao Simao (of the Ministério de Agricultura) located near Ribierao Preto, S.P., Brazil. All of the races studied came either from northern Argentina, Uruguay, southern Brazil, Paraguay, or Bolivia and were reasonably well-adapted to the growing conditions encountered in Piracicaba.

These races were grown during the summer of 1965-1966 at the Instituto de Genética, Piracicaba, Sao Paulo, Brazil, in a randomized complete block experiment with eight replications. Each plot consisted of a single 10-m row of (in case of perfect stand) 25 plants. Whenever possible nine competitive plants (i.e., plants not adjacent to missing hills) were measured for 16 characteristics which have been described in detail by Grant et al. (1963). (The measure of maturity used in this study differs from that used by Grant et al., however.) The characters studied were *(1)* plant height, *(2)* number of leaves per plant, *(3)* number of leaves above the uppermost ear, *(4)* leaf length, *(5)* leaf width, *(6)* number of primary tassel branches, *(7)* peduncle length, *(8)* length of branched part of tassel, *(9)* length of the central spike, *(10)* maturity, which was measured as the number of days from planting (November 9, 1965) until anthers had emerged on approximately 50% of the tassel, *(11)* ear length, *(12)* ear diameter, *(13)* number of rows of kernels on the ear, *(14)* kernel thickness, *(15)* kernel length, and *(16)* kernel width. The kernel characteristics were measured in millimeters, whereas the other characters were measured in centimeters.

A preliminary investigation, (Goodman, 1968a), had indicated that the within-race covariance matrices were heterogeneous (and that such heterogeneity was largely attributable to the histories of the races and samples used), so several simple transformations were used in attempts to eliminate some of the effects of such heterogeneity. (The means and standard errors [or variances] were correlated, thus logarithmic [or square root] transformations were used). In addition, two sampling experiments were conducted. From each plot a single plant was randomly chosen, and the plant with the most "typical" ear was also chosen (these were, on occasion, the same plant). Thus four sets of data were available for analysis: *(1)* the set of 120 untransformed plot means, *(2)* the set of 120 plot means from the transformed data, *(3)* the set of data from 120 "typical" plants, and *(4)* the set of data from 120 random plants.

Each of these sets of data was analyzed using the method of unweighted means (Anderson and Bancroft, 1952) to obtain the multivariate analysis of variance presented in Table 2. The distances between the races were then calculated using

$$D_{ij} = [\ (\bar{x}_i - \bar{x}_j)'\ \hat{\Sigma}_e^{-1}\ (\bar{x}_i - \bar{x}_j)]^{1/2},$$

where x_i is the vector of character means for race i, and Σ_e is the residual mean product matrix from the multivariate analysis of variance shown in Table 2.

When the mean vectors share a common residual covariance matrix and are multinormally distributed, then the D^2_{ij} follow essentially the noncentral F distribution. In the case of a

Table 1. The races, their population and/or collection numbers, and the numbers of plants used.

Racial code	Race name	Alternative name(s)	Collection no. or Population no.	No. of plants
CR	Cristal	Cristal Paulista	SP X	63
TU	Cristal Semi-dentado	Avati Tupi, C. Paraguay	NRC 2147	71
SS	Cateto São Simão			67
AD	Avati Djakaira	Ceremonial	NRC 2064	45
DP	Dente Paulista		SP III	70
MO	Moroti	Avati Moroti	NRC 2114	55
PP	Avati Pichinga	Pipoca Pontudo		62
CG	Caingang		PR III	66
CA	Cateto Sulino	Cateto Argentino	ARG III	70
MP	Moroti Precoce	Guarani Amarillo	BOL I	69
U3	Cateto Sulino Grosso	Amarillo Canario de Muchas Hileiras	URG III	70
C8	Canario de Ocho			67
CV	Cravo Rio Grandense		RGS VII	71
LH	Lenha		RGS XX	66
PR	Avati Pichinga Ihu	Pipoca Redondo	NRC 2169	68

Table 2. Multivariate analysis of variance.

Source of variation	df	Mean product matrix	E(MPM)
Blocks	7	S_1	$\Sigma_e + 15\Sigma_B$
Races	14	S_2	$\Sigma_e + 8\Sigma_R$
Error	98	S_e	Σ_e

$\Sigma_e = (1/k)\,\Sigma_w + \Sigma_p$, where Σ_w is the covariance matrix among plants in the same plot, Σ_p is the additional covariance matrix among plots within a replication, and k is the harmonic mean of plants per plot.

randomized complete block experiment with r blocks, n varieties, and p characters,

$$D_{ij} = ([2f_1(f_1+f_2-1)\ F^*]/rf_2)^{1/2},$$

where F^* is a noncentral F variable with $f_1 = p$ and $f_2 =$ rn-r-n-p+2 degrees of freedom and noncentrality parameter

$$\lambda_{ij} = (\mu_i - \mu_j)'\ \Sigma_e^{-1}\ (\mu_i - \mu_j),$$

where μ_i is the vector of character means for race i (Hotelling, 1954; Anderson, 1958). The relationship is exact only if the numbers of observations per race are equal. The mean of the noncentral F distribution with f_1 and f_2 degrees of freedom and noncentrality parameter λ is (Patnaik, 1949)

$$f_2\ (f_1 + \lambda)\ /\ [f_1(f_2 - 2)]$$

and its variance is

$$2f_2^2\ [(f_1+f_2-2)(f_1+2\lambda) + \lambda^2]/[f_1^2(f_2-2)^2\ (f_2-4)].$$

Thus $$\sigma^2_{D_{ij}} \doteq \frac{(f_1+f_2-1)[(f_1+f_2-2)(f_1+2\lambda) + \lambda^2]}{[r(f_1+\lambda)(f_2-2)(f_2-4)]},$$

since

$$\sigma^2_Y \doteq \left(\frac{dY}{dX}\right)^2_{X=\mu_X} \cdot \sigma^2_X$$

when $Y = f(X)$ [see Davies, 1957]. With the experimental conditions used in this experiment ($n=15$, $r=8$, $p=16$),

$$\sigma^2_{D_{ij}} \doteq (.00191436)\ (\lambda^2_{ij} + 194\lambda_{ij} + 1552)/(\lambda_{ij} + 16),$$

which can be estimated using $\hat{\lambda}_{ij} = D^2_{ij}$.

Anderson's (1958) multivariate extension[3] of Bartlett's well-known test of homogeneity of variances was used to test the hypothesis that the within-plot covariance matrices were identical for all the races studied both before and after the transformations were employed. The test criterion is

$$W = q\,(f_w\ \ln|\hat{\Sigma}_w| - \sum_{i=1}^{n} f_i\ \ln|\hat{\Sigma}_i|)$$

where $\hat{\Sigma}_w$ and f_w are the pooled within-plot covariance matrix and its degrees of freedom, respectively, and $\hat{\Sigma}_i$ and f_i are the estimated within-plot covariance matrix and its degrees of freedom, respectively, for race i.

$$q = 1 - [(\sum_{i=1}^{n}(1/f_i) - (1/f_w))\ (2p^2 + 3p - 1)/[6(p+1)(n-1)],$$

where p is the number of characters studied. The probability of a sample value of W larger than some particular value, W_0, when the covariance matrices are, in fact, identical is approximately $\Pr\left(X^2_v > W_0\right)$, where X^2_v is a chi-square with v degrees of freedom,

$$v = (1/2)\ (n-1)\ (p)\ (p+1),$$

although deviations from multinormality as well as deviations from homogeneity tend to lower such probabilities (Sokal, 1965; T. W. Anderson, personal communication.)

[3] The early printings of Anderson's book have a misprint in equation (7), p. 255. The equation should begin with $\Pr(-2\rho \log W_1 \leqslant z)$ rather than $\Pr(-\rho \log W_1 \leqslant z)$.

RESULTS

The means of the original, untransformed data for each race and the corresponding residual covariances and correlations from the unweighted means-multivariate analysis of variance are presented in Tables 3 and 4, respectively. Table 5 presents the resulting distances between the races along with their approximate standard errors.

The logarithms of the determinants of the within-row, within-race covariance matrices, before and after transformations were applied to the original data, are presented in Table 6. Despite the fact that the multivariate extension of Bartlett's test had indicated very significant heterogeneity ($\Pr < .00001$) without the use of such transformations and no heterogeneity with such transformations, the Spearman rank correlation between the distances obtained from the transformed and from the untransformed data was essentially one (0.991) and actual differences between corresponding distances were trivial. Similarly the Spearman rank correlations between the distances based upon plot means and the distances based upon "typical" and randomly chosen plants (all based on untransformed data) were 0.884 and 0.942, respectively.

Sokal and Sneath's (1963) unweighted average method of cluster analysis was used to construct the phenogram illustrated in Fig. 1 from the distances of Table 5. The vertical lines joining groups indicate the average distances between all pairs of races, where the pairs consist of one race from each group.

DISCUSSION

In certain respects these results support conclusions already reached (Brieger et al., 1958; Paterniani, op.

Table 3. Sample means (untransformed) for each character and each race.

Character no.	Race code														
	CR	TU	SS	AD	DP	MO	PP	CG	CA	MP	U3	C8	CV	LH	PR
1	266.0	244.0	276.0	229.0	258.0	242.0	226.0	236.0	189.0	191.0	168.0	172.0	227.0	197.0	205.0
2	18.5	17.8	18.3	16.3	18.3	18.5	17.5	17.1	14.1	14.9	13.8	12.2	17.1	17.0	17.3
3	6.2	6.2	6.3	6.1	6.1	6.0	6.1	6.2	4.9	5.9	5.4	4.7	6.2	6.0	5.4
4	85.0	86.0	84.0	80.0	80.0	88.0	79.0	76.0	72.0	76.0	75.0	80.0	73.0	71.0	70.0
5	9.4	9.2	9.2	8.2	9.4	10.3	9.1	8.5	8.5	8.5	8.7	8.7	8.7	8.1	9.6
6	21.8	25.5	24.9	17.9	21.9	33.3	22.0	20.4	19.0	20.6	19.7	16.0	21.2	21.2	19.9
7	18.1	18.5	18.8	18.5	19.2	15.1	14.4	17.5	23.0	17.3	21.6	25.9	17.9	15.0	16.0
8	18.0	17.0	18.5	14.3	17.1	19.5	16.8	15.9	14.5	15.4	11.0	13.0	13.1	13.2	10.9
9	28.7	28.8	26.0	28.6	25.6	24.9	27.9	24.4	21.1	20.4	22.0	24.5	24.0	24.0	25.4
10	75.3	75.3	72.0	71.6	71.3	77.0	74.7	66.9	56.0	59.2	54.4	54.6	67.0	64.9	71.8
11	20.7	20.5	19.6	19.4	18.6	20.0	18.5	20.3	17.0	16.5	14.3	19.9	15.5	14.8	15.1
12	3.9	4.2	3.8	4.1	4.6	4.4	3.8	4.2	4.0	4.0	4.6	3.7	5.1	5.3	3.0
13	12.5	13.3	13.9	13.5	14.2	14.0	14.3	11.4	11.5	13.8	17.1	8.2	19.6	20.6	15.7
14	4.3	4.0	3.7	4.7	3.8	4.6	4.0	4.4	3.8	4.7	3.4	4.0	3.3	3.9	3.0
15	10.1	10.0	10.2	9.4	12.5	11.1	10.4	11.6	10.4	10.1	10.6	9.9	13.6	10.6	8.9
16	8.7	8.7	7.7	8.9	8.8	9.2	6.7	10.0	9.3	8.5	7.6	10.8	7.2	7.8	5.4

Table 4. Residual covariance matrix from the multivariate analysis of variance using the untransformed plot means and the corresponding correlation coefficients. The variances are on the diagonal, the covariances above the diagonal and the correlations below the diagonal.

104.2	2.272	.8243	10.79	1.580	2.625	4.242	7.101	-.9121	-4.590	2.822	.2207	-.5160	-.0132	.8574	.6883
.44	.2559	.0680	-.2603	-.0058	.2293	-.0089	.1189	-.3525	-.0893	-.0306	.0029	.0384	-.0122	.0258	-.0042
.29	.48	.0771	.2154	.0232	.1319	-.0419	.0645	-.1473	-.1179	-.0103	.0037	.0281	-.0029	.0043	-.0009
.27	-.13	.20	15.60	1.051	.5196	-.6418	1.143	3.094	-.7323	1.549	.1493	.0972	.1474	.3987	.2159
.31	-.02	.17	.54	.2429	.2666	-.0488	.1535	.1455	-.2015	.0637	.0149	.0346	.0021	.0417	.0197
.13	.22	.24	.07	.27	4.060	-.3608	1.157	-1.268	-.8613	.0353	.0141	.1265	-.0494	.0235	-.0354
.33	-.01	-.12	-.13	-.08	-.14	1.551	.2207	-.0032	.1355	.1475	-.0020	-.1183	.0167	.0043	.0610
.57	.19	.19	.24	.25	.47	.14	1.494	-.6005	-.6668	.3739	-.0027	-.1137	.0036	.0297	.0499
-.05	-.40	-.31	.45	.17	-.36	.00	-.28	3.019	.8147	.5470	.0586	.0393	.1052	.0626	.1153
-.31	-.12	-.30	-.13	-.29	-.30	.08	-.38	.33	2.044	-.0311	.0071	.0630	.0086	.0363	.0178
.30	-.07	-.04	.43	.14	.02	.13	.33	.34	-.02	.8417	.0182	-.0051	.0394	.0257	.0493
.18	.05	.11	.31	.25	.06	-.01	-.02	.28	.04	.16	.0146	.0484	-.0003	.0272	.0083
-.07	.11	.14	.03	.10	.09	-.13	-.13	.03	.06	-.01	.57	.4967	.0051	.0656	-.0727
-.01	-.15	-.07	.24	.03	-.16	.09	.02	.39	.04	.27	-.02	.05	.0247	-.0077	.0031
.23	.14	.04	.28	.24	.03	.01	.07	.10	.07	.08	.63	.26	-.14	.1290	.0253
.27	-.13	-.01	.22	.16	-.07	.20	.16	.27	.05	.22	.28	-.42	.08	.28	.0613

Table 5. Distances between the races.

CR		TU		SS		AD		DP	
TU	5.3± .58	CR	5.3± .58	CR	7.6± .64	TU	8.5± .67	TU	9.7± .70
SS	7.6± .64	AD	8.5± .67	TU	8.7± .67	CR	8.9± .68	CR	10.5± .72
AD	8.9± .68	SS	8.7± .67	PP	10.0± .71	CG	10.3± .72	CG	11.8± .76
MO	10.3± .72	DP	9.7± .70	DP	11.9± .77	DP	11.9± .77	CV	11.8± .76
DP	10.5± .72	MO	9.8± .70	AD	12.0± .77	SS	12.0± .77	SS	11.9± .77
PP	10.8± .73	PP	11.6± .76	PR	15.0± .87	MO	13.0± .80	AD	11.9± .77
CG	13.8± .83	CG	14.0± .83	MO	15.3± .88	MP	13.1± .80	PP	12.1± .77
MP	19.5±1.03	CV	18.5± .99	CG	16.5± .92	PP	15.0± .87	MO	12.9± .80
PR	19.9±1.04	MP	18.8±1.00	CV	17.7± .96	LH	16.8± .93	MP	16.8± .93
CV	20.0±1.04	LH	18.9±1.00	MP	20.3±1.06	CV	18.5± .99	LH	16.8± .93
LH	21.4±1.10	PR	20.5±1.06	LH	21.8±1.11	CA	18.6± .99	CA	20.0±1.04
CA	24.0±1.19	CA	22.5±1.14	CA	23.7±1.18	PR	22.3±1.13	U3	22.3±1.13
U3	27.7±1.34	U3	25.3±1.24	U3	25.8±1.26	U3	22.4±1.13	PR	22.4±1.13
C8	30.3±1.44	C8	29.0±1.39	C8	30.4±1.45	C8	24.7±1.22	C8	28.2±1.36

MO		PP		CG		CA		MP	
TU	9.8± .70	SS	10.0± .71	AD	10.3± .72	MP	10.7± .73	CA	10.7± .73
CR	10.3± .72	CR	10.8± .73	DP	11.8± .76	U3	10.8± .73	CG	12.1± .77
DP	12.9± .80	TU	11.6± .76	MP	12.1± .77	C8	11.0± .74	AD	13.1± .80
AD	13.0± .80	DP	12.1± .77	CR	13.8± .83	CG	16.1± .90	U3	15.1± .87
SS	15.3± .88	AD	15.0± .87	TU	14.0± .83	AD	18.6± .99	LH	16.3± .91
CG	15.5± .88	MO	16.1± .90	MO	15.5± .88	LH	19.7±1.03	DP	16.8± .93
PP	16.1± .90	PR	16.3± .91	CA	16.1± .90	DP	20.0±1.04	CV	18.2± .98
MP	20.5±1.06	CV	16.9± .93	SS	16.5± .92	CV	20.5±1.06	C8	18.4± .99
LH	21.2±1.09	CG	18.7±1.00	CV	18.5± .99	TU	22.5±1.14	TU	18.8±1.00
CV	21.9±1.11	MP	21.1±1.08	PP	18.7±1.00	SS	23.7±1.18	CR	19.5±1.03
CA	26.1±1.28	LH	22.0±1.12	LH	20.0±1.04	CR	24.0±1.19	SS	20.3±1.06
PR	26.6±1.30	CA	25.9±1.27	C8	21.3±1.09	PP	25.9±1.27	MO	20.5±1.06
U3	29.7±1.42	U3	26.9±1.31	U3	22.6±1.14	MO	26.1±1.28	PP	21.1±1.08
C8	32.5±1.53	C8	33.8±1.58	PR	26.1±1.28	PR	27.6±1.34	PR	26.2±1.28

U3		C8		CV		LH		PR	
CA	10.8± .73	CA	11.0± .74	DP	11.8± .76	CV	14.7± .86	SS	15.0± .87
MP	15.1± .87	MP	18.4± .99	LH	14.7± .86	MP	16.3± .91	PP	16.3± .91
LH	16.6± .92	U3	19.2±1.09	PP	16.9± .93	U3	16.6± .92	CR	19.9±1.04
CV	17.7± .96	CG	21.3±1.01	SS	17.7± .96	AD	16.8± .93	TU	20.5±1.06
C8	19.2±1.01	AD	24.7±1.22	U3	17.7± .96	DP	16.8± .93	AD	22.3±1.13
DP	22.3±1.13	DP	28.2±1.36	MP	18.2± .98	TU	18.9±1.00	DP	22.4±1.13
AD	22.4±1.13	LH	28.9±1.39	TU	18.5± .99	CA	19.7±1.03	CV	22.9±1.15
CG	22.6±1.14	TU	29.0±1.39	AD	18.5± .99	CG	20.0±1.04	CG	26.1±1.28
TU	25.3±1.24	CV	29.9±1.43	CG	18.5± .99	MO	21.2±1.09	MP	26.2±1.28
SS	25.8±1.26	CR	30.3±1.44	CR	20.0±1.04	CR	21.4±1.10	MO	26.6±1.30
PP	26.9±1.31	SS	30.4±1.45	CA	20.5±1.06	SS	21.8±1.11	U3	27.3±1.32
PR	27.3±1.32	MO	32.5±1.53	MO	21.9±1.11	PP	22.0±1.12	CA	27.6±1.34
CR	27.7±1.34	PR	33.6±1.58	PR	22.9±1.15	PR	28.0±1.35	LH	28.0±1.35
MO	29.7±1.42	PP	33.8±1.58	C8	29.9±1.43	C8	28.9±1.39	C8	33.6±1.58

cit.) or reasonably clearly evident. For example, the two types of Cristal must be fairly closely related. Furthermore, with the exception of the apparently distinct round-seeded popcorn, Avati Pichinga Ihu, which seems to be very similar to the race Pororo from Bolivia (Raminez et al., 1961), all of the races known definitely to have been associated with Brazilian Indian tribes (Cristal, Avati Djakaira, Avati Pichinga, Moroti, and Caingang) form a relatively compact group. Although Dente Paulista and Cateto Sao Simao also fall within this group, these are relatively "modern" commercial types from Sao Paulo, Brazil, the first of which originated largely in the past century as a result of crosses between introduced and adapted types; the second originated as a result of mass selection from a local variety. With the exception of Caingang, grown by the Indians whose name it bears, all the indigenous races studied here were grown by the Tupi-Guarani Indians of southern Brazil and neighboring regions. In addition to the similarities among the indigenous

Table 6. Logarithms of the determinants of the individual within-row covariance matrices.

Race	d.f.	Original data Ln |S|	Transformed data* Ln |S|	Race	d.f.	Original data Ln |S|	Transformed data* Ln |S|
CR	55	13.24	13.88	CA	62	13.21	15.12
TU	63	7.94	8.96	MP	61	9.71	10.95
SS	59	12.88	13.82	U3	62	10.01	11.93
AD	37	8.94	9.93	C8	59	12.15	14.69
DP	62	13.55	13.76	CV	63	13.52	14.12
MO	47	10.50	10.50	LH	58	10.49	11.57
PP	54	9.85	11.06	PR	60	4.32	7.69
CG	58	10.43	11.20	Pooled	860	14.95	15.84

* Characters 1, 8, 12, and 16 were transformed using the transformation $Y = 10\sqrt{X}$. Characters 11, 13, 14, and 15 were transformed using the transformation $Y = 10 \ln X$. The other characters were not transformed.

races and the modern commercial varieties of Sao Paulo, there are definite similarities between Cravo and Lenha, both short-eared, many-rowed varieties from Rio Grande do Sul, Brazil. Cravo is actually most similar to Dente Paulista (Table 5), although this is not evident from Fig. 1. Both Cravo and Dente Paulista undoubtedly contain U.S. germplasm (Brieger et al., 1958). The similarities between Cateto Sulino, Cateto Sulino Grosso, and Canario de Ocho also were expected. These are races of yellow-orange colored flint corn from Uruguay and northern Argentina. The orange or yellow colored flint corns form a rather heterogeneous complex which extends (or, rather, extended) along the Atlantic coasts from Canada to Argentina. Very little is known about their origin(s) or even about the extent of heterogeneity present, although some hypotheses were presented by Brieger et al. (1958), who called them "catetos." While there seems to be no present disagreement over the meaning of this term, there are known to have been types of white (as well as yellow) flint called "cattete" or "catête" (earlier forms of the word "cateto") in Brazil as late as the end of the nineteenth century (Peckholt, 1878; Seixas, 1880).

The most surprising result obtained was the grouping of Moroti Precoce with Cateto Sulino. Moroti Precoce is an early maturing race of floury corn which has the characteristic, distinctive golden aleurone color of Moroti. It seems to be similar to the race Pojoso Chico described by Ramirez et al. (1961) from Bolivia, and, in fact, the population studied originated largely from samples collected near Santa Cruz, Bolivia. Other collections came from Rio Grande do Sul, Brazil. The grouping certainly appears to be an artificial one, which a non-numerical taxonomist would not have made. A previous study (Goodman, 1967) based

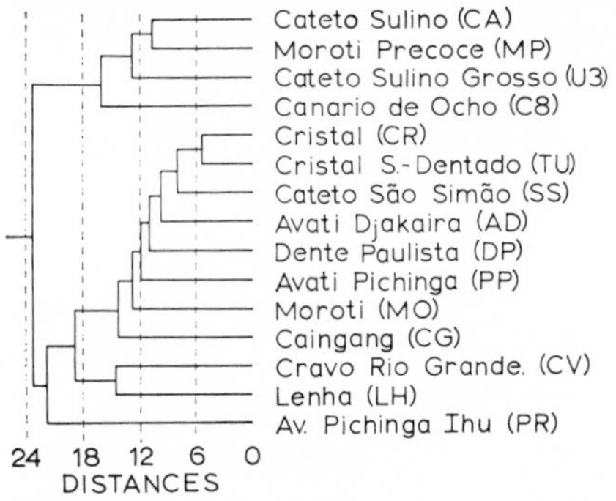

Fig. 1. Phenogram of 15 races of corn from southeastern South America.

upon the simple among-/within-races analysis generally employed in calculations of Mahalanobis' generalized distances also resulted in a similar grouping.

The multivariate analysis approach used in this study differs from the usual Mahalanobis' technique in one practical point. The field design used in this case is compatible with ordinary experimental studies; in the usual form of Mahalanobis' generalized distances, it is theoretically necessary to randomize *individual* plants. (Actually the author knows of no studies which have fulfilled this requirement). When the experimental area is not uniform, the unfavorable effects expected from any completely randomized design will also affect such generalized distances. However, the alternative approach presented here allows the choice of an appropriate field design (completely random, randomized block, or Latin square) where the experimental unit is the plot mean. The analysis of an experiment of this type, however, points out a very important aspect of the question of what are the appropriate choices of scales of measurement for the various characters. The error variances among plants within races, which determine the scales of measurement used in the ordinary form of Mahalanobis' generalized distances, can be written in the form

$$\sigma^2_a = \sigma^2_b + \sigma^2_p + \sigma^2_g + \sigma^2_e, \qquad [5]$$

in the case of a randomized block design, where the subscripts b and p indicate the contributions of blocks and plots, respectively. σ^2_g and σ^2_e are the genetic and environment variances within plots. The residual variances from the analyses of variance of plant means, which determine the scales of measurement in the analysis used herein, can be written in the form

$$\sigma^2_c = \sigma^2_p + (\sigma^2_g + \sigma^2_e)/k \qquad [6]$$

Appropriate choice of plot size can alter the proportions of the environmental and genetic within-race variances in [6]. If $\sigma^2_b > (k\text{-}1)\ \sigma^2_p$, then the relative genetic contribution to [6] exceeds that to [5] and vice versa. However, the largest genetic contribution to the overall variation among plants is ordinarily the component due to variation among races, which does not appear in either [5] or [6]. The question has been discussed elsewhere from a somewhat different point of view (Goodman, 1968b), but it is clear that morphological dissimilarity need not imply great genetic diversity. From [6] it can be seen that, if the heterogeneity among the within-row covariance matrices is of genetic and not environmental origin, then the effect upon the residual matrix is reduced by a factor of (1/k), which may explain, in part, the apparent lack of effect which such heterogeneity has upon the distance analysis.

The two sampling experiments demonstrated that proper experimental design combined with multivariate analysis can apparently reduce greatly the amount of data necessary for the calculation of generalized distances. The main disadvantage of the usual Mahalanobis approach has been the very large quantities of data necessary. Because of the problem of heterogeneity of the within-taxon covariance matrices, it has been widely accepted that at least 50 to 60 plants (and preferably more) should be used per taxon. If fewer plants are used, the individual within-taxon covariance matrices are poorly estimated, and decisions about the use of transformations and/or the pooling of matrices are obscured. The results obtained from these studies indicate that within reasonable limits the problem of heterogeneity is of relatively little importance, at least in the case of the races of maize, and that reasonable conclusions can be reached with relatively few plants per taxon. (The Spearman rank correlation between the distances calculated from the set of 120 randomly chosen plants – one per plot – and the distances calculated from the usual among-/ within-race analysis of 980 plants was 0.941. In the case of the 120 "typical" plants, the corresponding Spearman rank correlation was 0.907). The use of a smaller, more uniform experimental area (which the use of far fewer plants would easily permit) probably would lead to somewhat superior results.

LITERATURE CITED

Anderson, E. 1943. Races of *Zea mays*. II. A general survey of the problem. Acta Amer. 1:58-68.

———, and H. C. Cutler. 1942. Races of *Zea mays:* I. Their recognition and classification. Ann. Mo. Bot. Garden 29: 69-88.

Anderson, R. L., and T. A. Bancroft. 1952. Statistical theory in research. McGraw-Hill, New York. 399 p.

Anderson, T. W. 1958. An introduction to multivariate statistical analysis. John Wiley, New York. 374 p.

Brieger, F. G., J. T. A. Gurgel, E. Paterniani, A. Blumenschein, and M. R. Alleoni. 1958. Races of maize in Brazil and other eastern South American countries. Nat. Acad. Sci.-Nat. Res. Council Publ. 593. Washington, D. C. 283 p.

Committee on Preservation of Indigenous Strains of Maize. 1954. Collections of original strains of corn, I. Nat. Acad. Sci.-Nat. Res. Council. Washington, D. C. 300 p.

———. 1955. Collections of original strains of corn, II. Nat. Acad. Sci.-Nat. Res. Council. Washington, D. C. 298 p.

Davies, O. L. (ed.). 1957. Statistical methods in research and production. Oliver and Boyd, London. 396 p.

Goodman, M. M. 1967. The races of maize. I. The use of Mahalanobis' generalized distances to measure morphological similarity. Fitotecnia Lathinoamericana 4:1-22.

Goodman, M. M. 1968 a. A measure of "overall variability" in populations. Biometrics 24:189-192.

———. 1968b. Measuring evolutionary divergence between populations. 12th Int. Congr. Genetics, Vol. II. 212-213.

Grant, U. J., W. H. Hatheway, D. H. Timothy, C. Cassalett D., and L. M. Roberts. 1963. Races of maize in Venezuela. Nat. Acad. Sci.-Nat. Res. Council Publ. 1136. Washington, D. C. 92 p.

Hotelling, H. 1954. Multivariate analysis. P. 67-80 *in* O. Kempthorne, T. A. Bancroft, J. W. Gowan, and J. L. Lush (ed.). Statistics and mathematics in biology. Iowa State College Press, Ames.

Kendall, M. G. 1957. A course in multivariate analysis. Charles Griffin and Co. Ltd., London. 185 p.

Mahalanobis, P. C. 1936. On the generalized distance in statistics. Proc. Nat. Inst. Sci. India 2:49-55.

Murty, B. R., and V. Arunachalam. 1966. The nature of divergence in relation to breeding system in some crop plants. Indian J. Genet. Plant Breeding 26A:188-198.

Olson, E. C. 1964. Morphological integration and the meaning of characters in classification systems. P. 123-156 *in* V. H. Heywood and J. McNeill (ed.). Phenetic and phylogenetic classification. The Systematics Assoc. Publ. No. 6, London.

Patnaik, P. B. 1949. The noncentral and chi-square and F distributions and their applications. Biometrika 36:202-232.

Peckholt, T. 1878. Monographia do Milho e da Mandioca. Typographia Universal de E. and H. Laemmert. Rio de Janeiro. 169 p.

Pearson, K. 1926. On the coefficient of racial likeness. Biometrika 18:105-117.

Rao, C. R. 1952. Advanced statistical methods in biometric research. John Wiley, New York. 390 p.

Seixas, Silvino Adelino de. 1880. Historia Natural e Cultura do Milho. Typ. de Lopes Vellozo e C., Bahia. 50 p. (E. A. Thesis. Imperial School of Agriculture of Bahia, Salvador, Bahia, Brazil)

Snedecor, G. W. 1956. Statistical methods. 5th ed. Iowa State College Press, Ames. 534 p.

Sokal, R. R. 1965. Statistical methods in systematics. Biol. Rev. Cambridge Phil. Soc. 40:337-391.

Sokal, R. R., and P. H. A. Sneath. 1963. Principles of numerical taxonomy. W. H. Freeman, San Francisco. 359 p.

Wellhausen, E. J., L. M. Roberts, and E. Hernandex X. 1952. Races of maize in Mexico. The Bussey Institution, Harvard University, Cambridge, Mass. 223 p.

Part II

DISCRIMINANT ANALYSIS

Editors' Comments on Papers 8 Through 16

8 **FISHER**
The Use of Multiple Measurements in Taxonomic Problems

9 **MAHALANOBIS**
On the Generalized Distance in Statistics

10 **RAO**
The Utilization of Multiple Measurements in Problems of Biological Classification

11 **BRYAN**
The Generalized Discriminant Function: Mathematical Foundation and Computational Routine

12 **POREBSKI**
On the Interrelated Nature of the Multivariate Statistics Used in Discriminatory Analysis

13 **JOLICOEUR**
Multivariate Geographical Variation in the Wolf Canis lupus *L.*

14 **BAKER, ATCHLEY, and McDANIEL**
Karyology and Morphometrics of Peters' Tent-Making Bat, Uroderma bilobatum *Peters (Chiroptera, Phyllostomatidae)*

15 **BURNABY**
Growth-Invariant Discriminant Functions and Generalized Distances

16 **GREEN**
A Multivariate Statistical Approach to the Hutchinsonian Niche: Bivalve Molluscs of Central Canada

Under the general title of "discriminant analysis" are included some of the most important and useful multivariate statistical tools available to the biologist. Under the broad rubric of discriminant analysis, at least four pertinent questions might be asked when biologists apply the technique:

1. Are there significant differences among the *k* groups? While this question is the basic null hypothesis of multivariate analysis of variance, it is also the first step in discriminant analysis.

2. If the groups do exhibit statistical differences, in what *directions* do the population centroids differ? Are they collinear, coplanar, etc.? This question is relevant to situations where there are more than two groups and is the "typical" discriminant analysis question. It is answered by computing eigenvalues and associated eigenvectors of a specific characteristic equation that maximizes among-group variance. The relative spatial *configuration* of the groups in reduced subspace then provides information about directionality of group differences.

3. What are the relative *distances* among the *k* groups? Intergroup distance, which is obviously related to the question of whether significant differences exist, can be determined by one of several "distance statistics."

4. How are new or unknown specimens allocated to predetermined groups? Construction of probabilistic identification schemes generally based on Bayesian probability theory will provide answers on this question.

A study of the early literature on multivariate analysis indicates that the development of these four facets of discriminant analysis methodology result from the efforts of many different statisticians including, R. A. Fisher, P. C. Mahalanobis, M. S. Bartlett, and C. R. Rao.

It is generally felt that the idea of a discriminant function may have begun with Karl Pearson around 1920 when he devised the "coefficient of racial likeness" to ascertain the statistical distance between two samples. Dissatisfaction with Pearson's coefficient lead P. C. Mahalanobis to propose the D^2 statistic as an alternative. D^2, now commonly called the *Mahalanobis distance,* was first applied to a study of racial mixtures in India in 1925, but more recently has become an integral part of several multivariate procedures. In addition to developing the D^2 statistic, the early work of Mahalanobis probably signals the beginning of the flourishing Indian school of statistics.

The D^2 statistic has become widely used because it is an actual *measure* of metric distance between population centroids rather than primarily a criterion for testing the null hypothesis of zero distance. Subsequent work, however, has shown that D^2 and the T^2 statistic of Hotelling are proportional and, as a result, D^2 can be used as a test criterion. Research on identification procedures as discussed below have shown D^2 to have even wider uses than originally envisioned by the early workers. Mahalanobis published several papers on the D^2 statistic; however, we have selected his more general 1936 paper (Paper 9) for inclusion here.

The first published accounts of what we now know as discriminant analysis in the strictest sense were in craniometrics by Barnard (1935) and Martin (1936). These authors had the procedures suggested to them by Sir Ronald Fisher, who is rightfully given credit for developing the discriminant function technique. In 1936 (Paper 8), Fisher formally proposed the linear discriminant function as a solution to a practical problem of achieving optimal separation of two species of plants using a number of intercorrelated variables. Fisher's goal was simultaneously to consider several taxonomic characters in order to discern the probable genetic origin of a third allopolyploid species. This procedure would further facilitate the allocation of an unknown specimen into one or the other prior determined groups (species of plant). In two subsequent papers (1938, 1940), Fisher integrated the discriminant function with the multivariate T^2 test of significance of Hotelling and the D^2 distance statistic of Mahalanobis. In addition, in these later works Fisher discussed the extension of the two-group method to multiple groups. The latter two works of Fisher, together with his other published papers, have recently been reprinted by the University of Adelaide. The reprinted series, entitled *Collected Papers of R. A. Fisher,* is in five volumes.

Attempts to generalize Fisher's discriminant function to multiple groups have been made by several statisticians, including Bartlett (1938, 1947), Rao (Paper 10), and others. However, probably the first workable procedures were published by Bryan (Paper 11) in a short paper from a symposium on multiple discriminant functions. Bryan's solution involved the successive eigenvalues and associated eigenvectors of the characteristic equation, $(W^{-1}A - \lambda I)v = 0$, where W and A were within- and among-groups dispersion matrices, λ an eigenvalue, and v the matrix of discriminant coefficients. This procedure generated a set of orthogonal discriminant functions or canonical variates that maximized the ratio of among-groups residual variance to the within-groups variance.

The classification aspect of the two-group discriminant analysis alluded to by Fisher in his early papers received mathematical validity when Welch (1939) showed that the identification aspect of the Fisherian discriminant function was essentially an application of the Neyman–Pearson liklihood ratio principle. The rigorous mathematical expansion of the identification approach and its extension to several alternative groups was set forth in a series of papers by C. R. Rao, of which the 1948 paper (Paper 10) is now considered to be a classical work. The importance of these works of Rao is obvious. In addition to refining and generalizing the D^2 statistic, Rao formally defined the probability theory associated with the two-group case, discussed the problem of mixed series of samples, the problem of doubtful regions where definite decisions cannot be made, and the generalization of the identification problem to three or more groups.

The final paper on discriminant analysis "theory" by Porebski (Paper 12) has been included because of the excellent formulation of the discrimination problem with the use of standard symbolism and the clarification of the numerous interrelated test statistics associated with the technique.

Among the first critical applications of the multiple-group discriminant techniques in biology was Jolicoeur's 1959 study on geographic variation in the wolf *Canis lupus* (Paper 13). As is indicative of much of Jolicoeur's work, this paper includes a rather long and well-written introduction into the discriminant analysis technique, together with visual displays of the results.

A more recent use of these procedures in systematics is that of Baker et al. (Paper 14), who employed the procedure to determine if chromosomal races in the bat *Uroderma bilobatum* could be discriminated morphometrically. This paper was included because it is a "complete" discriminant analysis in that it examines all four of the questions posed earlier. We have also included at this point the important paper of Burnaby (Paper 15). Burnaby's work is of significance to systematics because it examines the problem of growth invariance and its relationship to discriminant functions and the Mahalanobis distance statistic.

Finally, we have included an application from the ecological literature. In an ecological study, Green (Paper 16) used discriminant analysis to identify the patterns of ecological variation in 10 species of bivalve mollusks and proposed a method for defining an ecological niche. In a sequel paper, Green (1974) has extended and generalized the methodology introduced in the 1971 paper and provided a solution to the problem of analysis of temporarily varying environmental parameters in ecology. Buzas (1967) and Cassie (1963) are further examples of the use of discriminant analysis in ecology, while Williams et al. (1970), Whitehouse (1971), and Riggs (1973) have used this technique in genetics.

Additional applications of discriminant analysis from the field of systematics include Atchley (1971), Atchley and Martin (1971), Blackith and Blackith (1969), Blackith and Kevan (1967), Clifford and Binet (1954), Delaney and Healy (1964), Delaney and Whittaker (1969), DuPraw (1965), Eyles and Blackith (1965), Lubischew (1962), Phillips et al., (1973), Rees (1969), Reyment (1973), Sailer and Flowers (1969), Smouse (1972), and Snyder and Jameson (1965). The recent book edited by Cacoullos (1973) includes a number of interesting papers on discriminant analysis methodology and, in addition, contains an extensive bibliography.

Before proceeding to the papers, it should be pointed out that some of the terminology in the discriminant analysis literature may be confusing to nonstatisticians. While the two-group case is usually simply called discriminant analysis, the multiple-group procedures have been given several names, including discriminant analysis, multiple-group or generalized discriminant analysis, canonical variate analysis, and canonical analysis. At first the latter two terms might seem particularly confusing since they are also used for canonical correlation analysis. This confusion can be clarified, however, when it is pointed out that both the canonical correlation and discriminant analysis procedures involve transformations of some matrix into "canonical form." Additionally, discriminant analysis is related to canonical correlation methodology because discriminant function analysis can be considered simply as a special application of canonical correlation. This is clarified by Glahn (1968) in the section on regression and correlation in Bryant and Atchley, 1975.

BIBLIOGRAPHY

Anderson, T. W. 1951. Classification by multivariate analysis. *Psychometrika 16:*31–50.

———, and R. R. Bahadur. 1962. Classification into two multivariate normal distributions with different covariance matrices. *Ann. Math. Stat. 33:*420–431.

Ashton, E. H., M. J. R. Healy, and S. Tipton. 1957. The descriptive use of discriminant functions in physical anthropology. *Proc. Roy. Soc. London B146:*552–572.

———, M. J. R. Healy, C. G. Oxnard, and T. F. Spence. 1965. The combination of locomotor features of the primate shoulder girdle by canonical analysis. *J. Zool. 147:*406–429.

Atchley, W. R. 1971. A statistical analysis of geographic variation in the pupae of three species of *Culicoides* (Diptera: Ceratopogonidae). *Evolution 25:*51–74.

———. 1973. A quantitative separation of the females of three species of

Culicoides (Diptera: Ceratopogonidae). *J. Med. Entomol. 10:*629–632.

———, and J. Martin. 1971. A morphometric analysis of differential sexual dimorphism in larvae of *Chironomus* (Diptera). *Canadian Entomol.* 319–327.

Bargmann, R. C. 1970. Interpretation and use of a generalized discriminant function, in *Essays in Probability and Statistics* (R. C. Bose et al., eds.). University of North Carolina Press, Chapel Hill, N.C. pp. 35–60.

Barnard, M. M. 1935. The secular varieties of skull characters in four series of Egyptian skulls. *Ann. Eugenics 6:*352–371.

Bartlett, M. S. 1938. Further aspects of theory of multiple regression. *Proc. Cambridge Phil. Soc. 34:*33–40.

———. 1939. A note on test of significance in multivariate analysis. *Proc. Cambridge Phil. Soc. 35:*180–185.

———. 1939. The standard errors of discriminant function coefficients. *J. Roy. Stat. Soc. Suppl. 6:*169–173.

———. 1947. Multivariate analysis. *J. Roy. Stat. Soc. Suppl. 9:*176–197.

———. 1951. The goodness of fit of a single hypothetical discriminant function in the case of several groups. *Ann. Eugenics 16:*199–214.

———. 1965. Multivariate statistics, in *Theoretical and Mathematical Biology* (T. H. Waterman and H. J. Morowitz, eds.). Xerox Publishing Company, Lexington, Mass. pp. 202–224.

———, and N. W. Please. 1963. Discrimination in the case of zero mean differences. *Biometrika 50:*17–21.

Beale, G. M. L., M. G. Kendall, and D. W. Main. 1967. The discarding of variables in multivariate analysis. *Biometrika 54:*357–366.

Blackith, R. E. 1960. A synthesis of multivariate techniques to distinguish patterns of growth in grasshoppers. *Biometrics 16:*28–40.

———, and R. M. Blackith. 1969. Variation of shape of discrete anatomical characters in the morabine grasshoppers. *Austral. J. Zool. 17:*697–718.

———, and D. K. M. Kevan. 1967. A study of the genus *Chrotogonus* (Orthoptera): VIII. Patterns of variation in external morphology. *Evolution 21:*76–84.

Bryant, E. H., and W. R. Atchley. 1975. *Multivariate Statistical Methods: Within-Groups Covariation.* Dowden, Hutchinson & Ross, Inc., Stroudsburg, Pa.

Buzas, M. A. 1967. An application of canonical analysis as a method for comparing faunal areas. *J. Anim. Ecol. 36:*563–577.

Cacoullos, T., ed. 1973. *Discriminant Analysis and Applications.* Academic Press, New York, 434 pp.

Cassie, R. M. 1963. Multivariate analysis in the interpretation of numerical plankton data. *New Zealand J. Sci. 6:*35–59.

Chaddha, R. L., and L. F. Marcus. 1968. An empirical comparison of distance statistics for populations with unequal covariance matrices. *Biometrics 24:*683–694.

Claringbold, P. J. 1958. Multivariate quantal analysis. *J. Roy. Stat. Soc. B20:*398–405.

Clifford, H. T., and F. E. Binet. 1954. A quantitative study of a presumed hybrid swarm between *Eucalyptus eleaphoria* and *E. goniocalyx*. *Aust. J. Bot. 2:*325–336.

Clunies-Ross, C. W., and R. H. Riffenburgh. 1960. Geometry and linear discrimination. *Biometrika 47:*185–189.

Cochran, W. G., and C. E. Hopkins. 1961. Some classification problems with multivariate qualitative data. *Biometrics 17:*19–32.

Cooper, P. W. 1963. Statistical classification with quadratic forms. *Biometrika 50:*439–448.

———. 1965. Quadratic discriminant functions in pattern recognition. *IEEE Trans. Inform. Theory II:*313–315.

Delaney, M. J., and M. J. R. Healy. 1964. Variation in the long-tailed field mouse (*Apodemus sylvaticus* L.) in north-west Scotland: II. Simultaneous examination of all the characters. *Proc. Roy. Soc. London B161:*200–207.

———, and H. M. Whittaker. 1969. Variation in the skull of the long-tailed field mouse *Apodemus sylvaticus* in mainland Britain. *J. Zool. 157:*147–157.

Dunn, O. J., and P. V. Varady. 1966. Probabilities of correct classification in discriminant analysis. *Biometrics 22:*908–924.

Dunsmore, I. R. 1966. A bayesian approach to classification. *J. Roy. Stat. Soc. B28:*568–577.

DuPraw, E. J. 1965. Non-Linnean taxonomy and the systematics of honey-bees. *Syst. Zool. 14:*1–24.

Eyles, A. C., and R. E. Blackith. 1965. Studies on hybridization in *Scolopostethus* Fieber (Heteroptera: Lygaeidae). *Evolution 19:*465–479.

Fisher, R. A. 1938. The statistical utilization of multiple measurements. *Ann. Eugenics. 8:*376–86.

———. 1940. The precision of discriminant functions. *Ann. Eugenics 10:*422–429.

Gilbert, E. S. 1968. On discrimination using qualitative variables. *J. Amer. Stat. Assoc. 63:*1399–1412.

———. 1969. The effect of unequal variance–covariance matrices on Fisher's linear discriminant function. *Biometrics 25:*505–515.

Glahn, H. R. 1968. Canonical correlation and its relationship to discrimiinant analysis and multiple regression. *J. Atmospheric Sci. 25:*23–31.

Gower, J. C. 1966. A *Q*-technique for the calculation of canonical variates. *Biometrika 53:*588–590.

Green, R. H. 1974. Multivariate niche analysis with temporally varying environmental factors. *Ecology 55:*73–83.

Healy, M. J. R. 1965. Descriptive uses of disciminant functions. *Math. Computer Sci. Biol. Med.* 93–102.

Horton, I. F., J. S. Russell, and A. W. Moore. 1968. Multivariate covariance and canonical analysis: a method for selecting the most effective discriminators in a multivariate situation. *Biometrics 24:*845–858.

John, S. 1960. On some classification problems: I, II. *Sankhya 22:*301–308, 309–316. Amended *Sankyha A23:*308.

Kendall, M. G. 1966. Discrimination and classification, in *Multivariate Analysis* (P. R. Krishnaiah, ed.). Academic Press, New York. pp. 165–185.

Kim, K. C., B. W. Brown, and E. F. Cook. 1963. A quantitative taxonomic study of the *Enderleinellus suturalis* complex (Anoplura: Hoplopeuridae). *Syst. Zool. 12:*134–148.

Lachenbruch, P. A. 1967. An almost unbiased method of obtaining confidence internals for the probability of misclassification in discriminant analysis. *Biometrics 23:*639–645.

———. 1968. On expected probabilities of misclassification in discriminant analysis, necessary sample size, and a relation with the multiple correlation coefficient. *Biometrics 24:*823–834.

———, and M. R. Mickey. 1968. Estimation of error rates in discriminant analysis. *Technometrics 10:*1–11.

Lohnes, P. R. 1961. Test space and discriminant space classification models and related significance tests. *Educ. Psychol. Measurement 21:*559–574.

Lubin, A. 1950. Linear and non-linear discriminating functions. *British J. Psychol. 3:*90–104.

Lubischew, A. A. 1962. On the use of discriminant functions in taxonomy. *Biometrics 18:*455–477.

Marcus, L. F. 1964. Measurement of natural selection in natural populations. *Nature 202:*1033–1034.

Martin, E. S. 1936. A study of the Egyptian series of mandibles with special reference to mathematical methods of sexing. *Biometrika 28:*149–178.

Maxwell, A. E. 1961. Canonical variate analysis when the variables are dichotomis. *Educ. Psychol. Measurement 21:*259–271.

Melton, R. S. 1963. Some remarks of failure to meet assumptions in discriminant analysis. *Psychometrika 28:*49–53.

Miller, R. G. 1962. Statistical prediction by discriminant analysis. *Meteorol. Monogr. 4(25):*54 pp.

Mirsa, R. K. Vectorial analysis for genetic clines in body dimensions in populations of *Drosphila subobscura* Coll. and a comparison with those of *D. robusta. Biometrics 22:*469–487.

Panse, V. G. 1946. An application of the discriminant function for selection in poultry. *J. Genetics 47:*242–248.

Penrose, L. S. 1947. Some notes on discrimination. *Ann. Eugenics 13:*228–237.

———. 1954. Distance, size and shape. *Ann. Eugenics 18:*337–343.

Philips, B. F., N. A. Campbell, and B. R. Wilson. 1973. A multivariate study of geographic variation in the whelk: *Dicathais. J. Exptl. Marine Biol. Ecol. 11:*27–69.

Quellette, R. P., and S. U. Qadro. 1968. The discriminatory power of taxonomic characteristics in separating salmonoid fishes. *Syst. Zool. 17:*70–75.

Quenouille, M. H. 1949. Note on the elimination of insignificant variates in discriminatory analysis. *Ann. Eugenics 14:*305–308.

Rao, C. R. 1946. Tests with discriminant functions in multivariate analysis. *Sankhya 7:*407–414.

———. 1949. On some problems arising out of discrimination with multiple characters. *Sankhya 9:*343.

———. 1953. Discriminant functions for genetic differentiation and selection. *Sankhya 12:*339–346.
———. 1954. On the use and interpretation of distance functions in statistics. *Bull. Inst. Intern. Stat. 34:*90–97.
———. 1962. Use of discriminant and allied functions in multivariate analysis. *Sankhya A22:*317–338.
———. 1966. Discriminant function between composite hypothesis and related problems. *Biometrika 53:*339–345.
———. 1970. Inference on discriminant function coefficients, in *Essays in Probability and Statistics* (R. C. Bose et al., eds.). University of North Carolina Press, Chapel Hill, N.C. pp. 587–602.
———, and P. Slater. 1949. Multivariate analysis applied to differences between neurotic groups. *British J. Psychol. 2:*17–29.
Rees, J. W. 1969. Morphologic variation in the mandible of the white-tailed deer (*Odococleus virginionus*): A study of populational skeletal variation by principal component and canonical analysis. *J. Morphol. 128:*113–130.
Reyment, R. A. 1962. Observations on homogeneity of covariance matrices in paleontologic biometry. *Biometrics 18:*1–11.
———. 1969. Covariance structure and morphometric analysis. *Math. Geol. 1:*185–197.
———. 1973. The discriminant function in systematic biology, in *Discriminant Analysis and Applications* (T. Cacoullos, ed.), pp. 311–337. Academic Press, New York, 434 pp.
Riffenburgh, R. H., and C. W. Clunies-Ross. 1969. Linear discriminant analysis. *Pacific Sci. 14:*251–256.
Riggs, T. J. 1973. The use of canonical analysis for selection within a population of spring barley. *Ann. Appl. Biol. 74:*249–258.
Rulon, P. J. 1951. Distinctions between discriminant and regression analysis and a geometric interpretation of the discriminant function. *Harvard Educ. Rev. 21:*80–90.
Sailer, S. B., and J. M. Flowers. 1969. Geographic variation in the American lobster. *Syst. Zool. 18:*330–338.
Schmid, J., Jr. 1950. A comparison of two procedures for calculating discriminant function coefficients. *Psychometrika 15:*431–434.
Smouse, P. G. 1972. The canonical analysis of multiple species hybridization. *Biometrics 28:*361–371.
Snyder, W. F., and D. L. Jameson. 1965. Multivariate geographical variation of mating call in populations of the Pacific tree frog (*Hyla regilla*). *Copeia 1965:* 129–142.
Symmons, P. M. 1967. A morphometric measure of phase in the desert locust *Schistocerca gregaria* (Forsk). *Bull. Entomol. Re . 58:*803–809.
Urbakh, V. Yu. 1971. Linear discriminant analysis: loss of discriminating power when a variate is omitted. *Biometrics 27:*531–534.
Vogt, W. G., and D. G. McPherson. 1972. The weighted separation index: a multivariate technique for separating members of closely related species using qualitative differences. *Syst. Zool. 21:*187–198.
Weiner, J. M., and O. J. Dunn. 1966. Elimination of variates in linear discrimination problems. *Biometrics 22:*268–275.

Welch, B. L. 1939. Note on discriminant functions. *Biometrika 31:*218–220.
Whitehouse, R. N. H. 1971. Canonical analysis as an aid in plant breeding, in *Barley Genetics II: Proc. 2nd Intern. Barley Genetics Symp.* (R. A. Nilan, ed.). pp. 269–282. Washington State University, Pullman, Wash. 621 pp.
Williams, D. A., R. A. Beatty, and P. S. Burgoyne. 1970. Multivariate analysis in the genetics of spermatozoan dimensions in mice. *Proc. Roy. Soc. London B175:*313–331.
Williams, G. J. 1955. Significance tests for discriminant functions and linear functional relationships. *Biometrika 42:*360–381.

8

Reprinted from *Ann. Eugenics, 7,* 179–188 (1936). Also appears in *Collected Papers of R. A. Fisher* (5 vols.), University of Adelaide, Australia (1971–1974)

THE USE OF MULTIPLE MEASUREMENTS IN TAXONOMIC PROBLEMS

By R. A. FISHER, Sc.D., F.R.S.

I. Discriminant functions

When two or more populations have been measured in several characters, $x_1, \ldots, x_s$, special interest attaches to certain linear functions of the measurements by which the populations are best discriminated. At the author's suggestion use has already been made of this fact in craniometry (*a*) by Mr E. S. Martin, who has applied the principle to the sex differences in measurements of the mandible, and (*b*) by Miss Mildred Barnard, who showed how to obtain from a series of dated series the particular compound of cranial measurements showing most distinctly a progressive or secular trend. In the present paper the application of the same principle will be illustrated on a taxonomic problem; some questions connected with the precision of the processes employed will also be discussed.

II. Arithmetical procedure

Table I shows measurements of the flowers of fifty plants each of the two species *Iris setosa* and *I. versicolor*, found growing together in the same colony and measured by Dr E. Anderson, to whom I am indebted for the use of the data. Four flower measurements are given. We shall first consider the question: What linear function of the four measurements

$$X = \lambda_1 x_1 + \lambda_2 x_2 + \lambda_3 x_3 + \lambda_4 x_4$$

will maximize the ratio of the difference between the specific means to the standard deviations within species? The observed means and their differences are shown in Table II. We may represent the differences by d_p, where $p = 1, 2, 3$ or 4 for the four measurements.

The sums of squares and products of deviations from the specific means are shown in Table III. Since fifty plants of each species were used these sums contain 98 degrees of freedom. We may represent these sums of squares or products by S_{pq}, where p and q take independently the values 1, 2, 3 and 4.

Then for any linear function, X, of the measurements, as defined above, the difference between the means of X in the two species is

$$D = \lambda_1 d_1 + \lambda_2 d_2 + \lambda_3 d_3 + \lambda_4 d_4,$$

while the variance of X within species is proportional to

$$S = \sum_{p=1}^{4} \sum_{q=1}^{4} \lambda_p \lambda_q S_{pq}.$$

The particular linear function which best discriminates the two species will be one for

Table I

Iris setosa				*Iris versicolor*				*Iris virginica*			
Sepal length	Sepal width	Petal length	Petal width	Sepal length	Sepal width	Petal length	Petal width	Sepal length	Sepal width	Petal length	Petal width
5·1	3·5	1·4	0·2	7·0	3·2	4·7	1·4	6·3	3·3	6·0	2·5
4·9	3·0	1·4	0·2	6·4	3·2	4·5	1·5	5·8	2·7	5·1	1·9
4·7	3·2	1·3	0·2	6·9	3·1	4·9	1·5	7·1	3·0	5·9	2·1
4·6	3·1	1·5	0·2	5·5	2·3	4·0	1·3	6·3	2·9	5·6	1·8
5·0	3·6	1·4	0·2	6·5	2·8	4·6	1·5	6·5	3·0	5·8	2·2
5·4	3·9	1·7	0·4	5·7	2·8	4·5	1·3	7·6	3·0	6·6	2·1
4·6	3·4	1·4	0·3	6·3	3·3	4·7	1·6	4·9	2·5	4·5	1·7
5·0	3·4	1·5	0·2	4·9	2·4	3·3	1·0	7·3	2·9	6·3	1·8
4·4	2·9	1·4	0·2	6·6	2·9	4·6	1·3	6·7	2·5	5·8	1·8
4·9	3·1	1·5	0·1	5·2	2·7	3·9	1·4	7·2	3·6	6·1	2·5
5·4	3·7	1·5	0·2	5·0	2·0	3·5	1·0	6·5	3·2	5·1	2·0
4·8	3·4	1·6	0·2	5·9	3·0	4·2	1·5	6·4	2·7	5·3	1·9
4·8	3·0	1·4	0·1	6·0	2·2	4·0	1·0	6·8	3·0	5·5	2·1
4·3	3·0	1·1	0·1	6·1	2·9	4·7	1·4	5·7	2·5	5·0	2·0
5·8	4·0	1·2	0·2	5·6	2·9	3·6	1·3	5·8	2·8	5·1	2·4
5·7	4·4	1·5	0·4	6·7	3·1	4·4	1·4	6·4	3·2	5·3	2·3
5·4	3·9	1·3	0·4	5·6	3·0	4·5	1·5	6·5	3·0	5·5	1·8
5·1	3·5	1·4	0·3	5·8	2·7	4·1	1·0	7·7	3·8	6·7	2·2
5·7	3·8	1·7	0·3	6·2	2·2	4·5	1·5	7·7	2·6	6·9	2·3
5·1	3·8	1·5	0·3	5·6	2·5	3·9	1·1	6·0	2·2	5·0	1·5
5·4	3·4	1·7	0·2	5·9	3·2	4·8	1·8	6·9	3·2	5·7	2·3
5·1	3·7	1·5	0·4	6·1	2·8	4·0	1·3	5·6	2·8	4·9	2·0
4·6	3·6	1·0	0·2	6·3	2·5	4·9	1·5	7·7	2·8	6·7	2·0
5·1	3·3	1·7	0·5	6·1	2·8	4·7	1·2	6·3	2·7	4·9	1·8
4·8	3·4	1·9	0·2	6·4	2·9	4·3	1·3	6·7	3·3	5·7	2·1
5·0	3·0	1·6	0·2	6·6	3·0	4·4	1·4	7·2	3·2	6·0	1·8
5·0	3·4	1·6	0·4	6·8	2·8	4·8	1·4	6·2	2·8	4·8	1·8
5·2	3·5	1·5	0·2	6·7	3·0	5·0	1·7	6·1	3·0	4·9	1·8
5·2	3·4	1·4	0·2	6·0	2·9	4·5	1·5	6·4	2·8	5·6	2·1
4·7	3·2	1·6	0·2	5·7	2·6	3·5	1·0	7·2	3·0	5·8	1·6
4·8	3·1	1·6	0·2	5·5	2·4	3·8	1·1	7·4	2·8	6·1	1·9
5·4	3·4	1·5	0·4	5·5	2·4	3·7	1·0	7·9	3·8	6·4	2·0
5·2	4·1	1·5	0·1	5·8	2·7	3·9	1·2	6·4	2·8	5·6	2·2
5·5	4·2	1·4	0·2	6·0	2·7	5·1	1·6	6·3	2·8	5·1	1·5
4·9	3·1	1·5	0·2	5·4	3·0	4·5	1·5	6·1	2·6	5·6	1·4
5·0	3·2	1·2	0·2	6·0	3·4	4·5	1·6	7·7	3·0	6·1	2·3
5·5	3·5	1·3	0·2	6·7	3·1	4·7	1·5	6·3	3·4	5·6	2·4
4·9	3·6	1·4	0·1	6·3	2·3	4·4	1·3	6·4	3·1	5·5	1·8
4·4	3·0	1·3	0·2	5·6	3·0	4·1	1·3	6·0	3·0	4·8	1·8
5·1	3·4	1·5	0·2	5·5	2·5	4·0	1·3	6·9	3·1	5·4	2·1
5·0	3·5	1·3	0·3	5·5	2·6	4·4	1·2	6·7	3·1	5·6	2·4
4·5	2·3	1·3	0·3	6·1	3·0	4·6	1·4	6·9	3·1	5·1	2·3
4·4	3·2	1·3	0·2	5·8	2·6	4·0	1·2	5·8	2·7	5·1	1·9
5·0	3·5	1·6	0·6	5·0	2·3	3·3	1·0	6·8	3·2	5·9	2·3
5·1	3·8	1·9	0·4	5·6	2·7	4·2	1·3	6·7	3·3	5·7	2·5
4·8	3·0	1·4	0·3	5·7	3·0	4·2	1·2	6·7	3·0	5·2	2·3
5·1	3·8	1·6	0·2	5·7	2·9	4·2	1·3	6·3	2·5	5·0	1·9
4·6	3·2	1·4	0·2	6·2	2·9	4·3	1·3	6·5	3·0	5·2	2·0
5·3	3·7	1·5	0·2	5·1	2·5	3·0	1·1	6·2	3·4	5·4	2·3
5·0	3·3	1·4	0·2	5·7	2·8	4·1	1·3	5·9	3·0	5·1	1·8

Table II. *Observed means for two species and their difference (cm.)*

	Versicolor	*Setosa*	Difference ($V-S$)
Sepal length (x_1)	5·936	5·006	0·930
Sepal width (x_2)	2·770	3·428	−0·658
Petal length (x_3)	4·260	1·462	2·798
Petal width (x_4)	1·326	0·246	1·080

Table III. *Sums of squares and products of four measurements, within species (cm.²)*

	Sepal length	Sepal width	Petal length	Petal width
Sepal length	19·1434	9·0356	9·7634	3·2394
Sepal width	9·0356	11·8658	4·6232	2·4746
Petal length	9·7634	4·6232	12·2978	3·8794
Petal width	3·2394	2·4746	3·8794	2·4604

which the ratio D^2/S is greatest, by variation of the four coefficients λ_1, λ_2, λ_3 and λ_4 independently. This gives for each λ

$$\frac{D}{S^2}\left\{2S\frac{\partial D}{\partial \lambda} - D\frac{\partial S}{\partial \lambda}\right\} = 0,$$

or

$$\frac{1}{2}\cdot\frac{\partial S}{\partial \lambda} = \frac{S}{D}\frac{\partial D}{\partial \lambda},$$

where it may be noticed that S/D is a factor constant for the four unknown coefficients. Consequently, the coefficients required are proportional to the solutions of the equations

$$\left.\begin{aligned} S_{11}\lambda_1 + S_{12}\lambda_2 + S_{13}\lambda_3 + S_{14}\lambda_4 = d_1, \\ S_{12}\lambda_1 + S_{22}\lambda_2 + S_{23}\lambda_3 + S_{24}\lambda_4 = d_2, \\ S_{13}\lambda_1 + S_{23}\lambda_2 + S_{33}\lambda_3 + S_{34}\lambda_4 = d_3, \\ S_{14}\lambda_1 + S_{24}\lambda_2 + S_{34}\lambda_3 + S_{44}\lambda_4 = d_4. \end{aligned}\right\} \quad \ldots\ldots(1)$$

If, in turn, unity is substituted for each of the differences and zero for the others, the solutions obtained constitute the matrix of multipliers reciprocal to the matrix of S; numerically we find:

Table IV. *Matrix of multipliers reciprocal to the sums of squares and products within species (cm.⁻²)*

	Sepal length	Sepal width	Petal length	Petal width
Sepal length	0·1187161	−0·0668666	−0·0816158	0·0396350
Sepal width	−0·0668666	0·1452736	0·0334101	−0·1107529
Petal length	−0·0816158	0·0334101	0·2193614	−0·2720206
Petal width	0·0396350	−0·1107529	−0·2720206	0·8945506

These values may be denoted by s_{pq} for values of p and q from 1 to 4.

Multiplying the columns of the matrix in Table IV by the observed differences, we have the solutions of the equation (1) in the form

$$\lambda = -0{\cdot}0311511, \quad \lambda_2 = -0{\cdot}1839075, \quad \lambda_3 = +0{\cdot}2221044, \quad \lambda_4 = +0{\cdot}3147370,$$

so that, if we choose to take the coefficient of sepal length to be unity, the compound measurement required is

$$X = x_1 + 5{\cdot}9037x_2 - 7{\cdot}1299x_3 - 10{\cdot}1036x_4.$$

If in this expression we substitute the values observed in *setosa* plants, the mean, as found from the values in Table I, is

$$5{\cdot}006 + (3{\cdot}428)\,(5{\cdot}9037) - (1{\cdot}462)\,(7{\cdot}1299) - (0{\cdot}246)\,(10{\cdot}1036) = 12{\cdot}3345 \text{ cm.};$$

for *versicolor*, on the contrary, we have

$$5{\cdot}936 + (2{\cdot}770)\,(5{\cdot}9037) - (4{\cdot}260)\,(7{\cdot}1299) - (1{\cdot}326)\,(10{\cdot}1036) = -21{\cdot}4815 \text{ cm.}$$

The difference between the average values of the compound measurements being thus 33·816 cm.

The distinctness of the metrical characters of the two species may now be gauged by comparing this difference between the average values with its standard error. Using the values of Table III, with the coefficients of our compound, we have

$$\begin{aligned}
19{\cdot}1434 + (9{\cdot}0356)\,(5{\cdot}9037) - (9{\cdot}7634)\,(7{\cdot}1299) - (3{\cdot}2394)\,(10{\cdot}1036) &= -29{\cdot}8508,\\
9{\cdot}0356 + (11{\cdot}8658)\,(5{\cdot}9037) - (4{\cdot}6232)\,(7{\cdot}1299) - (2{\cdot}4746)\,(10{\cdot}1036) &= 21{\cdot}1224,\\
9{\cdot}7634 + (4{\cdot}6232)\,(5{\cdot}9037) - (12{\cdot}2978)\,(7{\cdot}1299) - (3{\cdot}8794)\,(10{\cdot}1036) &= -89{\cdot}8206,\\
3{\cdot}2394 + (2{\cdot}4746)\,(5{\cdot}9037) - (3{\cdot}8794)\,(7{\cdot}1299) - (2{\cdot}4604)\,(10{\cdot}1036) &= -34{\cdot}6699,
\end{aligned}$$

and finally,

$$-29{\cdot}8508 + (21{\cdot}1224)\,(5{\cdot}9037) + (89{\cdot}8206)\,(7{\cdot}1299) + (34{\cdot}6699)\,(10{\cdot}1036) = 1085{\cdot}5522.$$

The average variance of the two species in respect of the compound measurements may be estimated by dividing this value (1085·5522) by 95; the variance of the difference between two means of fifty plants each, by dividing again by 25. For single plants the variance is 11·4269, so that the mean difference, 33·816 cm., between a pair of plants of different species has a standard deviation of 4·781 cm. For means of fifty the same average difference has the standard error 0·6761 cm., or only about one-fiftieth of its value.

III. Interpretation

The ratio of the difference between the means of the chosen compound measurement to its standard error in individual plants is of interest also in relation to the probability of misclassification, if the specific nature were judged wholly from the measurements. For reasons to be discussed later we shall estimate the variance of a single plant by dividing 1085·5522 by 95, giving 11·4269 cm.^{2} for the variance, and 3·3804 cm. for the standard deviation. Supposing that a plant is misclassified, if its deviation in the right direction exceeds half the difference, 33·816 cm., between the species, the ratio to the standard as estimated is 5·0018.

The table of the normal distribution (*Statistical Methods*, Table II) shows that a ratio 4·89164 is exceeded five times in a million, and 5·32672 only once in two million trials. By logarithmic interpolation the frequency appropriate to a ratio 5·0018 is about 2·79 per million. If the variances of the two species are unequal, this frequency is somewhat overestimated by this method, since we ought to divide the specific difference in proportion to the two standard deviations, and for constant sum of variances the sum of the standard deviations is greatest when they are equal. We may, therefore, at once conclude that if the measurements are nearly normally distributed the probability of misclassification, using the compound movement only is less than three per million.

The same ratio is of interest from another aspect. If the chosen compound X is analysed in respect to its variation within and between species, the sum of squares between species must be $25D^2$. Numerically we have, therefore,

Table V. *Analysis of variance of the chosen compound X, between and within species*

	Degrees of freedom	Sum of squares
Between species	4	28588·05
Within species	95	1085·55
Total	99	29673·60

Of the total only 3·6583 per cent. is within species, and 96·3417 per cent. between species. The compound has been chosen to maximize the latter percentage. Since, in addition to the specific means, we have used three adjustable ratios, the variation within species must contain only 95 degrees of freedom.

In making up the variate X, we have multiplied the original values of λ by $-32·1018$ in order to give to the measurement sepal length the coefficient unity. Had we used the original values, the analysis of Table V would have appeared as:

Table VI. *Analysis of variance of the crude compound X, between and within species*

	Degrees of freedom	Sum of squares	
Between species	4	27·74160	$=25D^2$
Within species	95	1·05341	$= \quad D = S$
Total	99	28·79501	$D(1+25D)$

On multiplying equations (1) by λ_1, λ_2, λ_3 and λ_4 and adding, it appears that $S = \Sigma\lambda d = D$, the specific difference in the crude compound X. The proportion (3·6 per cent.) of the sum of squares within species could therefore have been found simply as $1/(1+25D)$.

IV. The analogy of partial regression

The analysis of Table VI suggests an analogy of some interest. If to each plant were assigned a value of a variate y, the same for all members of each species, the analysis of variance of y, between the portions accountable by linear regression on the measurements $x_1, \ldots, x_4$, and the residual variation after fitting such a regression, would be identical with Table VI, if y were given appropriate equal and opposite values for the two species.

In general, with different numbers of representatives of the two species, n_1 and n_2, if the values of y assigned were

$$\frac{n_2}{n_1+n_2} \text{ and } \frac{-n_1}{n_1+n_2},$$

differing by unity, the right-hand sides of the equations for the regression coefficients, corresponding to equation (1), would have been

$$\frac{n_1 n_2}{n_1+n_2} d_p,$$

where d_p is the difference between the means of the two species in any one of the measurements. The typical coefficient of the left-hand side would be

$$S_{pq}+\frac{n_1 n_2}{n_1+n_2} d_p d_q.$$

Transferring the additional fractions to the right-hand side, we should have equations identical with (1), save that the right-hand sides are now

$$\frac{n_1 n_2}{n_1+n_2} d_p (1-\Sigma\lambda' d),$$

where λ' stands for a solution of the new equations; hence

$$\lambda' = \frac{n_1 n_2}{n_1+n_2} (1-\Sigma\lambda' d) \lambda,$$

multiply these equations by d and add, so that

$$\Sigma\lambda' d = \frac{n_1 n_2}{n_1+n_2} \Sigma\lambda d (1-\Sigma\lambda' d),$$

or

$$(1-\Sigma\lambda' d)\left(1+\frac{n_1 n_2}{n_1+n_2} \Sigma\lambda d\right) = 1,$$

and so in our example

$$1-\Sigma\lambda' d = \frac{1}{1+25D}.$$

The analysis of variance of y is, therefore,

Table VII. *Analysis of variance of a variate y determined exclusively by the species*

	Degrees of freedom	Sum of squares	
Regression	4	24·0854	$25^2D/1+25D$
Remainder	95	0·9146	$25/1+25D$
Total	99	25·0000	

The total $S\ (y^2)$ is clearly in general $\frac{n_1 n_2}{n_1+n_2}$; the portion ascribable to regression is

$$\frac{n_1 n_2}{n_1+n_2}\ \Sigma\lambda' d = \frac{25^2 D}{1+25D}.$$

In this method of presentation the appropriate allocation of the degrees of freedom is evident.

The multiple correlation of y with the measurements $x_1, \ldots, x_4$ is given by

$$R^2 = 25D/1+25D.$$

V. Test of significance

It is now clear in what manner the specific difference may be tested for significance, so as to allow for the fact that a variate has been chosen so as to maximise the distinctness of the species. The regression of y on the four measurements is given 4 degrees of freedom, and the residual variation 95; the value of z calculated from the sums of squares in any one of Tables V, VI or VII is 3·2183 or

$$\tfrac{1}{2}\ (\log 95 - \log 4 + \log 25 + \log D)$$

a very significant value for the number of degrees of freedom used.

VI. Applications to the theory of allopolyploidy

We may now consider one of the extensions of this procedure which are available when samples have been taken from more than two populations. The sample of the third species given in Table I, *Iris virginica*, differs from the two other samples in not being taken from the same natural colony as they were—a circumstance which might considerably disturb both the mean values and their variabilities. It is of interest in association with *I. setosa* and *I. versicolor* in that Randoph (1934) has ascertained and Anderson has confirmed that, whereas *I. setosa* is a "diploid" species with 38 chromosomes, *I. virginica* is "tetraploid", with 70, and *I. versicolor*, which is intermediate in three measurements, though not in sepal breadth, is hexaploid. He has suggested the interesting possibility that *I. versicolor* is a polyploid hybrid of the two other species. We shall, therefore, consider whether, when we use the linear compound of the four measurements most appropriate for discriminating three such species, the mean value for *I. versicolor* takes an intermediate value, and, if so, whether it differs twice as much from *I. setosa* as from *I. virginica*, as might be expected, if the effects of genes are simply additive, in a hybrid between a diploid and a tetraploid species.

If a third value lies two-thirds of the way from one value to another, the three deviations from their common mean must be in the ratio $4:1:-5$. To obtain values corresponding with the differences between the two species we may, therefore, form linear compounds of their mean measurements, using these numerical coefficients. The results are shown in Table VIII where, for example, the value 7·258 cm. for sepal length is four times the mean sepal length for *I. virginica* plus once the mean sepal length for *I. versicolor* minus five times the value for *I. setosa*.

Table VIII

Means	S_{pq}			
	Iris virginica. Fifty plants			
6·588	19·8128	4·5944	14·8612	2·4056
2·974	4·5944	5·0962	3·4976	2·3338
5·552	14·8612	3·4976	14·9248	2·3924
2·026	2·4056	2·3338	2·3924	3·6962
	Iris versicolor. Fifty plants			
5·936	13·0552	4·1740	8·9620	2·7332
2·770	4·1740	4·8250	4·0500	2·0190
4·260	8·9620	4·0500	10·8200	3·5820
1·326	2·7332	2·0190	3·5820	1·9162
	Iris setosa. Fifty plants			
5·006	6·0882	4·8616	0·8014	0·5062
3·428	4·8616	7·0408	0·5732	0·4556
1·462	0·8014	0·5732	1·4778	0·2974
0·246	0·5062	0·4556	0·2974	0·5442
	$4vi + ve - 5se$			
7·258	482·2650	199·2244	266·7762	53·8778
−2·474	199·2244	262·3842	74·3416	50·7498
19·158	266·7762	74·3416	286·5618	49·2954
8·200	53·8778	50·7498	49·2954	74·6604

Since the values for the sums of squares and products of deviations from the means within each of the three species are somewhat different, we may make an appropriate matrix corresponding with our chosen linear compound by multiplying the values for *I. virginica* by 16, those for *I. versicolor* by one and those for *I. setosa* by 25, and adding the values for the three species, as shown in Table VIII. The values so obtained will correspond with the matrix of sums of squares and products within species when only two populations have been sampled.

Using the rows of the matrix as the coefficients of four unknowns in an equation with our chosen compound of the mean measurements, e.g.

$$482{\cdot}2650\lambda_1 + 199{\cdot}2244\lambda_2 + 266{\cdot}7762\lambda_3 + 53{\cdot}8778\lambda_4 = 7{\cdot}258,$$

we find solutions which, when multiplied by 100, are

Coefficient of	sepal length	−3·308998
	sepal breadth	−2·759132
	petal length	8·866048
	petal breadth	9·392551

defining the compound measurement required.

It is now easy to find the means and variances of this compound measurement in the three species. These are shown in the table below (Table IX):

Table IX

	Mean	Sum of squares	Mean square	Standard deviation
I. virginica	38·24827	923·7958	18·8530	4·342
I. versicolor	22·93888	873·5119	17·8268	4·222
I. setosa	−10·75042	292·8958	5·9775	2·444

From this table it can be seen that, whereas the difference between *I. setosa* and *I. versicolor*, 33·69 of our units, is so great compared with the standard deviations that no appreciable overlapping of values can occur, the difference between *I. virginica* and *I. versicolor*, 15·31 units, is less than four times the standard deviation of each species.

The differences do seem, however, to be remarkably closely in the ratio 2 : 1. Compared with this standard, *I. virginica* would appear to have exerted a slightly preponderant influence. The departure from expectation is, however, small; and we have the material for making at least an approximate test of significance.

If the differences between the means were exactly in the ratio 2 : 1, then the linear function formed by adding the means with coefficients in the ratio 2 :−3 : 1 would be zero. Actually it has the value 3·07052. The sampling variance of this compound is found by multiplying the variances of the three species by 4, 9 and 1, adding them together and dividing by 50, since each mean is based on fifty plants. This gives 4·8365 for the variance and 2·199 for the standard error. Thus on this test the discrepancy, 3·071, is certainly not significant, though it somewhat exceeds its standard error.

In theory the test of significance is not wholly exact, since in estimating the sampling variance of each species we have divided the sum of squares of deviations from the mean by 49, as though these deviations had in all 147 degrees of freedom. Actually three degrees of freedom have been absorbed in adjusting the coefficients of the linear compound so as to discriminate the species as distinctly as possible. Had we divided by 48 instead of by 49 the standard error would have been raised by a trifle to the value 2·231, which would not have affected the interpretation of the data. This change, however, would certainly have been an over-correction, since it is the variances of the extreme species *I. virginica* and *I. setosa* which are most reduced in the choice of the compound measurement, while that of *I. versicolor* contributes the greater part of the sampling error in the test of significance.

The diagram, Fig. 1, shows the actual distributions of the compound measurement adopted in the individuals of the three species measured. It will be noticed, as was anticipated above, that there is some overlap of the distributions of *I. virginica* and *I. versicolor*, so that a certain diagnosis of these two species could not be based solely on these four measurements of a single flower taken on a plant growing wild. It is not, however, impossible that in culture the measurements alone should afford a more complete discrimination.

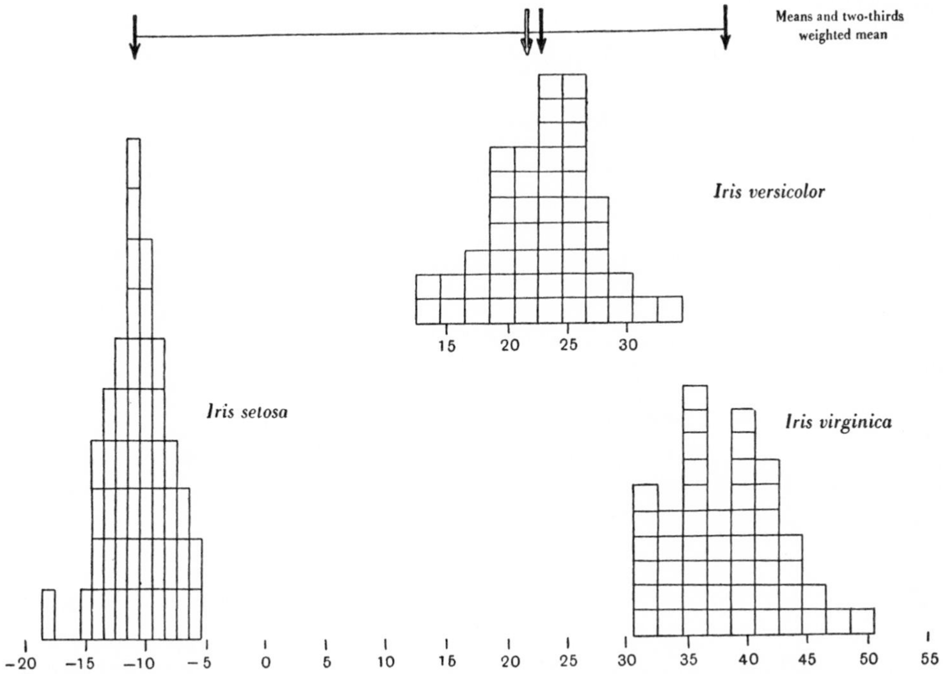

Fig. 1. Frequency histograms of the discriminating linear function, for three species of *Iris*.

REFERENCES

RANDOLPH, L. F. (1934). "Chromosome numbers in native American and introduced species and cultivated varieties of Iris." *Bull. Amer. Iris Soc.* **52**, 61–66.

ANDERSON, EDGAR (1935). "The irises of the Gaspe Peninsula." *Bull. Amer. Iris Soc.* **59**, 2–5.

—— (1936). "The species problem in *Iris*." *Ann. Mo. bot. Gdn.* (in the Press).

9

Reprinted from *Proc. Natl. Inst. Sci. India,* **2**(1), 49–55 (1936)

ON THE GENERALIZED DISTANCE IN STATISTICS.

By P. C. Mahalanobis.

(*Read January 4, 1936.*)

1. A normal (Gauss-Laplacian) statistical population in P-variates is usually described by a P-dimensional frequency distribution :—

$$df = \text{const.} \times e^{-\frac{1}{2\alpha}\left[A_{11}(x_1-\alpha_1)^2 + A_{22}(x_2-\alpha_2)^2 + \ldots\ldots + 2A_{12}(x_1-\alpha_1)(x_2-\alpha_2) + \ldots\ldots\right]} . dx_1 . dx_2 .. dx_P \qquad (1{\cdot}0)$$

where

$\alpha_1, \alpha_2 \ldots \alpha_P$ = the population (mean) values of the P-variates $x_1, x_2 \ldots x_P$ .. .. (1·1)

$\alpha_{ii} = \sigma_i^2$, are the respective variances .. .. (1·2)

$\alpha_{ij} = \sigma_i . \sigma_j . \rho_{ij}$, where ρ_{ij} = the coefficient of correlation between the ith and jth variates .. .. (1·3)

α is the determinant $| \alpha_{ij} |$ defined more fully in (2·2), and A_{ij} is the minor of α_{ij} in this determinant.

A P-variate normal population is thus completely specified by the set of $P(P+1)/2$ parameters* :—

$$(\alpha_1, \alpha_2 \ldots \alpha_P;\ \alpha_{11}, \alpha_{22} \ldots \alpha_{PP};\ \alpha_{12}, \alpha_{13} \ldots \alpha_{P-1,P}) \quad .. \qquad (1{\cdot}4)$$

A second P-variate normal population can be specified in the same way by the parameters :—

$$(\alpha'_1, \alpha'_2 \ldots \alpha'_P;\ \alpha'_{11}, \alpha'_{22} \ldots \alpha'_{PP};\ \alpha'_{12}, \alpha'_{13} \ldots \alpha'_{P-1,P}) \qquad (1{\cdot}5)$$

It is convenient to use the notation (α_i, α_{ij}) and $(\alpha'_i, \alpha'_{ij})$ to represent respectively the two sets of parameters given by (1·4) and (1·5). The α_i's or α'_i's are the mean values, while we may speak of (α_{ij}) or (α'_{ij}) as the respective dispersions.

2. It will be convenient at this stage to introduce the idea of a statistical field, such that at each point in this field there is a specified set of parameters (α_i, α_{ij}) which define a particular population. In other words, each point in a statistical field is the centre of a density-cluster belonging to a particular normal population completely specified by the value of (α_i, α_{ij}) at that particular point of the field.

---

* It may be noted here that such a population can be conveniently represented by a density-cluster in a P-dimensional space in which the position of the cluster is determined by the parameters $(\alpha_1, \alpha_2 \ldots \alpha_i \ldots \alpha_P)$, the size of the equi-frequential surfaces by $(\alpha_{11}, \alpha_{22} \ldots \alpha_{ii} \ldots \alpha_{PP})$, and the orientation of these surfaces by

$$(\alpha_{12}, \alpha_{13} \ldots \alpha_{ij} \ldots \alpha_{P-1,\ P}).$$

VOL. II—No. 1. [Published April 15th, 1936.

I shall first consider a field in which the dispersion is same at all points. That is

$$\alpha_{ij}=\alpha'_{ij} \quad \ldots \quad (2\cdot0)$$

for all values of i, j, and for all points in the field. In this case two populations can only differ in their mean values.

It has been shown elsewhere * that the distance between the two statistical populations can be conveniently measured by a Δ^2-statistics defined by

$$P\,.\,\Delta^2=\underset{i=1}{\overset{i=P}{S}}\frac{(\alpha_i-\alpha'_i)^2}{\alpha_{ii}} \quad \ldots \quad (2\cdot1)$$

for P independent variates.

This formula can be easily generalized for P correlated variates. Let us define the fundamental dispersion matrix :—

$$\alpha=\begin{vmatrix} \alpha_{11}, & \alpha_{12}, & . & . & . & \alpha_{1P} \\ a_{21}, & a_{22}, & . & . & . & a_{2P} \\ . & . & . & . & . & . \\ \alpha_{P1}, & \alpha_{P2}, & . & . & . & \alpha_{PP} \end{vmatrix} \quad \ldots \quad (2\cdot2)$$

$$\text{Let} \quad \alpha^{ij}=\frac{\text{Co-factor of } \alpha_{ij} \text{ in } \alpha}{\text{Determinant } \alpha} \quad \ldots \quad (2\cdot3)$$

The generalized Δ^2=statistics for P correlated variates can be now written in the form

$$P\,.\,\Delta^2=\underset{i,j=1}{\overset{i,j=P}{S}}\left[\alpha^{ij}\,.\,(\alpha_i-\alpha'_i)\,.\,(\alpha_j-\alpha'_j)\right] \quad \ldots \quad (2\cdot4)$$

Introducing the summation convention that a set of duplicated suffixes (say μ) will imply a summation for all values of $\mu=1, 2, \ldots. P$, we get

$$P\,.\,\Delta^2=\alpha^{\mu\nu}\,.\,(\alpha_\mu-\alpha'_\mu)\,.\,(\alpha_\nu-\alpha'_\nu) \quad \ldots \quad (2\cdot41)$$

If we write

$$d\alpha_\mu\equiv(\alpha_\mu-\alpha'_\mu), \quad d\alpha_\nu\equiv(\alpha_\nu-\alpha'_\nu) \quad \ldots \quad (2.42)$$

we get

$$P\,.\,\Delta^2=\alpha^{\mu\nu}\,.\,d\alpha_\mu\,.\,d\alpha_\nu \quad \ldots \quad (2\cdot5)$$

It is slightly more convenient to change the notation a little, and define

$$\alpha^{ij}\equiv\sigma_i\,.\,\sigma_j\,.\,\rho_{ij} \quad \ldots \quad (2\cdot61)$$

$$\alpha_{ij}=\frac{\text{Co-factor of } \alpha^{ij} \text{ in } \alpha}{\text{Determinant } \alpha} \quad \ldots \quad (2\cdot62)$$

and

$$(d\alpha)^i\equiv(\alpha_i-\alpha'_i), \quad (d\alpha)_j\equiv(\alpha_j-\alpha'_j) \quad \ldots \quad (2\cdot63)$$

* *Journ. Asiat. Soc. Bengal,* vol. XXVI, pp. 541–588, (1930).

Equation (2·5) can then be written in the form

$$P \,.\, \Delta^2 = \alpha_{\mu\nu} \,.\, (d\alpha)^\mu \,.\, (d\alpha)^\nu \qquad \ldots \qquad \ldots \qquad \ldots \qquad (2{\cdot}7)$$

Comparing with the formula for ds^2

$$ds^2 = g_{\mu\nu} \,.\, (dx)^\mu \,.\, (dx)^\nu \qquad \ldots \qquad \ldots \qquad \ldots \qquad (2{\cdot}71)$$

we notice that $P \,.\, \Delta^2$ in statistics is the exact analogue of ds^2 in the restricted theory of relativity.

This merely implies that a consistent geometrical representation is possible in both cases. It is possible, however, to use this formal equivalence to establish an exact correspondence between results in the two subjects.

3. We see therefore that a statistical field in which the dispersion is same everywhere (values of $\alpha_{\mu\nu}$'s same at all points of the field and independent of mean values) corresponds to the physical field in the restricted theory of relativity ($g_{\mu\nu}$'s same everywhere and independent of co-ordinate values). In fact $\alpha_{\mu\nu}$'s play the same part in statistics as $g_{\mu\nu}$'s in the theory of relativity, and all the results involving ds^2 can be formally obtained from the results for a statistical field in which the dispersion is constant by putting $P=4$; x_1, x_2, x_3 as the three space co-ordinates, and $x_4 = ct\sqrt{-1}$, where t is the time co-ordinate, c is the velocity of light, and $P_{44} = -1$.

The possibility of transformation to Galilean co-ordinates in physics is now seen to be a special case of a more general transformation from a set of correlated statistical variates to a set of independent statistical variates which is always possible when the dispersion is constant.

4. The expression for the statistical distance

$$P \,.\, \Delta^2 = \alpha_{\mu\nu} \,.\, (d\alpha)^\mu \,.\, (d\alpha)^\nu \qquad \ldots \qquad \ldots \qquad \ldots \qquad (2{\cdot}7)$$

is given in terms of the population parameters $\alpha_{\mu\nu}$ and $(d\alpha)^\mu$, and is therefore not subject to sampling fluctuations. In other words equation (2·7) is a functional and not a statistical equation.

In actual practice we are, however, obliged to work with values calculated from finite samples. It is therefore necessary to convert the functional equation (2·7) into a statistical equation by replacing the population parameters $\alpha_{\mu\nu}$ and $(\alpha)^\mu$, $(\alpha')^\mu$ by sample statistics $a_{\mu\nu}$ and $(a)^\mu$, $(a')^\mu$. Weakening the equation (2·7) by such substitution we thus get

$$P \,.\, D_1^2 = a_{\mu\nu} \,.\, (da)^\mu \,.\, (da)^\nu \qquad \ldots \qquad \ldots \qquad \ldots \qquad (4{\cdot}0)$$

where D_1^2 is the sample value of the Δ^2-statistic.

As the values of $a_{\mu\nu}$, or $(da)^\mu$ will in general fluctuate for samples drawn from the same two populations, the value of D_1^2 will also be subject to sampling fluctuations. We may consider the problem in two stages.

It is often possible to calculate the values of $a_{\mu\nu}$ by pooling together the variances and coefficients of correlation in a large number of samples on the assumption that the corresponding population values of $a_{\mu\nu}$ are identical. In such cases it is often possible to neglect the sampling fluctuations in $a_{\mu\nu}$ in comparison with the sampling fluctuations in the mean values $(a)^{\mu}$. We may then treat $a_{\mu\nu}$'s as constants and equal to $\alpha_{\mu\nu}$'s, and write :—

$$P \,.\, D_1^2 = \alpha_{\mu\nu} \,.\, (da)^{\mu} \,.\, (da)^{\nu} \quad \ldots \quad \ldots \quad \ldots \quad (4\cdot1)$$

in which $(da)^{\mu}$ or $(a)^{\mu}$ and $(a)^{\nu}$ are considered to be subject to sampling fluctuations.

I had shown * that the mathematical expectation of D_1^2 was

$$E(D_1^2) = \Delta^2 + \left(\frac{1}{n_1} + \frac{1}{n_2}\right) = \Delta^2 + \frac{2}{\bar{n}} \quad \ldots \quad \ldots \quad (4\cdot11)$$

where n_1, n_2 are the respective sizes of the two samples.

We may now define the sample value of the D^2-statistics in the following form :—

$$P \,.\, D^2 = P \,.\, D_1^2 - \frac{2 \,.\, P}{\bar{n}} \quad \ldots \quad \ldots \quad \ldots \quad (4\cdot12)$$

or

$$D^2 = D_1^2 - \frac{2}{\bar{n}} = \frac{1}{P} \,.\, [\alpha_{\mu\nu} \,.\, (da)^{\mu} \,.\, (da)^{\nu}] - \frac{2}{\bar{n}} \quad \ldots \quad \ldots \quad (4\cdot2)$$

with

$$E(D^2) = \Delta^2 \quad \ldots \quad \ldots \quad \ldots \quad \ldots \quad (4\cdot21)$$

that is the mathematical expectation of D^2 is the population value Δ^2.

The moment-coefficients of D^2 were also obtained by me † :—

$$\mu_2(D^2) = \frac{8}{P \,.\, \bar{n}} \left[\Delta^2 + \frac{1}{\bar{n}}\right] \quad \ldots \quad \ldots \quad \ldots \quad \ldots \quad (4\cdot22)$$

$$\mu_3(D^2) = \frac{32}{P \,.\, \bar{n}^2} \left[3\Delta^2 + \frac{1}{\bar{n}}\right] \quad \ldots \quad \ldots \quad \ldots \quad \ldots \quad (4\cdot23)$$

$$\mu_4(D^2) = \frac{192}{P \,.\, \bar{n}^2} \left[\left(\Delta^2 + \frac{2}{\bar{n}}\right)^2 + \frac{4}{P \,.\, \bar{n}} \left(2\Delta^2 + \frac{1}{\bar{n}}\right)\right] \quad \ldots \quad (4\cdot24)$$

---

* *Journ. Asiat. Soc. Bengal*, vol. XXVI, p. 557, (1930).

† *Journ. Asiat. Soc. Bengal*, vol. XXVI, p. 559, (1930).

The exact distribution of D_1 was found some time ago by Raj Chandra Bose,* and can be written in the following form :—

$$\text{Let} \quad \lambda^2=\tfrac{1}{2}\bar{n} \,.\, P \,.\, \Delta^2, \quad L^2=\tfrac{1}{2}\bar{n} \,.\, P \,.\, D_1^2 \quad \ldots \quad \ldots \quad \ldots \quad (4{\cdot}31)$$

Then the distribution of L is given by

$$f(L) \,.\, dL=\left(\frac{L}{\lambda}\right)^{\frac{P}{2}-1} \,.\, L \,.\, e^{\frac{1}{2}(L^2+\lambda^2)} \,.\, I_{\frac{P}{2}-1}(L\lambda) \,.\, dL \quad \ldots \quad (4{\cdot}32)$$

where I_z is the Bessel function of pure imaginary argument. It was also shown that the moment-coefficients previously given by me [equations (4·21) –(4·24)] were exact and remained valid for correlated variates. Five per cent. and one per cent. values of L for various values of λ have also been calculated by Raj Chandra Bose and Samarendra Nath Ray in conjunction with me and will be shortly published.†

Going a stage further we may consider both the dispersion $(a_{\mu\nu})$ as well as the mean values $(a)^\mu$ to be subject to sampling fluctuations, and relax the functional equation (2·7) completely into the statistical equation :—

$$P \,.\, D_2^2=a_{\mu\nu} \,.\, (da)^\mu \,.\, (da)^\nu \quad \ldots \quad \ldots \quad \ldots \quad (4{\cdot}4)$$

The exact distribution of D_2^2 is not known, but Raj Chandra Bose has recently succeeded in obtaining the exact moment-coefficients (for the case of P independent variates) which are quoted in an appendix.

5. We have been considering so far the case of a statistical field of populations for which the dispersions ($\alpha_{\mu\nu}$'s) are identical and independent of the mean values $(\alpha)^\mu$'s. We may now relax this condition. That is we shall now assume that $\alpha_{\mu\nu}$'s are functions of $(\alpha)^\mu$'s. For our purpose it is, however, not necessary that $\alpha_{\mu\nu}$ should be a mathematical function of $(\alpha)^\mu$. It is clearly sufficient that $\alpha_{\mu\nu}$ and $(\alpha)^\mu$, that is the dispersions and mean values should be statistically connected. This point is one of considerable importance. We assume that given $(\alpha)^\mu$, that is the mean values, we have knowledge of the distribution of $\alpha_{\mu\nu}$'s. This we may call a generalized statistical field in which the values of $\alpha_{\mu\nu}$'s will vary from point to point in the field. In such a field we may still continue to use

$$P \,.\, D_1^2=a_{\mu\nu} \,.\, (da)^\mu \,.\, (da)^\nu \quad \ldots \quad \ldots \quad \ldots \quad (5{\cdot}0)$$

as the expression for the line element. Here as also in the case where $\alpha_{\mu\nu}$

* *Science and Culture*, vol. I, p. 205, (1935).

† *Science and Culture*, vol. 1, (December, 1935). *Proc. Ind. Sc. Congress*, 1936.

are functionally connected with $(\alpha^{\mu}, \alpha^{\nu})$ the expression will be used only as a differential element the integral of which will depend on the path of integration. Equation (5·0) may be therefore considered to be a perfectly general expression* for the distance between two normal statistical populations. Before it can be used in practice it is, however, necessary to determine its exact distribution (or at least its moment-coefficients) which is now under investigation.

Appendix.

The moment-coefficients of the expression defined in (4·4) in the text are given below.

Let

$$D_i^2 = \frac{(a_i - a_i')^2}{(n_1\, a_{ii} + n_2\, a_{ii}')/(n_1+n_2-1)}, \quad \Delta_i^2 = \frac{(\alpha_i - \alpha_i')^2}{\alpha_{ii}} \quad .. \quad (6\cdot0)$$

$$D_0^2 = \frac{1}{P} \sum_{i=1}^{i=P} (D_i^2), \qquad \Delta^2 = \frac{1}{P} \sum_{i=1}^{i=P} (\Delta_i^2) \quad .. \quad (6\cdot1)$$

the summation extending over all the variates. It is also convenient to put

$$n = n_1 + n_2 - 1 \quad .. \quad .. \quad .. \quad (6\cdot2)$$

We then have

$$\mu_1(D_0^2) = \frac{n}{n-3}\left(\Delta^2 + \frac{2}{n}\right) \quad .. \quad .. \quad .. \quad (6\cdot3)$$

To compensate for the bias introduced by the finite size of the sample we now introduce a new statistics defined by

$$D^2 = \frac{n-3}{n} \cdot D_0^2 - \frac{2}{n} \quad .. \quad .. \quad .. \quad .. \quad (6\cdot4)$$

We then have the mathematical expectation of D^2

$$E(D^2) = \mu_1\,(D^2) = \Delta^2 \quad .. \quad .. \quad .. \quad (6\cdot5)$$

* I may just mention here that there is another alternative method of approaching the problem. A P-variate normal population can be represented by a density cluster in P-dimensional space, and can always be transformed to a set of independent variates. Consider two such clusters or populations. They may differ in their mean values or the position of the clusters. They may also differ in the size of the equi-frequential surfaces. Finally they may differ in the set of independent variates in terms of which they can be specified. In general two such non-identical normal populations can be completely superposed by a translation (equalizing the difference in mean values), a squeeze (equalizing differences in variances), and a rotation (equalizing differences in coefficients of correlation). Various quantitative forms are possible, some of which are under investigation at present.

The moment-coefficients of D^2 are given by the formulæ :—

$$\mu_2(D^2)=\frac{2}{n-5}\left[\overset{i=P}{\underset{i=1}{S}}\left(\frac{\Delta_i^4}{P^2}\right)+\frac{4(n-2)}{\bar{n}\,.\,P}\left\{\Delta^2+\frac{2}{\bar{n}}\right\}\right] \quad \ldots \quad (6\cdot6)$$

$$\mu_3(D^2)=\frac{16}{(n-5)(n-7)}\left[\overset{i=P}{\underset{i=1}{S}}\left(\frac{\Delta_i^6}{P^3}\right)+\frac{6(n-2)}{\bar{n}\,.\,P}\overset{i=P}{\underset{i=1}{S}}\left(\frac{\Delta_i^4}{P^2}\right)\right.$$
$$\left.+\frac{2(n-1)(n-2)}{\bar{n}^2\,.\,P^2}\left\{3\Delta^2+\frac{2}{\bar{n}}\right\}\right] \quad \ldots \quad (6\cdot7)$$

$$\mu_4(D^2)=\frac{12(n+9)}{(n-5)(n-7)(n-9)}\left[\overset{i=P}{\underset{i=1}{S}}\left(\frac{\Delta_i^8}{P^4}\right)\right.$$
$$+\frac{8(n-2)}{\bar{n}\,.\,P}\overset{i=P}{\underset{i=1}{S}}\left(\frac{\Delta_i^6}{P^3}\right)+\frac{12}{(n-5)^2}\left[\overset{i=P}{\underset{i=1}{S}}\left(\frac{2\Delta_i^4\Delta_j^4}{P^4}\right)\right.$$
$$\left.+\frac{8(n-2)}{\bar{n}\,.\,P}\overset{i=P}{\underset{i=1}{S}}\left(\frac{\Delta_i^4\Delta_j^2}{P^3}\right)+\frac{16(n-2)^2}{\bar{n}^2\,.\,P^2}\left(\Delta^2+\frac{2}{\bar{n}}\right)\right]$$
$$+\frac{384(n-2)}{(n-5)^2(n-7)(n-9)}\left[\frac{10n^2-69n+81}{\bar{n}^2\,.\,P^2}\overset{i=P}{\underset{i=1}{S}}\left(\frac{\Delta_i^4}{P^2}\right)\right.$$
$$\left.+\frac{2(n^3-6n^2+2n+9)}{\bar{n}^3\,.\,P^3}\left(2\Delta^2+\frac{1}{\bar{n}}\right)\right] \quad \ldots \quad (6\cdot8)$$

Statistical Laboratory,
Presidency College, Calcutta.

10

Reprinted from *J. Roy. Stat. Soc.* **B10**(2), 159–203 (1948)

THE UTILIZATION OF MULTIPLE MEASUREMENTS IN PROBLEMS OF BIOLOGICAL CLASSIFICATION*

By C. RADHAKRISHNA RAO

From the Duckworth Laboratory, University Museum of Archaeology and Ethnology, Cambridge

[Read before the RESEARCH SECTION of the ROYAL STATISTICAL SOCIETY, April 6, 1948, Dr. J. O. IRWIN in the Chair]

CONTENTS

INTRODUCTION

1. *Problems of Classification*

THE purpose of this paper is to present a statistical approach to two types of problems confronted in biological research. The first is that of specifying an individual as a member of one of many groups to which he can possibly belong, as when a taxonomist has to assign an organism to its proper species or subspecies, or an anthropologist is faced with the problem of sexing a skull or a jaw-bone. The second is the problem of classification of the groups themselves into some significant system based on the configuration of the various characteristics. The need of this is felt in the study of systematics and the evolution of species. A number of species or subspecies may have to be arranged in a hierarchical order showing the closeness of some and the distinctness of others. Such a representation superimposed on a geographical classification may, it is suggested, be of use in tracing the evolution of various species or subspecies.

The statistical methods developed in this article have been applied to determine the group

* I wish to express my sincere thanks to R. A. Fisher, P. C. Mahalanobis and J. C. Trevor for their valuable advice and suggestions.

constellations of 22 inbreeding Indian castes and tribes living in a compact geographical region. The classification so arrived at, while broadly agreeing with the existing social differences, has brought about new results which are of anthropological interest.

2. *Tests of Significance in Multivariate Analysis*

Tests of significance are fundamental to the interpretation of observed differences. Various contributions have appeared on this subject, the most important of which are by Bartlett (1934, 1938, 1947), Bose and Roy (1938), Fisher (1936*a*, 1938, 1939, 1940), Hotelling (1931), Hsu (1939*a*, 1939*b*), Mahalanobis (1936), Roy (1939), Wilks (1932) and Wishart (1928). Some of the tests of significance which arose out of the present work have been detailed in a separate paper due to appear in a forthcoming issue of *Biometrika*. Tests have been proposed for judging (*a*) the significance of differences in mean values of p correlated variables, (*b*) the asymmetry of organisms, (*c*) the adequacy of a selected panel of characters in discriminating a number of groups, and (*d*) the adequacy of certain linear combinations of characters to measure secular variations in the dimensions of organisms.

The analysis developed in this article is concerned with the interpretation of real differences among various groups as revealed by relevant preliminary tests of significance.

PART I: STATISTICAL CRITERIA TO DETERMINE THE GROUP TO WHICH AN INDIVIDUAL BELONGS

3. *The Problem of Two Groups and Discriminant Functions*

Let there be only two alternative groups with distributions specified by $f_1(x, \theta_1)$ and $f_2(x, \theta_2)$, where x stands for the available set of measurements on an individual and θ's for the population parameters. An individual belonging to one of the two groups has been observed, and the problem is to determine his group on the basis of the available set of measurements on him.

If there are p measurements $x_1, \ldots, x_p$, then the individual can be represented by an ordered set of elements $(x_1 \ldots x_p)$ or a point in a space of p dimensions. The problem requires the division of the space into two mutually exclusive regions R_1, R_2 with the rule of procedure of assigning an individual to the ith group if he falls in R_i.

If the individual is drawn at random from a population in which the two groups are mixed up in the ratio $\pi_1 : \pi_2 \, (\pi_1 + \pi_2 = 1)$, the chances of his being misclassified, by the above procedure, is

$$\alpha = \pi_2 \int_{R_1} f_2 \, dv + \pi_1 \int_{R_2} f_1 \, dv.$$

where dv is the volume element.

The frequency of misclassifications will be the lowest if the regions R_1 and R_2 are chosen such that α is minimum. It has been shown (Welch, 1939) that the best possible regions are defined by

$$R_1 \cap \pi_1 f_1 \geqslant \pi_2 f_2$$

$$R_2 \cap \pi_2 f_2 \geqslant \pi_1 f_1,$$

where the symbol $\cap$ stands for "defined by."

If π_1 and π_2 are not known, one may set up the logical requirement that the frequency of misclassifications should be the same for individuals of both the groups and that this should be the minimum possible. If R_1, R_2 are such regions then

$$\int_{R_2} f_1 \, dv = \int_{R_1} f_2 \, dv = \alpha,$$

in which case

$$\int_{R_1} f_1 \, dv = \int_{R_2} f_2 \, dv = 1 - \alpha,$$

which leads to the condition

$$\int_{R_2} (f_1 + f_2) \, dv = 1.$$

The region R_2 is, then, to be chosen such that

$$\int_{R_2} f_1 \, dv \text{ is a minimum}$$

subject to the above condition.

It is shown by Welch (1939) that the best regions are

$$R_2 \cap f_1 \leqslant a(f_1 + f_2) \text{ that is } f_2 \geqslant bf_1$$

$$R_1 \cap f_2 \leqslant bf_1 \text{ where } b \text{ is chosen such that}$$

$$\int_{R_2} f_1 \, dv = \int_{R_1} f_2 \, dv$$

If α is small then one can assert with some degree of confidence that an individual has been assigned to the right group. On the other hand it is possible to construct a region C_1 such that

$$\int_{C_1} f_1 \, dv \text{ is a maximum}$$

subject to the condition

$$\int_{C_1} f_2 \, dv = \alpha_1 \text{ (a small assigned quantity).}$$

The region C_1 is given by $f_1 \geqslant k f_2$ where k is properly chosen. An individual falling in C_1 can be taken to belong to the first group with some degree of confidence. An individual falling in $R_1 - C_1$ may be taken as likely to belong to the first group. Similarly the regions C_2 and $R_2 - C_2$ can be constructed for the individuals of the second group.

One thus has a division of the space into three mutually exclusive regions C_1, C_2 and the rest so that a decision is made only when the individual falls in C_1 or C_2 (Rao, 1947). This may be regarded as a footnote to Neyman and Pearson theory of testing of hypotheses where the problem requires the division of the space into two regions, one for rejecting a hypothesis and the other for keeping an open mind about it.

It is seen that the regions in all the cases discussed above are bounded by constant values of the likelihood ratio. If the two distributions are p variate normal populations with a common dispersion matrix, the surfaces of constant likelihood ratio are defined by

$$\sum_{i=1}^{p} (\alpha^{1i} d_1 + \alpha^{2i} d_2 + \ldots + \alpha^{pi} d_p) x_i = \text{constant},$$

where d_i is the difference in mean values of the ith character and (α^{ij}) is the matrix inverse to the dispersion matrix (α_{ij}). This is identical with the discriminant function introduced by Fisher. If the dispersion matrices are different, then these surfaces are defined by quadratic forms in the variables (Smith, 1947).

An illustrative example; *sources of the material*.—In August, 1939, a relatively complete male human skeleton was recovered from the ditch of an Iron Age camp on Highdown Hill, Goring by Sea, in the course of the excavations conducted under the auspices of the Worthing Archaeological Society. Fragments associated with the bones suggest that the burial could not have taken place later than the very beginning of Iron Age in Sussex, about 500 B.C. The camp went out of use not later than 250 B.C. and the remains themselves can be assigned to a 500 B.C. "invasion" horizon. It is doubtful, however, whether their owner was a Bronze Age "defender" or an Iron Age "invader." The principal question to be considered, in the present context, is whether the Highdown skull is more likely to have belonged to a Bronze Age or to an Iron Age population.

An attempt has been made here to answer this problem by utilizing the published data concerning Bronze Age and Iron Age + Romano-British (+ indicating a mixed series) crania. The characteristics of these groups have been computed from scanty material, so that the conclusion arrived at regarding the Highdown skull cannot be treated as final. This example has been chosen merely to illustrate the method.

Table 1 gives the mean values of male English Bronze Age (Morant, 1926) and Iron Age + Romano-British (Goodwin and Morant, 1940) cranial measurements and those on the Highdown skull for 7 characters, viz. maximum length (L), maximum breadth (B), minimum frontal breadth (B′), total sagittal arc (S), transverse biporial arc through bregma (βQ′), basi-bregmatic height (H′) and upper face height (G′H).

TABLE 1.—*Comparison of Highdown Skull and other Male English Skull Measurements*

Character	*Farringdon Street standard deviation*	*Mean values*		*Highdown skull*
		Bronze Age	*Iron Age + Romano-British*	
L	6·46	184·5	188·6	198·2
B	5·90	149·9	140·8	148·1
B′	4·58	99·7	97·1	101·2
S	14·20	376·6	383·5	407·2
βQ′	13·35*	320·5	316·7	333·9
H′	5·06	134·9	137·1	142·0
G′H	4·45	69·1	72·4	72·4

* This is for the transverse biporial arc measured through the apex and not the bregma, but the standard deviations for the two characters are likely to be very close.

Using the correlation matrix obtained from a long series of 17th century male English crania from Farringdon Street, London (Hooke, 1926), the following transformed characters (explained in section 7 below) are calculated. The lower-case letters indicate the measurements expressed in standard deviation units.

TABLE 2.—*Transformed Characters on the English Skull*

Transformed character	*Additional character used*	*Function of original characters*
α	L	l
β	B	$1{\cdot}0536\,b - {\cdot}3318\,l$
γ	B′	$1{\cdot}1802\,b' - {\cdot}5034\,b - {\cdot}2474\,l$
δ	S	$1{\cdot}6536\,s - {\cdot}1649\,b' - {\cdot}3262\,b - 1{\cdot}0948\,l$
ε	βQ′	$1{\cdot}5836\,\beta q' - {\cdot}4650\,s - {\cdot}1868\,b' - {\cdot}8903\,b + {\cdot}1177\,l$
ξ	H′	$2{\cdot}0840\,h' - 1{\cdot}8061\,\beta q' - 1{\cdot}5868\,s + {\cdot}4160\,b' + {\cdot}7843\,b + 1{\cdot}3341\,l$
η	G′H	$1{\cdot}0851\,g'h - {\cdot}1485\,h' + {\cdot}1549\,\beta q' - {\cdot}0511\,s - {\cdot}1238\,b' - {\cdot}3942\,b + {\cdot}0548\,l$

The affinities of the Highdown skull with English Bronze Age and Iron Age + Romano-British crania.—For the use of the discriminant function it must be known that an individual to be classified belongs to one or other of the two groups. This knowledge is, very often, available from external evidence. Association of grave findings with certain groups is one such piece of evidence in the case of skeletal finds. In some problems, as in the case of two sexes, there are only two possible groups. But in cases where the external evidence is meagre, the classification of an individual as a member of one of two groups may be subject to another kind of error, viz. the wrong assumption that he belongs to one or other of the groups when, in fact, he comes from a third unknown group. In the absence of definite knowledge, it may be necessary to examine by internal evidence supplied by the measurements whether an individual could be considered as belonging to either of the two groups. This leads to the evaluation of the following statistics and the probabilities associated with them.

What are the chances of the Highdown skull's belonging to the Bronze Age group?

	Transformed character		*Difference*
	Bronze Age mean	*Highdown skull*	
α .	28·5604	30·6811	. $-2\cdot1207$
β .	17·2929	16·2678	. 1·0251
γ .	5·8380	5·8537	. $-\cdot0157$
δ .	·7082	1·9954	. $-1\cdot2872$
ε .	2·3603	3·4884	. $-1\cdot1281$
ξ .	37·2023	37·6203	. ·4180
η .	4·1044	4·9418	. ·8374

$$\chi^2 = \Sigma\, d^2 = 9\cdot3539$$
Degrees of freedom, 7
$$P\,(\chi^2 \geqslant 9\cdot3539) > \cdot20$$

The chances are greater than 20 per cent.,* which is high enough not to invalidate his claims to belong to the Bronze Age group. For the Iron Age + Romano-British group the χ^2 is 3·7051, the chance of exceeding which on 7 degrees of freedom is over 78 per cent. The characteristics of the Highdown skull would thus be fairly common in the Iron Age + Romano-British group. While the chances in favour of his belonging to this group are great, the evidence is not sufficient to arrive at a definite conclusion, since the chances of his belonging to the Bronze Age group are not small. The discriminant function acts as a powerful tool in this situation. The differences between the Bronze Age and the Iron Age + Romano-British means of the transformed characters are the coefficients of the discriminant function. This is obtained as

$$\cdot6346\,\alpha - 1\cdot8354\,\beta - \cdot0505\,\gamma + \cdot7054\,\delta + \cdot8771\,\varepsilon + \cdot0502\,\xi + 1\cdot3841\,\eta$$

with its variance considered as a linear function of the variables as

$$(\cdot6346)^2 + (1\cdot8354)^2 + \ldots + (1\cdot3841)^2 = 6\cdot959106.$$

The values of the discriminant function for the Bronze Age and the Iron Age + Romano-British groups and the Highdown skull are, respectively, $-3\cdot7915$, 3·1676 and 2·5124. The chances of the Highdown skull's belonging to the Bronze Age and the Iron Age + Romano-British groups correspond to the areas beyond one side of the normal deviates

$$\frac{3\cdot7915 + 2\cdot5124}{\sqrt{6\cdot959106}} \text{ and } \frac{-3\cdot1676 + 2\cdot5124}{\sqrt{6\cdot959106}}$$
$$= 2\cdot389 \qquad\qquad = -\cdot247$$

The probabilities† are ·0084 and ·5987 which shows that the Highdown skull is less likely to have belonged to the Bronze Age group, thus establishing its claims to resemble the Iron Age + Romano-British one.

4. *Resolution of a Mixed Series into Two Gaussian Components*

In the last section we have considered the problem of determining the group to which an individual belongs when the distributions in the alternative groups are known. There may arise cases where a collection of individuals is observed but no information is available as to the distributions of the groups from which they have arisen and the proportion of mixture. The general

* In evaluating the probability of χ^2 it has been assumed that the means and the dispersion matrix for the Bronze Age group have been obtained from a large sample. On the other hand, if n represents the number of degrees of freedom for the variances and covariances and the mean values are based on a sample of size N, then the statistic $N\,(n + 1 - p)\,\chi^2 \,/\, np\,(N + 1)$ can be used as the variance ratio with p and $(n + 1 - p)$ degrees of freedom. Here $p = 7$.

† No method of evaluating these probabilities is available when the sample size and the degrees of freedom of the dispersion matrix are taken into account.

problem, in such cases, is to determine the characteristics of the various groups and also the proportion of mixture from the available set of measurements. This information may be finally used to specify the group of each individual if necessary.

Considering only two groups in which a certain character is distributed normally, the statistical problem reduces to that of estimating the two mean values μ_1, μ_2, standard deviations σ_1, σ_2 and the proportion of mixture ρ from the observed frequency distribution. The estimation of these five parameters by the method of moments was discussed by Pearson (1894). The estimates depend on a suitably chosen root of a nonic (ninth degree equation) constructed from the first five moments of the observed frequency distribution.

In many problems it is reasonable to suppose that $\sigma_1 = \sigma_2$, in which case there are only four parameters to be estimated. If the method of moments is followed only the first four moments are sufficient. In this case it has been shown that the estimates depend on the negative root of a simple cubic equation constructed from the first four moments. In practice, where large numbers are involved the estimates obtained by the method of moments, though not efficient, may serve the purpose at hand. In cases where higher efficiency is aimed at, the estimates have to be found by the method of maximum likelihood. The numerical computation involved in this method is set out in this section.

Whatever may be the method of estimation employed the numerical computation becomes much simpler when the standard deviations are assumed to be equal. The simplifying assumption may introduce bias in the estimates when, in fact, the standard deviations differ. Such estimates are, however, more accurate than those obtained without this assumption when the bias in any estimate is smaller than its standard error. If the mean values and the proportion of mixture are to be estimated with a higher precision small differences in the standard deviations can be ignored.

Estimation by the method of moments.—The method of estimating the parameters by the method of moments consists in equating the moments as calculated from the observations with functions of parameters representing the moments in the population. Since the expectations of calculated moments are not the same as the moments in the population this method might introduce a little bias in the estimating equations. This can be avoided by equating the calculated moments to their expected values. Instead of this, one can choose the system of k-statistics of Fisher (defined in *Statistical Methods for Research Workers*) and equate them to their expectations which are the cumulants of the distribution. If s_2, s_3 and s_4 are the second, third and fourth moments about the mean and s_1 the first moment about the origin as calculated by the usual method, the first four k-statistics derivable from them are given by

$$k_1 = s_1$$

$$k_2 = \frac{n}{(n-1)} s_2$$

$$k_3 = \frac{n^2}{(n-1)(n-2)} s_3$$

$$k_4 = \frac{n^2}{(n-1)(n-2)(n-3)} \left((n+1) s_4 - 3(n-1) s_2^2 \right).$$

If the moments are calculated from grouped data with class interval h the quantities $\frac{1}{12} h^2$ and $\frac{1}{120} h^4$ have to be subtracted from the expressions for k_2 and k_4 respectively.

If p, m_1, m_2 and s denote the estimates of ρ, μ_1, μ_2 and σ, the common standard deviation, then the estimating equations by this method are

$$1 = p + q$$

$$k_1 = pm_1 + qm_2$$

$$k_2 = s^2 + pd_1^2 + qd_2^2$$

$$k_3 = pd_1^3 + qd_2^3$$

$$k_4 = pd_1^4 + qd_2^4 - 3(pd_1^2 + qd_2^2)^2,$$

where $q = 1 - p$, $d_1 = m_1 - k$, $d_2 = m_2 - k_1$. Defining $x = d_1 d_2$, the equation giving x is obtained as the negative root of the cubic

$$x^3 + \frac{1}{2} k_4 x + \frac{1}{2} k_3^2 = 0.$$

If x is the required root, then d_1 is given by the negative root of the quadratic

$$d_1^2 + \frac{k_3}{x} d_1 + x = 0$$

and d_2 by $-(k_3/x) - d_1$. The estimates m_1, m_2, p and s are given by

$$m_1 = k_1 + d_1, \quad m_2 = k_1 + d_2,$$

$$p = d_2/(d_2 - d_1), \quad s^2 = k_2 + x.$$

Method of solving the cubic equation.—The fundamental cubic

$$x^3 + \frac{1}{2} k_4 x + \frac{1}{2} k_3^2 = 0$$

introduced above has always a single negative root greater than $(-k_2)$. The best method of determining the root is to start with a trial value and determine the correction by Newton's method of approximation. Since the equation is in a reduced form with the coefficient of x^2 being absent it is easy to guess the root correct to the nearest integer. If x_1 stands for the trial value then the additive correction δx_1 is given by

$$\left[3x_1^2 + \frac{k_4}{2}\right] \delta x_1 = -x_1^3 - \frac{1}{2} k_4 x_1 - \frac{1}{2} k_3^2.$$

The process is to be repeated until the expression on the right-hand side becomes very small. The data of Table 3 give the frequency distribution of heights of 454 plants of two different types grown on the same plot. The plants are indistinguishable except at the flowering stage. The problem is to estimate the mean heights of the two types of plants, their common standard deviation and the proportion of mixture.

TABLE 3.—*The Frequency Distribution of Heights in cm. for* 454 *Plants*

Class Interval	*Frequency*
7·5– 8·5	3
9·5	9
10·5	21
11·5	40
12·5	59
13·5	76
14·5	79
15·5	69
16·5	46
17·5	30
18·5	13
19·5	7
20·5	2
	454

The values of cumulants after adjustment for grouping are

$$k_1 = - \cdot 244493 \text{ (about 14 as the origin)}$$

$$k_2 = 4\cdot975963, \quad k_3 = \cdot728751, \quad k_4 = -5\cdot314741$$

$$\tfrac{1}{2} k_4 = -2\cdot657370, \quad \tfrac{1}{2} k_3^2 = \cdot265539.$$

The fundamental cubic is

$$x^3 - 2\cdot657370\, x + \cdot265539 = 0.$$

Taking $-1\cdot65$ as a trial root we find the correction δx is given by

$$[3\,(1\cdot65)^2 - 2\cdot657370]\, \delta x = -(-1\cdot65)^3 - 2\cdot657370\,(-1\cdot65) - \cdot265539$$
$$5\cdot510130\, \delta x = -\cdot158074, \quad \delta x = -\cdot0286878.$$

Similarly the second correction comes out as $+\cdot000707$ so that the second approximation is $-1\cdot678688 + \cdot000707 = -1\cdot677981$.

The quadratic giving d_1 is

$$d_1^2 - 0\cdot434302\, d_1 - 1\cdot677981 = 0,$$

which yields the negative root

$$d_1 = -1\cdot096293$$

$$d_2 = 1\cdot09629 + \cdot434302 = 1\cdot530595.$$

The estimates of m_1 and m_2 about 14 as the origin are

$$m_1 = -1\cdot096293 - \cdot244493 = -1\cdot340786$$

$$m_2 = 1\cdot530595 - \cdot244493 = 1\cdot286102$$

$$p = + d_2 / (d_2 - d_1) = \cdot582665$$

$$q = (1 - p) = \cdot417335$$

$$s^2 = 4\cdot975963 - 1\cdot677981 = 3\cdot297982$$

$$s = 1\cdot816035.$$

This completes the estimation of the four parameters by the method of moments.

Estimation by the method of maximum likelihood.—It is now established, under very general conditions, that the method of maximum likelihood which was originally introduced as a "primitive postulate" gives rise to best possible estimates in large samples. But the equations leading to these estimates are, usually, non linear, and considerable difficulty is experienced in solving these equations in practice. A certain amount of mechanization has been introduced by adopting a method of computation known as the scoring system, which is briefly recapitulated here. Further aspects of the scoring system are dealt with by Fisher (1946) and Rao (1948*b*).

If $f_1, \ldots, f_k$ are the observed frequencies with a total f in k classes with probabilities $P_1, \ldots P_k$ defined as functions of some unknown parameters $\theta_1, \ldots \theta_r$ then the score S_i of the data for θ_i is defined as

$$S_i = \sum_{s=1}^{k} \frac{f_s}{P_s} \frac{\partial P_s}{\partial \theta_i}.$$

The information matrix is denoted by (I_{ij}) where

$$I_{ij} = \sum_{s=1}^{k} \frac{1}{P_s} \frac{\partial P_s}{\partial \theta_i} \frac{\partial P_s}{\partial \theta_j}.$$

The maximum likelihood estimates are the values of $\theta_1, \ldots, \theta_r$ which make the scores $S_1, \ldots, S_r$ vanish simultaneously. In case the solutions are not directly calculable, they can be obtained by following the method of successive approximations as shown below.

If $\theta^o_1, \ldots, \theta^o_r$ denote the approximate values and the values of the scores and the elements of the information matrix calculated at these values of the parameters are represented with the index (o), then additive corrections $d\theta_1, \ldots, d\theta_r$ to the first approximations are obtained as solutions of the linear equations

$$I^o_{11}\, d\theta_1 + I^o_{12}\, d\theta_2 + \ldots + I_{1r}^o\, d\theta_r = S^o_1$$

$$I^o_{21}\, d\theta_1, + I^o_{22}\, d\theta_2 + \ldots + I_{2r}^o\, d\theta_r = S^o_2$$

$$\ldots \qquad \ldots \qquad \ldots$$

$$I^o_{r1}\, d\theta_1 + I^o_{r2}\, d\theta_2 + \ldots + I^o_{rr}\, d\theta_r = S^o_r.$$

This process can be repeated until the values on the right-hand side become negligible.

The method described above can be applied to the present problem for estimating ρ, μ_1, μ_2 and σ. The probability P_i of the ith class bounded by the values a_i and a_{i+1} is given by

$$P_i = 1/\sigma\sqrt{2\pi} \int_{a_i}^{a_{i+1}} \left[\rho \exp\left\{ -\tfrac{1}{2}\left(\frac{x-\mu_1}{\sigma}\right)^2\right\} + (1-\rho) \exp\left\{ -\tfrac{1}{2}\left(\frac{x-\mu_2}{\sigma}\right)^2\right\}\right] dx$$

$$\frac{\partial P_i}{\partial \rho} = A_1 - A_2,$$

where A_1 and A_2 are probabilities (areas under the normal curve) corresponding to the two groups for the ith class.

$$\frac{\partial P_i}{\partial \mu_1} = \frac{\rho}{\sigma}\{z(a_i) - z(a_{i+1})\}.$$

where $z(a_i)$ is the ordinate of the normal curve at a_i.

$$\frac{\partial P_i}{\partial \mu_2} = \frac{(1-\rho)}{\sigma}\{z'(a_i) - z'(a_{i+1})\}$$

where z' corresponds to the second group.

$$\frac{\partial P_i}{\partial \sigma} = \frac{\rho}{\sigma}[b_i\, z(a_i) - b_{i+1}\, z(a_{i+1})]$$

$$+ \frac{(1-\rho)}{\sigma}[b'_i\, z'(a_i) - b'_{i+}\, z'_1(a_{i+1})],$$

where

$$b_i = (a_i - \mu_1)\,/\,\sigma, \quad b'_i = (a_i - \mu_2)\,/\,\sigma.$$

For any given set of values of ρ, μ_1, μ_2 and σ the values of the derivatives can be obtained by the use of the ordinates and normal probabilities given in *Tables for Biometricians*, Part I, first edition, pages 2–10.

The values of the derivatives and scores at the values of the parameters found by the method of moments are set out in Table 4. The information matrix and the additive corrections to these estimates can be calculated with the help of this table.

TABLE 4.—*Scores and Information at the Estimates Obtained by the Method of Moments*

Probability for each Class (P)	Derivatives and Scores							
	$\frac{\partial P}{\partial \rho}$	$\frac{1}{P}\frac{\partial P}{\partial \rho}$	$\frac{\sigma}{\rho}\frac{\partial P}{\partial \mu_1}$	$\frac{\sigma}{\rho P}\frac{\partial P}{\partial \mu_1}$	$\frac{\sigma}{(1-\rho)}\frac{\partial P}{\partial \mu_2}$	$\frac{\sigma}{(1-\rho)P}\frac{\partial P}{\partial \mu_2}$	$\sigma\frac{\partial P}{\partial \sigma}$	$\frac{\sigma}{P}\frac{\partial P}{\partial \sigma}$
·006457	·010919	1·691032	—·028985	—4·488926	—·000366	—·056683	—·039267	—6·081307
·017701	·029300	1·655274	—·058811	—3·322468	—·002095	—·118355	—·052557	—2·969154
·045846	·072536	1·582166	—·108724	—2·371504	—·010097	—·220237	—·058287	—1·271365
·090033	·129569	1·439127	—·128542	—1·427720	—·033303	—·369898	—·010954	—·121666
·136833	·158811	1·160619	—·072267	—·528140	—·077902	—·569322	·061256	·447670
·166159	·112563	·677441	·038439	·231339	—·123047	—·740538	—·097298	—·585572
·167988	—·003545	—·021103	·119339	·710402	—·116904	—·695907	—·009165	—·054557
·144126	—·113329	—·786319	·121394	·842277	—·032367	—·224574	·079487	·551510
·109848	—·162826	—1·482285	·075090	·683581	·075216	·684574	—·015839	—·144190
·066345	—·126674	—1·909322	·031467	·474293	·131321	1·979365	·028207	·425156
·032277	—·069626	—2·157140	·009291	·287852	·106250	3·291817	·041672	1·291074
·011980	—·027340	—2·282137	·001969	·164357	·056245	4·694908	·045831	3·825626
·004404	—·010360	—2·352406	·000340	·077202	·027048	6·141689	·026913	6·111035

Instead of the actual information matrix one can calculate the sum of products from the derivatives and scores as listed in Table 4 and apply the corrections by the use of the multipliers σ / ρ, $\sigma / (1 - \rho)$, σ given at the column headings, at a later stage. This is obtained as

1·438129	−·855899	−·997251	−·606683
−·855899	1·070235	·138080	·534488
−·997251	·138080	1·330282	·613667
−·606683	·534488	·613667	1·006893

The reciprocal of this matrix, obtained by the C matrix method of Fisher, is

11·758460	10·304810	9·738131	−4·320342
10·304810	10·364783	8·796048	−4·653862
9·738131	8·796048	9·162066	−4·385647
−4·320342	−4·653862	−4·385647	3·533322

The total scores, obtained by taking the sum of products of individual scores and frequencies in the various class intervals, are

$$\cdot 612656, \quad 1\cdot 101496, \quad -2\cdot 179493, \quad 1\cdot 028987$$

The corrections to trial values are given by the sum of products of the scores and the elements in the rows of the reciprocal matrix divided by n, the total frequency.

Thus

$$\begin{aligned}\delta p &= [11\cdot 758460\,(\cdot 612656) + 10\cdot 304810\,(1\cdot 101496) \\ &\quad + 9\cdot 738131\,(-\,2\cdot 179493) - 4\cdot 320342\,(1\cdot 028987)] / 454 \\ &= -\,7\cdot 115166 / 454 = -\,\cdot 015672\end{aligned}$$

$$\frac{s}{p}\,\delta m_1 = -\frac{6\cdot 229605}{454} = -\,\cdot 01372$$

$$\delta m_1 = (-\,\cdot 013721)\,(\cdot 320844) = -\,\cdot 004402$$

$$\delta m_2 = (-\,\cdot 019441)\,(\cdot 229805) = -\,\cdot 004467$$

$$\delta s = (\cdot 011941)\,(\cdot 550650) = \cdot 006575$$

The first approximations to maximum likelihood estimates and their standard errors are

	Estimate	*Standard error*	
$p = p + \delta p =$	·566993,	$\sqrt{\frac{11\cdot 758460}{454}}$	$= \cdot 1609$
$m_1 = m_1 + \delta m_1 =$	−1·345188,	$\sqrt{\frac{10\cdot 364783}{454}}\left(\frac{p}{s}\right)$	$= \cdot 0485$
$m_2 = m_2 + \delta m_2 =$	1·281635,	$\sqrt{\frac{9\cdot 162066}{454}}\left(\frac{1-p}{s}\right)$	$= \cdot 0326$
$s = s + \delta s =$	1·822610,	$\sqrt{\frac{3\cdot 533322}{454}}\left(\frac{1}{s}\right)$	$= \cdot 0486.$

The variances are obtained by dividing the diagonal elements of the reciprocal matrix by n. In the particular example chosen above no further approximation is needed since the corrections are very small.

An important thing to be noted is that standard error of the proportion is as great as ·1609, which leads to a wide confidence range. On the other hand, the standard errors are small in comparison to the estimates for the other constants. These estimates can be safely used for interpreting differences in heights, but a certain amount of caution is necessary in specifying the proportion of mixture of the two types of plants.

Using these estimates the expected values are calculated and compared with the observed, as shown below:

Class Interval

	−9·5	−10·5	−11·5	−12·5	−13·5	−14·5	−15·5	−16·5	−17·5	−18·5	−19·5
Observed . .	12	21	40	59	76	79	69	46	30	13	9
Expected . .	11·12	20·34	40·31	61·06	74·50	75·95	66·69	49·94	31·05	14·98	7·90

The χ^2 measure of discrepancy is 1·30, which on 6 degrees of freedom yields a probability greater than ·95 so that the fit can be considered good.

Sexing of osteometric material.—The method of the discriminant function for assigning an individual to its proper group requires some basic material from which the characteristics of the two groups have to be estimated. For example, if a human mandible or lower jaw belonging to a particular group is to be assigned to the right sex, measurements on some mandibles of known sex from that group must be available for constructing the discriminant function. If such basic measurements are not available or if they are not sufficiently numerous, the best method is to treat the observed series (if it is large) as a mixture of two sexual groups and estimate the desired characteristics.

Thus if $x_1, \ldots, x_p$ represent measurements of p characters on an individual, then for each character treated separately the means for the two sexes, the common variance and the proportion of mixture can be estimated. This appears to be the best possible method with skeletal material where only a few measurements are possible on each individual bone. Correlations can also be estimated by considering pairs of characters, but the estimates will be extremely unreliable unless the series is very large.

A slightly different method which leads to the actual specification of the sex of an individual bone may be employed if the number of bones which supply measurements for a suitably chosen set of characters is fairly large.

The measurements for a chosen set of characters is replaced by a single linear function L. Treating the values of this function as a mixed series for a single variate the mean values L_1, L_2 of this function for males and females and the common variance could be evaluated. Any individual bone with a value L of this function is assigned to the male group if

$$|L - L_1| < |L - L_2|,$$

and to the female group if

$$|L - L_1| > |L - L_2|.$$

The success of this method largely depends on the nature of the linear function chosen for the above analysis. If there is no knowledge about the variabilities of these characters then a simple linear function which provldes a fair discrimination between the two sexes is

$$L = a_1 x_1 + \ldots + a_p x_p,$$

where $a = +1$, or -1 according as the male mean for the ith character is greater than or falls short of the female mean. This can, however, be decided without the knowledge of the actual mean values. All linear dimensions are usually smaller for females, whereas some angular measurements are smaller for males.

If the variabilities $\sigma_1, \ldots, \sigma_p$ of $x_1, \ldots, x_p$ are known at least approximately then a better discriminant is

$$L = a_1 \frac{x_1}{\sigma_1} + \ldots + a_p \frac{x_p}{\sigma_p}$$

where a's are chosen as stated above. If the differences $d_1, \ldots, d_p$ in mean values are known approximately then the following function

$$L = d_1 \frac{x_1}{\sigma_1^2} + \ldots + d_p \frac{x_p}{\sigma_p^2}$$

may be used. It may be noted that the above expression is the discriminant function if the p characters considered are uncorrelated. This has been termed the "shape factor" by Penrose (1947), who found that it is a good substitute for the discriminant function in some problems where the correlations are not all simultaneously zero.

If some measurements on individual bones of known sex are available one might construct the discriminant function in the usual way taking the correlations into consideration and choose this function for the above analysis. It may be emphasized that this analysis need be undertaken only when the material on which the discriminant function is based is inadequate and the series available is a random sample from a population of mixed sexes.

5. *The Problem of Three and More Groups*

In section 3 it is seen that, if measurements on a certain number of characters are available for two groups, it is possible to construct a discriminant function which affords the maximum discrimination between them. This function is useful in assigning an individual or individuals with a certain degree of confidence to one or other of the two groups to which they are known to belong. In taxonomic problems there arise cases where an individual specimen is known to belong to one of three or more groups and has to be assigned to its proper group. Thus a plant may have to be specified as *Iris versicolor*, *Iris setosa* or *Iris verginica*. This problem* is approached by the extension of the discriminant function analysis developed with special reference to two groups.

The general theory for three groups.—Let the probability densities in the three groups be represented by $f_1(x, \theta_1)$, $f_2(x, \theta_2)$, $f_3(x, \theta_3)$, where x stands for the available set of measurements and θ's for parameters. First we shall consider the general problem of classifying a collection of individuals drawn from a mixed population containing individuals of the three groups in the proportions π_1, π_2, π_3 $(\pi_1 + \pi_2 + \pi_3 = 1)$.

Any individual I characterized by p measurements can be represented by a point in a p-dimensional space. The problem of classifying an observed collection of individuals is the same as the division of the space into three mutually exclusive regions, R_1, R_2, R_3, with the rule of procedure of assigning an individual $I \subset R_i$ (represented by a point in R_i) to the ith group. If the probability of an individual of the ith group falling in R_i is β_i, then the expected value of the proportion of wrong classifications is

$$\alpha = 1 - (\pi_1 \beta_1 + \pi_2 \beta_2 + \pi_3 \beta_3).$$

The errors will be minimum for that choice of regions R_1, R_2, R_3 for which $\pi_1 \beta_1 + \pi_2 \beta_2 + \pi_3 \beta_3$ is maximum. Such regions, if they exist, may be termed the "best possible" regions.. The following theorem establishes the existence and nature of the best possible regions.

Theorem 1. *The regions defined by* $(\cap)$

$$R_1 \cap \pi_1 f_1 \geqslant \pi_2 f_2, \quad \pi_1 f_1 \geqslant \pi_3 f_3$$

$$R_2 \cap \pi_2 f_2 \geqslant \pi_3 f_3, \quad \pi_2 f_2 \geqslant \pi_1 f_1$$

$$R_3 \cap \pi_3 f_3 \geqslant \pi_1 f_1, \quad \pi_3 f_3 \geqslant \pi_2 f_2$$

constitute the best possible system of mutually exclusive regions.

Consider the function

$$\Delta = \pi_1 \int_{R_1} f_1 \, dv + \pi_2 \int_{R_2} f_2 \, dv + \pi_3 \int_{R_3} f_3 \, dv$$

* In a recent paper Brown (1947) considered this problem but did not suggest any solution. A formal solution is given by the author (Rao, 1948*a*).

B

and any other system R'_1, R'_2, R'_3 of mutually exclusive regions. Denoting by R_{ij} the intersection of R_i and R'_j and by $R_{ij} + R_{st}$ the region obtained by considering the two nonoverlapping regions R_{ij} and R_{st}, Δ may be written as

$$\begin{aligned}
\Delta &= \sum_i \pi_i \int_{R_{i1}+R_{i2}+R_{i3}} f_i \, dv \\
&= \sum_i \{ \int_{R_{1i}} \pi_1 f_1 \, dv + \int_{R_{2i}} \pi_2 f_2 \, dv + \int_{R_{3i}} \pi_3 f_3 \, dv \} \\
&\geqslant \sum_i \{ \int_{R_{1i}} \pi_i f_i \, dv + \int_{R_{2i}} \pi_i f_i \, dv + \int_{R_{3i}} \pi_i f_i dv \} \\
&= \pi_1 \int_{R'_1} f_1 \, dv + \pi_2 \int_{R'_2} f_2 \, dv + \pi_3 \int_{R'_3} f_3 \, dv
\end{aligned}$$

since the integrals satisfy the inequalities

$$\int_{R_{ij}} \pi_i f_i \, dv \geqslant \int_{R_{ij}} \pi_j f_j \, dv$$

which follows from the definition of the regions R_1, R_2, R_3. Hence the result of theorem 1.

If every individual is equally likely to be drawn from any group the best regions are

$$\begin{aligned}
&R_1 \cap f_1 \geqslant f_2,\ f_1 \geqslant f_3 \\
&R_2 \cap f_2 \geqslant f_3,\ f_2 \geqslant f_1 \\
&R_3 \cap f_3 \geqslant f_1,\ f_3 \geqslant f_2.
\end{aligned}$$

These regions may be used for classifying an observed collection of individuals when nothing is known about π_1, π_2, π_3 the proportions of mixture.

By adopting this procedure the probability of an individual of the first group being rightly assigned is $\int_{R_1} f_1 \, dv = \beta_1$, while the probabilities of the individuals of the second and third groups being wrongly assigned to the first group are

$$\alpha_{12} = \int_{R_1} f_2 \, dv \text{ and } \alpha_{13} = \int_{R_1} f_3 \, dv,$$

since in R_1, $f_1 \geqslant f_2$, $f_1 \geqslant f_3$ it follows that

$$\beta_1 \geqslant \text{the greater of } \alpha_{12} \text{ and } \alpha_{13}.$$

If α_{12} and α_{13} are small, one can assert with some confidence that an individual falling in R_1 is correctly classified. If they are not small it is pertinent to enquire whether there exists a region C_1 such that

$$\int_{C_1} f_1 \, dv \text{ is a maximum}$$

subject to the conditions that

$$\int_{C_1} f_2 \, dv \text{ and } \int_{C_1} f_3 \, dv$$

are both not greater than a quantity α_1, chosen to be small, say ·01 or ·05. If an observed individual specimen falls in such a region, then the hypothesis that it belongs to the second or the third groups may be rejected, in which case it is assigned to the first group. The existence and nature of such regions are established by theorem 2.

Theorem 2. *The region* C_1 *satisfying the condition*

$$\int_{C_1} f_1 \, dv \textit{ is a maximum}$$

subject to the restrictions

$$\alpha_1 \geqslant \int_{C_1} f_2\,dv,\ \int_{C_1} f_3\,dv$$

is defined by

$$f_1 \geqslant a f_2 + b f_3$$

where a *and* b *are suitably chosen.*

The proof of this theorem follows from the following lemma by Neyman and Pearson (1936). If $w_1, w_2, \ldots$ are a series of regions such that, with respect to the functions $\varphi_0, \varphi_1, \ldots$

$$\int_{w_j} \varphi_i\,dv = s_i \qquad (j = 1, 2, \ldots;\ i = 1, 2, \ldots)$$

then the region w_0 within which

$$\varphi_0 \geqslant a_1 \varphi_1 + a_2 \varphi_2 + \ldots$$

satisfies the condition that

$$\int_{w_0} \varphi_0\,dv \geqslant \int_{w_j} \varphi_0\,dv \qquad j = 1, 2, \ldots$$

when a's are chosen to satisfy the conditions

$$\int_{w_0} \varphi_i\,dv = s_i, \qquad i = 1, 2, \ldots .$$

To apply this lemma one needs to consider two quantities α_{12}, α_{13}, both less than the assigned quantity α_1 and choose a region such that

$$\int_{w} f_1\,dv \text{ is a maximum}$$

subject to the conditions

$$\int_{w} f_2 = \alpha_{12},\ \int_{w} f_3 = \alpha_{13}.$$

The inside of such a region is defined by

$$f_1 \geqslant a' f_2 + b' f_3$$

where a' and b' are properly chosen. Let the maximized value of $\int_w f_1\,dv$, which is evidently a function of α_{12}, α_{13}, be represented by $\beta(\alpha_{12}, \alpha_{13})$. This is not, in general, an increasing function of α_{12} and α_{13}, so that the maximum value is not necessarily attained when $\alpha_{12} = \alpha_{13} = \alpha_1$. If now the function $\beta(\alpha_{12}, \alpha_{13})$ is maximized with respect to α_{12}, α_{13} subject to the conditions α_{12}, $\alpha_{13} \leqslant \alpha_1$ one obtains two values, $\alpha^0{}_{12}$, $\alpha^0{}_{13}$, corresponding to the optimum solution. Denoting the values of a', b', corresponding to $\alpha^0{}_{12}$, $\alpha^0{}_{13}$, by a, b, the required region may be written as $f_1 \geqslant a f_2 + b f_3$, which proves theorem 2.

It is easy to see that at least one of the values $\alpha^0{}_{12}$, $\alpha^0{}_{13}$ coincides with the boundary value α_1. For, consider the best region corresponding to α_{12}, α_{13} both less than α_1. If $\alpha_{12} \geqslant \alpha_{13}$, it is always possible to add a region in which $f_2 \geqslant f_3$ so that α_{12} is increased to α_1 and α_{13} to a value $\leqslant \alpha_1$. If this is not possible a region in which $f_2 \leqslant f_3$ can be added such that at least one of α_{12} or α_{13} reaches the value α_1. The value of $\beta(\alpha_{12}, \alpha_{13})$ is increased in any case.

Having obtained the regions R_1 and C_1 as given in theorems 1 and 2, one may specify an individual falling in C_1 as 1 (i.e. belonging to the first group) and an individual falling in $R_1 - C_1$ as 1 ? (i.e. likely to belong to the first group). Similarly the regions R_2, C_2 and R_3, C_3 can be constructed.

If the proportions π_1, π_2, π_3 considered in theorem 1 are known then the region C_1 is determined by

$$f_1 \geqslant a(\pi_2 f_2 + \pi_3 f_3)$$

where a is chosen such that $\int_{C_1} (\pi_2 f_2 + \pi_3 f_3) = \alpha_1$ and so on. A detailed account of these latter regions is being published in a paper by Miss S. P. Vaswani.

It results in a certain amount of simplification if the best region C_1 is replaced by

$$C'_1 \cap f_1 \geqslant A f_2, f_1 \geqslant B f_3$$

where A and B are chosen such that

$$\int_{C'_1} f_1 \, dv \text{ is a maximum}$$

subject to the conditions

$$\int_{C'_1} f_2 \, dv \leqslant \alpha_1, \int_{C'_1} f_3 \, dv \leqslant \alpha_1.$$

This region is not, however, the best possible but is likely to be a good approximation.

In some situations it may be necessary to find regions R_1, R_2, R_3 such that the errors of classification are the same for each group or are to be in a given ratio $\rho_1 : \rho_2 : \rho_3$. The existence and nature of such regions are established by the following theorem.

Theorem 3. *The system of regions*

$$R_1 \cap a f_1 \geqslant b f_2, \; a f_1 \geqslant c f_3$$

$$R_2 \cap b f_2 \geqslant c f_3, \; b f_2 \geqslant a f_1$$

$$R_3 \cap c f_3 \geqslant a f_1, \; c f_3 \geqslant b f_2$$

where a, b, c *are suitably chosen are the best possible if the errors of classification for the three groups are to be in an assigned ratio.*

Let R'_1, R'_2, R'_3 be any other regions for which the errors of classification are in the assigned ratio. The region common to R'_i and R_j is represented by R_{ij}. Then

$$a \int_{R'_1} f_1 \, dv = a \int_{R_{11}} f_1 \, dv + \int_{R_{12}} a f_1 \, dv + \int_{R_{13}} a f_1 \, dv$$

$$\leqslant a \int_{R_{11}} f_1 \, dv + b \int_{R_{12}} f_2 \, dv + c \int_{R_{13}} f_3 \, dv.$$

Similar relationships can be set up starting with $b \int_{R'_2} f_2 \, dv$ and $c \int_{R'_3} f_2 \, dv$. The errors of classification with respect to the systems R_1, R_2, R_3 and R'_1, R'_2, R_3 may be represented by $\alpha\rho_1$, $\alpha\rho_2$, $\alpha\rho_3$ and $\alpha'\rho_1$, $\alpha'\rho_2$, $\alpha'\rho_3$ respectively, $\rho_1 : \rho_2 : \rho_3$ being the assigned ratio. Writing down the values of integrals in the above three relationships and adding one gets

$$(1 - \rho_1 \alpha')a + (1 - \rho_2 \alpha')b + (1 - \rho_3 \alpha')c$$

$$\leqslant (1 - \rho_1 \alpha)a + (1 - \rho_2 \alpha)b + (1 - \rho_3 \alpha)c$$

or

$$- \alpha' (a\rho_1 + b\rho_2 + c\rho_3) \leqslant - \alpha (a\rho_1 + b\rho_2 + c\rho_3)$$

or

$\alpha' \geqslant \alpha$, since a, b and c are positive quantities by definition. This proves the result of theorem 3.

The quantities a, b, c are to be determined from the equalities

$$\frac{1}{\rho_1} \int_{R_2 + R_3} f_1 \, dv = \frac{1}{\rho_2} \int_{R_3 + R_1} f_2 \, dv = \frac{1}{\rho_3} \int_{R_2 + R_1} f_3 \, dv.$$

Theorem 2 led us to the construction of 4 mutually exclusive regions with the help of which an observed specimen can either be assigned to a particular group or to none. In some problems it may be necessary to construct a system of 7 regions, 3 for assigning an observed specimen to particular groups, 3 others for specifying it as belonging to one of two of the groups, and the remaining one for making no decision. To construct these regions we set up three regions w_1, w_2, w_3 for not accepting respectively the first, the second and the third groups as the possible ones from which the observed specimen has arisen. The boundary surfaces of these regions determine by mutual intersections the required system of 7 regions. The region outside w_1, w_2, w_3 is the doubtful region, the intersection of w_i and w_j is the region for assigning an individual to the *k-th* group ($k \neq i \neq j$) and the region outside w_i, w_j but inside w_k is the region for specifying an individual as belonging to either the *i-th* or the *j-th* group. Some methods of constructing the regions w_1, w_2, w_3 are discussed below.

Regions w_i when π_1, π_2, π_3 are known. If π_1, π_2, π_3 considered in theorem 1 are known then the region w_1 is such that

$$\int_{w_1} \pi_1 f_1 \, dv = \alpha_1 \text{ (a small assigned quantity)}$$

and

$$\int_{w_1} (\pi_2 \; f_2 + \pi_3 \; f_3) \; dv \text{ is a maximum.}$$

The boundary surface of such a region is

$$\pi_1 \; f_1 \leqslant a \; (\pi_2 \; f_2 + \pi_3 \; f_3)$$

where the constant a is suitably determined. Similarly w_2 and w_3 can be constructed.

Regions w_i *independent of any aprori information.* Consider the region w'_1 such that

$$\int_{w'_1} f_1 \, dv = \alpha_1$$

and

$$\int_{w'_1} f_2 \, dv = \int_{w'_1} f_3 \, dv = \beta \text{ is a maximum.}$$

Having determined β it is possible to construct the region w_1 such that

$$\int_{w_1} f_1 \, dv = \alpha_1$$

$$\int_{w_1} f_i \, dv = \beta \quad i \neq 1$$

$$\int_{w_1} f_j \, dv \text{ is a maximum } j \neq i$$

where $i = 2$ if

$$\int_{w_1} f_2 \, dv \leqslant \int_{w_1} f_3 \, dv$$

and $i = 3$ otherwise. The boundary surface of such a region is of the form

$$f_1 \leqslant a f_2 + b f_3$$

where a and b are suitably determined.

There is an alternative method of constructing w_1, w_2, w_3 which may be useful in some practical situations. Let β_2 and β_3 be the maximum values of

$$\int_{u_2} f_2 \, dv \text{ and} \int_{u_3} f_3 \, dv$$

subject to the conditions

$$\int_{u_2} f_1 \, dv = \alpha_1 \, , \quad \int_{u_3} f_1 \, dv = \alpha_1$$

where u_2 and u_3 are the regions corresponding to the maxima. The region w_1 may be determined such that

$$\int_{w_1} f_1 \, dv = \alpha_1$$

and

$$\frac{1}{\beta_2}\int_{w_1} f_2 \, dv = \frac{1}{\beta_3}\int_{w_1} f_3 \, dv \text{ is a maximum.}$$

The boundary surface of such a region is again of the form

$$f_1 \leqslant a f_2 + b f_3$$

where a and b are suitably determined. The relative advantages of these two alternative systems require further investigation.

It is interesting to note that the above discussion is useful in the general theory of tests of statistical hypotheses outlined by Wald (1939). The problem of choice of one out of a given set of hypotheses is soluble with the help of the regions considered in the three theorems given above.

Applications to multivariate normal populations.—The probability densities in this case may be written as

$$f_r = \text{const} \times \exp\{ -\tfrac{1}{2} \Sigma \Sigma \alpha^{ij} (x_i - \mu_{ir})(x_j - \mu_{jr})$$

$$r = 1, 2, 3.$$

The surfaces of constant likelihood ratios f_r / f_s are defined by

$$\Sigma_j \{ \Sigma_i \alpha^{ij} (\mu_{ir} - \mu_{is})\} x_j = \text{constant.}$$

Defining the functions which may be called linear discriminant scores

$$L_r = \Sigma_j (\Sigma_i \alpha^{ij} \mu_{ir}) x_j - \tfrac{1}{2} \Sigma \Sigma \alpha^{ij} \mu_{ir} \mu_{jr}$$

$$r = 1, 2, 3$$

it is possible to express the above surfaces as

$$L_r - L_s = \text{constant.}$$

These functions can be directly used to get at the best possible regions. For instance, the region defined by $f_1 \geqslant f_2, f_1 \geqslant f_3$ is equivalent to the region $L_1 \geqslant L_2$, $L_1 \geqslant L_3$. If the three populations have different dispersion matrices then these functions will be quadratic in the x's.

Illustrative example.—The following Table 5 gives the mean values, correlations and standard deviations of 4 characters for three castes in the United Provinces (India).

TABLE 5.—*Statistical Constants of the Three Groups*

Group	*Stature* (*KH*)	*Sitting height* (*SH*)	*Nasal depth* (*NT*)	*Nasal height* (*NH*)
	Mean values			
Brahmin	164·51	86·43	25·49	51·24
Artisan	160·53	81·47	23·84	48·62
Korwa	158·17	81·16	21·44	46·72
	Correlations			
KH	—	·5849	·1774	·1974
SH	..	—	·2094	·2170
NT	..	..	—	·2910
NH	..	..	..	—
	Standard deviations			
	5·74	3·20	1·75	3·50

By inverting the correlation matrix one could easily calculate the functions L_1, L_2, L_3 defined above and use them for classification. In order to determine the probabilities of misclassifications it is necessary to go through a slightly complicated procedure. The four measurements can be replaced by two independent linear functions* (canonical variates explained in section 9) whose mean values differ from group to group. Omitting the computational procedure explained in section 10 the linear functions can be written as

$$X_1 = \cdot 0076\,kh + \cdot 4431\,sh + 1\cdot 0723\,nt + \cdot 1688\,nh$$

$$X_2 = -\cdot 5304\,kh + \cdot 8664\,sh + \cdot 5981\,nt + \cdot 0247\,nh$$

where the characters expressed in lower-case letters are in standard deviation units. The mean values of X_1 and X_2 for the three groups are:

	Mean values			$\frac{1}{2}(X^2_1 + X^2_2)$
	X_1	X_2		
Brahmin	30·2757	17·6628	.	614·2962
Artisan	28·4463	15·7153	.	528·0813
Korwa	26·8380	15·0156	.	472·8732

From these the following functions are constructed, the constant term being obtained by taking differences of the numbers in the last column of the table above.

$$Y_1 = (30\cdot 2757 - 28\cdot 4463)\,X_1 + (17\cdot 6628 - 15\cdot 7153)\,X_2 - (614\cdot 2962 - 528\cdot 0813)$$

$$= 1\cdot 8294\,X_1 + 1\cdot 9475\,X_2 - 86\cdot 2149$$

$$Y_2 = (30\cdot 2757 - 26\cdot 8380)\,X_1 + (17\cdot 6628 - 15\cdot 0156)\,X_2 - (614\cdot 2962 - 472\cdot 8732)$$

$$= 3\cdot 4377\,X_1 + 2\cdot 6472\,X_2 - 141\cdot 4230$$

$$Y_3 = Y_2 - Y_1 = 1\cdot 6083\,X_1 + \cdot 6997\,X_2 - 55\cdot 2081.$$

The regions for classifying individuals equally likely to have arisen from any group, are for

$$\text{Brahmin} \cap Y_1 \geqslant 0, \quad Y_2 \geqslant 0$$

$$\text{Artisan} \cap Y_1 \leqslant 0, \quad Y_3 \geqslant 0$$

$$\text{Korwa} \cap Y_2 \leqslant 0, \quad Y_3 \leqslant 0.$$

To find the errors of classification it is necessary to find the volume of the normal surface beyond regions specified above for each group. Making the transformation

$$z_1 = \frac{Y_1 + 86\cdot 2149}{\sqrt{(1\cdot 8294)^2 + (1\cdot 9475)^2}} = \frac{Y_1}{2\cdot 6720} + 32\cdot 2663$$

$$z_2 = \frac{Y_2 + 141\cdot 4230}{\sqrt{(3\cdot 4377)^2 + (2\cdot 6472)^2}} = \frac{Y_2}{4\cdot 3388} + 32\cdot 5947$$

$$z_3 = \frac{Y_3 + 55\cdot 2081}{\sqrt{(1\cdot 6083)^2 + (\cdot 6997)^2}} = \frac{Y_3}{1\cdot 7539} + 31\cdot 4771$$

* There are various methods of obtaining these functions. For instance the discriminant functions between any two pairs constitute two such independent linear functions. The advantage with canonical variates is that the two functions are uncorrelated and their between group variations give a clue as to the linearity or coplanarity of the groups. If they are linear then only the first canonical variate needs to be considered.

it is found that

$$V(z_1) = V(z_2) = V(z_3) = 1$$

$$\text{cov.}\,(z_1 z_2) = \cdot 9871,\ \text{cov.}\,(z_1 z_3) = \cdot 9186,\ \text{cov.}\,(z_2 z_3) = \cdot 9700.$$

To calculate the required volume say for Brahmins, the deviates

$$h = 32\cdot 2663 - \text{Mean } z_1, \quad k = 32\cdot 5947 - \text{Mean } z_2$$
$$= -1\cdot 33 \qquad\qquad = -2\cdot 17$$

are entered in tables for the normal bivariate surface* with $r = \cdot 9871$. The volume is about $\cdot 09$ so that the frequency of misclassification for Brahmin is 9 per cent., the corresponding figures for Artisans and Korwas being 20 per cent. and 19 per cent. respectively. Artisans and Korwas being close together there is expected to be greater overlap. Further, Artisans being mid-way between Brahmins and Korwas a greater percentage of Artisans rather than Korwas will be misclassified as Brahmins. This explains the higher frequency for Artisans.

An individual drawn at random has the values $z_1 = 33\cdot 0023$, $z_2 = 34\cdot 6642$. Since $z_1 - 32\cdot 2663$ and $z_2 - 32\cdot 5947$ are both positive he can be assigned to the Brahmin group. The chance of observing such an individual or a rarer type from Artisan is given by the probability of the inequalities $z_1 \geqslant 33\cdot 0023$, $z_2 \geqslant 34\cdot 6642$. The deviates are

$$h = 33\cdot 0023 - \text{mean } z_1, \quad k = 34\cdot 6642 - \text{mean } z_2$$
$$= 2\cdot 0720 \qquad\qquad = 2\cdot 5377.$$

The chance is about 6 in 1,000. The chances of the individual being a Korwa are rarer still, so that the observed individual may be taken as a Brahmin.

PART II: THE CONCEPT OF DISTANCE AND THE PROBLEM OF GROUP CONSTELLATIONS

6. *Generalized Distance between Two Populations*

In Part I, statistical criteria have been developed for specifying an individual as a member of one of many groups to which he can possibly belong. In this part we consider the problem of arriving at constellations of the groups themselves such that any two groups of a constellation are in some sense closer to each other than any two belonging to different constellations.

In order to answer this question, it is convenient to introduce the concept of a quantitative measure of "resemblance" or more conveniently of "separation" between two groups. Such a measure, when it satisfies some fundamental requirements, may be referred as the distance between two groups. This distance may be utilized in sorting out clusters or constellations of groups.

Mathematical concepts.—Let $f_1(x, \theta_1), f_2(x, \theta_2)$ denote the probability densities in two groups. It is shown in section 3 that there exist two regions R_1 and R_2 such that the probability

$$\int_{R_2} f_1\, dv = \int_{R_1} f_2\, dv = \alpha$$

is a minimum. This means that, by using the best possible criterion, a proportion α of the individuals belonging to any one group is liable to be misclassified; in other words, the two groups may be said to overlap to the extent of 100α per cent. The overlap is maximum when the two groups are identical and decreases with increase in the separation of the two groups. If the two groups are distinct in the sense that the regions of distributions are non-overlapping, then every individual can be specified to his group with certainty. In this case the percentage overlap is evidently zero. The extent of separation between two groups can, thus, be measured by α, the least proportion of overlapping individuals who are liable to be misclassified. One might choose a decreasing function of α as a measure of separation or distance between two groups so that this measure increases with the increase in separation. One such function is

* These are found in *Tables for Biometricians and Statisticians*, Part II, edited by Karl Pearson.

$(1 - \alpha)$. This satisfies the two fundamental postulates of distance as laid down in differential geometry*:

(i) The distance between two groups is not less than zero.

(ii) The sum of distances of a group from two other groups is not less than the distance between the two other groups. (Triangle law of distance.)

The first one follows since $\alpha \leqslant 1$, in which case the distance $(1 - \alpha) \nless 0$.

To prove the second, one may consider three groups, G_1, G_2, G_3. Let $R_1(1, 2)$, $R_2(1, 2)$ be the best divisions of the space corresponding to G_1 and G_2. Similar definitions hold for $R_1(1, 3)$, $R_3(1, 3)$, $R_2(2, 3)$, $R_3(2, 3)$. Defining

$$\int_{R_i(i,j)} f_i\, dv = 1 - \alpha_{ij}$$

the proposition required to be proved may be stated as follows:

$$(1 - \alpha_{12}) \ngtr (1 - \alpha_{13}) + (1 - \alpha_{23})$$

From definition it follows that

$$\int_{R_i(i,j)} f_i\, dv \nless \int_{R_i(i,j)} f_j\, dv$$

so that

$$\int_{R_i(i,j)} f_i\, dv + \int_{R_j(i,j)} f_j\, dv \nless 1.$$

Using this, one gets

$$\int_{R_1(1,2)} f_1\, dv \ngtr 1 \ngtr \int_{R_1(1,3)} f_1\, dv + \int_{R_3(1,3)} f_3\, dv$$

and

$$\int_{R_2(1,2)} f_2\, dv \ngtr 1 \ngtr \int_{R_2(2,3)} f_2\, dv + \int_{R_3(2,3)} f_3\, dv.$$

Adding, one gets $2(1 - \alpha_{12}) \ngtr 2(1 - \alpha_{13}) + 2(1 - \alpha_{23})$, which proves the desired result.

Empirical requirements of the distance function.—The distance function defined above has to satisfy some further logical requirements if it has to be of any value in biological classifications.

(i) The distance must not decrease when additional characters are considered.

(ii) The increase in distance by the addition of some characters to a suitably chosen set must be relatively small so that the group constellations arrived at on the basis of the latter set are not distorted when such additional characters are considered.

The first requirement is reasonable since adding characters to a basic set must necessarily reduce the errors of classification. This is, in fact, satisfied when the distance function is as chosen above. Let $P_1(x_1, \ldots, x_p)$ and $P_2(x_1, \ldots, x_p)$ denote the probability densities of two groups with $R_1(p)$ and $R_2(p)$ as the best divisions of the p-space. When an additional character is considered the probability densities can be written as $P_1(x_1, \ldots, x_{p+1})$ and $P_2(x_1, \ldots, x_{p+1})$ with $R_1(p+1)$ and $R_2(p+1)$ as the best division of the $(p+1)$-space. If Ω denotes the region obtained by considering $R_1(p)$ and the complete range for x_{p+1} then by definition it follows that

$$\int_{R_1(p+1)} P_1(x_1, \ldots, x_{p+1})\, dv' \leqslant \int_{\Omega} P_1(x_1, \ldots, x_{p+1})\, dv'$$

$$= \int_{R_1(p)} P_1(x_1, \ldots, x_p)\, dv$$

so that if α and α' represent the proportions of overlapping individuals in the two cases it follows that $\alpha' \leqslant \alpha$, which proves the result.

* By definition, the distance from the population A to the population B is the same as that from B to A. In problems where social relations are measured a new definition of distance may be necessary. Thus if Johnny (J_1) likes Jean (J_2) and Jean does not like Johnny, then the distances $J_1 J_2$ and $J_2 J_1$ should be considered different.

The second requirement has been introduced merely as a practical necessity. There must be some limit, due to the considerations of cost in obtaining the information and the numerical reduction of data, to the number of characters used in order to arrive at stable judgments. This can be empirically verified in any situation.

Application to multivariate normal populations with common dispersion matrices.—Denoting the common dispersion matrix by (α_{ij}) and its reciprocal by (α^{ij}) and the mean value of the ith character in the $_j$th group by μ_{ij}, the probability densities $f_1(x, \theta_1), f_2(x, \theta_2)$ can be written as

$$C \times \exp\{-\tfrac{1}{2}\Sigma\Sigma\alpha^{ij}(x_i - \mu_{i2})(x_j - \mu_{j1})\}$$

and

$$C \times \exp\{-\tfrac{1}{2}\Sigma\Sigma\alpha^{ij}(x_i - \mu_{i2})(x_j - \mu_{j2})\}$$

where C is a constant depending only on α_{ij}. The relation $f_2 \geqslant k f_1$ defining the region R_2 described in section 3 is given by

$$\sum_i(\sum_j \alpha^{ij}\mu_{j2})x_i \geqslant a + \sum_i(\sum_j \alpha^{ij}\mu_{j1})x_i$$

or

$$L(x) = \sum_i(\sum_j \alpha^{ij} d_j)x_i \geqslant a$$

where $d_i = \mu_{i1} - \mu_{ij}$ and $L(x)$ is as defined above. The quantity α is the common value of

$$\int_{L(x) \geqslant a} f_1\,dv = \int_{L(x) \leqslant a} f_2\,dv.$$

This can be determined by considering only the distribution of $L(x)$ which being a linear function of the x's is normally distributed. The mean $L(x)$ in the first group is

$$L(\mu_1) = \sum_i(\Sigma\alpha_{ij} d_j)\mu_{i1}$$

and in the second group

$$L(\mu_2) = \sum_i(\Sigma\alpha^{ij} d_j)\mu_{i2}$$

with the common variance as

$$D^2 = \Sigma\Sigma\alpha^{ij} d_i d_j$$

$$\alpha = \int_{L(x) \geqslant a} \exp -\tfrac{1}{2}\left\{\frac{L(x) - L(\mu_1)}{D}\right\}^2 d\,L(x)\,/\,D\sqrt{2\pi}.$$

Because of the similarity of the distributions of $L(x)$ in the first and second groups it is easy to see that

$$a = \frac{L(\mu_1) + L(\mu_2)}{2}.$$

Changing over to the variable $y = \{L(x) - L(\mu_1)\}\,/\,D$ the above integral becomes

$$\alpha = \frac{1}{\sqrt{2\pi}}\int_{D/2}^{\infty} e^{-\frac{1}{2}y^2}\,dy.$$

Hence

$$(1 - \alpha) = \frac{1}{\sqrt{2\pi}}\int_{-\infty}^{D/2} e^{-\frac{1}{2}y^2}\,dy$$

which shows that $(1 - \alpha)$ is an increasing function of D. The measure $(1 - \alpha)$ may, then, be conveniently replaced by the functionally dependent quantity D which is the same as the Mahalanobis' generalized distance. The above analysis supplies a logical derivation of this tool suggested on intuitive considerations. This also shows that Mahalanobis' distance function is strictly applicable to groups in which the characters are normally distributed.

Coefficient of racial likeness.—In 1921, in a paper by Miss M. L. Tildesley, Karl Pearson proposed a measure of racial likeness (*C.R.L.*), and since then it has been used by anthropologists of the Biometric school for purposes of classifying skeletal remains (Pearson, 1926; Morant, 1923, etc.). If n_{1i}, n_{2i} denote the number of observations on which the means m_{1i}, m_{2i} of the ith character for the first and second groups are based, and s_i the standard deviation of the ith character, and p, the number of characters used, then the *C.R.L* may be written as

$$\frac{1}{p}\sum_{i=1}^{p}\frac{n_{1i}\,n_{2i}}{n_{1i}+n_{2i}}\left(\frac{m_{1i}-m_{2i}}{s_i}\right)^2.$$

Apart from the multiplying factors depending on the sample sizes the *C.R.L.* differs from D^2 in an important aspect. In the former the various characters are treated as independent. All biological experience shows that a certain amount of correlation exists between any two characters, and the effect of this on the second expression in the *C.R.L.* is to make it increase rapidly with the increase in the number of characters used. It is difficult to say from the changes in the *C.R.L.* whether some newly added characters supply additional information for the purposes of discrimination. For instance, when a character highly correlated with the characters in a se is used in addition to those in the set the *C.R.L.* may be considerably altered while the change may be inappreciable when the high correlations are taken into account. One can be on safe grounds in using D^2 with some superfluous characters. If any character produces an appreciable change in D^2 it is taken to be of additional value in discrimination. In view of the fact that D^2 satisfies some logical requirements it may be safely used in preference to the *C.R.L.*

The use of indices.—Anthropologists and systematists make extensive use of the indices which are percentage ratios of pairs of measurements. It is believed that these functions of the measurements show greater variation between groups than the individual measurements themselves. This may be true, but the best functions that show maximum discrimination are certain linear combinations of the measurements. The best possible criterion is, then, a linear function of the measurements, and such a criterion can be based on any number of characters while the use of the index is confined to two only. In the case of two groups Mahalanobis' D^2 is the square of the difference in mean values of the best linear function.

It has sometimes been the practice to use the measurements in combination with the indices calculated from them in constructing a suitable measure of divergence. For example, the cephalic index which is the percentage of the ratio of head breadth to head length is used as a separate character in addition to the latter two. There appears to be no theoretical justification in using three such functionally related variables.

7. *An Illustrative Example**

Sources of the material.—The data consist of measurements of 12 characters taken by Dr. D. N. Mazumdar in 1941 on about 4,000 individuals in the United Provinces (India). They belong to 22 castes and tribes living in the same environment but without any social intermixture for a long period of time. The material has been analyzed by me under the guidance of Professor P. C. Mahalanobis, partly at the Indian Statistical Institute, Calcutta, and partly at Cambridge. The relevant results are summarized below to illustrate the use of D^2 in arriving at group constellations. The anthropological significance of the classification so arrived at is given in the report mentioned in the footnote.

The mean values of 9 out of the 12 characters for the 22 groups are given in Table 6 and the pooled correlations and standard deviations in Table 7. These latter estimates are based on a large number of degrees of freedom, and hence their variations can be neglected in tests of significance concerning the means.

* This is taken from the Report on the United Provinces Anthropometric Survey, 1941, by P. C. Mahalanobis, D. N. Majumdar and C. R. Rao. Some parts of this report relating to the biometric analysis are being published in *Sankhyā*, vol. 9, part 1.

TABLE 6.—*Mean Values by Groups and Characters*

	Group	Stature		Sitting height		Nasal depth		Nasal height		Head length		Frontal breadth		Bizygomatic breadth		Head breadth		Nasal breadth	
		N	*Mean*	N	*Mean*	N	*Mean*	N	*Mean*	N	*Mean*	N	*Mean*	N	*Mean*	N	*Mean*	N	*Mean*
1	Brahmin (Basti)	85	164·51	85	86·43	86	25·49	86	51·24	86	191·92	86	104·74	86	133·36	86	139·88	85	36·55
2	,, (Others)	92	165·07	92	86·25	91	24·74	91	50·40	92	191·35	92	104·46	92	132·68	92	139·50	90	36·13
3	Agharia	107	167·47	107	84·03	107	23·63	107	49·50	107	191·02	..	..	107	131·90	107	140·04	107	35·87
4	Chatri	139	163·33	139	82·25	139	24·73	137	52·72	139	192·58	139	103·98	139	131·70	139	131·72	137	35·64
5	Muslim	168	162·45	131	81·83	168	24·49	168	51·38	167	190·78	72	103·28	168	131·52	168	137·40	168	36·36
6	Bhatu	149	163·38	57	84·49	150	25·09	150	52·06	148	186·10	150	99·34	150	133·55	150	138·58	150	35·65
7	Habru	124	164·91	71	85·53	124	24·19	124	50·30	124	186·94	124	100·18	124	131·16	123	137·40	124	35·82
8	Bhil	186	162·92	..	..	186	24·05	187	48·60	187	181·87	187	103·36	186	131·18	186	137·62	187	37·49
9	Dom	113	166·53	102	84·19	112	25·33	113	50·34	113	186·40	113	104·16	113	132·64	113	137·52	113	38·11
10	Artisan (Ahir)	67	161·37	68	84·35	68	24·29	68	48·98	68	187·45	68	102·76	68	131·70	67	138·12	68	35·60
11	,, (Kurmi)	94	161·35	94	83·41	94	24·03	94	49·22	94	188·86	94	102·62	94	131·82	94	137·86	94	36·21
12	,, (Others)	172	161·34	172	83·09	171	23·73	173	48·72	173	187·69	173	102·44	173	131·30	173	136·84	173	36·27
13	,, (Kahar)	57	160·53	57	81·47	57	23·84	57	48·62	57	188·83	56	101·68	57	130·70	57	136·28	57	36·61
14	Tharu	191	163·33	191	83·57	191	21·72	191	49·94	191	187·78	191	100·00	191	131·68	191	135·90	191	37·90
15	Chamar	158	161·88	159	80·39	159	23·27	159	48·50	159	186·85	159	102·62	159	130·28	159	136·40	159	36·35
16	Chero	100	162·03	100	79·67	100	22·64	99	47·88	100	187·09	100	101·40	100	130·00	100	135·48	100	38·88
17	Majhi	153	162·96	156	79·89	156	22·37	156	46·70	156	187·12	155	102·56	155	131·34	155	135·92	156	39·24
18	Panika	156	159·76	157	78·71	157	22·40	157	47·22	158	186·34	158	102·14	158	130·44	158	135·10	157	38·05
19	Kharwar	197	160·79	188	80·31	197	22·19	197	48·34	197	185·92	192	102·34	197	131·22	196	136·74	196	38·47
20	Oraon	99	161·45	99	81·01	100	20·60	100	47·54	100	189·43	..	..	100	130·14	100	136·88	100	37·66
21	Rajwar	105	159·71	105	81·25	105	21·10	105	48·12	105	186·73	..	..	105	129·48	105	137·16	105	35·41
22	Korwa	101	158·17	98	81·16	101	21·44	101	46·72	101	187·48	..	..	101	130·34	101	134·62	101	39·78

TABLE 7.—*Pooled Estimates of Correlations and Standard Deviations*

	KH	*SH*	*NT*	*NH*	*HL*	*FB*	B_zB	*HB*	*NB*
KH	..	·5849	·1774	·1974	·2698	·2173	·2891	·1412	·2103
SH	..	..	·2094	·2170	·2651	·3012	·2995	·2069	·1182
NT	..	..	..	·2910	·1537	·1243	·1575	·1308	·1139
NH	..	..	..	..	·1758	·1139	·1852	·1735	·0438
HL	..	..	..	..	..	·2270	·2792	·1982	·1930
FB	..	..	..	..	..	..	·4930	·4461	·1831
B_zB	..	..	..	..	..	..	..	·5407	·2729
HB	..	..	..	..	..	..	..	..	·1413
NB	..	..	..	..	..	..	..	..	..
Standard deviation	5·74	3·20	1·75	3·50	6·60	3·92	4·58	4·50	2·57

Characters used.—The following characters have been used for the general analysis:

(1) Stature (*KH*). (2) Sitting height (*SH*). (3) Head length (*HL*). (4) Head breadth (*HB*). (5) Bizygomatic breadth (B_zB). (6) Nasal height (*NH*). (7) Nasal breadth (*NB*). (8) Nasal depth (*NT*). (9) Frontal breadth (*FB*)

Since the mean values of *FB* are missing for 4 out of 22 groups, only the first 8 characters have been used for comparing all the 22 groups. All nine have been used for classifying a subset of the groups.

Values of the square of the generalized distance, D^2.—The square of the generalized distance, D^2, between two groups as based on p characters is

$$D^2 = \sum_{i,j=1}^{p}\sum \alpha^{ij}\, d_i\, d_j$$

where (α_{ij}) is the matrix inverse to the pooled dispersion matrix (α_{ij}) and d_i is the difference in mean values for the ith character. This formula is not useful in practice, as the calculation of the inverse matrix and the further computation of the quadratic form in the differences of mean values become extremely laborious when the number of characters exceed 4 or 5. If the characters had been independent then the formula for D^2 would be simply the sum of squares of differences in mean values for the various characters. But when the characters are correlated, as is usually the case with biological populations, they may be replaced by a set of *transformed characters* which are linear functions of the observed characters and are mutually uncorrelated. When once these transformed characters are obtained, the calculation of D^2 is reduced to finding the simple sum of squares.

A very simple method of constructing the transformed characters is explained in Appendix 6 of the report *loc. cit.* If $x_1, \ldots, x_p$ are the values of the original characters in standard deviation units, then the transformed characters $y_1, \ldots, y_p$ with unit variances are defined by

$$y_1 = x_1$$

$$y_2 = a_{21}\, x_1 + a_{22}\, x_2$$

$$\ldots \qquad \ldots$$

$$y_p = a_{p1}\, x_1 + \ldots + a_{pp}\, x_p.$$

The advantages in obtaining such a system of uncorrelated variables by a proper choice of the coefficients is two-fold. The first is that D^2 can be calculated as a simple sum of squares. The second is that the sum of squares arising out of r of the variables $y_1, \ldots, y_r$ $(r < p)$ gives the D^2 with respect to the r original characters $x_1, \ldots, x_r$, so that the increase in D^2 by the addition of the characters $x_{r+1}, \ldots, x_p$ can be simply measured by the sum of squares corresponding to $y_{r+1}, \ldots, y_p$. The amount contributed by these additional characters enables one to judge their importance in the problem of classification. In the present case the transformed characters $y_1, \ldots, y_9$ are listed below. The original characters expressed in standard deviation units are written in lower-case letters. Thus (*hl*) stands for (*HL*) expressed in standard deviation units for head length.

$$y_1 = Y_1 = hl$$

$$\cdot 980162\ y_2 = Y_2 = hb - \cdot 198200\ Y_1$$

$$\cdot 822702\ y_3 = Y_3 = b_z b - \cdot 505209\ Y_2 - \cdot 279200\ Y_1$$

$$\cdot 970893\ y_4 = Y_4 = nh - \cdot 097610\ Y_3 - \cdot 144326\ Y_2 - \cdot 175800\ Y_1$$

$$\cdot 953943\ y_5 = Y_5 = nb + \cdot 023239\ Y_4 - \cdot 246667\ Y_3 - \cdot 107261\ Y_2 - \cdot 193000\ Y_1$$

$$\cdot 944814\ y_6 = Y_6 = nd - \cdot 069643\ Y_5 - \cdot 258066\ Y_4 - \cdot 094404\ Y_3 - \cdot 104439\ Y_2 - \cdot 153700\ Y_1$$

$$\cdot 924222\ y_7 = Y_7 = kh - \cdot 084080\ Y_6 - \cdot 045612\ Y_5 - \cdot 122926\ Y_4 - \cdot 211917\ Y_3 - \cdot 144918\ Y_2 - \cdot 269800\ Y_1$$

$$\cdot 788187\ y_8 = Y_8 = sh - \cdot 507938\ Y_7 - \cdot 115274\ Y_6 - \cdot 018904\ Y_5 - \cdot 141858\ Y_4 - \cdot 217927\ Y_3 - \cdot 160669\ Y_2 - \cdot 265100\ Y_1$$

$$\cdot 826753\ y_9 = Y_9 = fb - \cdot 162016\ Y_8 - \cdot 054612\ Y_7 - \cdot 027845\ Y_6 - \cdot 044106\ Y_5 + \cdot 006423\ Y_4 - \cdot 335334\ Y_3 - \cdot 417533\ Y_2 - \cdot 227000\ Y_1$$

The actual value of D^2 between any two groups, based on the 8 characters *HL, HB, B_zB, NH, NB, NT, KH, SH*, is obtained by taking the differences of the values of $y_1, y_2, \ldots, y_8$ for the two groups and taking the sum of their squares. The 231 values of D^2, thus calculated by considering all possible pairs from the 22 groups are given in Table 8. In this table the name of the group from which the D^2 values are measured is given at the top of each column and the names of the groups are arranged in each column in increasing order of D^2. The following symbols have been used to denote the different groups for convenience of reference:

A_1 = Ahir (Artisan), A_2 = Kurmi (Artisan), A_3 = Other Artisan, A_4 = Kahar (Artisan), B_1 = Basti Brahmin, B_2 = Other Brahmin, C_1 = Bhatu, C_2 = Habru, Ch = Chatri, Cm = Chamar, D = Dom, M = Muslim, Th = Tharu Male, T_1 = Chero, T_2 = Majhi, T_3 = Panika, T_4 = Kharwar, T_5 = Oraon, T_6 = Rajwar, T_7 = Korwa.

By classification is meant, in the first place, a division of all the groups into a number of clusters such that the groups within any cluster are close together, i.e. have a smaller D^2 among themselves than those from groups belonging to two different clusters. No formal rules can, however, be laid down for this purpose, and the classification must proceed by actual scrutiny of D^2 values as given in Table 8. The classification thus arrived at may be checked up by a study of the differences in individual characters for any two graphs within a cluster. A slightly different method of classification known as the method of *canonical* variates is developed in a later section. The procedure by the method of D^2 values is explained below.

It would be noticed that all the 22 groups fall into two distinct clusters which may be distinguished by T (tribal) and G (general) classes with Chamar and Tharu forming the borderline cases. The members of the T group Chero (T_1), Majhi (T_2), Panika (T_3), Kharwar (T_4), Oraon (T_5), Rajwar (T_6) and Korwa (T_7) are sharply differentiated from the rest with the generalized distances between any two of them being in general less than that between a member of this class and any other group. There is also an internal patterning of these seven members noticeable from the table of D^2's with T_1, T_2, T_3, T_4 being closer to one another and so also T_5 and T_6, and T_7 coming in between these two sub-clusters.

TABLE 8.—*Values of* D^2 *(Based on Eight Characters) Arranged in Increasing Order of Magnitude with Respect to each Group*

Brahmin Basti (B_1)		Brahmin Others (B_2)		Agharia (Ag)		Chatri (Ch)		Muslim (M)		Bhatu (C_1)		Habru (C_2)		Dome (D)		Bhil (Bh)		Ahir (A_1)		Kurmi (A_2)	
B_2	·27	B_1	·27	B_2	1·94	M	·39	Ch	·39	C_2	1·12	B_2	·69	Bh	1·15	D	1·15	A_2	·30	A_3	·11
A_1	1·13	C_2	·69	A_2	1·95	A_4	2·03	A_4	·78	A_1	1·26	A_1	·71	C_2	1·95	A_3	1·65	A_3	·48	A_1	·30
C_2	1·32	A_1	·73	C_2	2·00	A_2	2·11	A_2	1·21	B_1	1·53	A_2	·98	C_1	1·98	Cm	1·91	C_2	·71	A_4	·58
A_2	1·44	A_2	·98	A_3	2·07	Ag	2·33	Cm	1·23	A_2	1·56	A_3	·99	A_3	2·20	A_2	2·07	B_2	·73	B_2	·98
C_1	1·53	A_3	1·45	Cm	2·12	C_1	2·50	A_3	1·37	B_2	1·63	C_1	1·12	A_2	2·34	A_4	2·09	B_1	1·13	C_2	·98
A_3	2·11	C_1	1·63	M	2·18	Cm	2·50	C_1	2·07	A_3	1·73	B_1	1·32	M	2·47	C_2	2·17	C_1	1·26	M	1·21
D	2·82	Ag	1·94	Ch	2·33	A_3	2·58	Ag	2·18	D	1·98	D	1·95	A_4	2·50	A_1	2·38	A_4	1·52	B_1	1·44
M	2·83	M	2·61	A_4	2·40	B_2	2·81	A_1	2·33	M	2·07	Ag	2·00	A_1	2·75	T_4	2·50	M	2·33	C_1	1·56
Ch	2·98	A_4	2·68	A_1	2·58	B_1	2·98	D	2·47	Ch	2·50	Bh	2·17	B_2	2·78	C_1	2·56	Bh	2·38	Cm	1·69
Ag	3·17	D	2·78	B_1	3·17	A_1	3·18	C_2	2·53	Bh	2·56	A_4	2·31	B_1	2·82	T_1	3·10	Ag	2·58	Ag	1·95
A_4	3·26	Ch	2·81	D	3·21	C_2	3·23	B_2	2·61	A_4	2·77	M	2·53	Cm	2·90	M	3·15	D	2·75	Bh	2·07
Bh	4·41	Bh	3·79	Bh	3·46	D	3·84	B_1	2·83	Ag	3·50	Cm	3·22	Ag	3·21	Ag	3·46	Cm	2·94	Ch	2·11
Cm	5·40	Cm	4·32	C_1	3·50	Bh	5·01	Bh	3·15	Cm	3·75	Ch	3·23	Ch	3·84	T_3	3·49	Ch	3·18	D	2·34
Th	6·51	Th	5·04	T_6	3·53	T_6	5·36	T_4	3·42	Th	5·50	Th	3·51	T_1	4·35	T_2	3·58	T_6	4·16	T_6	3·12
T_6	7·69	T_6	5·71	Th	4·16	T_4	5·46	T_1	3·58	T_4	6·22	T_6	4·47	T_4	4·41	B_2	3·79	Th	4·49	Th	3·26
T_4	7·97	T_4	6·72	T_4	4·39	T_1	5·75	T_3	3·61	T_6	6·49	T_4	5·22	T_2	4·97	Th	3·90	T_4	5·11	T_4	3·35
T_1	9·10	T_1	7·88	T_1	4·79	Th	5·83	Th	4·23	T_3	7·60	T_1	6·48	Th	5·06	B_1	4·41	T_3	6·20	T_1	3·65
T_3	9·63	T_5	8·23	T_5	4·87	T_3	5·87	T_6	4·31	T_1	7·66	T_3	6·85	T_3	5·29	T_6	4·75	T_1	6·45	T_3	4·17
T_2	10·21	T_3	8·37	T_2	5·29	T_2	7·39	T_2	4·88	T_2	8·83	T_5	7·10	T_6	7·71	Ch	5·01	T_5	7·16	T_2	5·16
T_7	10·39	T_2	8·77	T_3	5·41	T_5	7·50	T_5	5·88	T_5	9·93	T_2	7·31	T_7	7·85	T_7	5·56	T_7	7·20	T_5	5·22
T_5	10·41	T_7	9·07	T_7	8·30	T_7	9·91	T_7	7·01	T_7	10·18	T_7	7·79	T_5	8·78	T_5	6·30	T_2	7·26	T_7	5·43

		Kahar		Tharu		Chamar (Cm)		Chero (T_1)		Majhi (T_2)		Panika (T_3)		Kharwar (T_4)		Oraon (T_5)		Rajwar (T_6)		Korwa (T_7)	
A_2	·11	A_3	·43	T_5	1·72	A_4	·59	T_2	·29	T_1	·29	T_1	·37	T_1	·48	T_6	1·20	T_5	1·20	T_4	1·64
A_4	·43	A_2	·58	T_4	1·79	T_3	1·15	T_3	·37	T_3	·56	T_2	·56	T_3	·56	T_4	1·60	Th	2·01	T_5	2·13
A_1	·48	Cm	·59	T_6	2·01	M	1·23	T_4	·48	T_4	·77	T_4	·56	T_2	·77	Th	1·72	T_4	2·31	Th	2·29
C_2	·99	M	·78	T_7	2·29	A_3	1·28	Cm	1·36	Cm	1·99	Cm	1·15	Cm	1·33	T_7	2·13	Cm	2·39	T_1	2·41
Cm	1·28	A_1	1·52	A_3	2·65	T_4	1·33	A_4	2·31	T_7	2·43	A_4	2·02	T_5	1·60	T_1	2·48	A_3	2·67	T_2	2·43
M	1·37	T_3	2·02	Cm	2·75	T_1	1·36	T_7	2·41	T_5	2·57	T_7	2·44	T_7	1·64	T_3	2·53	A_4	2·93	T_3	2·44
B_2	1·45	Ch	2·03	T_1	3·02	A_2	1·69	T_5	2·48	A_4	3·09	T_5	2·53	Th	1·79	T_2	2·57	A_2	3·12	T_6	3·92
Bh	1·65	T_4	2·04	A_4	3·24	Bh	1·91	Th	3·02	Th	3·42	A_3	3·35	T_4	2·04	Cm	3·25	T_3	3·36	A_4	4·21
C_1	1·73	Bh	2·09	A_2	3·26	T_2	1·99	Bh	3·10	Bh	3·58	T_6	3·36	T_6	2·31	A_4	4·17	Ag	3·53	Cm	4·50
Ag	2·07	T_1	2·31	T_2	3·42	Ag	2·12	A_3	3·58	A_3	4·25	Th	3·42	Bh	2·50	A_3	4·53	T_1	3·89	A_3	4·56
B_1	2·11	C_2	2·31	T_3	3·42	T_6	2·39	M	3·58	T_6	4·47	Bh	3·49	A_3	2·67	Ag	4·87	T_7	3·92	A_2	5·43
D	2·20	Ag	2·40	C_2	3·51	Ch	2·50	A_2	3·65	M	4·88	M	3·61	A_2	3·35	A_2	5·22	A_1	4·16	Bh	5·56
Ch	2·58	D	2·50	Bh	3·90	Th	2·75	T_6	3·89	D	4·97	A_2	4·17	M	3·42	M	5·88	M	4·31	M	7·01
Th	2·65	B_2	2·68	Ag	4·16	D	2·90	D	4·35	A_2	5·16	D	5·29	Ag	4·39	Bh	6·30	T_2	4·47	A_1	7·20
T_4	2·67	C_1	2·77	M	4·23	A_1	2·94	Ag	4·79	Ag	5·29	Ag	5·41	D	4·41	C_2	7·10	C_2	4·47	C_2	7·79
T_6	2·67	T_6	2·93	A_1	4·49	C_2	3·22	Ch	5·75	A_1	7·26	Ch	5·87	A_1	5·11	A_1	7·16	Bh	4·75	D	7·85
T_3	3·35	T_2	3·09	B_2	5·04	T_5	3·25	A_1	6·45	C_2	7·31	A_1	6·20	C_2	5·22	Ch	7·50	Ch	5·36	Ag	8·30
T_1	3·58	Th	3·24	D	5·06	C_1	3·75	C_2	6·48	Ch	7·39	C_2	6·85	Ch	5·46	B_2	8·23	B_2	5·71	B_2	9·07
T_2	4·25	B_1	3·26	C_1	5·50	B_2	4·32	C_1	7·66	B_2	8·77	C_1	7·60	C_1	6·22	D	8·78	C_1	6·49	Ch	9·91
T_5	4·53	T_5	4·17	Ch	5·83	T_7	4·50	B_2	7·88	C_1	8·83	B_2	8·37	B_2	6·72	C_1	9·93	D	7·71	C_1	10·18
T_7	4·56	T_7	4·21	B_1	6·51	B_1	5·40	B_1	9·10	B_1	10·21	B_1	9·63	B_1	7·97	B_1	10·41	B_1	7·69	B_1	10·39

The G-cluster shows distinct evidence of patterning within itself. To get a clearer view of the inter-relationships the D^2 for all pairs of groups in the G-cluster except Agharia (Ag), for which the mean value of FB is not available, have been calculated by using all the 9 transformed variates, i.e. using the characters *HL, HB, B_zB, NL, NB, NT, KH, SH, FB*, and tabulated in Table 9 in a manner similar to that adopted in Table 8. As observed before, the D^2's based on 9 characters can be obtained from those of Table 8 based on the 8 characters by adding the square of difference in the ninth transformed character y_9. The mean values of y_9 for groups other than those in the T cluster are given below.

Name of the Group	B_1	B_2	Ch	M	C_1	C_2	D
Mean y_9	·126	·143	·378	·276	−1·270	−·823	·325
Name of the Group	**Bh**	**A_1**	**A_2**	**A_3**	**A_4**	**Th**	**Cm**
Mean y_9	·318	−·078	−·079	·001	−·075	−·737	·254

TABLE 9.—*Values of* D^2 (*Based on* 9 *Characters*) *Arranged as in Table* 6

Brahmin Basti (B_1)		Brahmin Others (B_2)		Bhatu (C_1)		Habru (C_2)		Dome (D)		Bhil (Bh)		Chamar (Cm)	
B_2	·27	B_1	·27	C_2	1·32	A_1	1·26	Bh	1·15	D	1·15	A_4	·69
A_1	1·17	A_1	·78	A_1	2·68	C_1	1·32	C_2	2·11	A_3	1·75	M	1·22
A_2	1·48	A_2	1·03	A_2	2·98	A_2	1·53	A_3	2·31	A_2	2·23	A_3	1·35
A_3	2·13	A_3	1·47	A_3	3·35	B_2	1·63	A_2	2·41	Cm	2·23	A_2	1·81
C_2	2·23	C_2	1·63	B_1	3·48	A_3	1·67	M	2·47	A_4	2·24	Bh	2·23
M	2·86	M	2·62	B_2	3·61	D	2·11	A_4	2·66	C_2	2·43	Ch	2·52
D	2·86	A_4	2·72	A_4	4·20	B_1	2·23	B_2	2·81	A_1	2·53	D	2·91
Ch	3·05	D	2·81	M	4·46	Bh	2·43	B_1	2·86	M	3·16	A_1	3·05
A_4	3·30	Ch	2·87	D	4·52	A_4	2·87	Cm	2·91	B_2	3·82	Th	3·74
C_1	3·48	C_1	3·61	Bh	5·08	Th	3·51	A_1	2·91	B_1	4·45	B_2	4·34
Bh	4·45	Bh	3·82	Ch	5·25	M	3·74	Ch	3·84	Th	5·01	C_2	4·38
Cm	5·42	Cm	4·34	Th	5·68	Cm	4·38	C_1	4·52	Ch	5·02	B_1	5·42
Th	7·25	Th	5·82	Cm	6·07	Ch	4·68	Th	6·18	C_1	5·08	C_1	6·07

Chatri (Ch)		Muslim (M)		Ahir (A_1)		Kurmi (A_2)		Other Artisan (A_3)		Kahar (A_4)		Tharu (Th)	
M	·40	Ch	·40	A_2	·30	A_3	·12	A_2	·12	A_3	·43	A_3	3·19
A_2	2·12	A_4	·90	A_3	·49	A_1	·30	A_4	·43	A_2	·58	C_2	3·51
A_4	2·24	A_2	1·34	B_2	·78	A_4	·58	A_1	·49	Cm	·69	A_2	3·72
Cm	2·52	A_3	1·45	B_1	1·17	B_2	1·03	Cm	1·35	M	·90	Cm	3·74
A_3	2·72	Cm	1·22	C_2	1·26	M	1·34	M	1·45	A_1	1·52	A_4	3·78
B_2	2·87	A_1	2·45	A_4	1·52	B_1	1·48	B_2	1·47	Bh	2·24	A_1	4·93
B_1	3·05	D	2·47	M	2·45	C_2	1·53	C_2	1·67	Ch	2·24	Bh	5·01
A_1	3·38	B_2	2·62	Bh	2·53	Cm	1·81	Bh	1·75	D	2·66	M	5·25
D	3·84	B_1	2·86	C_1	2·68	Ch	2·12	B_1	2·13	B_2	2·72	C_1	5·68
C_2	4·68	Bh	3·16	D	2·91	Bh	2·23	D	2·31	C_2	2·87	B_2	5·82
Bh	5·02	C_2	3·74	Cm	3·05	D	2·41	Ch	2·72	B_1	3·30	D	6·18
C_1	5·25	C_1	4·46	Ch	3·38	C_1	2·98	Th	3·19	Th	3·78	Ch	7·07
Th	7·07	Th	5·25	Th	4·93	Th	3·72	C_1	3·35	C_1	4·20	B_1	7·25

The internal patterning has become more definite by the use of FB as an additional character. The two Brahmins Basti (B_1) and Others (B_2) cluster together and so also the four Artisans Ahir (A_1), Kuroni (A_2), Others (A_3) and Kurmi (A_4), who appear to be linearly arranged in the order A_1, A_2, A_3 and A_4, with A_4 being more distant from A_1. Bhatu (C_1) and Habru (C_2) gc together with C_2, being nearer to the Brahmin and Artisan clusters and C_1 far removed from them. Bhil (Bh) and Dome (D) are closer and so also Muslim (M) and Chatri (Ch). Tharu (Th) occupy a distinct position, being far away from the groups considered above. In fact they occupy an intermediate position between the tribal and the general clusters with more affinities towards the former. Chamar (Cm) also occupy a distinct position far removed from the Brahmins (B_1, B_2), Bhatu (C_1), Habru (C_2) and Tharu (Th), but closer to the rest of the members of the G-cluster.

The classification arrived at by a proper scrutiny of the Tables 8 and 9 may be summarized as follows:

1. There are two distinct clusters, one comprising the tribal groups ($T_1, T_2, \ldots, T_7$) and the other general groups ($A_1, \ldots, A_4$, Ag, B_1, B_2, Bh, C_1, C_2, Ch, D).
2. Tharu (Th) and Chamar (Cm) form border-line cases with Th having more affinities with the T cluster and Cm with the G cluster.
3. There is an internal patterning in the T cluster with T_1, T_2, T_3, T_4 and T_5, T_6, T_7 forming sub-clusters, T_7 in the latter being nearest to the members of the former.

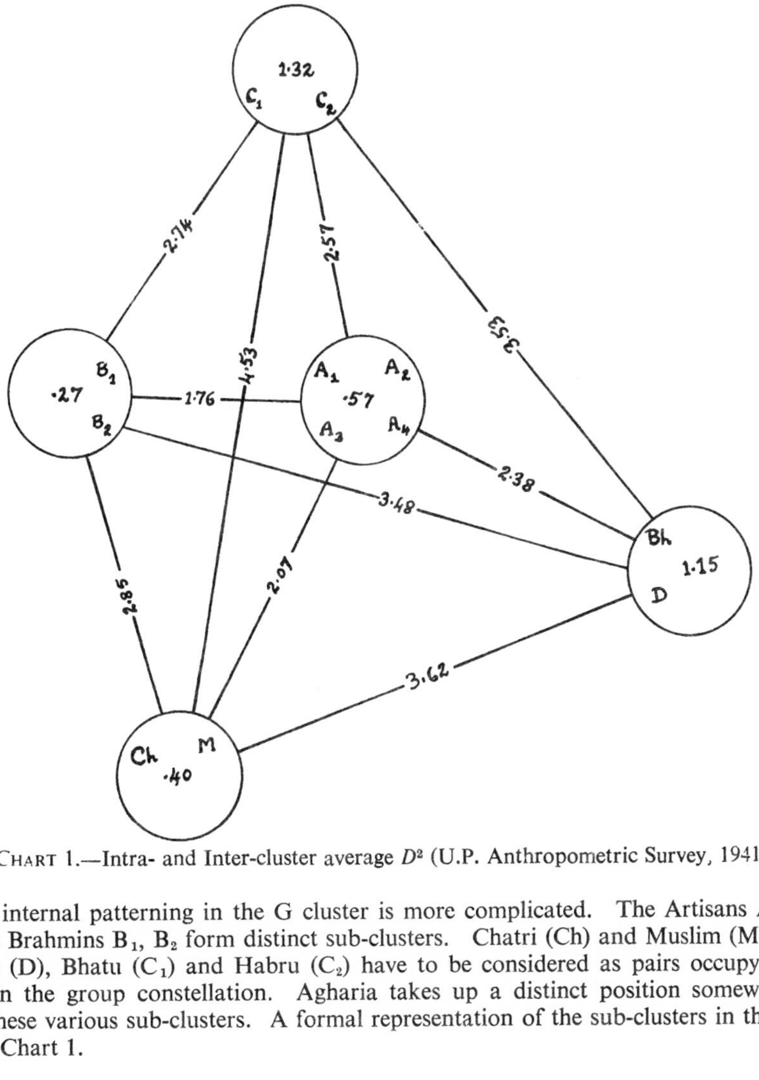

CHART 1.—Intra- and Inter-cluster average D^2 (U.P. Anthropometric Survey, 1941).

4. The internal patterning in the G cluster is more complicated. The Artisans A_1, A_2, A_3, A_4 and the Brahmins B_1, B_2 form distinct sub-clusters. Chatri (Ch) and Muslim (M), Bhil (Bh) and Dome (D), Bhatu (C_1) and Habru (C_2) have to be considered as pairs occupying distinct positions in the group constellation. Agharia takes up a distinct position somewhere in the centre of these various sub-clusters. A formal representation of the sub-clusters in the G cluster is given in Chart 1.

PART III: THE USE OF CANONICAL VARIATES IN DERIVING GROUP CONSTELLATIONS

8. *Graphical Methods of Representing the Groups*

The method of finding the group constellations as described in Part II becomes considerably simple if the groups are characterized by two or three measurements. In the case of two characters x_1 and x_2, one can represent the mean values, expressed in standard deviation units, of the k populations under consideration on a two-dimensional chart with axes inclined at an angle

$\cos^{-1}r$, where r is the correlation between x_1 and x_2. In such a chart, the distance between two points is equivalent, apart from a constant multiplier to Mahalanobis's D between two populations represented by the two points. This is valuable as it facilitates the study of group constellations, and also serves as a pictorial representation of the configuration of various groups.

In the case of three characters one can construct a three-dimensional model representing the characters along three mutually inclined axes. In order that the distance between two points might be equal to the D, apart from a constant multiplier, between two populations represented by them, the angle between the axes corresponding to x_1 and x_2 should be chosen as $\cos^{-1}(r_{12} - r_{13}\, r_{23})$ where r_{ij} represents the correlation between x_i and x_j.

If the object of representation is only to measure D by the actual distance in space, one can transform the characters to independent variables, in which case they can be represented along three mutually orthogonal axes. This method of representation fails when one is dealing with more than three characters. In such cases it might be useful to examine whether the configuration of mean values with respect to $p > 3$ characters can be preserved, so far as is possible, by representing the groups with respect to two or three suitably chosen functions of the p characters. A convenient measure for examining the adequacy of such a simpler representation is given by the ratio of the sum of squares of all possible $k\,(k-1)/2$ distances arising out of the k populations in the simpler to that of the p-dimensional representation. The former sum of squares is not greater than the latter, and the two representations are identical when the ratio is unity. When this ratio is close to unity the simpler model might be considered as a fair representation of the groups in the total character space.

9. *The Problem of Maximal Average D^2*

Restricting only to linear functions of the characters the general problem may be stated as follows:

What are the best t ($<$ p) *linear combinations of the characters which make the sum of all possible* D^2*'s arising out of* k *populations, as calculated with these functions alone, a maximum?*

Let M denote the matrix

$$M = \begin{pmatrix} m_{11} & \cdots & m_{1p} \\ \cdot & \cdots & \cdot \\ m_{k1} & \cdots & m_{kp} \end{pmatrix}$$

where m_{ij} stands for the mean of the jth character in the ith population. Let S denote the common dispersion matrix of the k populations. I have shown elsewhere* that if R denotes the matrix

$$R = ((m_{ij} - \bar{m}_{.j}))$$

where $k\bar{m}_{.j} = m_{1j} + \ldots + m_{kj}$, the best t functions are supplied by the first t canonical vectors of the matrix $R'RS^{-1}$. Either the canonical variates corresponding to the canonical vectors or their linear combinations supply the solution to the above problem.

10. *An Illustrative Example*

Using the transformed characters given in section 7, the variances and covariances between 22 groups are calculated and given in Table 10. Since all the transformed characters are independent, the determinantal equation giving the canonical variances is

$$\begin{vmatrix} a_{11}-\lambda & a_{12} & \ldots & a_{18} \\ a_{21} & a_{22}-\lambda & \ldots & a_{28} \\ \cdot & \cdots & \cdots & \cdot \\ a_{81} & a_{82} & \ldots & a_{88}-\lambda \end{vmatrix} = 0$$

where a_{ij} are as given in Table

* Report on Anthropometric Survey of the United Provinces, 1941, by P. C. Mahalanobis, D. N. Mazumdar and C. R. Rao (Appendix 5).

The process of obtaining the canonical vectors and the canonical variances by numerical methods is very laborious, especially when the number of characters exceeds 4 or 5. In the present case, solutions were obtained with the help of the Mallock's Machine* housed in the Mathematical Laboratory of Cambridge University.

TABLE 10.—*The Dispersion Matrix between Groups*

y_1	y_2	y_3	y_4	y_5	y_6	y_7	y_8
·1304	·0313	−·0337	·0547	−·0873	·0443	−·0001	·0433
..	·0884	−·0070	·0758	−·1183	·1416	·0469	·1051
..	..	·0467	·0144	·0417	·0473	−·0003	·0187
..	..	..	·1780	−·1494	·2049	·0308	·1090
..	..	..	..	·3046	−·2141	−·0196	−·1767
..	..	..	..	..	·5500	·0610	·1931
..	..	..	..	..	..	·1056	·0270
..	..	..	..	..	..	..	·3800

TABLE 11.—*The First Three Canonical Variances and Vectors*

	Variance	Canonical Vector							
		y_1	y_2	y_3	y_4	y_5	y_6	y_7	y_8
λ_1	1·0029	·128	·244	·020	·333	−·439	·648	·091	·438
λ_2	·2745	−·214	−·055	·183	·066	·329	·648	·094	−·614
λ_3	·2087	−·445	−·057	·312	−·200	·506	·116	·049	·624

TABLE 12.—*Mean Values of the Canonical Variates for the 22 Groups Considered in Section 7*

Group	Mean Values		
	λ_1	λ_2	λ_3
B_1	1·596	−·213	·226
B_2	1·318	−·448	·150
Ag	·496	−·209	−·592
Ch	·814	·214	−1·129
M	·457	·374	−·698
C_1	1·323	·205	·178
C_2	1·000	−·319	·384
Bh	·681	·955	·556
D	·129	·628	·701
A_1	·966	−·323	·299
A_2	·582	−·194	·052
A_3	·358	−·150	·161
A_4	·051	·209	−·209
Th	−·671	−·732	·382
Cm	−·331	·396	−·457
T_1	−1·240	·629	−·081
T_2	−1·425	·632	·099
T_3	−1·291	·550	−·198
T_4	−1·112	·168	·105
T_5	−1·461	−·890	−·298
T_6	−·699	−1·061	−·410
T_7	−1·541	−·421	·779

* I am indebted to Dr. M. V. Wilkes for his valuable help in working this machine.

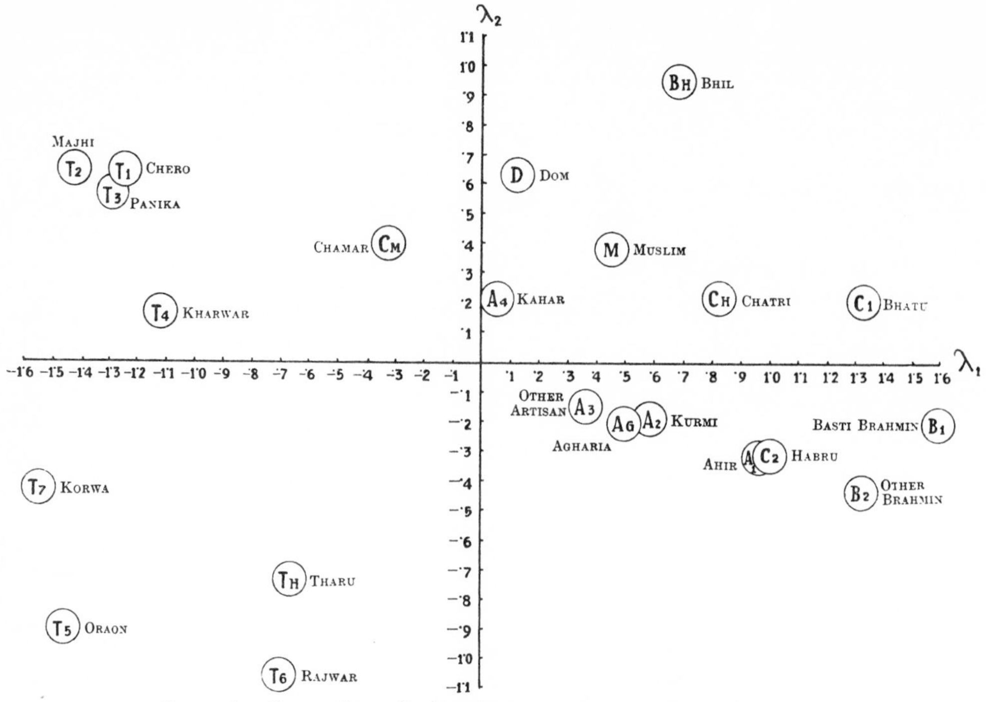

CHART 2.—Group Constellations (U.P. Anthropometric Survey, 1941).

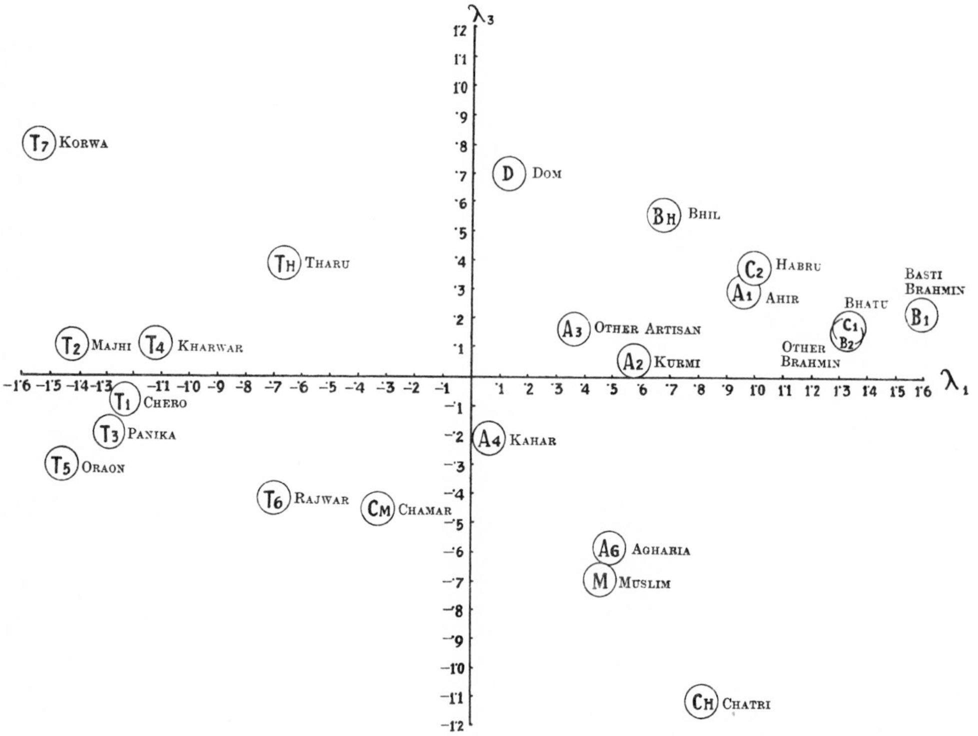

CHART 3.—Group Constellations (U.P. Anthropometric Survey, 1941).

Out of the 8 roots with a total of 1·7837, the first three are $\lambda_1 = 1 \cdot 0029$, $\lambda_2 = \cdot 2745$, $\lambda_3 = 2087$, adding up to 1·4862. The other five roots are all less than ·07, and individually contribute very little to the discrimination of the various groups. Two or three dimensional models giving the positions of the various groups with respect to two or three canonical variates can, therefore, give a good idea of the configuration in the total character space.

The first three canonical vectors and the mean values of the canonical variates for all the groups are given in Tables 11 and 12 respectively. A glance at Table 12 shows that the first canonical variate corresponds to a general factor indicating a sharp differentiation of the tribal from the general cluster. It is also the best single linear function with which all the groups can be differentiated. The second canonical variate has got greater variance for members of the tribal cluster, and hence is more useful for determining its sub-clusters. On the other hand, the third canonical variate has greater variance for members of the general cluster and is, therefore, more useful in classifying its members. The position is explained in charts 2 and 3, giving the representations of the variates (λ_1, λ_2) and (λ_1, λ_3) respectively. These charts, besides offering pictorial representations of the groups, are useful in studying the relative positions of the groups in sub-clusters.

Summary

Attempts have been made in this article to supply statistical criteria for assigning an individual specimen to its proper group and for arranging the groups themselves in a hierarchical system. The first problem was solved by extending the theory of discriminant functions which is appropriate for cases involving two groups. The second problem was solved by the introduction of a distance function which is capable of a simple interpretation. The scope of these methods is very general, and useful applications can be found in economic and psychological and other studies. Recently Patrick Slater and myself had the opportunity of examining the data collected by Dr. Maya Gross where-in five "neurotic groups" and one "normal group" had to be classified on the basis of three measurements. All the "neurotic" groups, while differing widely from the "normal," had an internal pattern of their own. The results of this investigation will be published in a forthcoming issue of the *British Journal of Psychology*.

So far as the first problem is concerned no serious difficulty is experienced in the choice of characters. Usually all the available or some of the best discriminating characters on the individual may be used. As for the second problem, the characters used should satisfy some conditions if useful classifications are to be obtained.

The aim of anthropological classifications in particular is, as Morant (1939) says, "to unravel the course of human evolution, and it may be taken for granted to-day that the proper study of the natural history of man is concerned essentially with the mode and path of his descent."

For arriving at such a classification it is necessary to choose those characters which are indicators of racial affinity. Very little work has been done in this direction. Fisher (1936*b*), writing on the future of craniometry, says, "Broadly speaking and by analogy with the progress of other sciences it may be suggested that the fundamental problems needed for the ethnographic interpretation of cranial remains must be advanced many steps further than the present state of knowledge before they can contribute appreciably to our knowledge of racial history."

This problem is, essentially, connected with the advances in human genetics and the effects of "nature and nurture" on morphological characters. It is to be hoped that these advances will bring to light the complex factors effecting the course of evolution, and that the study of skeletal remains will be of real evidential value in determining the mode and path of descent of our own species. On the other hand, it is difficult to maintain that the tentative classifications obtained on the basis of available sets of measurements are mere gratifications of idle curiosity. One may recall the following statement by Karl Pearson (1898):

"No scientific investigation can be final; it merely represents the most probable conclusion which can be drawn from the data at the disposal of the writer. A wider range of facts, or more refined analysis, experiment, and observation will lead to new formulae and new theories. This is the essence of scientific progress."

APPENDIX 1: GENERALIZATION OF A LEMMA OF NEYMAN AND PEARSON

(A) Let $R'_1, R'_2, \ldots$ be a set of mutually exclusive regions such that with respect to functions $g_1, g_2, \ldots$ the values

$$\int_{R_i} g_j \, dv = s_{ji}$$

are constant,

(B) Consider the system of regions

$$R_k \cap F_k \leqslant F_s \quad s = 1, 2, \ldots$$
$$k = 1, 2, \ldots$$

where $$F_k = \varphi_k + \lambda_{k1} g_1 + \lambda_{k2} g_2 + \ldots$$

φ_k being some assigned functions.

We now prove that the value of the integral

$$\int_{R'_1} \varphi_1 \, dv + \int_{R'_2} \varphi_2 \, dv + \ldots$$

Subject to the conditions (A) is a minimum for the set of regions defined in (B).

The intersection of the regions R_i and R'_j is represented by R_{ij}. It follows from definition that

$$\int_{R_1} F_1 \, dv \leqslant \int_{R_{11}} F_1 \, dv + \int_{R_{12}} F_2 \, dv + \ldots$$

Writing down the above relationships for all R_1 and adding, we get

$$\int_{R_1} F_1 \, dv + \int_{R_2} F_2 \, dv + \ldots \quad \leqslant \int_{R'_1} F_1 \, dv + \int_{R'_2} F_2 \, dv + \ldots$$

or

$$\int_{R_1} \varphi_1 \, dv + \int_{R_2} \varphi_2 \, dv + \ldots \quad + \Sigma \Sigma \lambda_{ij} s_{ji}$$

$$\leqslant \int_{R'_1} \varphi_1 \, dv + \int_{R'_2} \varphi_2 \, dv + \ldots \quad + \Sigma \Sigma \lambda_{ij} s_{ji}$$

which establishes the result stated above. All the theorems given in Section 5 are special cases of the above lemma.

APPENDIX 2: MECHANICAL TABULATION OF DATA

A word or two may be said about the computational aspects of the formulae developed in this article. Biologists may feel that these formulae are too complicated to be of any use to them in practice. Much of the labour can be simplified by the use of mechanical tabulation. A rough idea may be given as to the amount of computation involved in the biometric analysis of about 4,000 observations on 22 castes and tribes described in this article.

The first step was to arrive at the dispersion matrix. This is not a difficult task if the punched card technique is used to calculate the variances and covariances. In such a connection, reference may be made to the paper read before this Society by Dr. H. O. Hartley (1946).

The second step was to construct the transformed characters as explained in section 7. If the mean values of these transformed characters are used, the calculation of D^2 is reduced to finding a simple sum of squares. In the present case, using 9 characters it took about 30 computing hours to calculate the transformed characters and 231 D^2 values.

The third step was calculate the canonical variates described in section 9 of this article. This was done with the help of the Mallock Machine in the Mathematical Laboratory of Cambridge University. In fact, this machine can be used to achieve a full canonical analysis (with three digital accuracy) up to 15 characters in about 6 hours' time. Higher accuracy, if necessary, can be achieved by a proper use of this machine.

Bibliography

(1) Bartlett, M. S. (1934), "The vector representation of a sample," *Proc. Cam. Phil. Soc.*, **30**, 327.
(2) —— (1938), "Further aspects of the theory of multiple regression," *ibid.*, **34**, 33.
(3) —— (1947), "Multivariate analysis," *J.R.S.S.*, **9**, 176.
(4) Brown, G. W. (1947), "Discriminant functions," *Ann. Math. Stat.*, **18**, 514.
(5) Bose, R. C., and Roy, S. N. (1938), "The distribution of the Studentised D^2-statistic," *Sankhyā*, **4**, 19.
(6) Fisher, R. A. (1936*a*), "The use of multiple measurements in taxonomic problems," *Ann. Eugen.*, **7**, 179.
(7) —— (1936*b*), "The coefficient of racial likeness," *J. Roy. Anthrop. Inst.*, **66**, 57.
(8) —— (1938) "The statistical utilisation of multiple measurements," *Ann. Eugen.*, **8**, 376.
(9) —— (1939), "The sampling distribution of some statistics obtained from non-linear regression," *ibid.*, **9**, 238.
(10) —— (1940), "The precision of the discriminant function," *ibid.*, **10**, 422.
(11) —— (1946), "A system of scoring linkage data with special reference to pied factors in mice," *Amer. Nat.*, **80**, 568.
(12) Goodwin, C. N., and Morant, G. M. (1940), "The human remains of Iron Age and other periods from Maiden Castle, Dorset," *Biom.*, **31**, 295.
(13) Hartley, H. O. (1946), "The application of some commercial calculating machines to certain statistical calculations," *J.R.S.S. Supp.*, **8**, 154.
(14) Hotelling, H. (1931), "The generalization of Student's ratio," *Ann. Math. Stat.*, **2**, 360.
(15) Hooke, B. G. E. (1926), "A third study of the English skull with special reference to the Farringdon Street crania," *Biom.*, **18**, 1.
(16) Hsu, P. (1939*a*), "On the generalised analysis of variance (1)," *ibid.*, **31**, 221.
(17) —— (1939*b*), "On the distribution of the roots of certain determinantal equations," *Ann. Eugen.*, **9**, 250.
(18) Mahalanobis, P. C. (1936), "On the generalized distance in statistics," *Proc. Nat. Inst. Sc.* (*India*), **12**, 49.
(19) Morant, G. M. (1923), "A first study of the Tibetan skull," *Biom.*, **14**, 193.
(20) —— (1926), "A first study of craniology of England and Scotland from Neolithic to early historic times, with special reference to Anglo-Saxon skulls in London Museums," *ibid.*, **18**, 56.
(21) —— (1939), "The use of statistical methods in the investigation of problems of classification in Anthropology," *ibid.*, **31**, 72.
(22) Neyman, J., and Pearson, E. S. (1936), "Contributions to the theory of testing statistical hypotheses," *Stat. Res. Memoirs*, **1**, 1.
(23) Pearson, K. (1894), "Contributions to the mathematical theory of evolution. I. Dissection of frequency curves," *Philos. Trans.*, A, **185**, 71.
(24) —— (1898), "Mathematical contributions to the theory of evolution. V. On the reconstruction of stature of prehistoric races," *ibid.*, **192**, 169.
(25) —— (1914), "On sexing osteometric material," *Biom.*, **10**, 479.
(26) —— (1926), "On the coefficient of racial likeness," *ibid.*, **18**, 105.
(27) Penrose, L. S. (1947), "Some notes on discrimination," *Ann. Eugen.*, **13**, 228.
(28) Rao, C. R. (1947), "The problem of classification and distance between two populations," *Nature*, **159**, 30.
(29) —— (1948*a*), "A statistical criterion to determine the group to which an individual belongs," *ibid.*, **160**, 835.
(30) —— (1948*b*), "Large sample tests of statistical hypotheses concerning several parameters with applications to problems of estimation," *Proc. Cam. Phil. Soc.*, **44**, 50.
(31) —— (1948*c*), "Tests of significance in multivariate analysis," *Biom.* (in press).
(32) Roy, S. N. (1939), "*p*-Statistics or some generalisations in the analysis of variance appropriate to multivariate problems," *Sankhyā*, **4**, 381.
(33) Smith, C. A. B. (1947), "Some examples of discrimination," *Ann. Eugen.*, **13**, 272.
(34) Tildesley, M. L. (1921), "A first study of the Burmese skull," *Biom.*, **13**, 176.
(35) Wald, A. (1939), "Contributions to the theory of statistical estimation and testing hypotheses," *Ann. Math. Stat.*, **10**, 299.
(36) Welch, B. L. (1939), "Note on discriminant functions," *ibid.*, **31**, 218.
(37) Wilks, S. S. (1932), "Certain generalisations in analysis of variance," *ibid.*, **24**, 471.
(38) Wishart, J. (1928), "The generalised product moment distribution in samples from a normal multivariate population," *ibid.*, **20**A, 32.

DISCUSSION ON MR. RAO'S PAPER

DR. B. L. WELCH: When Mr. Rao came to this country about two years ago, few of us were aware of the nature and extent of the contributions he had made to mathematical statistics. Since then we have seen accounts of his work, appearing regularly in a wide variety of statistical journals, and have come to look to him as a writer who always has something fundamental and interesting to add to the discussion of current problems. It has, therefore, given the Society very great satisfaction that he has been able to address a paper to one of their meetings, before his return to India.

The subject Mr. Rao has chosen is not an easy one. The computations involved can be difficult, whether one has access to a Mallock machine or not. The complexity of the calculations increases rapidly with the number of variables, and, this being the case, the newcomer to the subject may wonder how the number and choice of variables should be made. My own experience in the closely allied subject of multivariate regression analysis is that it is seldom profitable to introduce more than two *independent* variables to explain the variation observed in a third variable. One can see, however, that there are special features in discriminant analysis which might make the simultaneous introduction of more variables than this worth while. I would be interested to know whether Mr. Rao, from his own experience, has come to any conclusions as to the number of variables beyond which it is seldom profitable to go. In some sections of the paper he has, of course, made comments which are relevant to this question. In one example, for instance, he starts with eight variables and shows that most of their discriminating value may, be contained in three canonical vectors. However, these three vectors are still, themselves linear functions of the whole set of eight variables. At the top of p. 184 Mr. Rao presents a somewhat different approach. He starts there with one variable and then brings in the others successively, one at a time. What I do not fully appreciate, however, is how the choice of variables at each stage is made and how one decides where to stop.

I have only one point to make which might possibly be interpreted as a criticism of anything in Mr. Rao's paper and this, perhaps, should rather be described merely as an interrogatory than as a criticism. It is concerned with the equations on p. 164, which must be solved when a distribution is resolved into two Gaussian components by the method of moments. I wonder whether it is really best always to use the k-statistics simply because their expectations are equal to the population cumulants. One knows that the k-statistics have some advantage in that methods exist for writing down more concisely some of their associated sampling formulae. But there is another consideration which, I think, was first alluded to by Bertelsen in 1927, and this is that certain consistency relationships between the cumulants may not hold between the corresponding k's. For instance it is well-known that the relation

$$\kappa_4 + 2\kappa_2^2 - \kappa_3^2/\kappa_2 \geqslant 0$$

must subsist between the cumulants of any real distribution. If, from a sample of n values, we calculate the k-statistics, it is, however, not necessarily true (although it will usually be true) that

$$k_4 + 2k_2^2 - k_3^2/k_2 \geqslant 0.$$

This may be seen, for instance, by considering a sample of 4, having two values equal to a (say) and two values equal to b. It follows that if we accept the k's as presumptive values of the κ's we may in fact be presuming values for the κ's which can belong to no real distribution.

I do not wish to stress this fact unduly, particularly as it is unlikely to arise in any practical problem of resolving distributions into normal components. This type of resolution is scarcely an exercise one would indulge if n were not reasonably large and then the consistency relation referred to will almost certainly be satisfied by the k's. But, of course, if n is large, biasses are scarcely likely to be important, whatever the way in which the presumptive values of the κ's are defined. As Mr. Rao has himself raised the question of bias, perhaps I may be forgiven for raising this other which is little more irrelevant. This latter question is moreover one which arises also in other connections. I should therefore be interested to know whether Mr. Rao has any views about it.

In conclusion I should like to propose, on behalf of all present, a cordial vote of thanks to Mr. Rao for his paper. It will, I am sure, become a constant source of reference for anyone wishing in future to employ the techniques of discriminant analysis.

DR. H. O. HARTLEY: I think we all agree that Mr. Rao's paper makes extremely interesting and stimulating reading. There is a great deal of incitement to further thinking and study. The range of subject matter treated is extremely wide: Starting from the abstract logic of the Problem of two groups and passing through Discriminant Functions, Maximum Liklihood, Generalized distance D^2, Mr. Rao, in a final appendix, gives a description of a mechanical computing programme, a terrific descent from the sublime to the mundane.

It is a most gratifying feature of the paper that at all stages the theory is illustrated with examples. These are mainly taken from anthropology, and I hope that later on, we shall have contributions from the expert anthropologists on the applications of Mr. Rao's methods. I am not qualified to speak about these and will confine myself to a few remarks on technique.

Mr. Rao's method of discrimination is derived from a most general principle—that of minimizing the chance of misplacement—and his method must, therefore, have this optimum property of minimizing misjudgment. As so often happens, the very fact that the method is the "best" under certain conditions implies that it often involves laborious calculations. The question that must therefore be asked in general is this: does the gain in statistical efficiency always warrant the extra computational labour? And most emphatically do we ask this question if there is some doubt as to whether the basic assumptions are strictly justified, because if they are not, the optimum property of Mr. Rao's method becomes doubtful.

Now, to the assumptions made: Mr. Rao's theory is, of course, perfectly general, but when he comes to execution he makes the assumption of characters distributed in a multinormal distribution with a common dispersion matrix for all groups. Nobody will blame him for the assumption of normality, but the condition of homoscedasticity is often not satisfied as has been pointed out in recent work by Penrose and Smith, although their examples are not anthropological ones.

It is perhaps ungrateful to ask for more where so much has been given, but the user of these methods would like to know when he can be absolved from using Mr. Rao's most efficient methods in favour of some simpler, less efficient substitutes. For example on the author mentions a short cut discriminant function used by Professor Penrose: the "size" and "shape" factor, and states that it has been found a good substitute in some problems where the correlations are not all simultaneously zero. What one would like to know is under what conditions (if any) is this substitute definitely not good enough.

Again, in the problem of splitting a mixed up series into two Gaussian components the author himself provides first a simple solution using the method of moments, he then gives a statistically more efficient method based on maximum likelihood and Fisher's method of scoring. In particular, Mr. Rao uses the moment solution to start off the iteration process leading to the maximum likelihood solution which is found to differ from the former by corrections δm_1, δm_2 and δs. These are all about 1/3 per cent. of the original moment solution, about 1/8 of their respective standard errors and, therefore, obviously unnecessary. The example is, perhaps, an unfortunate one from the author's point of view and a lucky one for the method of moments, but I should like to know whether the author has worked out standard errors of his moment estimates and compared them with those of the maximum likelihood solution.

One final minor point concerning the number of figures quoted. One knows, of course, that in Fisher's method of scoring it is necessary to compute certain quantities to a high accuracy as differentials are obtained as differences between two almost identical quantities, and no objection is raised on this score. However, take, for example, the system of equations where coefficients are given to 6 places of decimals. The idea of 0·023 239 × Nasal Height is somewhat ridiculous. How accurately can one determine a nasal height? The system of equations is, of course, recurrent, but does the accumulation of rounding off errors warrant 6– decimal accuracy in the coefficients?

My remarks may, perhaps, appear a little critical but should in no way detract from the excellence of the paper and the high standard of the work. I have much pleasure in seconding the vote of thanks.

Dr. G. M. Morant said that the problems discussed by the author had been treated by a number of distinguished statisticians during the past fifty years, chiefly by Karl Pearson and R. A. Fisher. The pioneer work of Pearson in the field aimed at providing the solutions of certain anthropological problems. His successors extended the theory and applied it principally to the same kind of problems, as the present author had done in most of his examples. It had been pointed out that the methods might be applied to material of various kinds. His own practical experience of the matter had been limited to Pearsonian methods and to anthropological applications.

With that kind of background he desired to make a few comments of a severely practical kind. The first class of problem treated by the author was that of assigning an individual to one of two or more populations on the basis of multiple characters. Judging from experience of less refined methods, this was often not a profitable line of inquiry. The answer in the case of anthropological populations was quite likely to be that there are highest and almost equal probabilities that an individual belonged to two or more groups not closely related to one another. The matter could be examined experimentally by choosing individuals at random from a known series and comparing them with other series. It might be argued that if any method was likely to assign brothers to different communities then the question being asked was not of the right kind. Anthropology was essentially the study of groups, and the student need not despair if he was unable to place an individual.

As a variant of the same class of problems, the author considered the resolution of mixed-up series into two Gaussian components. After many unsuccessful attempts to do this, the speaker was obliged to conclude that the anthropologist could hope to derive little benefit by applying the Pearsonian method, either because the samples were seldom large enough, or because the populations were not sufficiently far apart to give effective results. In these days of easier computing, by those who have the facilities, the matter could be investigated by treating mixtures of samples. On the practical issue of sexing bones, could any mathematical method of sexing give a more accurate result than the anatomical? In 1936 Dr. E. S. Martin published in *Biometrika* the results of an examination of 430 human lower jaws. A discriminant function method suggested by Prof. Fisher was found to give the best results, but it was little more effective than a simple method of scoring. Again there was scope for the experimental testing of the practical utiilty of the statistical methods proposed.

The second class of problem discussed by the author was that of assessing the relative degrees of resemblance of a number of populations on the basis of mean measurements of samples for a number of characters. The only method of this kind widely applied was that of Karl Pearson, and experience of its use suggested the following conclusions. In any comparison of two series, distinctions will depend on a minority of the characters used, and for comparisons in general a considerable number—20 to 30, say— should be used when possible. If it was desired that the method should be one of wide applications, at least 20 characters would be needed. A less precise method applied to a larger number of characters might give better results than a more precise method applied to a smaller number of characters.

Secondly, the value of the method and the way in which its results should be interpreted could be judged only by the criterion of reasonableness. In the case of anthropological material the statistical measures of resemblance could only be accepted as measures of relationships of the populations if there was conformity in a general way with historical and archaeological evidence and with considerations regarding geographical and secular contiguity.

It was soon discovered that the coefficient of racial likeness gave some results which were obviously unreasonable. A more distant degree of resemblance might be found between series representing two western European populations, say, than between one of them and a non-European series from a remote part of the world. This kind of situation was encountered repeatedly. It was found, however, that reasonable and suggestive results were always given if use was made only of coefficients of racial likeness indicating a close order of resemblance, leaving all others out of account. A classification of populations could then be reached by taking short jumps and ignoring different degrees of distant relationship. Anthropological evidence of other kinds suggested that this was a proper and safe way to proceed. These conclusions were derived from experience of the coefficient of racial likeness. The maker of that tool, Karl Pearson, was the first to point out that it was an imperfect one because it neglected correlations between characters. It seemed to the speaker extremelv probable that the conclusions referred to would still apply to the results given by methods which were theoretically more correct.

The anthropological situation was such that ten characters might not be enough to give reasonable results. Another point was that the new methods used the evidence of all degrees of relationship, and it was possible that this was not justified. One would like to see an experimental inquiry carried out applying the three methods he had mentioned to give comparisons of the same set of series. It might be that the results would be in close agreement.

He desired to comment on a point which Dr. Hartley had raised regarding Mr. Rao's assumption of equality in variances and normality of distributions. It seemed to him there that Mr. Rao was entirely justified, as experience had shown that general rules for all populations in the world are that anthropometric characters are normally distributed, and that their variances are very close, if there had been no special selection of the series.

Mr. Rao's paper was a valuable contribution to these problems. In his own brief comments he had stressed the difficulties, as they appeared to him, in connection with the statistical tools which were proposed. In recent years the most notable work regarding the specialized topics here considered had been carried out by Indian statisticians, and it was to be hoped that Mr. Rao and his colleagues would continue to pursue these questions in theory and practice.

Dr. J. M. Tanner said that as a "consumer" of Mr. Rao's product he wished very much to welcome this paper and to point out, what was obvious, that its use extended far beyond anthropology. The technique described would be of immediate use in physiology, medicine and psychology. For example, many measurements had been made on patients suffering from various constitutional diseases, but the lack of a suitable statistical technique for handling this data had been partly responsible for the limited success of this approach.

A good deal of the activity of physical anthropologists could be classified under two headings:

(1) Differentiating ethnic and other groups by means of measurements in order to classify the *groups*.

(2) Classifying the measurements themselves, in order to discover something about the genesis of the measurements; that is to find out about the growth of the owner of the measurements, and through this, to detail his genetic constitution and environmental mouldings.

It was because of his admiration for the way Mr. Rao had tackled the first of these problems that he would like to ask him if he had yet turned his attention to the second.

In particular, a discussion of the subject of the relations between factor analysis, the technique specifically concerned with the second of the two problems, and Mr. Rao's present tests would be most helpful to anthropologists at the present time. The serious application of factor analysis to bodily measurements dated back to 1935 and had a considerable, though not apparently widely known literature.

Clearly there were close relations between factor analysis and Mr. Rao s techniques. Specifically he desired to ask two questions on this point:

(1) What was the relation between Mr. Rao's canonical variates λ_1, λ_2, λ_3, and the first three principal factors of a Hotelling solution of the correlation matrix of Table 7? Perhaps Mr. Rao had made such an analysis of Table 7, and if so, could he say whether the principal components were similar to the canonical variates in terms of their anthropometric composition? It seemed to him that such might be the case from biological considerations.

(2) Could a good approximation to D^2 be obtained by summing the squares of the estimated mean score of each group in the first few factors of the factor analysis?

Mr. Patrick Slater congratulated Mr. Rao on a paper which had very valuable psychological applications, particularly to problems of selection. In his remarks and in personal discussions afterwards he raised the following questions:

1. He had experimented with a different approach to the same problem while working at the Directorate of Selection of Personnel, War Office. A set of measures, 1, . . . , m, could be used to find the means of a representative sample of the whole army, say $X_1, \ldots, X_m$, and of a sample of men performing duty A (excluding any who were considered by their officers definitely too good or not good enough for it), say $a_1, \ldots a_m$. Using $[d_a]$ for the differences $X_1 - a_1$, etc., written as a one-column matrix, and $[M]$ for the matrix of mean square variances and covariances of the measurements in the representative sample, the weights providing the discriminant function to use in selecting men for A would be

$$[M^{-1}]\ [d_a] = [w_a],\ \text{say},$$

where $[M^{-1}]$ is the reciprocal of $[M]$. In terms of the discriminant function the army sample mean would be 0 and the duty sample mean

$$[d_a]'[w_a] = a,\ \text{say},$$

where $[d_a]'$ is the transpose of $[d_a]$. Corrections for homogeneity in the duty sample being usually too small to be worth making, $\bar{a}$ could also be taken as the mean square variance about the mean, i.e. assuming that the matrix of mean square variances and covariances in the duty sample can be put equal to $[M]$

$$S_a^2 = [d_a]'[M^{-1}]\ [M]\ [M^{-1}]\ [d_a] = \bar{a}$$

Now let $x_1, \ldots, x_m$ be an individual's measurements in the same terms, and $[d_x]$ the column matrix of differences, $X_1 - x_1$, etc., corresponding to $[d_a]$. His score on the discriminant function measuring his suitability for A would be

$$[d_x]'[w_a] = i_a,\ \text{say},$$

and his probability of forming a homogeneous member of the group in the duty could be found by entering $\dfrac{i_a - \bar{a}}{S_a}$ as an abscissa in a table of the normal probability integral. Finding probabilities for other occupations similarly would show in which his chance of successful adaptation was greatest.

Mr. Slater did not claim that this method was preferable to Mr. Rao's, but he asked whether some convenient simplifications might not be obtained, when many samples were to be compared, if each could be treated, as in his instance, as drawn from the same population. Comparing each sample with every other must involve a larger number of comparisons than comparing each with a parent population.

2. Is canonical transformation of the variables a necessary step in finding the likelihood that an individual of unknown origin belongs to one or other of the groups under consideration? Would Mr. Rao explicitly confirm, if he agrees, that this is obtained more directly and no less precisely from the original observations?

3. Could Mr. Rao give a rule of procedure to be followed when out of an undistributed population with the parameters $X_1, \ldots, X_m$ and $[M]$, n_a have to be selected for duty A, n_b for duty B, etc., given the means for each duty? This supposes that the whole population must be distributed among the duties and the quota and the desired standards for each duty are laid down.

4. Given the parameters for the undistributed population and the means for group A, could Mr. Rao give a method for estimating the probability of finding a member of group A in the population? In selection procedures it is often possible to say what are the average requirements for admission to a selected sample, e.g. of university students or officer cadets, but not how many an efficient selection procedure should be able to discover in the population from which selection is to be made.

Mr. K. D. Tocher added his tribute to Mr. Rao's paper. He admired Mr. Rao's determination to give a clear exposition of a method of handling data and not to obscure this by the introduction of significance tests. Statisticians always had to avoid the danger of devoting their lives to proving that the other fellow had proved nothing at all.

He confessed a disappointment that Mr. Rao had not dealt with a difficulty which worried him. The best discriminant function in the multivariate normal case involved the elements of the covariance matrix. If the wrong covariance matrix was used the function was not the best one and the probability of misclassification was larger than necessary. If estimates of the elements from samples are substituted then the probability of misclassification has a sampling distribution, but all possible values are greater than the real minimum value. How do we allow for this? The calculated probability of misclassification will also be based on the sample covariance matrix and so will have a distribution. What is relation between these two distributions? He realized that such problems would involve very great mathematical difficulties, but had hoped to see a better solution than Mr. Rao's, which consisted of asserting that if the degrees of freedom of the covariance matrix estimates were large enough this effect would be negligible, and if they were small the method of analysis of mixed-up series should be used.

The general theory of the paper deserved praise. It was encouraging to see the idea of a division of the sample space into more than two parts, allowing, in addition to the assertions of belonging to one of the two alternative populations, an expression of doubt. This had first been introduced by Neyman and Pearson in the early thirties, but dropped to accelerate the development of the more important ideas of their theory of testing hypotheses. It was essential to the idea of sequential analysis and had been raised again by Barnard in his review of Wald's book *Sequential Analysis* in the *Journal of the American Statistical Association*. In that same review, Barnard had stressed the importance of the likelihood ratio as the sole criterion between two alternative simple hypotheses. Mr. Rao's work added weight to this and sketched the extension when several hypotheses were under consideration.

The iterative solution of the maximum likelihood equation for the constants of two mixed-up series followed standard lines. The last paper before the Society, that of Mr. Steven's, had used the same technique. It was not as well known as it should be that the information matrix which played an essential part in the solution need not be known exactly. In particular, if it is not changed at each step of the iteration, this will still converge. However, the matrix at the maximum likelihood solution is needed to give reasonable approximations to the standard errors of the constants.

The problem of group constellations reminded him of the problem of confluence analysis as tackled by Ragnar Frisch, who had evolved a "cluster" technique whereby variables were added to a concordant group until the value of a cluster coefficient associated with such a group "exploded." It would seem possible to combine the D^2 of Table 8 and use this in place of the cluster coefficients.

The paper contained many incidental remarks of great value to statisticians. In particular, the transformation from one set of variables to an orthogonal set should obtain better circulation as the best practical method of making such a transformation proposed to date.

Mr. Tocher added the following comment in writing after the meeting, enlarging on his suggestion for classifying the populations into groups:

The definition of a group being vague, we will not be able to introduce probabilistic concepts or make tests of the significance of our groupings. In these circumstances the simplest appropriate function of the D^2's should be used. Suppose we take the sum of D^2 for every pair of populations in any group and denote it by $\triangle^2$. Choose that group of two with the lowest $\triangle^2$. In Rao's data, this is A_2 and A_3 with $\triangle^2 = \cdot 11$. Now add each of the other populations in turn to this group and select that of lowest $\triangle^2$. This is A_1 A_2, A_3 with $\triangle^2 = \cdot 89$. Continuing in this way we obtain the following table:

Population Added to Group	$\triangle^2$	No. of Terms n	$\frac{\text{Increase in } \triangle^2}{\text{Increase in } n}$
A_2 A_3	·11	1	—
A_1	·89	3	·390
A_4	3·42	6	·843
C_2	8·41	10	1·248
B_2	14·94	15	1·306
B_1	24·47	21	1·588
C_1	36·07	28	1·657
M	51·80	36	1·966
Ag, D	95·38	55	2·294
Cp	123·36	66	2·544
Cm	155·30	78	2·662
Bh	191·00	91	2·746
Th	251·13	105	4·295
T_4	313·73	120	4·173
T_6	383·34	136	4·351
T_1	460·77	153	4·555
T_3	541·49	171	4·484
T_2	633·96	190	4·867
T_5	739·39	210	5·272
T_7	858·61	231	5·677

After adding M, the value of Δ^2 increases equally if either Ag or D is added. In these circumstances both have been added together. The number of terms in each Δ^2 is tabulated in the third column. The ratio of these two would give an estimate of D^2, if the populations were all alike, but this would not be very sensitive to an increase in any one D^2. By differencing the Δ^2 and dividing by the differences of n we obtain the fourth column, giving estimates of D^2 for equal populations, and more sensitive to increases in D^2. This clearly indicates that the group is completed with the addition of Bh, as the addition of Th, the most favourable of the remaining populations, inflates this measure from 2·746 to 4·295, a markedly larger increase than experienced before.

The remaining members can be treated in the same way. We obtain the following table:

Population Added to Group	$\triangle^2$	No. of Terms n	$\frac{\text{Increase in } \triangle^2}{\text{Increase in } n}$
T_1, T_2	·29	1	—
T_3	1·22	3	·465
T_4	3·03	6	·603
T_7	11·95	10	2·230
T_8	23·26	15	2·262
Th	38·92	21	2·610
T_6	60·08	28	3·023

This clearly gives T_1, T_2, T_3 and T_4 in one group and the rest in a third group. This grouping agrees with that of Rao except that it definitely commits Th to the tribal group and Cm to the general group.

Dr. C. C. SPICER wished to comment on a remark of Dr. Tanner who had said that the relevant data for calculating discriminant functions in some cases of medical interest already existed in the literature. Some years ago he had thought so himself, but had not been able to find any satisfactory figures as the authors did not give the correlation coefficients or covariances of their measurements so that it was impossible to calculate a discriminant function. However, the material collected by Draper and his colleagues in this connection was all preserved in the U.S.A. and would almost certainly be analysed using discriminant function technique in the future. He wished to say that the overlap in the measurements of body build of various pathological types was very great, and it would be interesting to see how efficiently the discriminant function was able to classify individuals suffering, for example, from cholecystitis and peptic ulcer on the basis of body-build measurements alone. In an application of discriminants to a war time problem of classifying wounded as incapacitated for further combat or not, he had found that the variability of the material was too great for this technique to be of real value. The discriminant sorted out about 40 per cent. correctly, but it was the 40 per cent. that could be sorted out without using statistics. He did not wish to decry discrminants on this account as they could not be expected to turn bad data into good.

He wished to thank Mr. Rao very much for a most interesting paper.

The following contributions were received in writing:

Professor M. S. BARTLETT expressed his regret that absence abroad prevented his being present at the meeting, and congratulated the author on his extremely valuable paper, both for its theoretical contributions to the general problems of discrimination and for its particular applications to Indian anthropometric material. On the theoretical side Dr. Rao's development of the discriminatory problem to three or more groups in Section 5 and his definition of a generalized distance between populations in Section 6 were especially important.

It might be noted that the particular procedure adopted when more than two groups are present, i.e. more than one group contrast, will depend to some extent on the problem. For example, in the psychological example referred to in the Summary, it might be of interest to consider separately the discriminatory problems between normal and abnormal persons, and between particular types of abnormals. Alternatively a canonical analysis would indicate from the data themselves which group contrasts were most marked.

Dr. C. A. B. SMITH: This is a very interesting paper. An especially important part is that of the resolution of a mixed up series into two Gaussian components. This might have application to genetical problems, where a population might be separated into two overlapping parts corresponding to the presence or absence of some gene—as, for example, with the threshold of tasting phenylketonuria in man. However, Mr. Rao's solution depends on the two assumptions of equality of variance and exact Gaussian form, and neither of these are likely to be exactly fulfilled in practice. I would, therefore, like to ask how much his solution is likely to be affected by the failure of these assumptions.

Professor L. S. PENROSE: In Section 6 of his paper, Mr. Rao refers to Pearson's C.R.L., which is inadequate because it assumes the correlation between characters to be zero. I would like to know whether he has considered—and, if so, what he thinks about—Zarapkin's criterion, say, Z. This is, in effect, the standard deviation of the differences between the groups expressed in terms of their own standard measurements, i.e.

$$Z^2 = \frac{1}{(p-1)}\left[\overset{p}{\underset{1}{S}}\left(\frac{d_i}{\sigma_i}\right)^2 - \frac{1}{p}\left\{\overset{p}{\underset{1}{S}}\left(\frac{d_i}{\sigma_i}\right)\right\}^2\right]$$

This measurement is an approximation to a kind of D^2 and is very easy to calculate. It has been investigated by Cavalli. I have (1947) pointed out that there is justification for using Z^2, when different groups drawn from the same species (e.g. different groups of men and women) are compared by taking several measurements of intercorrelated variables.

The presentation of the data in Sections 7 and 8 in full is most welcome. I have not had time to do so, but it will be easy to examine the results of treating the data by much simpler methods, e.g. by tabulating size against shape for each class. *N.B.* (see p. 7):

$$P\text{ (shape)} \equiv K_1\frac{x_1}{\sigma_1} + K_2\frac{x_2}{\sigma_2} + \ldots + K_p\frac{x_p}{\sigma_p},$$

where $K_1 = \frac{d_1}{\sigma_1} - \frac{1}{p}\overset{p}{\underset{1}{S}}\left(\frac{d_i}{\sigma_i}\right)$.

$$Q \text{ (Size)} = \frac{x_1}{\sigma_1} + \frac{x_2}{\sigma_2} + \ldots + \frac{x_p}{\sigma_p}.$$

These multiple measurements are uncorrelated if all intercorrelations are equal. This is, of course, not quite true. But, after all, even the precise calculation of the best linear discriminant assumes that the variances and covariances are the same in all classes and this is probably not true either.

Mr. Rao subsequently replied as follows:

The question raised by Dr. Welch regarding the number of characters cannot be easily answered. It depends on the nature of the groups compared. For anthropological classification about 10 well chosen characters may be sufficient if the groups belong to a compact geographical region. If a number of groups living under different environmental conditions are considered, the procedure would be to obtain first a broad classification based on a small number of characters and then use possibly different panels of characters to distinguish the groups classed alike by the above method. If, for instance, the ethnic composition of a country like India is to be studied, one might use the four characters, stature, sitting height, head length and head breadth to obtain a broad classification into clusters and then use a panel of about 10 characters for studying the relative positions within a cluster. If the above procedure is adopted one need not use about 30 characters as suggested by Dr. Morant.

The reasons for constructing the transformed characters are threefold: (1) They supply a set of uncorrelated linear functions of the characters easily derivable from the correlation matrix. (2) The discriminant function based on a subset of the characters chosen in the order L, B, B′, S, βQ′, H′, G′H can be constructed by using only the transformed characters involving them. Thus the discriminant function based on (L, B, B′) is simply

$$\alpha\, d_\alpha + \beta d_\beta + \gamma\, d_\gamma$$

where d_α, d_β and d_γ are the differences in mean values of α, β and γ. The discriminant function based on L, B, B[1], S

$$\alpha d_\alpha + \beta d_\beta + \gamma d_\gamma + \delta d_\delta$$

is obtained by adding δd_δ to the above. This is useful in classifying a collection of skulls not all of which supply measurements on the seven characters. In all the available series it has been found that the numbers of skulls supplying the measurements L, B, B′, S, βQ′, H′, G′H are in decreasing order. Also a skull supplying the measurements S, βQ′, H′ admits the measurements L, B, B′. A relatively smaller number of skulls supply G′H in addition to the above. The order of characters chosen in Table 2 is based on these considerations. (3) The importance of the addition of any character to the set above it can be judged by the magnitude of the difference in the mean values of the transformed character corresponding to it. If this is small the newly added character is dropped and another one is considered.

As for Dr. Welch's last point, the only assertion made in the paper is that at least the estimating equations are unbiased. I presume no difficulty would be experienced in practice by using the k-statistics if the samples are large. The situation is similar to that where the use of Sheppard's corrections may lead to a correlation coefficient which sometimes exceeds unity. Should Sheppard's corrections be ignored as not *fully appropriate*?

Dr. Hartley thought that the calculation of the coefficients to 6 decimal places in the transformed characters is unnecessary. All anthropological measurements be it nasal height or stature are recorded to three significant figures, and perhaps the mean values will be correct to four figures. I believe it is a good arithmetical discipline to calculate some functions of the observations to a higher number of significant figures than that of the observations. In the present case, it is relevant, for the method of computation adopted, to retain 6 decimal places at each stage in order that the final values may be correct to four significant figures.

Dr. Morant raised many problems which are of importance to a practical worker in the field of anthropology. I may mention, at the outset, that the theoretical methods discussed in the paper provide some basic formulae with which any empirically derived formula may be compared. It is not suggested that more refined techniques should be adopted when simpler methods can very nearly yield the same results. In cases where adequate empirical formulae are not available and the situations demand an accurate analysis as in the case of the Highdown Skull (Section 2 of the paper) it may be desirable to use a theoretically justifiable formula.

Dr. Morant pointed out that the method of resolution of a mixed series is not effective when the samples are not large and the populations are not far apart. But the aim of the statistical methods is only to extract as much information as is possible from the data, and the method of the maximum likelihood should be adequate for this purpose.

As for the application of D^2 in problems of anthropological classification I agree with Dr. Morant that much caution is necessary in the interpretation of the observed relationships. Smaller values of D^2 would mean a closer organic resemblance only when the characters chosen are real indicators of racial affinity. It is, however, safer to restrict the use of D^2 to groups of persons living under the same environmental conditions. If groups living in widely different regions are to be compared the interpretation of the observed relationship would depend on the "nature and nurture" effects on the characters used. To argue that any measure of distance is useless because the results derived from it conflict some uncertain historical evidence is to deny the possibilities of measuring organic relationships between two groups and to evade the fundamental problem of interpreting the observed differences.

Professor Penrose suggested the simple method of tabulating the shape and size factors for each group. These are not necessarily the best linear functions and also, as shown in Section 8, at least three dimensions are necessary to represent the various groups. I think the simpler method is not profitable in the present study.

Zarapkin's criterion is easier to calculate, but its validity depends on the nature of the correlation matrix. If all the correlations are nearly of the same magnitude, then using the average correlation ρ, one can construct

$$z_1 = \frac{1}{1-\rho}\{\Sigma d^2 - \rho(\Sigma d)^2/(1 + \overline{p-1}\,\rho)\}$$

where the d's are differences in mean values expressed in standard deviation units. This formula is simple to evaluate and may be a good approximation to D^2. Zarapkin's criterion is appropriate only when the characters are all highly correlated. Similar formulae are available when the characters can be thrown into two equally correlated sets. The usefulness of these formulae will be discussed elsewhere.

Mr. Patrick Slater's solution of the first problem is approximately correct. A direct and a more simpler method would be to use the results of theorem 1 of Section 5.

The discriminant scores obtained from the original observations are simpler to evaluate and are necessarily more accurate than those derivable from the canonical variates.

The third problem raised by Mr. Slater can be dealt with as follows: Let us consider the three groups, army, navy and air force with the probability densities f_1, f_2, f_3 and the proportions in the general population Π_1, Π_2, Π_3. If n_1 posts are to be filled in the army then for each individual we calculate the score

$$\lambda = \frac{\Pi_1\, f_1 + \Pi_2\, f_2 + \Pi_3\, f_3}{r_{21}\;\Pi_2\; f_2 + r_{31}\;\Pi_3\; f_3}.$$

Where r_{21} and r_{31} are the ratios of the losses incurred in accepting in the army individuals suitable for the navy and air force. The risk will be a minimum for the army if the individuals corresponding to the highest n_1 values of λ are chosen.

If the problem is that of distributing N individuals into groups of n_1, n_2, n_3 for the army, navy and air force a slightly complicated procedure is needed. If R_1, R_2, R_3 are the regions for assigning the individuals, then the expected loss is

$$\sum_{i=1}^{3} \int_{R_i} (\Pi_j r_{ji} f_j + \Pi_k r_{ki} f_k)\,dv \qquad i \neq j \neq k.$$

We choose R_i such that the above value is a minimum subject to the conditions

$$\int_{R_i} (\Pi_1 f_1 + \Pi_2 f_2 + \Pi_3 f_3)\,dv = \rho_i,\; i = 1, 2, 3.$$

Where ρ_1, ρ_2, ρ_3 are the assigned ratios in which the N individuals have to be distributed. Using the result of Appendix 1 we find

$$R_i \;\cap\; F_i \leqslant F_j\,,\;\; j = 1, 2, 3 : i = 1, 2, 3$$

where $$F_i = \Pi_j r_{ji} f_j + \Pi_k r_{ki} f_k + \lambda_i\,(\Pi_1 f_1 + \Pi_2 f_2 + \Pi_3 f_3),$$

the λ's being suitably chosen constants. The three regions so determined will not necessarily contain n_1, n_2 and n_3 individuals because of sampling errors. In such a case the expected values of ρ_1, ρ_2, ρ_3 may be altered in such a manner as to admit the requisite numbers and to minimize the total loss. This solution, though theoretically the best is very complicated. Some approximate methods are worth investigating.

I have no ready solution for Mr. Slater's fourth problem.

The problem of the sampling distribution of misclassifications raised by Mr. Tocher is extremely difficult. The discriminant function method supplies the best possible use of the data and in practice only a rough idea of the errors is needed.

Mr. Tocher's suggestion of some short cuts in iteration is very valuable.

The transformed characters as expressed in Section 7 are analogous to orthogonal polynomials. The coefficients in any transformed character can be calculated independently of each other. The actual computational procedure is discussed in the report cited in the paper.

The method of deriving group constellations as indicated by Mr. Tocher is quite useful, but I am not in favour of laying down any formal rules. I have used various devices such as finding the change in average D^2 etc., in arriving at the clusters given in Section 7. A proper scrutiny of the D^2 table always helps. I am inclined to treat Tharu as an intermediate group between the General and Tribal clusters with more affinities towards the latter. If Th has to be classified in the Tribal cluster, one would expect Th and T_5, T_6, T_7, the members of the T cluster closer to Th, to move together in columns of D^2 corresponding to the members of the G cluster. But this is not so.

Dr. Tanner referred to an important problem in his first question. The canonical analysis refers to the factorization of the between group correlation matrix, while the method of principal components to the within group correlation matrix. The computational techniques are similar. They both give rise to uncorrelated functions of characters; the former being used for describing the configuration of the means of various groups while the latter for explaining the differences in measurements of the individuals within a group. They serve different purposes and need not have any explicit relation.

As Dr. Tanner observed a good approximation to D^2 can be obtained by taking the sum of squares of differences in mean values of the canonical variates. It is necessary for this purpose that the sum of the variances of the canonical variates retained account for a high proportion of the total variance. It might, however, happen that the variance of a neglected canonical variate is concentrated among a few groups in which case some of the D^2's may be considerably altered.

I cannot think of any alternative method for Dr. Spicer's problem.

As for Dr. Smith's query, I believe the assumption of equality of variances does not invalidate the estimates of means provided the ratio of the variances is close to unity. I have not investigated the effects of non-normality in the method of resolution discussed in Section 4.

I have used the methods suggested by Dr. Bartlett in arriving at the classification of neurotic groups mentioned in the Summary of the paper.

In conclusion, I wish to thank all those who contributed to the discussion and have given me so much encouragement.

11

Reprinted from *Harvard Educ. Rev.*, **21**(2), 90–95 (1951)

THE GENERALIZED DISCRIMINANT FUNCTION: MATHEMATICAL FOUNDATION AND COMPUTATIONAL ROUTINE

Joseph G. Bryan

Introduction

When the number of groups G exceeds two, discriminative analysis involves more than one dimension. The coordinates of the group means may be regarded as defining G points in the multidimensional space represented by the test variables used, and if the number of test variables n is at least as great as G-1, the group means will define a subspace of G-1 dimensions. If the number of test variables is less than G-1, it will be mathematically impossible to determine a space of G-1 dimensions, and in that case there will be no measure at all of some of the potentially distinguishing characteristics of the groups. For this reason, it will be assumed that n is at least as great as G-1. Furthermore, it is wise not to put all the eggs in one basket, and since only a portion of the variability of any test is relevant to the contrasts between groups, it will be advisable to utilize more than just the bare mathematical minimum of estimates of possible group differences. In other words, it would be a good idea to take n substantially larger than G-1.

In any event, if we let r denote the smaller of the two numbers G-1, n and work out the immediate mathematical consequences of the discriminative criterion (namely, that the ratio of the sum of squares among groups to that within groups shall be maximized), we arrive at the following facts:

1. The entire linear discriminative capacity of the test battery is exhausted by r linear functions of the test variables.
2. These functions turn out to be mutually uncorrelated.
3. They are independent of the origin of coordinates and the units of measurement.
4. They can be identified with the latent vectors of a certain matrix.

Mathematical Statement

Given n variates $x_1, x_2, \ldots, x_n$ representing measurements on G groups of persons, there being N_g $(g = 1, 2, \ldots, G)$ persons within group g. An individual score, for example that of the pth person in group g on test j, would then be x_{pgj}. Because the contemplated mathematical development involves quantities which are independent of the mean test scores, we may assume without loss of generality that $\sum_{p,g} x_{pgj} = O$ for all values of j. However the separate group means, such as $\bar{x}_{gj} = \frac{1}{N_g}\sum_p x_{pgj}$ would not in general vanish. Consider a linear function

$$y = v_1x_1 + v_2x_2 + \ldots + v_nx_n$$

an individual value being y_{pg}. The group mean of y, of course, is

$$\bar{y}_g = \frac{1}{N_g}\sum_p y_{pg}$$

the sum of squares of y among groups is

$$\Sigma N_g \bar{y}_g^{\,2}$$

and the sum of squares of y within groups is

$$\sum_{p,g} (y_{pg} - \bar{y}_g)^2$$

The problem is to determine the coefficients $v_1, v_2, \ldots, v_n$ so that the ratio λ of the among-groups sum of squares to the within-groups sum of squares is maximized. It will be found that the function λ has, in fact, several extrema, each of which is indicative of a distinct dimension* of the subspace defined by the group means. To carry out the process of maximization, we express the sums of squares as quadratic forms in the test variates and then apply the ordinary techniques of differential calculus. The manipulation is greatly simplified by the use of matrix algebra. Accordingly we introduce the symmetrical matrices,

$$A = \|a_{ij}\|, \quad a_{ij} = \Sigma N_g \bar{x}_{gi} \bar{x}_{gj} = a_{ji} \quad (i, j = 1, 2, \ldots, n)$$

$$W = \|w_{ij}\|$$

$$w_{ij} = \sum_{p,g} (x_{pgi} - \bar{x}_{gi})(x_{pgj} - \bar{x}_{gj}) = w_{ji}$$

and the column vector $v = \|v_j\|$. Then the two sums of squares become, respectively,

$$v'Av \text{ and } v'Wv$$

and their ratio is

$$\lambda = \frac{v'Av}{v'Wv} \qquad (1)$$

Upon setting the partial derivatives of λ with respect $v_1, v_2, \ldots, v_n$ equal to zero, we arrive at the matrix equation

$$(v'Wv)Av - (v'Av)Wv = 0 \qquad (2)$$

Now dividing through by $v'Wv$ and collecting terms, we get

$$(A - \lambda W)\, v = 0 \qquad (3)$$

and this simplifies to

$$(R - \lambda I)\, v = 0 \qquad (4)$$

where I is the unit matrix and $R = W^{-1}A$. Therefore the coefficients of the discriminant functions are determined by the latent vectors of R, and the corresponding latent roots of R equal the respective ratios of among-groups to within-groups sums of squares. By considering the rank of the matrix A, it is a simple matter[1] to show that the number of solutions of (4) such that $\lambda \neq 0$ is at most equal to the smaller of the two integers G-1,n. Consequently, letting r stand for the smaller number, the total discriminative power of the test battery is exhausted by r linear functions defined in the manner stated. Among these, all functions corresponding to distinct values of λ are mutually uncorrelated[2] as they stand. Repeated roots other than zero are possible but unlikely to occur. If one or more multiple roots should occur, however, the vectors corresponding to any one of them are already uncorrelated with the vectors corresponding to all different roots and can be chosen in such a way as to be uncorrelated also among them-

*The question of statistical significance is not considered here.

[1] J. G. Bryan, "A Method for the Exact Determination of the Characteristic Equation and Latent Vectors of a Matrix with Applications to the Discriminant Function for More Than Two Groups" Ch. 5, p. 131, 145 (Unpublished Ed. D Thesis, Graduate School of Ed., Harvard Univ., 1950).

[2] J. G. Bryan, Op. Cit., Chapter 5, pp. 134-138 incl.

selves. The numerical values of these functions are independent of the origin of coordinates, the units of measurement, and in fact independent of any non-singular linear transformaton of the test variates.[2]

Numerical Illustration

The simplest situation which gives an idea of the general case is that with three groups and four tests. Although the methods described in this paper have been applied successfully to real data[3], it seems preferable from the pedagogical standpoint to present a fictitious example so constructed that the number work will be easy to follow. Let us suppose that the matrices A and W obtained from three groups and four tests are as follows:

$$A = \begin{Vmatrix} 1 & -1 & -2 & 3 \\ -1 & 2 & 5 & -4 \\ -2 & 5 & 13 & -9 \\ 3 & -4 & -9 & 10 \end{Vmatrix}$$

$$W = \begin{Vmatrix} 2 & 3 & 4 & 3 \\ 3 & 5 & 8 & 6 \\ 4 & 8 & 18 & 15 \\ 3 & 6 & 15 & 14 \end{Vmatrix}$$

We then find the inverse of W, which is

$$W^{-1} = \begin{Vmatrix} 31 & -29 & 11 & -6 \\ -29 & 28 & -11 & 6 \\ 11 & -11 & 5 & -3 \\ -6 & 6 & -3 & 2 \end{Vmatrix}$$

and afterwards, the product R, which is

$$R = W^{-1}A = \begin{Vmatrix} 20 & -10 & -10 & 50 \\ -17 & 6 & 1 & -40 \\ 3 & 4 & 15 & 2 \\ 0 & -5 & -15 & 5 \end{Vmatrix}$$

Following the writer's method[4] for determining the latent roots and vectors of any matrix, we construct the key matrix B_4. This is

$$B_4 = \begin{Vmatrix} 0 & -50 & 670 & 0 \\ 0 & 40 & -552 & 0 \\ 0 & -2 & 62 & 0 \\ -1 & 41 & -366 & 0 \end{Vmatrix}$$

Next we obtain a certain row vector f_4 given by

$$f_4 = \begin{Vmatrix} 1 & -46 & 401 & 0 & 0 \end{Vmatrix}$$

The *characteristic equation* of R can now be written down at once from f_4 and is as follows

Characteristic Equation of R $\lambda^4 - 46\lambda^3 + 401\lambda^2 = 0$
$= \lambda^2(\lambda^2 - 46\lambda + 401)$

The non-vanishing roots of the characteristic equation are $\lambda = 23 \pm 8\sqrt{2}$. We thus put $\lambda_1 = 23 + 8\sqrt{2} = 34.313708$, $\lambda_2 = 23 - 8\sqrt{2} = 11.686292$. To obtain the latent vectors, we construct the matrix L (the elements of which are the powers of the latent roots) and form the product B_4L. The columns of B_4L are the latent vectors. Here we are not interested in the vanishing latent roots and so we neglect them in constructing L. The result is

$$L = \begin{Vmatrix} \lambda_1^3 & \lambda_2^3 \\ \lambda_1^2 & \lambda_2^2 \\ \lambda_1 & \lambda_2 \\ 1 & 1 \end{Vmatrix}$$

$$= \begin{Vmatrix} 40402.010071 & 1595.989929 \\ 1177.430591 & 136.569409 \\ 34.313708 & 11.686292 \\ 1.000000 & 1.000000 \end{Vmatrix}$$

$$B_4L = \begin{Vmatrix} -35881.345190 & 1001.345190 \\ 28156.056824 & -988.056824 \\ -227.411286 & 451.411286 \\ -4686.172968 & -273.827032 \end{Vmatrix}$$

Now since the essential properties of

[3] J. G. Bryan, Op. Cit., Chapter 6.
[4] J. G. Bryan, Op. Cit., Chapter 4 esp. pp. 80-90 incl.

latent vectors are preserved under arbitrary changes of scale, it is conventional to choose the scale such that the magnitude of the largest element is unity. Denoting by V the matrix thus obtained from B_4L we have

$$V = \begin{Vmatrix} 1.000000 & 1.000000 \\ -.784699 & -.986729 \\ .006338 & .450805 \\ .130602 & -.273459 \end{Vmatrix}$$

Therefore the two discriminant functions are

$$y_1 = 1.000000x_1 - .784699x_2 + .006338x_3 + .130602x_4$$

$$y_2 = 1.000000x_1 - .986729x_2 + .450805x_3 - .273459x_4$$

and the ratios of the sums of squares are 34.313708 and 11.686292 respectively.

Computational Short Cut

If the number of test variates n is substantially greater than G-1, a spectacular* saving in labor can be realized through the special routine outlined in this section. The rank of A cannot exceed G-1, and from sampling considerations it is almost out of the question for the rank to be less than G-1 when $n \geqslant$ G-1. Therefore we shall assume that the rank r of A equals G-1. The matrix R can then be condensed into a matrix of lower order having r = G-1 rows and columns. The latent roots of this matrix are the same as the nonvanishing roots of R, and the corresponding latent vectors are simply related to those of R.

Partition A into four parts:

$$A = \begin{Vmatrix} A_{11} & A_{12} \\ A'_{12} & A_{22} \end{Vmatrix}$$

Where A_{11} is r x r, A_{12} is r x (n-r), and A_{22} is (n-r) x (n-r). In the foregoing numerical example r = 2, so that

$$A = \left\Vert \begin{array}{cc|cc} 1 & -1 & -2 & 3 \\ -1 & 2 & 5 & -4 \\ \hline -2 & 5 & 13 & -9 \\ 3 & -4 & -9 & 10 \end{array} \right\Vert$$

That is

$$A_{11} = \begin{Vmatrix} 1 & -1 \\ -1 & 2 \end{Vmatrix} \quad A_{12} = \begin{Vmatrix} -2 & 3 \\ 5 & -4 \end{Vmatrix}$$

$$A'_{12} = \begin{Vmatrix} -2 & 5 \\ 3 & -4 \end{Vmatrix} \quad A_{22} = \begin{Vmatrix} 13 & -9 \\ -9 & 10 \end{Vmatrix}$$

Now the fact that the rank of A is r implies linear dependence of the following sort.

$$A_{12} = A_{11}B, \quad A'_{12} = B'A_{11}, \quad A_{22} = B'A_{11}B$$

and by solving the first of these equations for B we obtain

$$B = A_{11}^{-1}A_{12}$$

The matrix B can be used to effect the desired transformation of R. The algebraic demonstration[5] need not concern us here; the process is as follows

Step 1—Derivation of B

Take as initial matrix

$$\left\Vert \begin{array}{c|c} A_{11} & A_{12} \\ \hline I_r & O \end{array} \right\Vert$$

and compute the modified Crout Auxiliary[6]. The lower right hand quadrant of the modified Crout Auxiliary is B. Thus in the example considered

$$\text{Initial Matrix} = \left\Vert \begin{array}{cc|cc} 1 & -1 & -2 & 3 \\ -1 & 2 & 5 & -4 \\ \hline 1 & 0 & 0 & 0 \\ 0 & 1 & 0 & 0 \end{array} \right\Vert$$

*The writer does not know whether or not this statement would hold true for high speed electronic calculators.

[5]J. G. Bryan, Op. Cit., Chapter 5, pp. 140-147 incl.

[6]J. G. Bryan, Op. Cit., Appendix A.

$$\text{Modified Auxiliary Matrix} = \begin{Vmatrix} 1 & -1 & -2 & 3 \\ -1 & 1 & 3 & -1 \\ 1 & 1 & 1 & 2 \\ 0 & 1 & 3 & -1 \end{Vmatrix}$$

Accordingly,

$$B = \begin{Vmatrix} 1 & 2 \\ 3 & -1 \end{Vmatrix} \qquad B' = \begin{Vmatrix} 1 & 3 \\ 2 & -1 \end{Vmatrix}$$

Step 2—Define the Transformation Matrices

These are J, J′, H, H′, and H* where

$$J = \begin{Vmatrix} I_r & O_{12} \\ -B' & I_{n-r} \end{Vmatrix} \qquad H = \begin{Vmatrix} I_r & O_{12} \\ B' & I_{n-r} \end{Vmatrix}$$

$$H^* = \begin{Vmatrix} I_r \\ B' \end{Vmatrix}$$

Of course J′ and H′ are merely the transposes of J and H. Here

$$J = \begin{Vmatrix} 1 & 0 & 0 & 0 \\ 0 & 1 & 0 & 0 \\ -1 & -3 & 1 & 0 \\ -2 & 1 & 0 & 1 \end{Vmatrix} \qquad H = \begin{Vmatrix} 1 & 0 & 0 & 0 \\ 0 & 1 & 0 & 0 \\ 1 & 3 & 1 & 0 \\ 2 & -1 & 0 & 1 \end{Vmatrix}$$

$$H^* = \begin{Vmatrix} 1 & 0 \\ 0 & 1 \\ 1 & 3 \\ 2 & -1 \end{Vmatrix}$$

Step 3—Derive an Intermediate Matrix $P^* = H'W^{-1}H^*$

Denoting the initial matrix by M and the modified Crout Auxiliary by K, we set up M thus.

$$M = \begin{Vmatrix} W & H^* \\ H' & O \end{Vmatrix}$$

Then P* will emerge as the lower right hand partition of K. The numerical results are

$$M = \begin{Vmatrix} 2 & 3 & 4 & 3 & 1 & 0 \\ 3 & 5 & 8 & 6 & 0 & 1 \\ 4 & 8 & 18 & 15 & 1 & 3 \\ 3 & 6 & 15 & 14 & 2 & -1 \\ 0 & 1 & 3 & -1 & 0 & 0 \\ 1 & 0 & 1 & 2 & 0 & 0 \\ 0 & 0 & 1 & 0 & 0 & 0 \\ 0 & 0 & 0 & 1 & 0 & 0 \end{Vmatrix}$$

$$K = \begin{Vmatrix} 2 & 1.5 & 2.0 & 1.5 & 0.5 & 0 \\ 3 & 0.5 & 4.0 & 3.0 & -3.0 & 2.0 \\ 4 & 2.0 & 2.0 & 1.5 & 2.5 & -0.5 \\ 3 & 1.5 & 3.0 & 0.5 & -5.0 & -5.0 \\ 1 & -1.5 & 5.0 & -2.5 & 30.0 & 7.0 \\ 0 & 1.0 & -1.0 & -2.5 & 10.0 & 7.0 \\ 0 & 0 & 1.0 & -1.5 & 7.0 & 15.0 \\ 0 & 0 & 0 & 1.0 & -5.0 & -5.0 \end{Vmatrix}$$

Thus

$$P^* = \begin{Vmatrix} 30 & 7 \\ 7 & 15 \\ 10 & 7 \\ -5 & -5 \end{Vmatrix}$$

Step 4—Compute $Q^* = P^*A_{11}$

$$Q^* = \begin{Vmatrix} 30 & 7 \\ 7 & 15 \\ 10 & 7 \\ -5 & -5 \end{Vmatrix} \begin{Vmatrix} 1 & -1 \\ -1 & 2 \end{Vmatrix}$$

$$= \begin{Vmatrix} 23 & -16 \\ 3 & 4 \\ -8 & 23 \\ 0 & -5 \end{Vmatrix} = \begin{Vmatrix} Q_{11} \\ Q_{21} \end{Vmatrix}$$

Step 5—Derive the Characteristic Equation and Latent Vectors of Q_{11}

$$B_2 = \begin{Vmatrix} 0 & 16 \\ -1 & 23 \end{Vmatrix}$$

$f_2 = (1 \quad -46 \quad 401)$

Ch. Eqn: $\lambda^2 - 46\lambda + 401 = 0$

The latent roots are, as before, $\lambda_1 = 23 + 8\sqrt{2}$, $\lambda_2 = 23 - 8\sqrt{2}$. We now find L and B_2L. These matrices are

$$L = \begin{Vmatrix} \lambda_1 & \lambda_2 \\ 1 & 1 \end{Vmatrix}$$

$$B_2L = \begin{Vmatrix} 16 & 16 \\ -8\sqrt{2} & 8\sqrt{2} \end{Vmatrix}$$

Hence, simplifying B_2L by dividing by 8 and calling the result S* we have

$$S^* = \begin{Vmatrix} 2 & 2 \\ -\sqrt{2} & \sqrt{2} \end{Vmatrix}$$

Step 6—Restore Dimensionality to Obtain Discriminant Functions

To this end compute J′Q*S* and then conventionalize the result.

$$J'Q^*S^* =$$

$$\begin{Vmatrix} 1 & 0 & -1 & -2 \\ 0 & 1 & -3 & 1 \\ 0 & 0 & 1 & 0 \\ 0 & 0 & 0 & 1 \end{Vmatrix} \begin{Vmatrix} 23 & -16 \\ -8 & 23 \\ 3 & 4 \\ 0 & -5 \end{Vmatrix} \begin{Vmatrix} 2 & 2 \\ -\sqrt{2} & \sqrt{2} \end{Vmatrix}$$

$$= \begin{Vmatrix} 54.142136 & 25.857364 \\ -42.485281 & -25.514719 \\ 0.343146 & 11.656854 \\ 7.071068 & -7.071068 \end{Vmatrix}$$

Conventionalizing this result, we find

$$V = \begin{Vmatrix} 1.000000 & 1.000000 \\ -.784699 & -.986730 \\ .006338 & .450805 \\ .130602 & -.273459 \end{Vmatrix}$$

12

Reprinted from *British J. Math. Stat. Psychol.*, **19**, p. 2, 197–214 (Nov. 1966)

ON THE INTERRELATED NATURE OF THE MULTIVARIATE STATISTICS USED IN DISCRIMINATORY ANALYSIS

By Olgierd R. Porebski
University of New South Wales, Australia

This paper is concerned with discriminatory analysis involving two populations having equal *a priori* probabilities and a common covariance matrix and normal distribution. It is thus limited to a simple problem of discrimination to which a number of different solutions were proposed during the early period in the development of the subject. These well-known results are presented here under one heading and in common notation. The paper begins with a discussion of some alternative logical ways of formulating the problem; then it shows the various statistical methods of deriving a discriminant function; finally, it exhibits the interrelated nature of the statistics which may be used in the test of significance. Each statistic is expressed as a scalar product of the same vector of differences between p pairs of means and a vector of discriminatory coefficients which varies only in length.

1. The Fundamental Lemma

In discriminatory analysis appropriate for two populations we test a hypothesis (H_1) that an observation $\mathbf{X}$, specified by p measurements

$$(X_1, X_2, \ldots, X_p),$$

has originated from a certain p-variate population π_1, against an alternative hypothesis (H_2) that it has originated from a p-variate population π_2. We admit H_1 and H_2 as the only hypotheses and thus assume that the sampling space of $\mathbf{X}$ is restricted to π, the ' logical sum ' of π_1 and π_2.

The origin of $\mathbf{X}$ may be tested by comparing the posterior probabilities

$$P(H_i|\mathbf{X}, \pi) \propto P(H_i|\pi)P(\mathbf{X}|H_i, \pi) \qquad (i=1, 2) \tag{1}$$

acccording to Bayes' theorem. If the *a priori* probabilities are equal this is reduced simply to the examination of a ratio of the corresponding likelihoods. We give preference to H_1 when $P(\mathbf{X}|H_1, \pi) > P(\mathbf{X}|H_2, \pi)$. (For the relationship between Bayes' theorem or postulate and the principle of maximum likelihood, see Kendall (1940).)

If the *a priori* probabilities are different, due to different frequencies with which cases are sampled from the two populations, we choose H_1 when

$$\frac{P(\mathbf{X}|H_1, \pi)}{P(\mathbf{X}|H_2, \pi)} > \frac{P(H_2|\pi)}{P(H_1|\pi)} = \text{a constant.} \tag{2}$$

All solutions involve the determination of a likelihood ratio and differ only with respect to the constant. This applies also to a risk function, if available, in which those probabilities are further weighted by various psychological, economic and other considerations. A decision in favour of H_1 is based then on the inequality

$$\frac{P(\mathbf{X}|H_1, \pi)}{P(\mathbf{X}|H_2, \pi)} > \frac{P(H_2|\pi)W_2}{P(H_1|\pi)W_1} = K, \tag{3}$$

in which W_1 and W_2 are any weights whatever applied to H_1 and H_2 respectively, and K is a constant. Therefore, writing $p_i(\mathbf{X})$ for $P(\mathbf{X}|H_i, \pi)$ in eqns. (2) or (3) we have

$$\frac{p_1(\mathbf{X})}{p_2(\mathbf{X})} > K \tag{4}$$

generally.

This fundamental result is known as the lemma of Neyman and Pearson (1928, 1932). It was originally derived by the method of maximum likelihood, involving a recognition of two kinds of error in relation to H_1. Either the hypothesis is rejected when it is true or it is accepted when an alternative hypothesis (H_2) is true.

The method of maximum likelihood leads to a study of certain critical regions.

If we choose a region R_2 in the sample space π as a critical rejection region to test H_1 then the probability that $\mathbf{X}$ falls into R_2, if H_1 is true, is $P_1(R_2) = \epsilon$, and depends simply on the size of the critical region chosen. The probability that $\mathbf{X}$ falls into R_1, the region mutually exclusive to R_2, is $P_1(R_1) = 1 - P_1(R_2)$. The occurrence of this event involves an error of the second kind, if H_2 is true, and its probability is $P_2(R_1) = 1 - P_2(R_2)$.

To minimize these two sources of error in a given ratio α we must satisfy the condition

$$P_1(R_2) = \alpha P_2(R_1) = \epsilon. \tag{5}$$

This is equivalent to finding a minimum of ϵ subject to the restriction that

$$P_2(R_2) + P_1(R_2)/\alpha = 1. \tag{6}$$

Therefore, in the case of a continuous distribution of $\mathbf{X}$, we have to find a minimum of

$$\int_{R_2} [p_1(\mathbf{X}) - \beta\{p_2(\mathbf{X}) + p_1(\mathbf{X})/\alpha\}]d\mathbf{X}, \tag{7}$$

where β is a Lagrange multiplier. Solving and absorbing all constants in a new constant c, we obtain $cp_1(\mathbf{X}) - p_2(\mathbf{X}) = 0$ for the boundary condition, and $p_2(\mathbf{X})/p_1(\mathbf{X}) \geqslant c$ for the critical rejection region R_2.

The same result can be obtained from a slightly different approach, discussed by Neyman and Pearson (1932). We may consider which is the best critical rejection region for H_1, with respect to a single alternative H_2. It is clear that, out of many possible regions for which the error of the first kind is of size ϵ, we have to choose one, say R_2, for which $P_2(R_2)$ is a maximum. But a maximum of $P_2(R_2)$, subject to the restriction that $P_1(R_2)=\epsilon$, involves finding a stationary value of

$$\int_{R_2} \{p_2(\mathbf{X}) - cp_1(\mathbf{X})\} d\mathbf{X},$$

where c is a constant chosen so that the critical rejection region has the required size. This is achieved by taking into R_2 all those points for which $p_2(\mathbf{X}) - cp_1(\mathbf{X}) \geqslant 0$.

Under certain conditions, the method of Neyman and Pearson and Bayes' theorem provide the same result. This was first pointed out by Welch (1939), who also recognized the relationship between the result so obtained and Fisher's discriminant function.

The discriminatory problem may also be solved by a minimax method. An explicit example of this for a univariate case was given by Von Mises (1945). His procedure involves proving first that the greatest risk of making a false decision is $(1-P_{\min})$. In the case of two populations, this is made minimum when the smaller of the two expressions

$$P_i = \int_{R_i} p_i(X)\,dX \qquad (i=1, 2)$$

is given its possibly greatest value. It follows almost immediately that the probability for an observation from π_i to be allocated to R_i must be the same for the two regions. Thus, the condition $P_1=P_2$ must be satisfied. In a contrary case, no solution is possible unless there is some prior (probabilistic) knowledge concerning the origin of the observation. When a sample of such observations is to be classified then an additional condition is required. This condition is derived by considering an infinitesimal variation of the boundary containing any two points X_1 and X_2. For it can be shown that the determinant $p_1(X_1)p_2(X_2) - p_2(X_1)p_1(X)_2$ must be zero for the solution to be possible. Hence, along the border between R_1 and R_2 the ratio $p_1(X)/p_2(X)$ must be constant.

The minimax method may be extended to a p-variate case by considering the critical regions in a p-dimensional space in which an observation is represented by a point. Its further extension may involve the application of differential weights $W_i(i=1, 2)$ to the error according to whether H_1 or H_2 is true. The procedure is then concerned with a minimization of the maximum loss and leads to the choice of a constant for which the two risks are equal (Neyman and Pearson, 1933; Wald, 1944, 1947, 1950; Anderson, 1951).

A minimax solution so obtained is identical with a maximum likelihood solution for a given ratio at which the errors on the two hypotheses are minimized. For, if the necessary minimax condition is written as

$$W_1P_1(R_2) = W_2P_2(R_1), \tag{8}$$

then clearly it is also expressible by eqn. (5) in which W_2/W_1 is substituted for α.

The same solution can also be derived from Bayes' theorem by inserting in eqn. (2) certain *a priori* probabilities whose ratio is equal to the required constant.

For this reason, as pointed out by Anderson (1951, p. 27), " every minimax solution is a Bayes' solution for some *a priori* probabilities ".

When the *a priori* probabilities are not available and the errors of misclassification are considered of the same importance, the constants in eqns. (2) and (3) are both equal to unity. Such a solution may be regarded as a Bayes' solution with equal *a priori* probabilities. It involves the assumption that $P_1(R_1+R_2)=P_2(R_1+R_2)=1$. It also specifies that the relationship $p_2(\mathbf{X})>p_1(\mathbf{X})$ must be satisfied for all observations $\mathbf{X}$ allocated to π_2 and that the relationship $p_1(\mathbf{X})>p_2(\mathbf{X})$ must be satisfied for all those allocated to π_1.

If there is only one region of overlap between the two distributions, a continuous sampling and classification of observations to π_1 or π_2 must satisfy the inequalities

$$\begin{aligned} P_2(R_2) &> P_1(R_2), \\ P_1(R_1) &> P_2(R_1), \end{aligned} \tag{9}$$

when the probabilities of the observations so classified are summed or integrated over the respective regions. If the two distributions are not identical the common region will, as a rule, be unequally divided. However, as a special case, it is possible that

$$\begin{aligned} P_1(R_1) &= P_2(R_2), \\ P_1(R_2) &= P_2(R_1), \end{aligned} \tag{10}$$

which is the necessary condition for a solution by the method of maximum likelihood (or minimax) when the errors are minimized at an equal ratio.

Similarly the acceptance of eqn. (10) and a subsequent maximization of the inequalities of eqn. (9) will not necessarily result in $K=1$ (consider in eqn. (7) $\alpha=1$, $\beta\neq\frac{1}{2}$). Such maximization is performed without any prior knowledge of the specific value of the constant, and condition (10) alone is insufficient to determine it. Therefore, in general, the equal *a priori* probabilities do not imply the same ratio of success or error on the two hypotheses, nor does the same ratio of success or error imply that the *a priori* probabilities are equal ($K=1$).

A specific solution with $K=1$ may, however, be obtained on assuming the appropriate distributions. In this paper it will be assumed that the two distributions are multivariate normal and that they differ only as to their means. With this assumption it will be shown that a maximum likelihood solution leads directly to Fisher's discriminant function and is equivalent to a Bayes' solution with equal *a priori* probabilities.

2. Discriminant Functions

To obtain a numerical expression for the boundary conditions we consider the probability densities of $\mathbf{X}$ on the two hypotheses, assuming in each case a multivariate normal distribution. Since the population parameters are usually unknown, the optimum estimates of these parameters are obtained from samples of size N_1 and N_2 drawn respectively from the two populations. The means are given then by the means of the samples and the variances and covariances by the pooled within classes estimates common to the two groups (Fisher, 1936). (For a case of unequal covariance matrices in the two groups see Penrose (1947), Smith (1947) and Kullback (1952); for a case of unspecified covariances see James (1954).)

If $\mathbf{m}_i$ is the estimated $p \times 1$ column vector of means ($i=1, 2$) and $\mathbf{C}$ is the common covariance matrix within classes, the estimated probability density on the hypothesis H_i is

$$\hat{p}_i(\mathbf{X}) = \frac{1}{(2\pi)^{p/2}|\mathbf{C}|^{1/2}} \exp\left(-\tfrac{1}{2}\mathbf{x}_i'\mathbf{C}^{-1}\mathbf{x}_i\right) \tag{11}$$

where $\mathbf{x}_i = \mathbf{X} - \mathbf{m}_i$ is a column vector.

On taking the logarithms of both sides of eqn. (4) and substituting from eqn. (11) we obtain

$$\mathbf{x}_2'\mathbf{C}^{-1}\mathbf{x}_2 - \mathbf{x}_1'\mathbf{C}^{-1}\mathbf{x}_1 > 2 \log K \tag{12}$$

as a condition for accepting H_1 rather than H_2. Developing now the left side of eqn. (12) and substituting in turn $\mathbf{d} = \mathbf{m}_1 - \mathbf{m}_2$ we have

$$\mathbf{d}'\mathbf{C}^{-1}\mathbf{X} > \log K + \tfrac{1}{2}(\mathbf{m}_1'\mathbf{C}^{-1}\mathbf{m}_1 - \mathbf{m}_2'\mathbf{C}^{-1}\mathbf{m}_2) = \bar{Z} = \text{constant}, \tag{13}$$

since the right side of the inequality does not depend on $\mathbf{X}$. The statistic

$$Z = \mathbf{d}'\mathbf{C}^{-1}\mathbf{X} = \mathbf{t}'\mathbf{X} \tag{14}$$

is equivalent to Fisher's discriminant function, the latter being usually derived so that its coefficients $\mathbf{l}$ are $(N_1 + N_2 - 2)$ times smaller than those in $\mathbf{t}$.

Following Fisher's original approach (1936) we choose, out of all possible sets of arbitrary discriminant coefficients $\mathbf{l}$, those which maximize the ratio of the sum of squares between classes to the sum of squares within classes. We denote discriminant functions corresponding to the means of the two groups by $\bar{Z}_1 = \mathbf{m}_1'\mathbf{l}$ and $\bar{Z}_2' = \mathbf{m}'_2\mathbf{l}$, and obtain

$$\bar{Z}_1 - \bar{Z}_2 = (\mathbf{m}_1' - \mathbf{m}_2')\mathbf{l} = \mathbf{d}'\mathbf{l}. \tag{15}$$

Then, we write $\mathbf{B}$ and $\mathbf{W}$ for the between and within classes dispersion matrices (matrices of sum of squares and products), and B^* and W^* for the corresponding sums of squares of the discriminant function. B^* and W^* can now be expressed as

$$B^* = \lambda^2(\mathbf{l}'\mathbf{d}\mathbf{d}'\mathbf{l}) = \mathbf{l}'\mathbf{B}\mathbf{l} \tag{16}$$

where $\lambda^2 = N_1N_2/(N_1 + N_2)$ and

$$W^* = \mathbf{l}'\mathbf{W}\mathbf{l}. \tag{17}$$

To maximize

$$\phi = B^*/W^* \tag{18}$$

(originally Fisher (1936) used $(\mathbf{d}'\mathbf{l})^2$ instead of B^*, since λ^2 was to be absorbed in the constant; he also wrote S for our W^*) we first reduce $\partial(\phi)/\mathbf{l}'=0$ to

$$\frac{\phi\partial\langle W^*\rangle}{\partial\mathbf{l}'}=\frac{\partial\langle B^*\rangle}{\partial\mathbf{l}'} \tag{19}$$

and then solve the differential equation. This results in

$$\mathbf{Wl}=c\mathbf{d}, \tag{20}$$

where $c=\lambda^2(\mathbf{d}'\mathbf{l}\,|\phi)$ is a constant. Putting now $c=1$, and noting (14), we find that

$$\mathbf{l}=\mathbf{W}^{-1}\mathbf{d}=\mathbf{t}/(N_1+N_2-2). \tag{21}$$

Fisher (1936, 1938) has also expressed the discriminant weights as regression coefficients predicting a dummy variate with a constant value $N_2/(N_1+N_2)$ for the first sample and a constant value $-N_1/(N_1+N_2)$ for the second sample. The sum of squares of the dummy variate is then $\lambda^2=N_1N_2/(N_1+N_2)$ and the sum of products of any variate j with the dummy variate is $\lambda^2 d_j$ (where d_j is the difference between the two means on variate j).

Following the usual formulation of the multiple regression problem based on the *total* dispersion matrix we obtain

$$(\mathbf{W}+\mathbf{B})\mathbf{b}=\mathbf{d}\lambda^2 \tag{22}$$

and hence

$$\mathbf{b}=(\mathbf{W}+\mathbf{B})^{-1}\mathbf{d}\lambda^2. \tag{23}$$

The regression weights $\mathbf{b}$ are proportional to $\mathbf{l}$ since from eqns. (22) and (16) we have $\mathbf{Wb}=\mathbf{d}\lambda^2(1-\mathbf{d}'\mathbf{b})$, which in virtue of eqn. (21) leads to

$$\mathbf{b}=\mathbf{W}^{-1}\,\mathbf{d}\lambda^2(1-\mathbf{d}'\mathbf{b})=\mathbf{l}c, \tag{24}$$

where c is a constant.

Yet another formulation of the problem may be considered. It is based on the theory of canonical correlation, originally developed by Hotelling (1935, 1936) and has the additional merit of being applicable to more than two groups. Such a formulation was provided by Bartlett (1938). The underlying problem of distribution was, in principle, solved by Fisher (1939) and Hsu (1939). Additional contributions were made later by Bartlett (1947, 1951), Rao (1948b, 1952), Lubin (1950), Williams (1955), Kendall (1957), Anderson (1958) and others.

We may consider the problem of discrimination within the framework of finding a canonical correlation between a set of q dummy variates corresponding to $k=(q+1)$ samples and a set of p usual statistical variates (which may also be represented by the within classes dispersion matrix). Then the best single discriminant function is given as a latent vector corresponding to the largest root of the characteristic equation

$$(\mathbf{B}-\phi\mathbf{W})\mathbf{a}=0, \tag{25}$$

and is obtained from the solution of

$$|\mathbf{B}-\phi\mathbf{W}|=0. \tag{26}$$

Fisher's analysis of variance solution may be regarded as a particular case of the solution of the determinantal equation (26), and can be expressed in the same form.

In either case we maximize

$$\phi=\frac{\mathbf{l}'\mathbf{B}\mathbf{l}}{\mathbf{l}'\mathbf{W}\mathbf{l}} \tag{27}$$

which in a general form results in eqn. (25) and, in the particular case of two groups, leads to eqn. (20). Substituting in the latter case for the constant c we obtain

$$\phi\mathbf{W}\mathbf{l}=\lambda^2\mathbf{d}\mathbf{d}'\mathbf{l}$$

$$\text{or} \tag{28}$$

$$(\mathbf{B}-\phi\mathbf{W})\mathbf{l}=0.$$

Alternatively, we can maximize

$$\phi'=\frac{\mathbf{h}'\mathbf{B}\mathbf{h}}{\mathbf{h}'\mathbf{W}\mathbf{h}+\mathbf{h}'\mathbf{B}\mathbf{h}}$$

(where $\mathbf{h}$ is a column vector of coefficients), so that in the determinantal equation

$$|\mathbf{B}-\phi'(\mathbf{W}+\mathbf{B})|=0 \tag{29}$$

the new largest root is defined as

$$\phi'=\frac{\phi}{1+\phi}. \tag{30}$$

In both eqns. (26) and (29), owing to the degeneracy of $\mathbf{B}$, which is simply $\lambda^2\mathbf{d}\mathbf{d}'$, there is only one non-zero root.

To determine ϕ and ϕ' we postmultiply eqns. (26) and (29) by $\mathbf{W}^{-1}$ and $(\mathbf{W}+\mathbf{B})^{-1}$ respectively. Developing the resulting determinants and substituting $\phi=tr(\mathbf{B}\mathbf{W}^{-1})$ in the first and $\phi'=tr[\mathbf{B}(\mathbf{W}+\mathbf{B})^{-1}]$ in the second, we reduce them to $|\mathbf{B}\mathbf{W}^{-1}|$ and $|\mathbf{B}(\mathbf{W}+\mathbf{B})^{-1}|$ respectively. Both these determinants are equal then to zero, since $\mathbf{B}$ is of rank one.

We find the latent vector in eqn. (25) by first writing the trace of $(\mathbf{B}\mathbf{W}^{-1})$ as

$$\phi=tr\,[\lambda^2\mathbf{d}(\mathbf{d}'\mathrm{W}^{-1})]=\lambda^2\mathbf{d}'\mathbf{W}^{-1}\mathbf{d}. \tag{31}$$

On substituting then for ϕ in eqn. (25) we solve the equation with

$$\mathbf{a}=\mathbf{W}^{-1}\mathbf{d}c, \tag{32}$$

since c is a constant and $\lambda^2\mathbf{d}(\mathbf{d}'\mathbf{W}^{-1}\mathbf{d})c-\lambda^2(\mathbf{d}'\mathbf{W}^{-1}\mathbf{d})\mathbf{d}c$ is identically zero. Thus, putting $c=1$, as in eqn. (20), we obtain $\mathbf{a}=\mathbf{l}$. Similarly, it can be found that

$$\phi'=\lambda^2\mathbf{d}'(\mathbf{W}+\mathbf{B})^{-1}\mathbf{d}, \tag{33}$$

S.P. N

and

$$\mathbf{h} = (\mathbf{W}+\mathbf{B})^{-1}\mathbf{d}c. \tag{34}$$

When $c=1$, $\mathbf{h}$ may also be expressed as $\mathbf{h}=\mathbf{W}^{-1}\mathbf{d}(1-\lambda^2\mathbf{d}'\mathbf{h})$ and hence as $\mathbf{h}=\mathbf{a}(1-\phi')$ or $\mathbf{h}(1+\phi)=\mathbf{a}$. This indicates that $\mathbf{h}$ is always proportional to $\mathbf{a}$ and, therefore, is also proportional to $\mathbf{l}$ and $\mathbf{b}$. By an arbitrary choice of $c=1+\phi$, $\mathbf{h}$ is made identical with $\mathbf{a}$ and, by a similar choice of $c=\lambda^2$, it is made identical with $\mathbf{b}$.

To allocate an observation into one of the two populations without the consideration of a doubtful region, we need to determine only one critical point. For large values of N_1 and N_2 we assume that a discriminant function Z is normally distributed with a standard deviation $\sigma_1=\sigma_2$ and with the mean value $\bar{Z}_1$ or $\bar{Z}_2$, according to whether H_1 or H_2 is true. When the errors on the two hypotheses are considered of equal importance, it is reasonable to make their probabilities equal. Denoting the required critical value by $\bar{Z}$, and assuming $\bar{Z}_1 > \bar{Z}_2$, we write

$$\int_{-\infty}^{\bar{Z}} p_1(z)dz = \int_{\bar{Z}}^{\infty} p_2(z)dz.$$

From this we immediately obtain

$$\int_{\bar{Z}}^{\bar{Z}_1} p_1(z)dz = \int_{\bar{Z}_2}^{\bar{Z}} p_2(z)dz,$$

and hence

$$\frac{\bar{Z}_1+\bar{Z}_2}{2} = \bar{Z}. \tag{35}$$

(For a more complete proof, see Wald (1944).)

Although the numerical value of Z is dependent on the arbitrary constant c, this does not affect the problem of discrimination. If, for instance, in eqn. (14) we choose a value of $\mathbf{X}$, classified on the basis $Z=\mathbf{t}'\mathbf{X}>\bar{Z}$, its classification is not affected when it is made according to $Z'=c\mathbf{t}'\mathbf{X}>\bar{Z}'$, which is equivalent to

$$Z' = cZ > c\bar{Z} = c(\bar{Z}_1+\bar{Z}_2)/2 = \bar{Z}'. \tag{36}$$

In addition to the constant c, which is entirely arbitrary, and constant $\bar{Z}$, which is partly arbitrary, we have to consider the constant K which, in this solution, has a specific value. K expresses the ratio of the *a priori* probabilities of the two hypotheses according to Bayes' theorem. Its value can be obtained by writing $\bar{Z}_1=\mathbf{d}'\mathbf{C}^{-1}\mathbf{m}_1$ and $\bar{Z}_2=\mathbf{d}'\mathbf{C}^{-1}\mathbf{m}_2$, in virtue of eqn. (14), and substituting these values in the right-hand side of the inequality (13). The substitution results in $\log K+\bar{Z}=\bar{Z}$, and hence in $K=1$.

The solution considered here is a maximum likelihood solution based on the equal probability of error on the two hypotheses. It is found to be equivalent to a Bayes' solution with equal *a priori* probabilities. This equivalence is specifically due to the additional assumptions concerning the two distributions, but, as pointed out in the previous section, it does not hold generally.

3. Tests of Significance

A multivariate test of significance in discriminatory analysis was originally developed by Hotelling (1931) and involved the determination of T^2. This statistic can be expressed in terms of Fisher's z, but it requires a small adjustment before it can be applied directly to two samples. It may, therefore, be convenient to begin with the equivalent test due to Fisher (1936). The latter is based on a familar form of F-ratio and is derived from the analysis of variance of a discriminant function.

W^*—Fisher's dispersion statistic

The F-ratio in the case of a discriminant function departs from a simple analysis of variance only in that it requires further removing of $p-1$ degrees of freedom from the residual. The between classes sum of squares is now tested with p degrees of freedom, corresponding to the p hypothetical constraints $\mu_{j1}=\mu_{j2}(j=1, \ldots, p)$. We have then

$$F=\frac{N_1+N_2-p-1}{p}\frac{B^*}{W^*} \tag{37}$$

which, in virtue of eqns. (16), (17) and (21) may be expressed as

$$F=\frac{N_1+N_2-p-1}{p}\lambda^2\frac{(\mathbf{l}'\mathbf{d})^2}{\mathbf{l}'\mathbf{d}} \tag{38}$$

or

$$F=\frac{N_1+N_2-p-1}{p}\frac{N_1N_2}{N_1+N_2}W^*. \tag{39}$$

W^* denotes the within classes sum of squares of a discriminant function. It also measures the difference between the two composite means derived with the aid of $\mathbf{l}$ coefficients. This is evident from the identity

$$W^*=\mathbf{l}'\mathbf{d} \tag{40}$$

resulting from eqns. (17) and (21).

R^2—Fisher's multiple (discriminatory) correlation

On substituting the result of eqn. (24) in eqn. (40) and then in eqn. (39) we obtain

$$F=\frac{N_1+N_2-p-1}{p}\frac{\mathbf{b}'\mathbf{d}}{1-\mathbf{b}'\mathbf{d}}=\frac{N_1+N_2-p-1}{p}\frac{R^2}{1-R^2}, \tag{41}$$

the usual test of significance for multiple correlation.

This test can also be established directly in at least two ways discussed by Fisher (1940). The analysis of variance of a dummy variate distinguishing the two contrasted samples may be considered as separating the portion expressible in terms of $\mathbf{X}$ as independent variables from a residue not so expressible. This leads to the subdivision of the total sum of squares λ^2 in terms of

$\lambda^2 R^2/\lambda^2(1-R^2) = R^2/(1-R^2)$ as shown in eqn. (41). Alternatively, we may regard the discriminant function as a variate and analyse its variation between and within the samples. The total sum of squares is then $\lambda^2 R^2$, and its analysis can be expressed as a ratio $\lambda^2 R^4/\lambda^2 R^2(1-R^2)$, since the between classes sum of squares is $\lambda^2 \mathbf{b}'\mathbf{dd}'\mathbf{b} = \lambda^2 R^4$. Apart from a constant factor R^2, we obtain again the same ratio.

R^2 measures the difference between the ' predicted ' means of the two samples since

$$R^2 = \mathbf{b}'\mathbf{d}. \tag{42}$$

This difference is not identical with W^*, for we find from eqns. (21) and (40) that

$$W^* = \mathbf{d}'\mathbf{W}^{-1}\mathbf{d}, \tag{43}$$

whereas, from eqns. (23) and (42) we have

$$R^2 = \lambda^2 \mathbf{d}'(\mathbf{W}+\mathbf{B})^{-1}\mathbf{d}. \tag{44}$$

D^2—Mahalanobis' generalized distance

The D^2 statistic[1] was developed by Mahalanobis to study the metric distance between two populations, whereas the purpose of Fisher's discriminant function was to study the direction of such a difference. In the case of uncorrelated variables (Mahalanobis, 1930), D^2 represents the square of the actual distance between two ' centroids ' corresponding to two samples. In the case of correlated variables (Mahalanobis, 1936), D^2 refers directly to the distance only if the oblique system of coordinates is used. A direct derivation of the ' Studentized ' test of significance for D^2 was originally due to Bose and Roy (1938). Rao (1948 b, 1952) generalized the use of D^2 to several large samples and did most of the statistical work related to systematization and practical application of this statistic. He also pointed out the possibility of obtaining a discriminant function during the process of determining D^2 in the case of two samples (for an alternative derivation of D^2, see Jambunathan (1954) or Kullback (1952)).

The D^2 statistic may be defined in matrix notation as

$$D^2 = \mathbf{d}'\mathbf{C}^{-1}\mathbf{d}. \tag{45}$$

Its comparison with eqn. (43) establishes the identity

$$D^2 = (N_1+N_2-2)W^* \tag{46}$$

which, in view of eqn. (39), leads immediately to

$$F = \frac{N_1+N_2-p-1}{p} \, \frac{N_1N_2}{(N_1+N_2)} \, \frac{D^2}{(N_1+N_2-2)}. \tag{47}$$

[1] Mahalanobis originally defined his statistic as D^2/p, but, subsequently, it has been more usual to define it as in (45). D^2 must not be confused with the same symbol used by Garrett (1943) who, following Fisher's original notation (1936), expressed it as $(\mathbf{l}'\mathbf{d})^2$, or, as equivalent to our W^{*2}. Fisher subsequently (1938) discontinued the use of D^2 for any other purpose than to denote Mahalanobis' statistic.

D^2, like W^* and R^2, represents the difference between two composite means and can therefore be represented as a scalar product of a vector of discriminant coefficients, say $\mathbf{t}$, and the vector $\mathbf{d}$.

From eqn. (45) we find that the vector

$$\mathbf{t}=\mathbf{C}^{-1}\mathbf{d} \tag{48}$$

defines the discriminant function in eqn. (14), which was derived directly from the lemma of Neyman and Pearson.

T^2—Hotelling's generalized t-ratio

The generalization of Student's ratio by Hotelling (1931) may be expressed in matrix notation as

$$T^2=\boldsymbol{\xi}'\mathbf{C}^{-1}\boldsymbol{\xi}, \tag{49}$$

where

$$\boldsymbol{\xi}=(\mathbf{m}_i-\mathbf{m})\sqrt{(N_i)} \tag{50}$$

is a random vector, and $\mathbf{m}_i$ and $\mathbf{m}$ are vectors of p means of the sample i and of p population means respectively. Since T^2 has an additive property comparable to that of independent values of χ^2, its use can be extended to two samples by means of the statistic

$$T^{*2}=T_1{}^2+T_2{}^2=\boldsymbol{\xi}_1{}'\mathbf{C}^{-1}\boldsymbol{\xi}_1+\boldsymbol{\xi}_2{}'\mathbf{C}^{-1}\boldsymbol{\xi}_2 \tag{51}$$

with

$$\boldsymbol{\xi}_1=(\mathbf{m}_1-\mathbf{m})\sqrt{(N_1)}$$

and

$$\boldsymbol{\xi}_2=(\mathbf{m}_2-\mathbf{m})\sqrt{(N_2)}.$$

We can also express eqn. (51) as

$$T^{*2}=\gamma(\boldsymbol{\xi}_1-\boldsymbol{\xi}_2)'\mathbf{C}^{-1}(\boldsymbol{\xi}_1-\boldsymbol{\xi}_2)\gamma, \tag{52}$$

where $\gamma=\sqrt{(N_1+N_2)}/\{\sqrt{(N_1)}+\sqrt{(N_2)}\}$. Alternatively, we can redefine $\boldsymbol{\xi}$ as

$$\boldsymbol{\xi}=(\boldsymbol{\xi}_1\boldsymbol{\xi}_2)$$

and obtain yet another form

$$T^{*2}=tr(\boldsymbol{\xi}'\mathbf{C}^{-1}\boldsymbol{\xi}). \tag{53}$$

Eqns. (51), (52) and (53) are each reducible to

$$T^{*2}=\lambda^2\mathbf{d}'\mathbf{C}^{-1}\mathbf{d} \qquad (\lambda^2=(N_1N_2)/(N_1+N_2)). \tag{54}$$

This expression can also be derived directly (see Fisher (1938) and Hsu (1938)) by writing $\boldsymbol{\xi}=\lambda\mathbf{d}$ in (49).

On considering eqns. (45) and (47) it is immediately evident that the appropriate test of significance for T^{*2} is

$$F=\frac{N_1+N_2-p-1}{p}\,\frac{T^{*2}}{N_1+N_2-2}. \tag{55}$$

This test is in agreement with the original result of Hotelling (1931, p. 377) when the required degrees of freedom for the within classes covariance is substituted in eqn. (55), that is, making $n=N_1+N_2-2$.

Λ—Wilks' likelihood ratio

The test of significance of a likelihood ratio in a univariate case was first derived by Pearson and Neyman (1930; Neyman and Pearson, 1931). A generalization of the test to multivariate populations was developed by Wilks (1932). Among Wilks' results were also exact tests for certain specific cases, including the case of differences between two multivariate samples on the assumption that covariance matrices are equal and that the elements of the common covariance matrix follow Wishart's distribution (1928).

The likelihood ratio associated with this may be expressed as a fractional power of the original ratio (l_2) and denoted by W_2, so that asymptotically

$$-2\log l_2=-N\log W_2\to\chi^2 \text{ (approx.)} \tag{56}$$

with qp d.f., where $q=k-1$, and k is the number of samples. (For a more detailed treatment see Bartlett (1938), Wilks (1938), Hsu (1940), Rao (1948b) and Lawley (1956).)

The division of the numerator and denominator of W_2 by N^p leads to an equivalent (or nearly equivalent) expression suggested by Bartlett (1934) and defined by him as

$$\Lambda=\frac{|\mathbf{W}|}{|\mathbf{W}+\mathbf{B}|}. \tag{57}$$

It is this ratio that we shall examine here (see also Bartlett (1938)). Since we consider only two samples we can write (57) as

$$\lambda=\frac{1}{|1+\mathbf{BW}^{-1}|}=\frac{1}{|1+\lambda^2\mathbf{dd}'\mathbf{W}^{-1}|}. \tag{58}$$

$\mathbf{BW}^{-1}$ is of rank one and, therefore, all its minors vanish when the determinant in eqn. (58) is expanded. This leads to the identity $|1+\mathbf{BW}^{-1}|=1+tr(\mathbf{BW}^{-1})$. On substituting the value of the trace of $\lambda^2\mathbf{dd}'\mathbf{W}^{-1}$, already considered in eqn. (31), we find that

$$\Lambda=\frac{1}{1+\lambda^2\mathbf{d}'\mathbf{W}^{-1}\mathbf{d}}. \tag{59}$$

Now, in view of eqn. (43) we note that

$$1/\Lambda=1+\lambda^2W^*,$$

which, on considering eqn. (39), leads immediately to

$$F = \frac{N_1 + N_2 - p - 1}{p} \frac{1 - \Lambda}{\Lambda}. \tag{60}$$

The comparison of eqn. (59) with (40), evaluated in terms of eqns. (24) and (42), gives another useful result $1 - \Lambda = R^2$, which on substituting for R^2 from eqn. (44) takes the form

$$1 - \Lambda = \lambda^2 \mathbf{d}'(\mathbf{W} + \mathbf{B})^{-1}\mathbf{d}. \tag{61}$$

The test of significance in eqn. (60) is in agreement with Wilks' specific result for two samples (1932). For a complete list of Wilks' specific results expressed in terms of Λ, see Rao (1948 b). For the discussion of other aspects of likelihood tests, see Pearson and Wilks (1933), Wilks (1938, 1946), Bishop (1939), Bartlett (1937, 1947, 1951), Williams (1955) and Lawley (1956). For a more comprehensive treatment refer to the book by Anderson (1958).

V—*Hsu's generalized correlation ratio*

A multivariate generalization of the correlation ratio (η^2) in the form $|\mathbf{B}|/|\mathbf{B} + \mathbf{W}|$ was found to be unsatisfactory for a number of reasons considered by Pearson and Wilks (1933). This determinantal ratio is equivalently given by the product of the roots of the equation $|\mathbf{B} - \phi'(\mathbf{B} + \mathbf{W})| = 0$. Wilks–Bartlett likelihood ratio $\Lambda = |\mathbf{W}|/|\mathbf{B} + \mathbf{W}|$, which is generally preferred, is given on the other hand by the product of $(1 - \phi')$ values (Bartlett, 1947). It may thus be regarded as a generalization of $(1 - \eta^2)$. Another useful generalization of the correlation ratio, this time involving $\eta^2/(1 - \eta^2)$, was proposed by Hsu (1940). It was given as the V statistic and found to be equivalent to the sum of the roots of the equation $|\mathbf{B} - \phi\mathbf{W}| = 0$. It should be noted that $\phi = \phi'/(1 - \phi')$.

Hsu's V statistic was originally defined as

$$V = \sum_{ij=i}^{p} w^{ij} b_{ij}, \tag{62}$$

where b_{ij} is a typical element of matrix $\mathbf{B}$ and w^{ij} is a typical element of matrix $\mathbf{W}^{-1}$. Since eqn. (62) may also be written as $V = tr(\mathbf{BW}^{-1})$, it is expressible as the sum of quadratic forms $\boldsymbol{\xi}'\mathbf{W}^{-1}\boldsymbol{\xi}$, in which $\boldsymbol{\xi}$ is a random vector defined in eqn. (50). We have then an alternative expression

$$V = tr(\mathbf{BW}^{-1}) = tr(\mathbf{Y}'\mathbf{W}^{-1}\mathbf{Y}), \tag{63}$$

with $\mathbf{Y}$ being a matrix $(\boldsymbol{\xi}_1\boldsymbol{\xi}_2 \ldots \boldsymbol{\xi}_k)$ assembled from k samples. On comparing eqn. (63) with eqns. (51) and (53) we find that V is also expressible as a function of Hotelling's T^2 statistics, since

$$V = tr(\mathbf{Y}'\mathbf{C}^{-1}\mathbf{Y})/n = \sum_{i=1}^{k} T_i^2/n \tag{64}$$

for n degrees of freedom within classes.

When $k = 2$, V is further reduced to

$$V = (T_1^2 + T_2^2)/n = T^{*2}/(N_1 + N_2 - 2) = \lambda^2 \mathbf{d}'\mathbf{W}^{-1}\mathbf{d} \tag{65}$$

as a consequence of eqns. (51) and (54). In view of eqn. (55) its test of significance is then obtained from

$$F = \frac{N_1 + N_2 - p - 1}{p} V. \tag{66}$$

T_0^2—Hotelling's generalized T measure of dispersion

A direct generalization of T^2 to k samples was given by Hotelling (1947, 1950) in the form of the T_0^2 statistic. This statistic was defined (1950, p. 31) as

$$T_0^2 = T_1^2 + T_2^2 + \ldots + T_k^2. \tag{67}$$

On comparing eqn. (67) with eqns. (64) and (63) it is clear that Hotelling's T_0^2 is closely related to Hsu's V. We have simply

$$T_0^2 = tr(\mathbf{Y}'\mathbf{C}^{-1}\mathbf{Y}) = n\{tr(\mathbf{BW}^{-1})\} = nV, \tag{68}$$

where n are the degrees of freedom of matrix $\mathbf{W}$. If $q = k - 1$ are the degrees of freedom corresponding to matrix $\mathbf{B}$, T_0^2 may also be written as $q\{tr(\mathbf{SC}^{-1})\}$, where $\mathbf{S}$ and $\mathbf{C}$ are the covariance matrices between and within classes respectively. T_0^2/q is then the sum of the roots of the equation $|\mathbf{S} - \lambda\mathbf{C}| = 0$. We note that these roots are also obtainable from the equation $|\mathbf{B} - \phi\mathbf{W}| = 0$, since $\lambda = n\phi/q$.

The distribution of T_0^2 for $p = 2$ was given by Hotelling (1950), and for $p > 2$ by Ito (1956) to whom Hotelling suggested the problem. As n tends to infinity, T_0^2 (similarly as nV and $-n \log \Lambda$ and $-n \log W_2$) tends to χ^2 with qp d.f., but with small n gives an exaggerated value of χ^2 which requires a correction.

In the case of two samples T_0^2 can be ' Studentized ' and referred to the F-distribution, for it is known (e.g. Hotelling (1931), Bartlett (1934)) that the distribution with $q = 1$, $p = (\text{any})\, r$, is the same as with $p = 1$ and $q = (\text{any})\, r$. This also follows from the comparison of eqn. (67) with (51), leading to $T_0^2 = T^{*2}$ for $k = 2$. Therefore, in view of eqn. (55) we have

$$F = \frac{N_1 + N_2 - p - 1}{p} \frac{T_0^2}{N_1 + N_2 - 2}. \tag{69}$$

ϕ, ϕ'—Roots of certain determinantal equations

All general tests of hypotheses (i.e. Λ, V, T_0^2) are dependent on the characteristic roots of the matrix $\mathbf{B}(\mathbf{B} + \mathbf{W})^{-1}$ or $\mathbf{BW}^{-1}$ and can be equivalently expressed in terms of those roots. The sum of these roots or the logarithm of their product is also a simple function of χ^2 in the limit, but requires a correction in finite samples to be so related.

Since the number of roots cannot exceed p or $k - 1$, whichever is the smaller, there can only be one root when the observations are limited to two samples. In such case the root of the matrix $\mathbf{BW}^{-1}$, or equivalently of the equation $|\mathbf{B} - \phi\mathbf{W}| = 0$, has the value given in eqn. (31). This value can be tested for significance from

$$F = \frac{N_1 + N_2 - p - 1}{p} \phi, \tag{70}$$

since on comparing eqn. (31) with eqn. (65) it is found that $\phi = V$. Similarly, by examining the relationship between eqns. (33) and (44), we find that the root of $|\mathbf{B} - \phi'(\mathbf{W}+\mathbf{B})| = 0$ is identical with Fisher's R^2. Therefore, in view of eqn. (41) we have

$$F = \frac{N_1 + N_2 - p - 1}{p} \frac{\phi'}{1-\phi'}. \tag{71}$$

Table 1 summarizes the statistical conclusions reached in this paper.

TABLE 1. COMPARISON OF THE MULTIVARIATE STATISTICS USED IN DISCRIMINATORY ANALYSIS OF TWO SAMPLES

Nature of test	*Equivalent value of statistics*	*Difference between composite means*	*which equals* *Vector of discriminatory coefficients*	*Vector of differences between means*	*Statistic originally introduced by*
	$\frac{T^{*2}}{N_1+N_2-2}$	T^{*2}	$=\lambda^2\mathbf{d}'\mathbf{C}^{-1}$		Hotelling (1931) Fisher (1938)
	$\frac{1-\Lambda}{\Lambda}$	$1-\Lambda$	$=\lambda^2\mathbf{d}'(\mathbf{W}+\mathbf{B})^{-1}$		Wilks (1932) Bartlett (1934)
	$\frac{N_1N^2}{(N_1+N_2)}\frac{D^2}{(N_1+N_2-2)}$	D^2	$=\mathbf{d}'\mathbf{C}^{-1}$		Mahalanobis (1936) Fisher (1938)
	$\frac{N_1N_2}{N_1+N_2}W^*$	W^*	$=\mathbf{d}'\mathbf{W}^{-1}$		Fisher (1936)
$F_{N_1+N_2-p-1,\, p}$ for all methods	$\frac{R^2}{1-R^2}$	R^2	$=\lambda^2\mathbf{d}'(\mathbf{W}+\mathbf{B})^{-1}$	$\mathbf{d}$ for all methods	Fisher (1936) Fisher (1938)
	ϕ	ϕ	$=\lambda^2\mathbf{d}'\mathbf{W}^{-1}$		Hotelling (1935, 1936) Bartlett (1938)
	$\frac{\phi'}{1-\phi'}$	ϕ'	$=\lambda^2\mathbf{d}'(\mathbf{W}+\mathbf{B})^{-1}$		Hotelling (1935, 1936) Fisher (1939)
	V	V	$=\lambda^2\mathbf{d}'\mathbf{W}^{-1}$		Hsu (1940)
	$\frac{T_0^2}{N_1+N_2-2}$	T_0^2	$=\lambda^2\mathbf{d}'\mathbf{C}^{-1}$		Hotelling (1947) Hotelling (1950)

The table shows clearly how the multivariate statistics used in discriminatory analysis are interrelated and how, in the case of two samples, they can be tested for significance. Associated with each statistic is a discriminant function involving a different constant. The differences between these constants account entirely for the differences in the statistics. For we find that the latter in each case is constituted of a product of the same vector of differences between p pairs of means and a vector of discriminatory coefficients which varies only in length.

Each scalar value of such a product represents a difference between two composite means or ' centroids ' along the same direction, but evaluated in different units. It also represents a measure of the squared distance between the two populations, since each vector of discriminatory coefficients includes **d** as its element. These measures are, on the whole, arbitrarily dependent on N. D^2 is, however, an exception to this. It has the required metric property as a direct distance statistic.

R^2, which is identical with ϕ' and $1-\Lambda$, has also some special properties. As a canonical correlation it measures the degree of association between composite vectors corresponding to the ' total ' and ' between ' dispersion matrices respectively. Its measure of distance between the two ' centroids ' is, therefore, a function of this correlation. It is expressed with reference to the maximum distance of unity and varies as a cosine of the angle between the two vectors and attains the maximum only when the observations are perfectly homogeneous within the two samples. The discriminant function is then not unlike an artificial variate considered by Fisher, for its two means corresponding to the two samples differ by unity and its within sample variance is zero. R^2 may, therefore, be regarded as a measure of correlation with such a dummy variate and as a relative measure of distance.

The remaining statistics, although less suitable as distance statistics may, however, be used to obtain a discriminant function and to test its significance.

References

Anderson, T. W. (1951). Classification by multivariate analysis. *Psychometrika*, **16**, 31–50.

Anderson, T. W. (1958). *An Introduction to Multivariate Statistical Analysis*. Wiley: New York.

Bartlett, M. S. (1934). The vector representation of a sample. *Proc. Camb. Phil. Soc.* **30,** 327–340.

Bartlett, M. S. (1937). Properties of sufficiency and statistical tests. *Proc. Roy. Soc.* A, **160,** 268–282.

Bartlett, M. S. (1938). Further aspects of the theory of multiple regression. *Proc. Camb. Phil. Soc.* **34,** 33–40.

Bartlett, M. S. (1939). The standard errors of discriminant function coefficients. *J. Roy. Statist. Soc., Supplement*, **6,** 169–173.

Bartlett, M. S. (1947). Multivariate analysis. *J. Roy. Statist. Soc., Supplement*, **9,** 176–197.

Bartlett, M. S. (1951). The goodness of fit of a single hypothetical discriminant function in the case of several groups. *Ann. Eugen.* **16,** 199–214.

Bartlett, M. S. (1954). A note on the multiplying factors for various χ^2 approximations. *J. Roy. Statist. Soc.* B, 296–298.

Bartlett, M. S. (1962). *Essays on Probability and Statistics*. Wiley: New York.

Bayes, Thomas (1763). Essay towards solving a problem in the doctrine of chances. *Phil. Trans. Roy. Soc.* **53,** 370–418 (reprinted with a note by Barnard, G. A. (1958). *Biometrika*, **45,** 293–315).

Beal, G. (1945). Approximate methods in calculating discriminant functions. *Psychometrika*, **10,** 205–218.

Bishop, D. J. (1939). On a comprehensive test of the homogeneity of variances and covariances in multivariate problems. *Biometrika*, **31**, 31–55.

Bose, R. C. (1936). On the exact distribution and moment-coefficients of the D^2 statistic. *Sankhya*, **2**, 143–54.

Bose, R. C. and Roy, S. N. (1938). The exact distribution of the Studentized D^2 statistic. *Sankhya*, **4**, 19–38.

Cochran, W. G. and Bliss, C. I. (1948). Discriminant functions with covariance. *Ann. math. Statist.* **19**, 151–176.

Dwyer, P. S. and MacPhail, M. S. (1948). Symbolic matrix derivatives. *Ann. math. Statist.* **19**, 517–534.

Fisher, R. A. (1936). The use of multiple measurements in taxonomic problems. *Ann. Eugen.* **7**, 179–188.

Fisher, R. A. (1938). The statistical utilization of multiple measurements. *Ann. Eugen.* **8**, 376–386.

Fisher, R. A. (1939). The sampling distribution of some statistics obtained from non-linear equations. *Ann. Eugen.* **9**, 238–249.

Fisher, R. A. (1940). The precision of discriminant functions. *Ann. Eugen.* **10**, 422–429.

Garrett, H. E. (1943). The discriminant function and its use in psychology. *Psychometrika*, **8**, 65–79.

Horst, P. and Smith, S. (1950) The discrimination of two racial samples. *Psychometrika*, **15**, 271-290.

Hotelling, H. (1931). The generalization of Student's ratio. *Ann. math. Statist.* **2**, 360–378.

Hotelling, H. (1935). The most predictable criterion. *J. educ. Psychol.* **26**, 139–142.

Hotelling, H. (1936). Relations between two sets of variates. *Biometrika*, **28**, 321–377.

Hotelling, H. (1947). A generalized T measure of multivariate dispersion. (Abstract), *Ann. math. Statist.* **18**, 298.

Hotelling, H. (1950). A generalized T test and measure of multivariate dispersion. *Proc. Second Berkeley Symposium on Math. Statist. and Probability*, 23–41. California: University Press. London: Cambridge University Press (published 1951).

Hsu, P. L. (1938). Notes on Hotelling's generalized T. *Ann. math. Statist.* **9**, 231–243.

Hsu, P. L. (1939). On the distribution of roots of certain determinantal equations. *Ann. Eugen.* **9**, 250–258.

Hsu, P. L. (1940). On the generalized analysis of variance. *Biometrika*, **31**, 221–237.

Ito, K. (1956). Asymptotic formulae for the distribution of Hotelling's generalized T_0^2 statistic. *Ann. math. Statist.*, **27**, 1091–1105.

Jambunathan, M. V. (1954). Some properties of Beta and Gamma distributions. *Ann. math. Statist.* **25**, 401–405.

James, G. S. (1954). Tests of linear hypotheses in univariate and multivariate analysis when the ratios of the population variances are unknown. *Biometrika*, **41**, 19–43.

Kendall, M. G. (1940). On the method of maximum likelihood. *J. Roy. Statist. Soc.* **103**, 388–399.

Kendall, M. G. (1957). *A Course in Multivariate Analysis*. Griffin: London.

Kullback, S. (1952). An application of information theory to multivariate analysis. *Ann. math. Statist.* **23**, 88–102.

Lawley, D. N. (1956). A general method for approximating to the distribution of likelihood ratio criteria. *Biometrika*, **43**, 295–303.

Lubin, A. (1950). Linear and non-linear discriminating functions. *Brit. J. Psychol. Statist. Sec.* **3**, 90–104.

Mahalanobis, P. C. (1930). On tests and measures of group divergence. I. Theoretical formulae. *J. Asiat. Soc. Beng.* **26**, 541–588.

Mahalanobis, P. C. (1936). On the generalized distance in statistics. *Proc. Nat. Inst. Sci. (India)* **12**, 49–55.

MISES, R. VON (1945). On the classification of observation data into distinct groups. *Ann. math. Statist.* **16,** 68–73.
NEYMAN, J. and PEARSON, E. S. (1928). On the use of interpretation of certain test criteria for purposes of statistical inference. *Biometrika,* **20,** A, 175–240, 263–294.
NEYMAN, J. and PEARSON, E. S. (1931). On the problem of *k* samples. *Bull. Intern. Acad. Pol. Sci. Lettres* A, 460–481, Cracow Univ. Press.
NEYMAN, J. and PEARSON, E. S. (1932). On the problem of the most efficient tests of statistical hypotheses. *Phil. Trans. Roy. Soc.* London, A, **231,** 281–337.
NEYMAN, J. and PEARSON, E. S. (1933). On the testing of statistical hypotheses in relation to probability *a priori*. *Proc. Camb. Phil. Soc.* **29,** 492–510.
PEARSON, E. S. and NEYMAN, J. (1930). On the problem of two samples. *Bull. Intern. Acad. Pol. Sci. Lettres* A, 73–96. Cracow Univ. Press.
PEARSON, E. S. and WILKS, S. S. (1933). Method of statistical analysis appropriate for *k* samples of two variables. *Biometrika,* **25,** 353–378.
PENROSE, L. S. (1947). Some notes on discrimination. *Ann. Eugen.* **13,** 228–237.
RAO, C. R. (1948 a). The utilization of multiple measurements in problems of biological classification. *J. Roy. Statist. Soc.* B, **10,** 159–193.
RAO, C. R. (1948 b). Tests of significance in multivariate analysis. *Biometrika,* **35,** 58–79.
RAO, C. R. (1952). *Advanced Statistical Methods in Biometric Research.* Wiley: New York.
RAO, C. R. and SLATER, P. (1949). Multivariate analysis applied to differences between neurotic groups. *Brit. J. Psychol., statist. Sec.,* **2,** 17–29.
SMITH, C. A. (1947). Some examples of discrimination. *Ann Eugen.* **13,** 272–282.
WALD, A. (1944). On a statistical problem arising in the classification of an individual into one of two groups. *Ann. math. Statist.* **15,** 145–163.
WALD, A. (1947). An essentially complete class of admissible decision functions. *Ann. math. Statist.* **18,** 549–555.
WALD, A. (1950). *Statistical Decision Functions.* Wiley: New York.
WALD, A. and BROOKNER, R. J. (1941). On the distribution of Wilks' statistic for testing the independence of several groups of variates. *Ann. math. Statist.* **12,** 137–152.
WELCH, B. L. (1939). Note on discriminant functions. *Biometrika,* **31,** 218–220.
WILKS, S. S. (1932). Certain generalizations in the analysis of variance, *Biometrika,* **24,** 471–494.
WILKS, S. S. (1938). The large-sample distribution of the likelihood ratio for testing composite hypotheses. *Ann. math. Statist.* **9,** 60–62.
WILKS, S. S. (1946). Sample criteria for testing equality of means, equality of variances and equality of covariances in a normal multivariate distribution. *Ann. math. Statist.* **17,** 257–281.
WILLIAMS, E. J. (1955). Significance tests for discriminant functions and linear functional relationship. *Biometrika,* **42,** 360–381.
WISHART, J. (1928). The generalized product moment distribution in samples from a normal multivariate population. *Biometrika,* **20,** A, 32–58.

13

Reprinted from *Evolution,* **13**(3), 283–299 (1959)

MULTIVARIATE GEOGRAPHICAL VARIATION IN THE WOLF *CANIS LUPUS* L.[1]

PIERRE JOLICOEUR[2]

Department of Zoology, University of British Columbia, Vancouver, Canada

Received July 7, 1958

INTRODUCTION

Modern biology is subdivided in numerous disciplines dealing with particular aspects of life. But what lives and evolves in reality is not isolated characteristics, it is whole organisms or, even more exactly, populations of organisms. Nowhere else perhaps as much as in evolutionary studies has a synthetic approach become important. While this makes the theory of evolution the focus of biology, it also creates intricate problems. The student of evolution often has to compare groups of organisms with respect to multiple characters. Multivariate statistical techniques should be most efficient in such problems but comparatively little use has been made of them. The following analysis of geographical variation in the wolf (*Canis lupus* L.) is an attempt to evaluate these techniques in practice. Although only morphological characters are considered here, physiological, behavioral and ecological data could be analysed in the same manner.

The last comprehensive taxonomic study of North American wolves was that of Goldman (Young and Goldman, 1944). It consisted primarily of qualitative skull and pelage descriptions and failed to show clearly the nature and the extent of geographical variation in the species as a whole. During the present study groups of specimens have been compared with respect to complexes of metrical characters. This has disclosed multivariate trends of variation between several Nearctic wolf populations. The techniques used and the results obtained are exposed and discussed in this paper.

MATERIAL AND DATA

Numerous wolf specimens were collected in northwestern Canada by the Canadian Wildlife Service and the Manitoba Game Department during recent predator control operations. This material was deposited in the Museum of Zoology of the University of British Columbia. In this study it has been compared with British Columbia, Alaska and Arctic material, some of which was bor-

[1] Based on a thesis submitted in partial fulfillment for the degree of Master of Arts, Department of Zoology, University of British Columbia, April 1958.

[2] Present address: Walker Museum, University of Chicago, Chicago 37, Illinois.

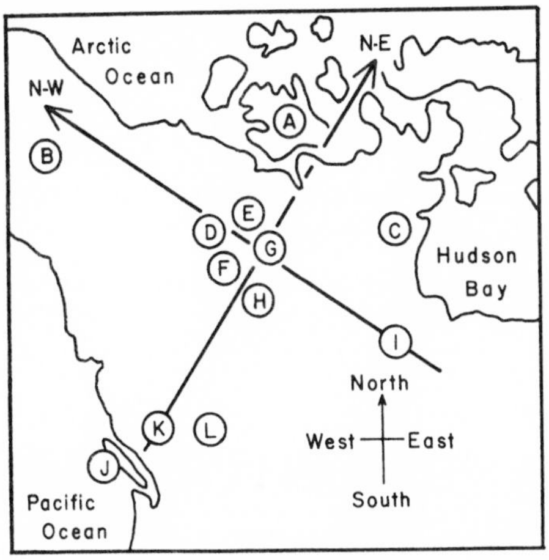

FIG. 1. Geographical origin of the samples.

rowed from the British Columbia Provincial Museum, the National Museum of Canada and Dr. R. Rausch, Anchorage, Alaska. Approximately five hundred specimens were studied. They were grouped according to sex and geographical origin as shown in figure 1 and table 1. Juvenile specimens were excluded as there was only half a dozen of them. Four general areas are represented by large samples: British Columbia (group K), Manitoba (I), and the Northwest Territories between Great Slave Lake and Great Bear Lake (groups D + E and group G). Two arrows have been lined up on these large samples in the map (fig. 1) and in a subsequent graph (fig. 10). They point approximately northeastward and northwestward and help to refer biometrical differences to their geographical context.

Skulls were available for most specimens while there were pelage and body data for only part of the material. The analysis of geographical variation was therefore based primarily on skull dimensions. Photographic transparencies of the carcasses were available for four samples of the Northwest Territories and they were examined with respect to pelage coloration. Twelve skull dimensions were measured, in the manner illustrated (fig. 2). They are referred to hereafter by the following coded designations:

L_1: Condylobasal length
L_2: Palatal length
L_3: Postpalatal length
W_1: Zygomatic width
W_2: Palatal width outside the first upper molars M_1
W_3: Palatal width inside the second upper premolars Pm_2
W_4: Width between the postglenoid foramina
W_5: Interorbital width
C_1: Least width of the braincase
C_2: Width between the auditory bullae
T_1: Alveolar length of the upper carnassial Pm_4
T_2: Crown length of the first upper molar M_1

TECHNIQUES OF ANALYSIS

The advantages of multivariate statistical techniques for evolutionary problems have been discussed by Anderson (1954). Typically evolutionary studies lead to comparisons of groups of organisms with respect to numerous charac-

TABLE 1. *Size and sex-composition of the samples*

Locality	Group	Males	Females	Undetermined	Total
Arctic	A	11	8	—	19
Alaska	B	3	3	3	9
Keewatin	C	5	3	6	14
Northwest	D	41	39	—	80
Territories	E	41	40	—	81
	F	12	8	—	20
	G	33	33	—	66
	H	—	—	9	9
Manitoba	I	73	64	—	137
Vancouver Island	J	5	5	—	10
Mainland of B. C.	K	15	12	18	45
Rocky Mountains	L	6	3	—	9
					499

teristics. But in reality, as stressed by Olson and Miller (1958), the various aspects of living organisms distinguished by man are intimately associated. When comparing evolutionary groups of organisms it is therefore best to consider as many characters as possible simultaneously. Theoretically such joint comparisons call for multivariate analysis.

Practical reasons as well as theoretical ones make multivariate techniques desirable in evolutionary studies. Groups of organisms may be entirely distinct with respect to several characters jointly and yet overlap with respect to every one of the same characters separately. Hypothetical examples of such cases are illustrated here. In figure 3 for instance two groups of specimens, represented respectively by open circles and solid dots, occupy readily separable portions of a bivariate scatter diagram. But the complete separation of these two sam-

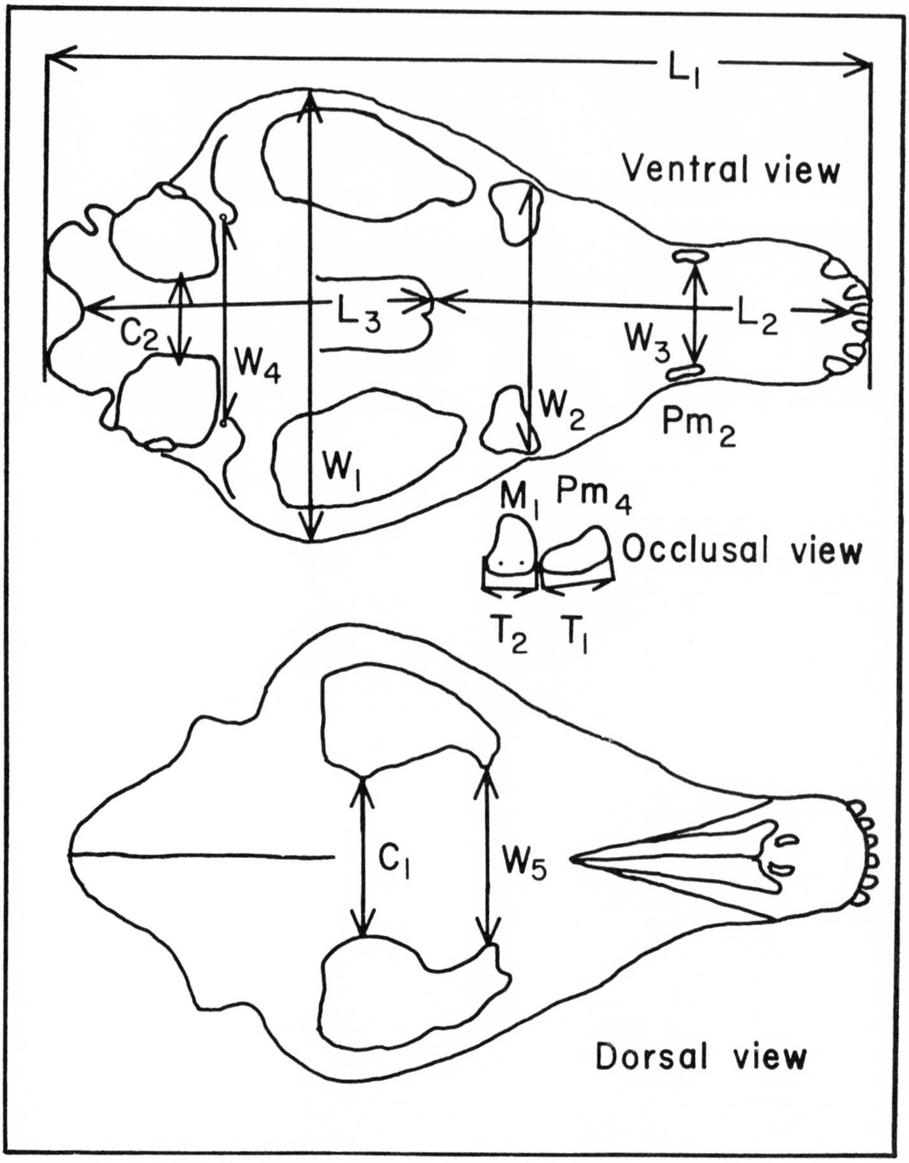

FIG. 2. Skull dimensions measured and coded designations; dorsal and ventral views of the skull and occlusal view of left upper teeth.

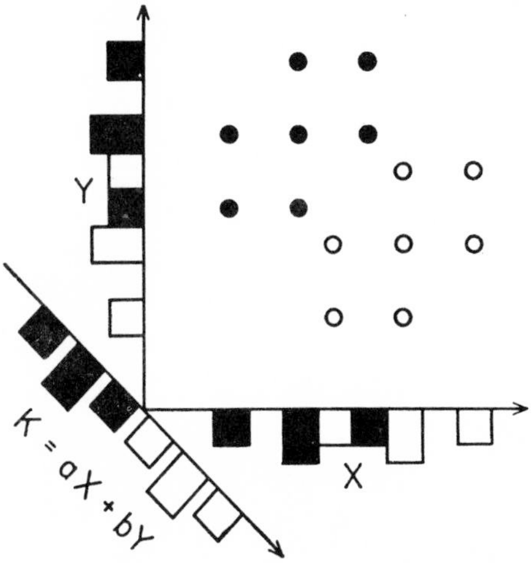

FIG. 3. Hypothetical example of discriminatory analysis for two characters. Scatter diagram and frequency histograms of characters X and Y and of discriminant function K = aX + bY.

ples would fail to appear if the two metrical characters X and Y were considered one after the other. This can be seen by projecting the bivariate diagram upon axes X and Y respectively; the black and the white frequency bars are interspersed in the resulting frequency histograms. The difference of the two samples could be fully expressed by a single variable though, the so-called *discriminant function* $K = aX + bY$; it corresponds to the projection of the graph upon axis K and its frequency histograms of the two samples are completely separated.

For three characters a scatter diagram assumes the form of a box within which groups of dots, circles and crosses are dispersed (fig. 4). The three edges OX, OY and OZ of the box are the coordinate axes of characters X, Y and Z. Bivariate diagrams of these data would correspond to perpendicular views of the tridimensional graph through faces XOY, XOZ and YOZ. However the true nature and magnitude of the differences between the groups would not show up well if the characters were considered one by one or two by two. The relative position of the samples appears much more clearly if the graph is viewed through the discriminant plane K_1OK_2 or projected upon it. This optimum two-dimensional representation of between-group differences is obtained by calculating the discriminant functions $K_1 = a_1X + b_1Y + c_1Z$ and $K_2 = a_2X + b_2Y + c_2Z$ and making their scatter diagram (fig. 5). The projections X, Y and Z of coordinate axes OX, OY and OZ upon the discriminant plane (fig. 5) indicate its signification in terms of the original variables themselves, the three metrical characters.

But it is in joint studies of four or more characters that the efficiency of discriminant functions is unequalled for ordinary graphing methods are impossible in more than three dimensions. Nevertheless multivariate procedures permit the analyst to examine optimum two-dimensional slices of the unexisting multidimensional graph (fig. 10, 11, 12). Much time can also thus be saved; no lengthy univariate or bivariate search is necessary for the most informative combinations of characters. There would be sixty-six different manners to associate twelve characters two by two for instance; but one or two discriminant graphs bring out almost all available information. Discriminant functions then are indispensable not only to disclose the true degree of distinctness of multivariate samples but also to condense the information relative to large numbers of characters in very few graphs.

Many samples of biometrical data follow the normal distribution at least approximately. Graphically they appear as elliptical clusters of points (fig. 8). In practice much of the information contained by such samples should be explicit in the scatter diagrams themselves. But comparisons involving several large samples would be likely to engender confusion. To avoid this, statistical summarization is necessary. *Equal frequency ellipses* are particularly effective for normal bivariate data; they

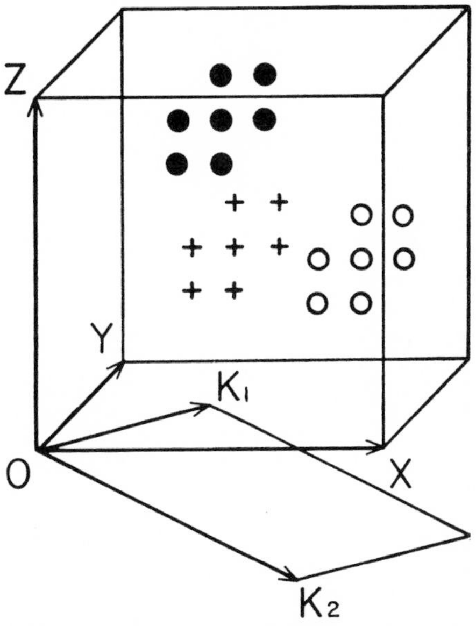

FIG. 4. Hypothetical example of discriminatory analysis for three characters. Scatter diagram of characters X, Y and Z and discriminant plane K_1OK_2 (see fig. 5).

have been used for instance by Kotaka (1953) on *Anadara granosa* (Pleistocene pelecypod from Japan). Their biometrical utilization has been thoroughly discussed by Defrise-Gussenhoven (1955). Confidence belts about regression lines and quadrilateral range diagrams have also been used to represent bivariate dispersion. However a regression line yields an estimate of the value of one character from others and this is not of primary importance in studies of evolution. As for the range diagrams utilized by Imbrie in a study of Triassic amphibians (Colbert and Imbrie, 1956), they would be preferable to ellipses only for certain abnormal types of data; but departures from normality are often negligible in practice and, when important, they can frequently be corrected by using logarithmic instead of arithmetic values. In general then equal frequency ellipses are more suitable than any other current statistical device to outline the range of variation of bivariate specimens. And, used in conjunction with discriminant scatter diagrams, they are applicable to studies of any number of characters.

Multivariate calculations follow the same basic plan however many variables are analyzed jointly. The average value of metrical characters in a group of specimens is the mean; it is the center of the sample of points. The variation of the characters within this group of organisms is summarized numerically by a set of mean squared deviations and cross products, the so-called *within-group covariance matrix* W. Matrix W is generated by the elliptical dispersion of the individuals around their group mean and it figures in the equation of the equal frequency ellipses (or multidimensional ellipsoids) circumscribing the sample. The multivariate analysis of within-group variation consists in calculating the *principal axes* of such ellipses; to these directions of maximum variation (or *characteristic vectors*) correspond *principal variance components* (or *characteristic roots*). It is when biometrical samples are examined with respect to these principal trends of variation that their information content is the most explicit. Principal component analysis is closely similar in principle to factor analysis as applied by psychologists and, more recently, by biometricians.

As noted by Yates (1950), the analysis of differences between groups of specimens is analogous to that of within-group variation. The dispersion of group means around their grand mean is expressed by the *between-group covariance matrix* B. The so-called *discriminant functions* are the principal axes of matrix B after standardization by matrix W of within-group variation (pooled over all samples). Geometrically speaking the ellipsoid of within-group variation is taken as a yardstick for between-group variation. But, within ordinary

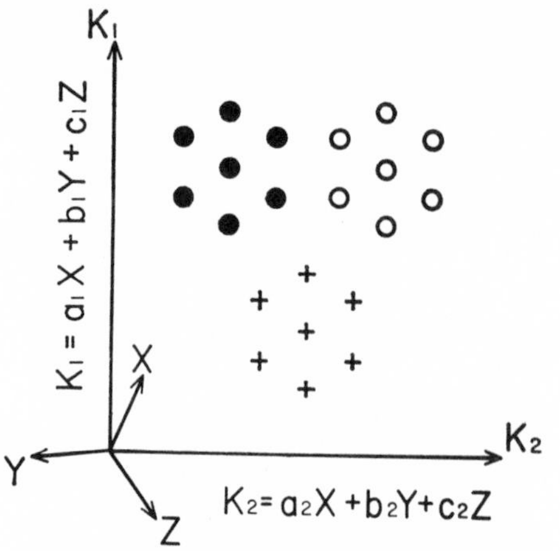

FIG. 5. Scatter diagram of discriminant functions K_1 and K_2 corresponding to plane K_1OK_2 of figure 4 and projections thereupon of axes OX, OY and OZ.

samples of living organisms, the greatest variation component of body dimensions is generally age and size (Teissier, 1955; Wright, 1932). Consequently the standardization of matrix B by matrix W emphasizes other factors of variation than age and size. The purpose formerly served by age classes and size-independent ratios is thus automatically fulfilled in discriminatory analysis. In multivariate group comparisons as in principal component analysis it is in the direction of the principal axes that the observations differ the most and that most of the information lies.

Several expositions of multivariate methods are now available (Hotelling, 1954; Kendall, 1957; Quenouille, 1952; Rao, 1952); to understand them elements of multidimensional algebra and geometry are indispensable; Murdoch's excellent introduction (1957) provides a relatively easy way to the latter. Most recent applications of multivariate analysis have unfortunately been too abstract; the graphical means of presentation utilized in this study make multivariate results just as easy to visualize as univariate or bivariate ones. As for the length of computations, it is no longer prohibitive thanks to the growing availability of electronic digital computers. Multivariate statistical techniques should be used more and more along with simpler methods when problems call for them. This appears to be the case in biometrical studies of evolution.

Pelage Color Variation

The pelage coloration of wolves is very variable in intensity, in hue and in pattern. As detailed verbal descriptions are not

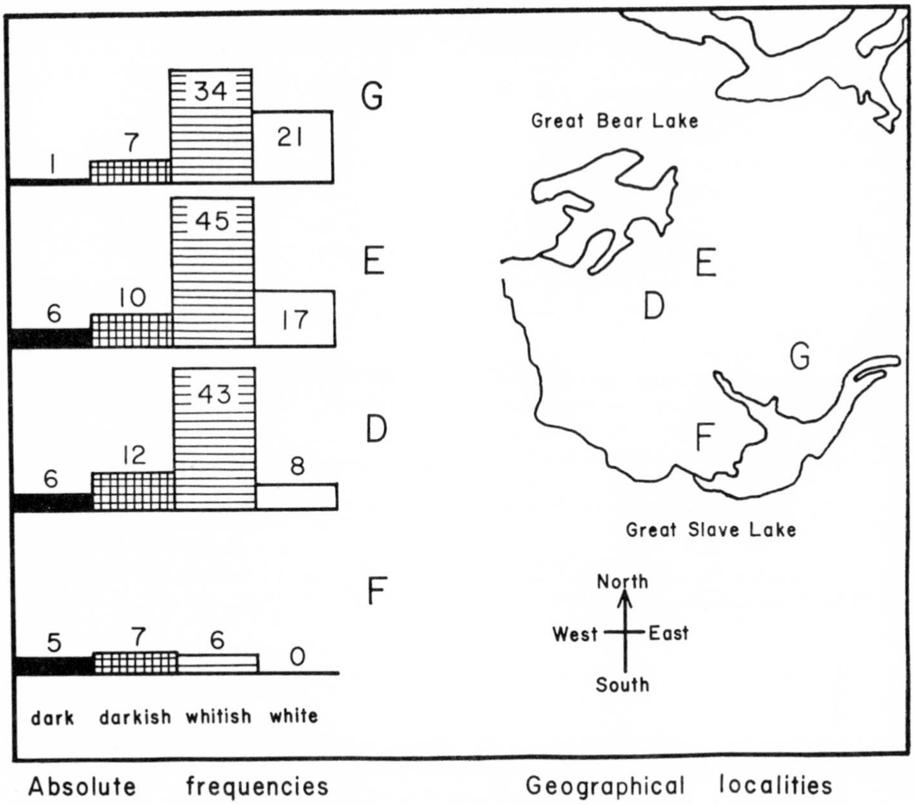

FIG. 6. Northeastward increase in the relative frequency of pale wolves toward the tundra between Great Slave Lake and Great Bear Lake.

suitable for large samples, the photographic transparencies examined were classified into four arbitrary classes according to the general darkness of pigmentation: dark, darkish, whitish and white. Such arbitrary classes do not correspond in the wolf to actually discrete color phases as in some polymorphic species. Such a classification is also only approximate and fits adequately only the present material. It does disclose however gradual differences in color-class frequencies analogous to the clines in color-phase frequencies of the red fox and the black bear (Cowan, 1938).

The relative frequency of pale wolves increases in a northeastward direction (toward the tundra) between Great Slave Lake and Great Bear Lake in the Northwest Territories (fig. 6). There are gradually more white and whitish and fewer dark and darkish individuals in samples F, D, E and G successively. Samples D and E differ little from each other but differ significantly from the two extreme groups (95% chi-square). Seasonal variation is apparently not involved since specimens were collected at comparable dates. On the other hand, the environmental conditions of these four localities appear too similar for such pronounced differences to be phenotypical. Consequently the color frequency shift probably expresses a cline in gene frequencies. The extreme whiteness of tundra in winter is of course very well known and the higher frequency of pale wolves there is clearly a case of homochromy. The selective value of concealing coloration for a predator would presumably lie in the corresponding ease to approach preys. Pale pelage coloration may be just one facet of a physiological complex however and it is possibly related to other factors than homochromy.

Recent barren-ground caribou studies (Banfield, 1954; Kelsall, 1957) have shown caribou to migrate more intensely through areas D and E than through areas F and G. The similarity of pelage coloration between wolf populations appears to be proportional thus to the local intensity of caribou migrations. This relationship could be expected if the movements of wolves were correlated with those of caribou. The latter correlation is effectively suggested by field observations. Wolves are often seen with caribou herds (Rausch, 1951) and, statistically speaking at least, they probably follow them at migration (Banfield, 1951).

A relatively higher frequency of dark individuals has been reported in the Rocky Mountains (Cowan, 1947). The short-distance cline exhibited by the present material may therefore be part of a long-distance cline going at least from the Rockies to the Northwest Territories.

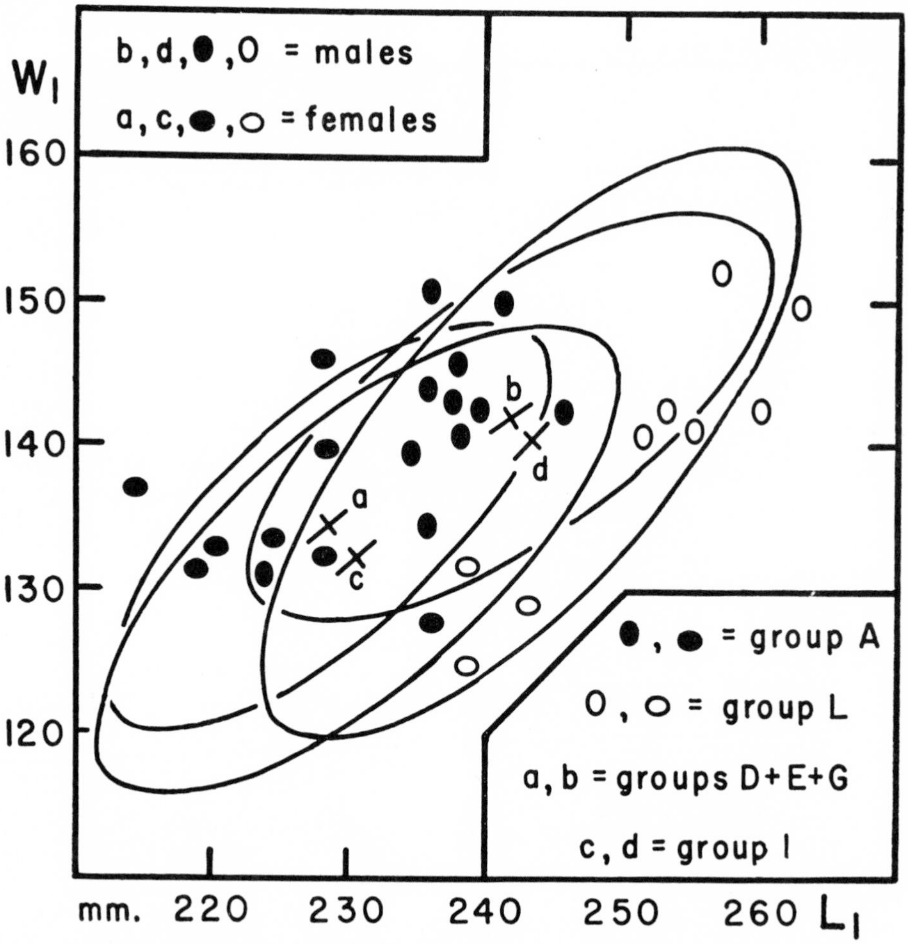

FIG. 7. Condylobasal length (L_1) and zygomatic width (W_1). Dots represent individual specimens; crosses and ellipses represent means and 95% frequency intervals respectively. Males have a larger skull than females and northeastern individuals are shorter and relatively broader-skulled than southwestern ones.

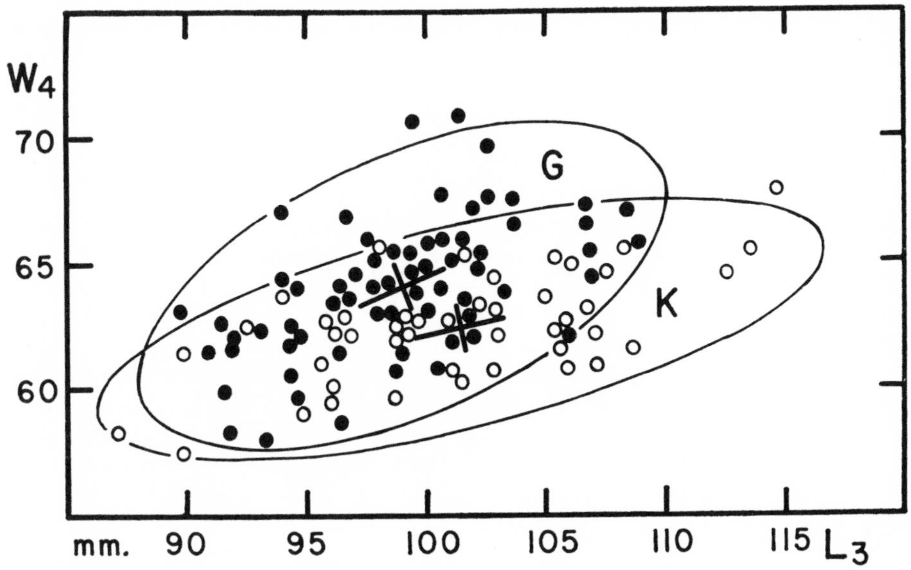

FIG. 8. The wolves of the Northwest Territories (G) have a higher growth rate of postglenoid width (W_4) relatively to postpalatal length (L_3) than those of British Columbia (K).

More data on the pelage coloration of wolves would perhaps show analogies with the geographical distribution of the color phases of the red fox and the black bear (Butler, 1947; Cowan, 1938).

Bivariate Skull Variation

Overall skull size can be satisfactorily described by condylobasal length (L_1) and zygomatic width (W_1). Bivariate scatter diagrams of these two dimensions were made and 95% equal frequency ellipses were calculated following Defrise-Gussenhoven's procedure (1955). Figure 7 summarizes the most important information. Males are approximately 4% larger than females in linear skull dimensions. This agrees with Hildebrand's conclusions on the body size of Canidae (1952). Other facts brought out are the small skull size (close to the left lower corner of the graph) and the great relative skull breadth (close to the left upper corner) of northeastern wolves. Groups L, I, D + E + G, and A are successively closer to the left side of the graph. This ordering of samples according to skull size and relative breadth is strikingly similar to the ordering of the geographical localities of origin projected upon a line of northeastward direction. Such gradual geographic variation was termed *clines* by Huxley (1938).

The shortness and greater relative breadth of skull of northeastern wolves also shows up in a scatter diagram (fig. 8) of postglenoid breath (W_4) and postpalatal length (L_3). The wolves of the Northwest Territories (G) are shorter and broader-skulled than those of British Columbia (K) with respect to these two dimensions. But here the difference of proportion increases with size; there is a difference of relative growth rate. Equal frequency ellipses fit the data satisfactorily; there is no obvious curvature of trend and no need for a logarithmic transformation.

A third bivariate association shows geographical variation (fig. 9): interbullar breadth (C_2) with carnassial length (T_1). The specimens of Manitoba (I) and the Northwest Territories (D + E) are at the center of this graph and constitute the average. The wolves of British Columbia (K) have shorter upper carnassial teeth than the average and those of Vancouver Island (J) a narrower interbullar space. Simple examination of the skulls confirms what the graphical analysis summarizes. Distinct spaces show in between the small teeth of British Columbia wolves and the ten Vancouver Island specimens have markedly inflated auditory bullae with a narrow interval. Surprisingly in this graph the Vancouver Island wolves differ the most from those to which they are the closest geographically. Further discussion of this will follow the joint multivariate analysis of all twelve skull dimensions.

Multivariate Skull Variation

Comparing biometrical samples with respect to complexes of characters requires the calculation of discriminant functions. The coefficients K of these functions and the components D of between-group variation for which they account are the solution of the following matrix equation: KB = DKW. Matrices B and W are the *between-group* and the pooled *within-group covariance matrices* respectively and their signification has already been explained in these pages. As geographical variation in skull proportions was of primary interest here, sexes were kept together; sexual skull differences were mostly size differences (fig. 7) and separating males and females here would have spread the group means too much in the direction of size. The equation KB = DKW was solved on an electronic digital computer by matrix operations (Murdoch, 1957: 165, 166) corresponding to the transformations suggested by Rao (1952: 357, 367). Matrices were diagonalized following the Jacobi method. The within-group variances and covariances of the discriminant functions were calculated to verify the computations; they were exact to two or three significant digits (KWK′ = I) and this was considered acceptable.

The sum of the components D_i of between-group variance was 75.464; of this

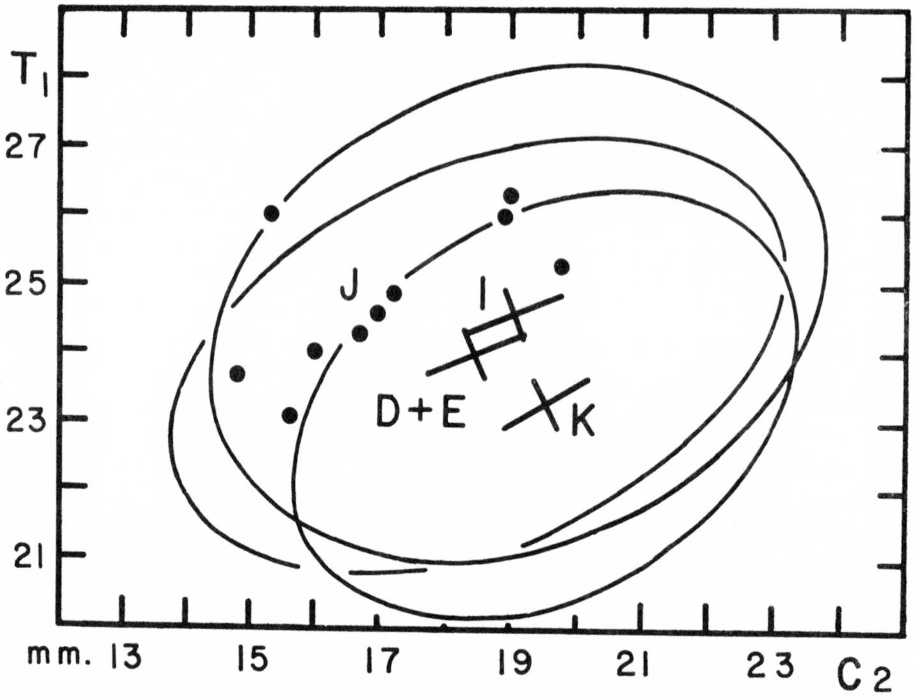

Fig. 9. The Vancouver Island specimens (J) have a narrow interbullar space (C_2) and those of British Columbia (K) short upper carnassial teeth (T_1).

TABLE 2. *Discriminatory analysis*

Variance component	D_1	D_2	D_3	D_4	D_5
Magnitude	39.274	13.474	9.008	5.958	3.206
Per cent of total	52%	18%	12%	8%	4%
Probability	$<.01$	$<.01$	$<.01$	$<.01$	$<.05$

94% was accounted for by the first five functions (table 2). The statistical significance of these variances was tested as prescribed by Rao (1952: 372) for large samples. If the various samples came from a same population the probability of such large components would be less than 1% for the first four and less than 5% for the fifth. To each component correspond twelve coefficients for the skull dimensions in the discriminant functions; the numerical values of these coefficients are not given here for they could be deduced from the vectors of the discriminant graphs (figs. 10 and 12). The group means, the grand mean and the pooled within-group standard deviation of the twelve skull dimensions are listed for reference (table 3); inspecting them rapidly confirms the reality of the multidimensional trends of variation disclosed by discriminatory analysis. Tabulating other statistics or raw data here would consume too much space without making anything explicit.

The pattern of dispersion of the groups in the plane of the first and second discriminant axes (fig. 10) presents similarities to the geographical disposition of the localities of origin. Northern samples tend to congregate in the left upper corner of the graph, eastern samples in the right upper corner and contrarywise for southern and western samples. The two ar-

TABLE 3. *Multivariate statistics (in mm.)*

Characters..		L_1	L_2	L_3	W_1	W_2	W_3	W_4	W_5	C_1	C_2	T_1	T_2
Grand Mean		236.91	118.02	99.69	137.67	78.50	33.57	63.88	45.68	41.29	18.90	24.33	17.28
Pooled Standard Deviation		10.14	5.017	5.022	8.012	3.524	1.987	2.675	3.503	2.835	1.871	1.347	.8787
Group Means													
Group	N												
A	19	231.63	113.84	98.84	139.37	80.32	32.90	65.84	45.44	40.08	19.20	25.77	17.54
B	9	245.67	123.11	102.22	140.45	81.33	35.36	65.65	46.86	43.06	19.46	24.40	17.70
C	14	234.14	116.79	98.50	135.29	78.72	33.67	64.01	45.71	40.52	18.70	24.50	17.33
D + E	161	234.73	117.30	98.49	138.56	78.37	33.58	63.78	46.37	41.28	18.45	23.99	17.28
F	20	242.15	119.55	103.15	140.75	79.63	33.79	65.07	45.64	41.64	19.67	25.13	17.42
G	66	235.98	118.35	99.00	137.76	78.13	33.18	64.05	46.08	41.03	18.59	24.42	17.36
H	9	243.33	120.89	102.89	140.22	79.59	34.39	64.34	46.27	40.49	20.54	23.91	17.19
I	137	237.20	117.75	99.88	136.52	78.63	34.01	63.82	45.11	41.15	19.03	24.61	17.34
J	10	236.30	119.60	98.30	136.70	77.73	31.85	61.15	44.13	41.94	17.03	24.82	16.70
K	45	240.18	119.36	101.40	135.27	76.92	32.97	62.49	44.07	42.16	19.50	23.28	16.80
L	9	251.00	123.45	106.33	139.67	79.91	32.73	66.10	47.51	42.61	22.23	25.23	17.76

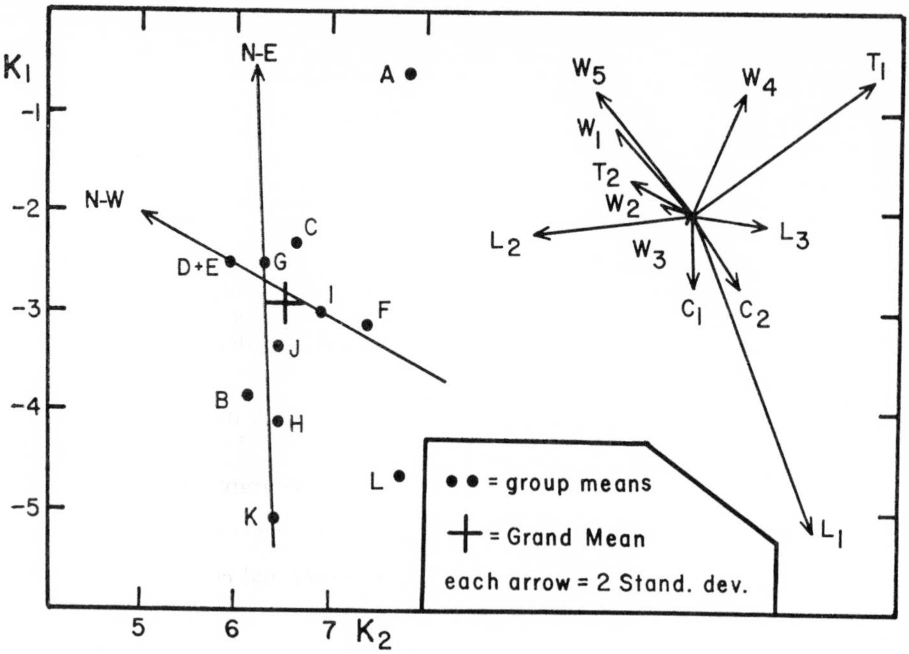

FIG. 10. Group dispersion (left) in discriminant functions K_1 and K_2 and variation of the skull dimensions (right); arrows N–E and N–W correspond to those of the map (fig. 1).

rows marked N–E and N–W correspond to those of the map (fig. 1) and help to evaluate the similarity of the pattern of biometrical dispersion to the pattern of geographical origin. Discrepancies come mostly from small samples; the most obvious one is the relative position of Alaska (B) and Vancouver Island (J) wolves. But this discrepancy is largely compensated for by the dispersion pattern of the third and fourth discriminant functions (fig. 12); here the average of the Alaska specimens diverges from the southern groups and the Vancouver Island individuals contrast sharply with all others. The first component of multivariate variance ($D_1 = 52\%$ of total), which corresponds closely to a northeastward direction, is markedly larger than the next one ($D_2 = 18\%$ of total). Geographical variation in skull dimensions would thus appear to be most pronounced northeastward. Ascertaining this last point would require more material however for the present samples are far from evenly distributed with respect to latitude and longitude.

As discussed previously, the full significance of discriminant scatter diagrams would not be explicit unless the coordinate axes of the original variables were projected thereupon. Sets of *vectors* (= arrows) bearing the coded designations of the skull dimensions indicate these projections in the discriminant graphs (figs. 10 and 12). Each vector shows the change in the discriminant functions that the corresponding dimension would generate if it varied independently. All skull dimensions are intercorrelated of course and these vectors must be considered jointly rather than separately. Northeastern wolves differ generally from southwestern ones (fig. 10) by a decrease in skull length (L_1) and in braincase development (C_1 and C_2) opposed to an increase in skull breadth (W_1, W_4 and W_5). Eastern individuals have longer upper carnassial teeth (T_1) and a shorter palate (L_2) than western ones. The wolves of

Vancouver Island (J) differ very much from the others (fig. 12) by six skull dimensions (greater T_1 and C_1; smaller T_2, C_2, W_4 and W_5) and very little with respect to the six others. The role of these two groups of dimensions is contrasted not only by the directions but also by the lengths of their vectors. The Vancouver Island specimens are much further from the grand mean than the arrows (one

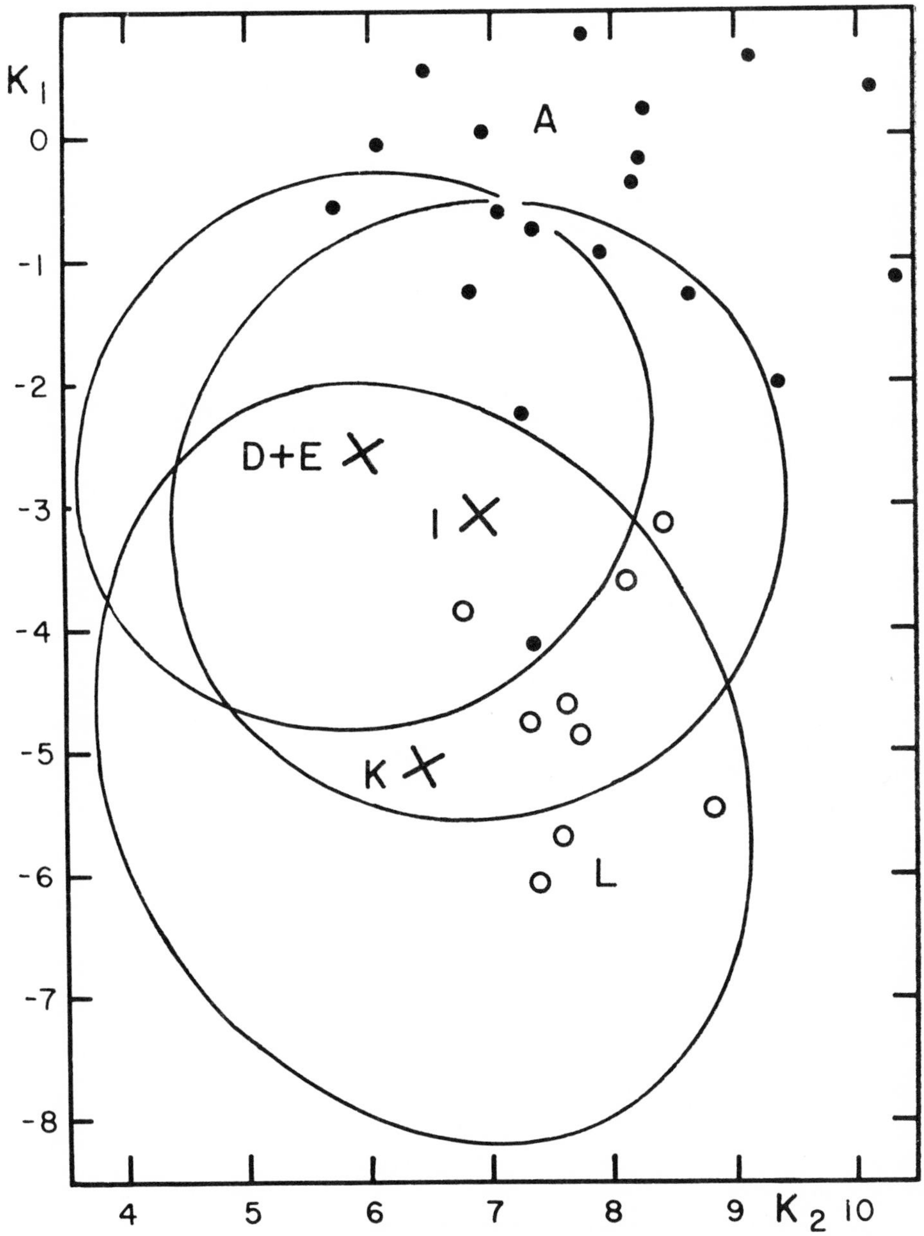

FIG. 11. Overlapping of groups in discriminant functions K_1 and K_2; dots represent individuals, mean crosses and 95% equal frequency ellipses represent samples; letters refer to closest symbols.

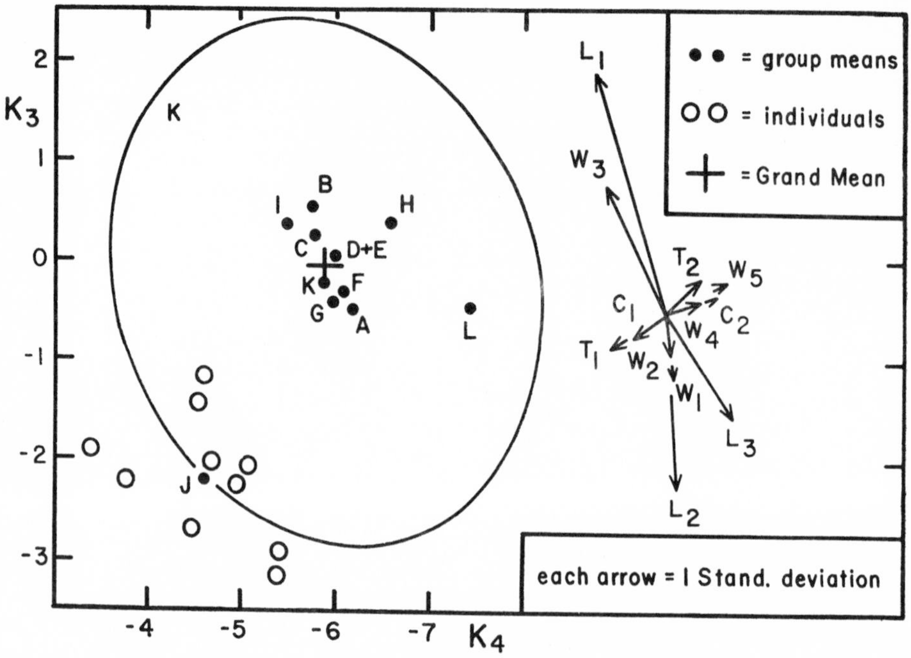

Fig. 12. Group dispersion (left) in discriminant functions K_3 and K_4 and variation of the skull dimensions (right); 95% equal frequency ellipse of group K.

standard deviation) of their discriminators are long.

The amount of overlapping of groups of specimens in discriminant graphs can be illustrated adequately with individual dots for small samples and equal-frequency ellipses for large samples. To evaluate overlapping exactly all discriminant dimensions should be considered jointly and the peripheral decrease in frequency within the samples should be taken into account. But for the sake of simplicity *biometrical overlapping* is defined here as the *percentage of common area between samples in a two-dimensional scatter diagram.* Following this definition the British Columbia wolves (K) overlap by approximately 50% (fig. 11) with those of Manitoba (I) and the Northwest Territories (D + E). The individuals of the Rocky Mountains (L) are intermediary and overlap largely both with those from British Columbia and those from Manitoba. The Arctic specimens (A) overlap by approximately 50% with those of the mainland. The lowermost point of sample A represents a subadult female from Coronation Gulf (mainland) which should apparently have been grouped with mainland specimens and is relatively narrow-skulled. This individual excepted, the Arctic wolves do not overlap with those of the Rockies. Larger samples would probably do so to some extent however. The Vancouver Island specimens (J) overlap by approximately 50% with the others (fig. 12).

To sum up, the present material shows northeastern wolves to have shorter and relatively broader skulls than southwestern ones and eastern wolves to have a shorter palate and longer upper carnassial teeth than western ones. Such a generalization is only approximate however; the pattern of biometrical differences does not correspond exactly to the pattern of geographical origin and the first two discriminant functions account for only 70% of the variance between the groups. More variance is associated with a northeastward direction than with any other one. The Vancouver Island wolves differ markedly from the others with respect to six skull dimensions but very little with respect to the six others. The amount of biometrical overlapping and separation between all groups corresponds approximately to the degree of geographical separation by distance, insularity, etc.

Interpretations and Conclusions

The approximate correspondence of biometrical differences to geographical separation could readily be interpreted in terms of population genetics. Theoretically genetic differentiation within an incompletely panmictic population should increase with geographical distance and other factors of isolation (Li, 1955: 306–310). In the case of wolves, genetic differentiation should have been promoted by the immensity of their area of distribution and its subdivision by Pleistocene glaciers (Rand, 1954); but these influences should have been counteracted by the high mobility of such a large cursorial predator. The magnitude of biometrical differences between populations could be taken as indications of genetical differentiation.

The genetical interpretation of geographical variation is not the only possible one however. The specimens studied represent phenotypes and the latter have been exposed to highly varied environmental conditions. In fact environmental conditions vary throughout the area of distribution of the wolf. Arctic winters are gradually colder, darker and longer northward or northeastward for instance; and taiga is gradually replaced by tundra in a general northeastward direction. Such environmental differences might induce gradual phenotypical variation directly as well as genotypical variation indirectly. Direct effects of the environment on physiological processes and morphological characteristics are well known. Seasonal changes in temperature, illumination, humidity and food supply affect the reproductive cycles and migrations of various vertebrates. Molts and coat-color changes of weasels were controlled photoperiodically by Bissonnette and Bailey (1944), apparently through the pituitary gland. Prairie and tundra wolves could be particularly affected by seasonal periodicity as they cannot evade the climatic extremes of their habitat by seeking microclimates as small mammals do. The circumstances in which local populations live might induce modifications of growth processes, including those involved in skull development.

The growth of the various parts of the mammalian skull is neither perfectly synchronous nor isometric. The facial region reaches full development much later than the braincase (Baer, 1954). This was noted in Canidae at least as early as 1880 (Huxley). Because of this asynchronism and anisometry, temporary physiological disturbances might have very different effects according to their time of occurrence. Young northeastern wolves enter their first cold and dark arctic winter at an age (5 or 6 months) at which their facial growth potential is presumably still very high; their undersized face as adults is perhaps largely due to a metabolic and/or hormonal imbalance during this period of low temperature and illumination. Mice with the pituitary dwarf genotype differ from normal ones by *"their smaller size, their shorter snout, the short fleshy ears, and the relatively shorter tail"* (Grüneberg, 1952: 122). As for juvenile sheep with thyroid deficiencies, they grow skulls with normal braincase and teeth but with an underdeveloped facial region (Todd and Wharton, 1934); the descriptions of the latter would fit surprisingly well indeed the skulls of northeastern wolves with large teeth cramped in a short palate. The possible influence of low winter illumination on endocrine balance and growth is suggested by the effects of darkness on the teleost *Astyanax mexicanus* (Rasquin and Rosenbloom, 1954); darkness led to a loss of calcium from the bones and resulted in shorter and deeper-bodied fish than normal; skeletal changes were not repaired after return of the survivors to light.

Baer (1954) was led to postulate the existence of two major and relatively independent developmental processes in the postnatal mammalian skull; the expansion of the braincase on one hand, and the elongation of the face and the base of the skull on the other hand. It would be of great interest to see how much of variation in relative skull breadth in the present material could be ascribed to alterations of the equilibrium between two such growth processes. Very much of skull variation between the various breeds of domestic dogs (Stockard, 1941) could probably also be accounted for in this way. According to Scott's discussion (1957), the elongation of the mammalian skeleton is due to cartilage growth and promoted by pituitary growth hormones while its massiveness is due to subperiosteal growth and depends primarily on function and robustness factors. Although such a distinction seems most applicable to the post-cranial skeleton, it appears to correspond also with trends of skull variation.

Whether or not the biometrical characteristics of wolves can be ascribed to specific growth processes however, the most important evolutionary problem is the extent to which such characters are genotypical or merely phenotypical. Bringing up and breeding captive animals from various localities under controlled conditions would be informative in this respect but such data are not available at present. Gradual phenotypical variation induced directly by environmental conditions is possibly present along with genotypical variation in the skull dimensions of wolves.

Taxonomical conclusions can be based on geographical variation only inasmuch as the latter is hereditary. How much of biometrical variation is inherited in the present material is unknown. Even if the biometrical differences were entirely genetical however, the overall pattern of variation between the populations sampled is more suggestive of an incompletely panmictic continuum than of distinct subspecific units. Only one population seems sharply different from its immediate neighbours, probably as a result of insular isolation, that of Vancouver Island. In fact, it resembles northern wolves more than those presently on the mainland of British Columbia. It is perhaps with northern populations that Vancouver Island had its most intense recent biotic contacts. It seems doubtful that such a small insular population would have a major long-term evolutionary significance however. Ascertaining the taxonomic relationships of northwestern Nearctic wolves would require more material from Vancouver Island, Alaska, Alberta and the regions in between. An adequate analysis of variation is of course desirable for the species as a whole or at least for all its North American representatives. There are most likely far too many subspecific designations in use (Miller and Kellog, 1955).

As far as multivariate statistical techniques are concerned, their application to the present problem has probably already made their value apparent to many readers. One last comment is made in conclusion. Discriminant scatter diagrams of contemporaneous groups of organisms are closely analogous to what sections of phylogenetic trees at given time levels should be like. Discriminatory analysis may well prove an ideal tool in phylogenetic studies. Multivariate statistical techniques may make eventually possible the exact calculation of horizontal slices of the tree of Evolution.

Summary

An analysis is made of pelage and skull variation in five hundred wolf specimens from northwestern Nearctic localities. Pale wolves are relatively more frequent toward the tundra. Males are 4% larger than females in linear skull dimensions. Northeastern specimens are shorter- but relatively broader-skulled than southwestern ones. Multivariate biometrical differences between populations appear approximately proportional to geographical separation. This may express genetic differentiation through imperfect panmictia. But direct environmental influences may be involved. Formal taxonomic conclusions are postponed but it seems probable that far too many subspecific designations are now in use. Multivariate statistical techniques are very efficient for evolutionary comparisons with respect to complexes of characters.

Acknowledgments

This study was done at the suggestion of Dr. I. McT. Cowan at the Department of Zoology of the University of British Columbia. Field aspects of the problem were discussed with several members of the Canadian Wildlife Service. Explanations and advice on multivariate statistical techniques were given by Dr. S. W. Nash, Department of Mathematics. Help in multidimensional algebra and geometry was received from Marcel Banville, Department of Physics, W. R. Knight and Bomshik Chang, Department of Mathematics, and several others. Most calculations were done at the University's Computing Center, with much assistance from its personnel. A Wildlife Conservation Fellowship from the Canadian Industries Limited was held during this work. Finally criticisms from Dr. E. C. Olson and H. Barghusen of the University of Chicago led to improvements of the manuscript. All of this is very gratefully acknowledged.

Appendix

Multivariate Statistical Formulae

Value of ith character measured on an individual specimen: X_i

Set of values of the p characters measured on an individual specimen: $X = (X_1, \cdots X_p)$

Mean of a sample: $\overline{X} = (\overline{X}_1, \cdots \overline{X}_p)$

Variance of ith character within a sample: W_{ii}

Covariance of ith and j^{th} characters within a sample: $W_{ij} = W_{ji}$

Variance-covariance matrix of a sample:

$$W = \begin{bmatrix} W_{11} & \cdots & W_{1p} \\ \cdots & \cdots & \cdots \\ W_{p1} & \cdots & W_{pp} \end{bmatrix}$$

Equation of 95% equal frequency ellipse: $(X - \overline{X})\, W^{-1} (X - \overline{X})' = 5.99$

X' = X transposed; W^{-1} = inverse matrix of W; only the elements of X and W corresponding to the two characters are included here in X and W; 5.99 = 95% chi-square with 2 degrees of freedom; see Defrise-Gussenhoven (1955).

Covariance matrix of sample means around their grand mean:

$$B = \begin{bmatrix} B_{11} & \cdots & B_{1p} \\ \cdots & \cdots & \cdots \\ B_{p1} & \cdots & B_{pp} \end{bmatrix}$$

Matrix of the coefficients K_{ij} of the discriminant functions:

$$K = \begin{bmatrix} K_{11} & \cdots & K_{1p} \\ \cdots & \cdots & \cdots \\ K_{p1} & \cdots & K_{pp} \end{bmatrix}$$

Matrix of the components D_i of between-group variance corresponding to the discriminant functions:

$$D = \begin{bmatrix} D_1 & \cdots & 0 \\ \cdots & \cdots & \cdots \\ 0 & \cdots & D_p \end{bmatrix}$$

Matrix equation yielding K and D: $KB = DKW$

Coordinate of any point $X = (X_1, \cdots X_p)$ upon the ith discriminant axis: $K_i = K_{i1}X_1 + \cdots + K_{ip}X_p$

Within-group variances and covariances of the discriminant functions: KWK'

K' = K transposed

$$KWK' = I = \begin{bmatrix} 1 & \cdots & 0 \\ \cdots & \cdots & \cdots \\ 0 & \cdots & 1 \end{bmatrix}$$ for the pooled within-group covariance matrix.

Literature Cited

Anderson, E. 1954. Efficient and inefficient methods of measuring specific differences. In Kempthorne and others, 1954: 93–106.

Baer, M. J. 1954. Patterns of growth of the skull as revealed by vital staining. Human Biology, **26**: 80–126.

Banfield, A. W. F. 1951. Notes on the mammals of the Mackenzie District, Northwest Territories. Arctic, **4**: 112–121.

——. 1954. Preliminary investigation of the barren ground caribou. Wildl. Man. Bul. 1, 10 A and 10 B. Canadian Wildlife Service, Ottawa.

Bissonnette, T. H., and E. E. Bailey. 1944. Experimental modification and control of molts and changes of coat-color in weasels by controlled lighting. Ann. N. Y. Acad. Sci., **45**: 221–260.

Butler, L. 1947. The genetics of the colour phases of the red fox in the Mackenzie River locality. Can. Jour. Res. D, **25**: 190–215.

COLBERT, E. H., AND J. IMBRIE. 1956. Triassic metoposaurid amphibians. Bull. Amer. Mus. Nat. Hist., **110**: 399–452.

COWAN, I. McT. 1938. Geographic distribution of color phases of the red fox and black bear in the Pacific Northwest. Jour. Mamm., **19**: 202–206.

——. 1947. The timber wolf in the Rocky Mountain National Parks of Canada. Can. Jour. Res. D, **25**: 139–174.

DEFRISE-GUSSENHOVEN, E. 1955. Ellipses équiprobables et taux d'éloignement en biométrie. Bull. Inst. Royal Sci. Nat. Belgique, **XXXI** (26). Bruxelles.

GRÜNEBERG, HANS. 1952. The genetics of the mouse. (2nd ed.) xiv + 650 pp. Martinus Nijhoff, The Hague.

HILDEBRAND, M. 1952. An analysis of body proportions in the Canidae. Amer. Jour. Anat., **90**: 217–256.

HOTELLING, H. 1954. Multivariate analysis. In Kempthorne and others, 1954: 67–80.

HUXLEY, J. S. 1938. Clines: an auxiliary taxonomic principle. Nature, **142**: 219.

HUXLEY, T. H. 1880. Cranial and dental characters of the Canidae. Proc. Zool. Soc. London, 1880: 238–287.

KELSALL, J. P. 1957. Continued barren-ground caribou studies. Wildl. Man. Bull. 1, 12. Canadian Wildlife Service, Ottawa.

KEMPTHORNE, O., AND OTHERS (editors). 1954. Statistics and mathematics in biology. vii + 632 pp. Iowa State College Press, Ames.

KENDALL, M. G. 1957. A course in multivariate analysis. 185 pp. Charles Griffin & Co. Ltd., London.

KOTAKA, TAMIO. 1953. Variation of Japanese *Anadara granosa*. Trans. Proc. Paleont. Soc. Japan, N.S., No. **10**: 31–36.

LI, C. C. 1955. Population genetics. xi + 366 pp. The University of Chicago Press, Chicago.

MILLER, G. S., AND R. KELLOG. 1955. List of North American Recent mammals. U. S. Nat. Mus. Bull. 205: 954. Smithsonian Institution, Washington.

MURDOCH, D. C. 1957. Linear algebra for undergraduates. xi + 239 pp. John Wiley and Sons, Inc., New York.

OLSON, E. C., AND R. L. MILLER. 1958. Morphological integration. xv + 317 pp. The University of Chicago Press, Chicago.

QUENOUILLE, M. H. 1952. Associated measurements. x + 242 pp. Butterworths Scientific Publications, London.

RAND, A. L. 1954. The ice age and mammal speciation in North America. Arctic, **7**(1): 31–35.

RAO, C. R. 1952. Advanced statistical methods in biometric research. xvii + 390 pp. John Wiley and Sons, Inc., New York.

RASQUIN, P., AND L. ROSENBLOOM. 1954. Endocrine imbalance and tissue hyperplasia in teleosts maintained in darkness. Bull. Amer. Mus. Nat. Hist., **104**: 359–426.

RAUSCH, R. 1951. Notes on the Nunamiut Eskimo and mammals of the Anaktuvuk Pass Region, Brooks Range, Alaska. Arctic, **4** (3): 147–195.

SCOTT, J. H. 1957. Muscle growth and function in relation to skeletal morphology. Amer. Jour. Phys. Anthrop., N.S., **15** (2): 197–234.

STOCKARD, C. R., AND OTHERS. 1941. The genetic and endocrinic basis for differences in form and behavior as elucidated by studies of contrasted pure-line dog breeds and their hybrids. xx + 775 pp. Amer. Anat. Mem. 19. Wistar Inst. Anat. Biol., Philadelphia.

TEISSIER, G . 1955. Allométrie de taille et variabilité chez *Maia squinado*. Arch. Zool. Expér. et Gén., **92**: 221–264.

TODD, T. W., AND R. E. WHARTON. 1934. The effect of thyroid deficiency upon skull growth and form in the sheep. Amer. Jour. Anat., **55**: 97–115.

WRIGHT, S. 1932. General, group and special size factors. Genetics, **17**: 603–619.

YATES, F. 1950. The place of statistics in the study of growth and form. Proc. Royal Soc. B, **137**: 479–489.

YOUNG, S. P., AND E. A. GOLDMAN. 1944. The wolves of North America. xx + 636 pp. American Wildlife Institute, Washington.

14

Reprinted with permission from *Syst. Zool.*, **21**(4), 414–429 (1972)

KARYOLOGY AND MORPHOMETRICS OF PETERS' TENT-MAKING BAT, *URODERMA BILOBATUM* PETERS (CHIROPTERA, PHYLLOSTOMATIDAE)

ROBERT J. BAKER, WILLIAM R. ATCHLEY AND V. RICK MCDANIEL

Abstract

Baker, R. J., W. R. Atchley and V. R. McDaniel (Department of Biology and The Museum, Departments of Biology and Statistics, and Department of Biology and The Museum, Texas Tech University, Lubbock, Texas 79409. Present address of McDaniel. Division of Biological Sciences, Arkansas State University, Jonesboro, Arkansas 72467) 1972. Karyology and Morphometrics of Peters Tent-Making Bat, Uroderma bilobatum *Peters (Chiroptera, Phyllostomatidae). Syst. Zool., 21:414–429.*—There is more chromosomal variation in the tent-making bat than has been reported for any other species of bat. This is true in the number of chromosomal races (three), the minimum number of chromosomal rearrangements required to produce the respective races from a common ancestor, and the number of individuals surviving with atypical karyotypes. Some of the atypical karyotypes may be the result of interbreeding of cytotypes. Univariate analysis of variance indicated different patterns of secondary sexual dimorphism among the races. *U. b. molaris* was dimorphic in eight measurements whereas the others were dimorphic in only two. Multivariate statistical analysis were used to examine phenetic differentiation among races. Based on morphometrics, *U. b. davisi* was least divergent from the geographically adjacent *U. b. convexum.* Chromosomally, *U. b. davisi* was more similar to the geographically distant *U. b. bilobatum* and *U. b. trinitatum.* The Bayesian probability classification procedures indicated that at least 90% of all specimens were correctly classified (excepting *U. b. bilobatum*). The misclassified individuals appeared to represent a random geographic collection rather than clusters at potential zones of interbreeding. [Karyology; Morphometrics; *Uroderma.*]

The tent-making bat, *Uroderma bilobatum* Peters (Chiroptera, Phyllostomatidae) is distributed from southern Mexico southward throughout parts of tropical South America. The species has been divided into five subspecies (*bilobatum*, *molaris*, *convexum*, *trinitatum*, and *thomasi*) based on variations in external and cranial measurements and pelage color (Davis, 1968).

At the beginning of this study the karyological data available for three subspecies revealed chromosomal races with diploid numbers of 38 (involving part of *U. b. convexum*), 42 (including *bilobatum* and *trinitatum*), and 44 (involving part of *U. b. convexum*), and, as can be seen in some cases, these races did not correspond to existing subspecific boundaries (Baker, 1970A and Baker and Lopez, 1970A). In Colombia, specimens of *U. b. convexum* had a diploid number of 38 and a fundamental number of 44, whereas specimens of the same subspecies from Mexico had a diploid number of 44 and a fundamental number of 48 (Baker and Lopez, 1970A).

Extensive studies of chromosomal variation have failed to reveal another species of bats in which such chromosomal divergence occurs (Baker and Patton, 1967; Baker 1967, 1970A, 1970B; Hsu et al., 1968; Baker and Hsu, 1970; Capanna, 1968). In as much as locality records between the Colombian and Mexican samples of *U. b. convexum* suggested continuous populations between the two cytotypes, this situation provided an excellent opportunity to study the significance of karyotypic divergence in the speciation and raciation of Central American representatives of this species. Specifically, we wished to examine the following problems: 1) whether the two cytotypes interbreed or whether the chromosomal divergence signals completion of speciation; 2) what the chromosomal characteristics of intermediate populations might be, assuming that the two cytotypes were taken from a continuum of interbreeding populations (as suggested by their assignment to the same subspecies by Davis 1968); and 3) to what degree morphometric differentiation has paralleled chromosomal divergence. Toward this end, 126 specimens of *U. bilobatum* (including specimens of *U. b. molaris*) were karyotyped from localities in Central America. These, along with the 22 specimens (18 of these were used in the morphometric analysis) reported previously in the literature, form the basis for this report.

During the preparation of this manuscript it became apparent that the $2N = 44$ chromosomal race from Mexico and El Salvador was distinct from the described taxa from Central America. Although the $2N = 44$ populations may prove to be specifically distinct from *convexum* and *molaris* it seems best to recognize them as a subspecies (see discussion) until additional data are available. We feel that *Systematic Zoology* is not the proper place to introduce subspecific names therefore *U. b. davisi* Baker and McDaniel (1972) was described elsewhere.

METHODS AND MATERIALS

Specimens were collected from natural populations with mist-nets and karyotyped by the in vivo, bone marrow techniques, described by Baker (1970B). At least 10 somatic spreads were studied for each specimen. Metacentric, submetacentric, subtelocentric, acrocentric, and fundamental number are used as defined by Patton (1967).

Measurements of specimens were taken with dial calipers. The number before each measurement identifies that measurement in Tables 1, 2, 3, 6, 7, 9, and various places in the text. External measurements used are (1) length of forearm and (2) length of third metacarpal. Cranial measurements as shown in Figure 1 are (3) greatest length of skull, A-B; (4) condylobasal length, E-B; (5) zygomatic breadth, H-I; (6) braincase breadth, CC-DD; (7) mastoid breadth, AA-BB; (8) interorbital constriction width,

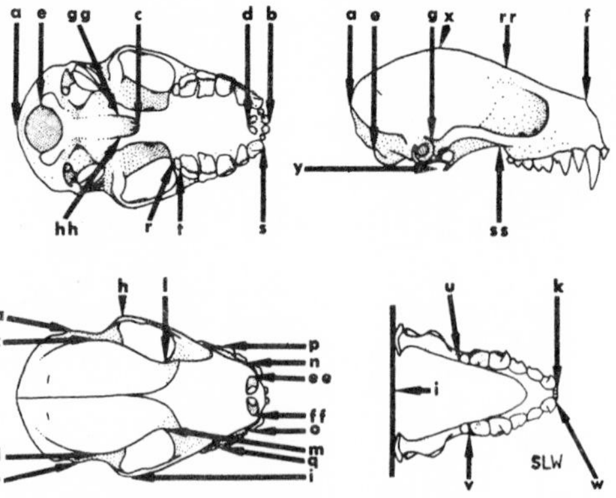

FIG. 1.—Skull of *Uroderma bilobatum* showing cranial measurements used in statistical studies. Names of measurements are given in text.

L-M; (9) breadth across P_4-P_4, P-G; (10) canine to molar two length (upper), T-S; (11) canine to molar two length (lower), V-W; (12) mandibular length, J-K; (13) occipitonasal length, A-F; (14) depth of braincase, S-Y; (15) palatal length, C-D; (16) postpalatal length, A-C; (17) condylonasal length, E-F; (18) zygorostral length, G-F; (19) canine to canine breadth, N-O; (20) width across the mesopterygoid fossa, GG-HH; (21) interorbital depth, RR-SS; (22) breadth of nasal opening, EE-FF; (23) length of maxillary tooth row, R-S; (24) length of mandibulary tooth row, U-K; (25) index M2 (described by Davis, 1968).

Specimens were divided into groups on the basis of karyotypes, as follows: a $2N = 44$ group, (*davisi* Baker and McDaniel, 1972) which was not further subdivided; specimens with $2N = 38$ divided between Pacific (referred to as the *convexum* group) and Atlantic (*molaris* group) versants along lines suggested by Davis (1968); and the $2N = 42$ type, which was divided into Trinidad (*trinitatum*) and mainland (*bilobatum*) groups.

Only adults (as determined by complete ossification of the phalanges) were used in the statistical analysis (sixteen karyotyped specimens were juveniles). Thirty-six specimens not karyotyped were used in the statistical analyses. These were from series containing specimens of known karyotype.

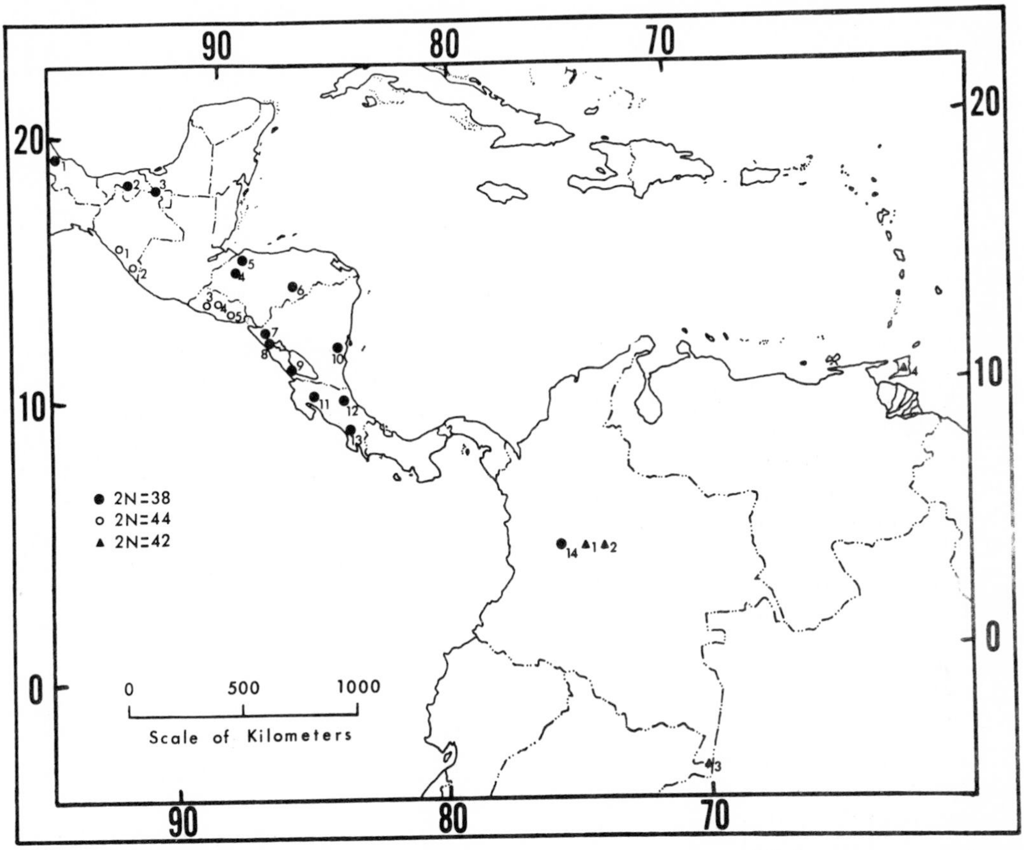

FIG. 2.—Geographic distribution of cytotypes of *Uroderma bilobatum*. Solid dots represent specimens with a $2N = 38$, open circles represent specimens with a $2N = 44$ and solid triangles represent specimens with a $2N = 42$. Numbers adjacent to each symbol identify that locality in "specimens examined."

Descriptive statistics were computed for the 25 morphometric variables on 162 specimens, divided into the 5 groups defined above. The sexes (90 males and 72 females) were treated separately. Univariate analyses of variance were used to determine the extent of secondary sexual dimorphism within each taxon and to ascertain the degree of divergence in each character among the various taxa. The extent of multivariate morphometric divergence was assessed by canonical variate analysis (Seal, 1964; Blackith and Reyment, 1971). The Texas Tech University version of the BMD07M program for canonical variate analysis was the primary source of these results.

For each taxon, we have pooled collections from several localities. We therefore are testing the null hypothesis of no morphometric divergence over and above the existing patterns of geographic variation. For this purpose, a multivariate analysis of variance was employed using the Wilks Lambda criterion (Morrison, 1967) as a test statistic. Further, a pairwise test for equality of mean vectors based on the Mahalanobis generalized distance statistic ($\mathbf{D}^2$) was employed using the formula

$$F_{ab} = \frac{(n-g-r+1)n_a n_b}{r(n-g)(n_a+n_b)} D^2_{ab}$$

with r and $n-g-r+1$ degrees of freedom. For this test, n represents the number of observations, g represents the number of groups, r represents the number of variables, and a and b represent the two groups.

In certain cases, a step-wise discriminant analysis was used to select an optimal subset of variables that would facilitate classification of unknown specimens and still maximize the separation of the groups. A "step-up" procedure was employed where the criterion for selection of the variables was the largest F-value from the univariate analyses of variance. At each step, the variable having the largest F-value was entered; the F-values for the remaining variables were then recomputed and the process continued. When the variance associated with the entered variable is partialled out, the data are readjusted such that the intercorrelation among the remaining variables is minimized.

The posterior probabilities based on the Bayesian probability theorem were used to construct a classification matrix to determine the amount of phenetic overlap among the taxa based on the morphometric variables.

Finally, a matrix of phenetic distance statistics is presented, which represents the square root of the Mahalanobis generalized distance statistic between pairs of taxa. This distance statistic is most useful in morphometric studies because the values are weighted by the inverse of the pooled covariance matrix, thereby eliminating the effects of correlation among characters.

RESULTS

Results of the karyotypic analysis are shown in Figures 2–7. Samples from the Atlantic versant of Mexico, Honduras, Nicaragua, and Costa Rica and the Pacific versant of Nicaragua and Costa Rica had a diploid number of 38 and a fundamental number of 44 (Fig. 3). The X chromosome was a medium-sized biarmed element and the Y chromosome was a small biarmed element. The four pairs of biarmed autosomes were indistinguishable from the X element. Only one of the 106 specimens examined from these samples had a karyotype that differed from the above description. A male from 2.5 mi. W Chinandega, Nicaragua, had a diploid number of 39 and fundamental number of 45 (Fig. 4). If the X and Y elements are assumed to be like those of other males from the Chinandega area, then the karyotype has an additional medium-sized acrocentric element added to the typical $2N = 38$ karyotype.

Specimens from the Pacific versant of Mexico and El Salvador had a karyotype of $2N = 44$ (in three specimens $2N = 43$) and a fundamental number of 48 (Figs. 5 and 6). This karyotype had X and Y chromosomes similar to those described for the $2N = 38$ type. Specimens with $2N = 44$ had four biarmed autosomes, which were indistinguishable in size from the X, and 36 acrocentric elements, whereas specimens with a diploid number of 43 had five such biarmed elements and 34 acrocentrics. All specimens with a $2N = 43$ or 44 had a small pair of metacentric autosomes. Specimens with a $2N = 43$ were taken only at La Herradura, El Salvador (three of sixteen specimens).

Analyses of variance for secondary sexual dimorphism in the 25 morphometric variables indicate peculiar patterns of secondary sexual variation. In *davisi*, only two characters, zygomatic breadth and breadth across canines, were dimorphic, with males having the largest values. In *molaris*, eight characters were sexually dimorphic including skull length, condylobasal length, mandible length, and occipitonasal length at $P < .01$ and depth of brain case, palatal length, zygorostral length, and breadth across canines at $P < .05$. Males are larger than females for all eight characters. Only two characters, postpalatal length, and breadth across canines were dimorphic in *convexum*. For *bilobatum* and *trinitatum*, we were hindered by the small sample sizes and, therefore, the statistical tests would not be meaningful.

For many of the succeeding analyses, we have first examined the divergence among all five taxa and then pooled *molaris*

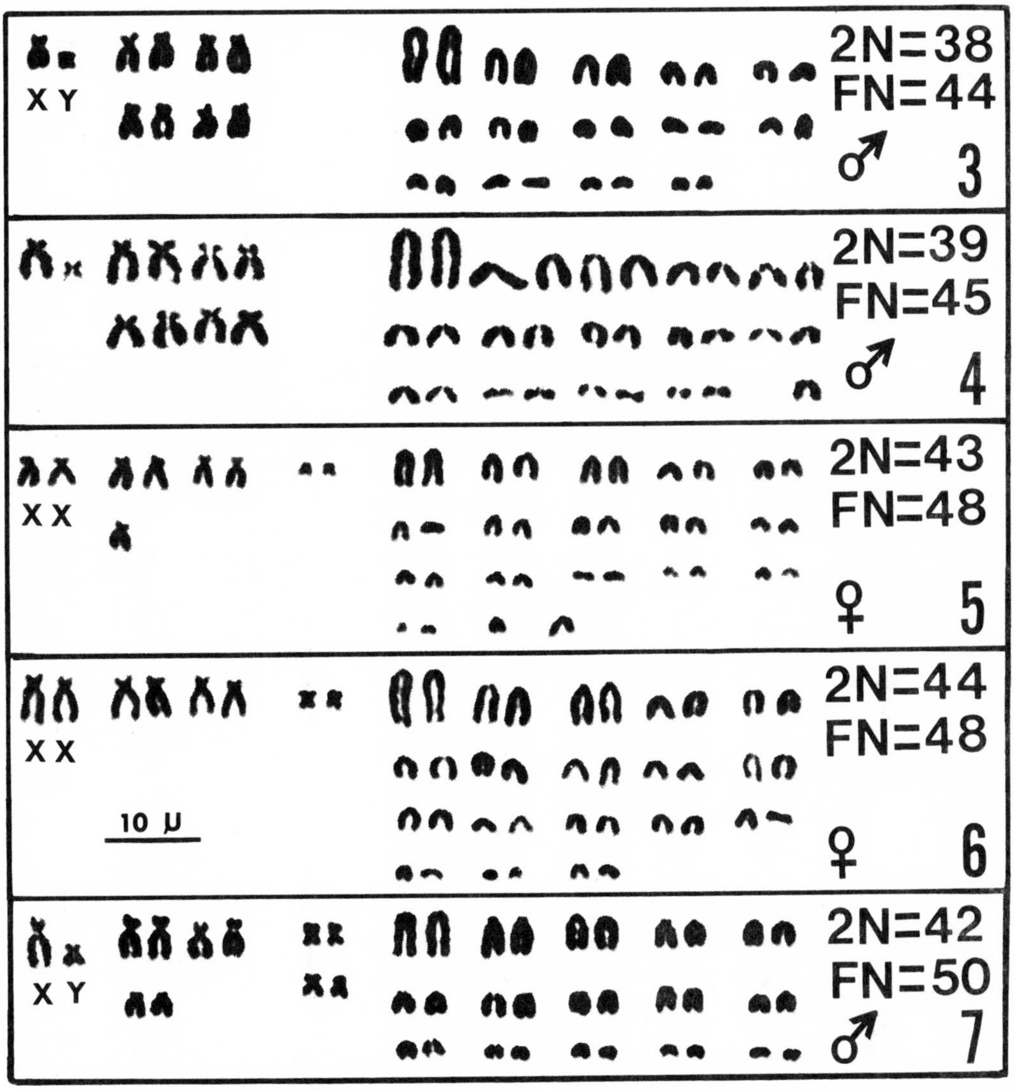

FIG. 3.—Karyotype representative of *U. b. molaris*. A karyotype indistinguishable from that of *U. b. molaris* is characteristic of *U. b. convexum*.

FIG. 4.—Karyotype of a male collected from near Chinandega, Nic. Only one of 26 specimens from the vicinity of Chinandega had this karyotype.

FIG. 5.—Karyotype of a female collected from La Herradura, El Salvador. Three of 16 specimens from this locality had this karyotype.

FIG. 6.—Karyotype representative of *U. b. davisi*.

FIG. 7.—Karyotype representative of *U. b. bilobatum*. A karyotype indistinguishable from that of *U. b. bilobatum* is characteristic of *U. b. trinitatum*.

and *convexum* (all specimens with a $2N =$ 38) and analyzed the resulting four groups.

In the univariate analyses of variance for intertaxa morphometric divergence in the five taxa, significant differences were found among means at $P < .01$ in all characters for females except numbers 5, 9 and 20, which were different at $P < .05$, and 7, 19 and 21,

TABLE 1.—CANONICAL VARIATE COEFFICIENTS FOR 25 MORPHOMETRIC VARIABLES IN 5 *Uroderma* TAXA. PERCENTAGE OF THE TOTAL VARIANCE ACCOUNTED FOR BY EACH VECTOR IS PRESENTED WITH TAXA MEANS ON EACH VECTOR.

	♂♂				♀♀			
	I	II	III	IV	I	II	III	IV
1.	−8.27	−0.17	−1.19	3.56	11.08	−1.92	1.72	−2.76
2.	−1.03	4.93	−0.07	−8.08	−9.47	−5.37	−5.08	3.49
3.	22.15	−1.19	17.97	16.09	−30.26	−0.16	14.76	−6.22
4.	0.81	21.70	−32.33	−5.47	8.39	−4.23	−11.37	−7.36
5.	−24.43	−13.25	8.16	2.55	42.26	3.78	3.26	−4.33
6.	−2.75	2.93	2.26	4.93	1.10	−8.77	−7.09	−18.78
7.	7.11	2.48	4.63	3.83	0.80	0.36	−11.54	27.43
8.	9.33	−1.76	−41.08	−7.16	−9.47	10.54	10.97	21.65
9.	18.20	27.68	29.29	30.74	−40.67	−20.01	38.36	5.25
10.	−8.83	−14.96	1.59	−47.50	9.27	−16.62	26.21	18.54
11.	−14.34	18.12	6.24	11.53	−35.60	4.13	1.51	−5.56
12.	−4.10	−3.39	−8.82	−7.22	−32.99	22.11	22.17	7.46
13.	−7.29	−5.24	6.37	−14.04	36.02	25.84	−23.44	12.09
14.	−2.10	6.42	13.70	−1.40	−21.58	−5.60	19.26	−7.51
15.	26.19	−0.31	−3.17	14.50	−15.36	−11.90	−11.00	−14.42
16.	−11.79	7.28	−41.79	18.94	11.54	−49.53	−18.91	−14.18
17.	−0.03	−0.47	0.31	−1.11	−37.86	6.05	9.35	−15.78
18.	15.92	−4.20	20.83	−11.84	21.57	−16.04	−5.36	14.71
19.	−7.49	−12.26	4.44	−46.14	−14.98	−7.31	6.95	−21.15
20.	27.28	−29.39	−5.82	−19.96	−3.40	47.66	19.51	3.63
21.	8.25	−11.70	−19.50	27.32	1.70	−5.11	−12.31	−4.68
22.	7.77	−2.87	−0.14	−33.43	9.29	8.32	−26.76	17.14
23.	10.21	5.60	27.99	48.79	−79.64	25.64	11.34	−41.79
24.	−23.66	−22.17	−25.36	12.06	−36.77	−16.79	−39.73	62.34
25.	0.94	−0.13	−1.08	−1.69	1.11	0.18	−0.83	−0.28
% Variance explained	61.1	17.4	14.6	6.9	61.1	24.2	10.4	4.3
			Group	Means				
davisi	−0.26	−1.60	1.69	0.12	0.03	2.26	1.86	−0.16
molaris	−1.26	1.59	0.56	0.07	−0.32	−1.78	0.39	−0.20
convexum	−1.07	−0.58	−1.15	0.03	1.66	0.50	−0.69	0.36
trinitatum	5.14	0.36	−0.29	0.80	−9.60	0.70	−0.53	1.47
bilobatum	2.92	0.15	0.03	−2.80	−2.86	1.75	−1.81	−2.02

which showed no statistically significant differentiation. In the males, characters 5, 7, 16, 18, 19, and 25 exhibited no significant differentiation, 10 and 23 differed at the $P < .05$ level, and the remainder at $P < .01$. When *molaris* and *convexum* were pooled and four groups examined, considerable change resulted in the patterns of variation. In females, characters 5, 7, 9, 13, 18, 19, 21, and 22 showed no significant differences, whereas 6, 10, 11, 17, and 20 differed at $P < .05$; and the remainder were significantly different at $P < .01$. In males, characters 4, 5, 8, 9, 10, 13, 16, 18, 18, 22, and 23 were not significant, character 7 was different at $P < .05$, and the remainder at $P < .01$.

The results of the canonical variate analysis for the five taxa are given in Tables 1–3. Table 1 gives the canonical variate coefficients for the four vectors, the amount of the variance accounted for by each vector, and the group means along each axis. Table 2 gives the standardized coefficients and indicates the relative contribution of each character to each vector. The coefficients are standardized by multi-

TABLE 2.—STANDARDIZED CANONICAL VARIATE COEFFICIENTS FOR 5 TAXA. Coefficients from Table 2 have been multiplied by pooled standard deviations of each character. Standardized coefficients indicate the contribution of each character to each vector.

	♀				♂			
	I	II	III	IV	I	II	III	IV
1.	1.24	–0.21	0.19	–0.30	0.98	–0.02	–0.14	0.42
2.	0.95	–0.54	–0.51	0.35	–0.12	0.59	0.00	0.97
3.	1.45	–0.00	0.71	–0.29	1.05	–0.05	0.85	0.76
4.	0.37	–0.19	–0.51	–0.33	0.03	0.88	–1.32	–0.22
5.	1.42	0.12	0.11	–0.14	–0.82	–0.44	0.27	0.08
6.	0.03	–0.26	–0.21	–0.57	–0.07	0.07	0.05	0.13
7.	0.01	0.00	–0.25	0.60	0.16	0.05	0.11	0.09
8.	–0.16	0.17	0.18	0.36	0.15	–0.02	–0.66	–0.11
9.	–1.06	–0.52	1.00	0.13	0.43	0.66	0.70	0.74
10.	0.21	–0.38	0.60	0.42	–0.18	–0.30	0.03	–0.97
11.	–0.71	0.08	0.03	–0.11	–0.31	0.39	0.13	0.25
12.	1.01	0.68	0.68	0.23	–0.13	–0.11	–0.28	–0.23
13.	0.57	0.41	–0.37	0.19	–0.37	–0.27	0.33	–0.72
14.	–0.50	–0.13	0.44	–0.17	–0.05	0.16	0.34	–0.03
15.	–0.54	–0.42	0.38	–0.51	0.81	–0.01	–0.09	0.45
16.	0.28	–1.23	–0.47	–0.35	–0.33	0.20	–1.19	0.54
17.	–0.52	0.08	0.12	–0.21	0.00	–0.09	0.06	–0.21
18.	0.75	–0.56	–0.18	0.51	0.50	–0.13	0.65	–0.37
19.	–0.25	–0.12	0.11	–0.35	–0.12	–0.20	0.07	–0.77
20.	–0.04	0.63	0.26	0.04	0.29	–0.32	–0.06	–0.21
21.	–0.04	–0.13	–0.31	–0.12	0.19	–0.27	–0.45	0.63
22.	0.12	0.11	–0.36	0.23	0.10	–0.04	0.00	–0.46
23.	1.93	0.62	0.27	–1.01	0.20	0.11	0.56	0.98
24.	–0.83	–0.38	0.90	1.42	–0.49	–0.46	0.53	0.25
25.	0.25	0.04	–0.18	–0.06	0.19	–0.02	–0.22	–0.34

plying each by the pooled standard deviation for that character.

For females, standardized coefficients for the first canonical variate indicate that variation in this vector relates to length of the maxillary tooth row, greatest length of skull, length at forearm, breadth across P4-P4, index M2, breadth of palate, and mandible length. The first eigenvector separates *bilobatum* and *trinitatum* from the remaining groups. The second vector, which discriminates *molaris* from *davisi* is primarily a reflection of variation in post-palatal length. *Uroderma b. molaris* has the greater palatal length, whereas *davisi* has the smaller. The third canonical vector distinguishes *davisi* and has the highest standardized coefficients for breadth across P4-P4, length of the mandibular tooth row, greatest length of skull, mandibular length, and C-M2 length. The final canonical variate, which has the largest coefficients for

TABLE 3.—POSTERIOR PROBABILITY CLASSIFICATION MATRICES FOR 5 TAXA. First values are for males followed by values for females in parentheses. Diagonal elements are those that are correctly classified. Off diagonal values are misclassified.

	davisi	*molaris*	*convexum*	*trinitatum*	*bilobatum*
davisi	16 (10)	0 (0)	0 (0)	0 (0)	0 (0)
molaris	1 (0)	20 (25)	4 (1)	0 (0)	0 (0)
convexum	3 (2)	1 (0)	29 (27)	0 (0)	0 (0)
trinitatum	0 (0)	0 (0)	0 (0)	11 (3)	0 (0)
bilobatum	0 (0)	0 (0)	0 (0)	1 (0)	4 (4)

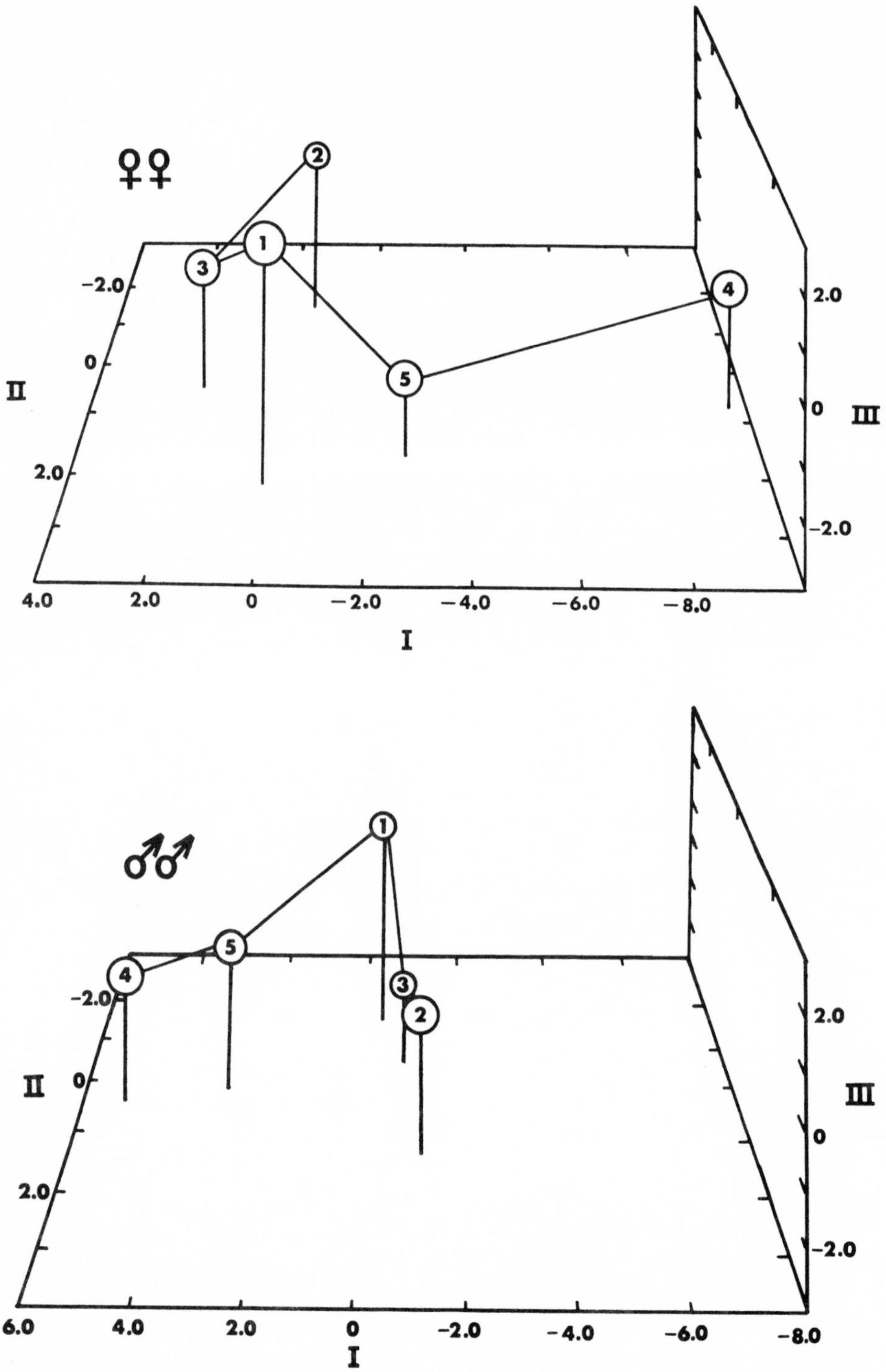

FIG. 8.—Projection of group means onto the first three canonical variates. Pairs of taxa are linked by smallest D value between pairs. 1. represents *U. b. davisi*, 2. *molaris*, 3. *convexum*, 4. *trinitatum*, 5. *bilobatum*.

TABLE 4.—DISTANCE MATRIX BASED ON THE SQUARE ROOT OF THE MAHALANOBIS GENERALIZED DISTANCE STATISTIC. FIRST VALUES ARE FOR MALES FOLLOWED BY VALUES FOR FEMALES IN PARENTHESES.

	davisi	*molaris*	*convexum*	*trinitatum*	*bilobatum*
molaris	5.643 (6.104)				
convexum	5.399 (5.581)	5.862 (5.978)			
trinitatum	7.543 (11.063)	8.396 (10.988)	7.955 (12.343)		
bilobatum	6.629 (6.658)	7.405 (7.226)	6.955 (7.287)	6.516 (8.679)	

the lengths of the mandibular and maxillary tooth rows, isolates *bilobatum*. The classification matrix for the females (Table 3) indicates that the five taxa are nearly discrete, with only three of the 72 females "misclassified."

The results for males are similar to those for females. The canonical coefficients (Tables 1, 2) for the first eigenvector reflect among group variation in greatest length of skull, length of forearm, and zygomatic breadth and palatal length. As in females, this first axis discriminates *bilobatum* and *trinitatum* from the other three taxa. The second canonical axis has highest coefficients for condylobasal length, length of third metacarpal, and breadth of P4-P4, and maximized the separation between *davisi* and *molaris*. The third canonical variate separates *davisi* from *convexum* and has the highest coefficients on condylobasal length, postpalatal length, and skull length. The final vector separates *trinitatum* primarily on length of the maxillary tooth row (both in total and C-M2) and length of third metacarpal. When the classification is examined (Table 3) 10 of 91 males were "misclassified."

The first three canonical variates have been projected into a three-dimensional model projection for each sex to illustrate phenetic relationships among the taxa. They are further linked by the smallest paired D value to summarize all of the phenetic information. Figure 8 illustrates the spatial configuration of the five groups for the first three canonical variates. These figures together with the distance matrices (Table 4) indicate that *bilobatum* is phenetically most distinct of the five. The D matrix further indicates that there has been less phenetic divergence among males than among females. Some slight discrepancies also were noted. For instance, *bilobatum* is most distinct from *molaris* in males, but in females is most distinct from *convexum*.

TABLE 5.—ORDER OF VARIABLES RANKED BY THEIR DISCRIMINATING POWER AS SHOWN BY STEPWISE DISCRIMINANT ANALYSIS. THE VARIABLES AT THE TOP OF EACH COLUMN HAVE THE GREATEST DISCRIMINATORY POWER.

5 Group Analysis		4 Group Analysis	
♂	♀	♂	♀
15	9	15	15
1	15	1	1
9	1	4	24
21	13	5	2
4	17	16	23
16	5	21	5
2	2	24	17
19	16	3	16
24	20	2	20
5	24	8	9
3	23	18	12
8	3	9	18
18	14	19	11
20	11	20	3
11	12	23	8
25	18	11	21
23	22	25	6
14	19	22	10
22	10	10	14
10	6	13	13
13	7	17	7
17	8	14	22
12	21	7	4
7	25	12	19
6	4	6	25

TABLE 6.—CANONICAL VARIATE COEFFICIENTS FOR 25 MORPHOMETRIC VARIABLES FOR 4 TAXA.

	♂			♀		
	I	II	III	I	II	III
1.	8.35	0.81	3.60	–10.81	–1.88	2.03
2.	0.86	3.28	–8.20	6.51	8.47	–2.83
3.	–22.71	–14.61	15.94	23.23	–9.48	–2.11
4.	–0.73	38.89	–5.33	–8.00	9.41	10.39
5.	24.81	–14.82	2.62	–35.07	–9.66	5.96
6.	2.58	0.17	4.89	–5.28	7.53	17.30
7.	–7.37	–1.95	3.75	2.93	12.99	–19.41
8.	–8.16	30.30	–6.57	12.89	–10.30	–21.43
9.	–20.15	–4.52	30.10	20.06	–14.78	–26.00
10.	9.40	–10.84	–47.59	–18.36	–10.13	–29.15
11.	13.55	7.02	11.25	31.78	–0.59	2.95
12.	4.50	4.57	–7.09	35.91	–25.18	–12.33
13.	7.35	–8.25	–14.14	–14.35	2.75	–6.11
14.	1.49	–6.31	–1.69	12.51	–11.21	–3.32
15.	–26.21	2.14	14.64	8.15	14.69	13.31
16.	12.72	36.73	19.55	–31.46	38.68	9.79
17.	0.04	–0.55	–1.11	32.83	–9.17	9.14
18.	–16.41	–18.70	–12.16	–24.28	12.92	–12.87
19.	7.87	–11.28	–46.30	6.96	–3.17	13.33
20.	–26.13	–14.68	–19.57	22.85	–40.54	–0.70
21.	–7.32	7.25	27.92	0.51	12.09	7.60
22.	–7.68	–1.70	–33.57	0.98	18.42	–2.40
23.	–11.27	–17.90	48.58	–60.26	–37.38	42.53
24.	25.30	5.02	12.80	33.27	53.85	45.10
25.	–0.91	0.74	–1.68	–0.76	0.41	0.67
% Variance explained	72.9	18.8	8.3	71.9	20.1	8.0
			Group Means			
davisi	0.27	–2.34	0.12	0.75	–2.74	–0.09
molaris + *convexum*	1.15	0.55	0.05	–0.86	0.39	–0.09
trinitatum	–5.17	0.45	0.80	8.70	1.25	–1.52
bilobatum	–2.94	0.07	–2.82	3.41	0.49	2.65

The null hypothesis of no morphometric divergence among the taxa was rejected at $P < .001$ for both sexes using the Wilks Lambda criterion. In the pair-wise F-tests for females, all pairs were significantly different at $P < .001$ except one and five which differed at the $P < .05$ level. In males, all pairs were significantly different at $P < .001$ except 4 and 5, which were significant at $P < .05$.

A list of characters in order of their ability to discriminate all taxa is given in Table 5. For the analysis of the five groups, the three best discriminating variables were the same for both sexes—palatal length, length of forearm, and breadth across P4-P4. After the first three characters, however, there was little correlation between the characters in the two sexes with regard to their discriminatory ability.

The analyses were repeated on the data with *molaris* and *convexum* pooled to give four groups. This was done to compare the $2N = 38$ and the $2N = 44$ cytotypes. Coefficients for the canonical variates are given in Table 6 together with group means projected onto each axis. Table 7 gives the classification matrix for the four groups.

To facilitate the allocation of new material to the correct taxon, we selected a small subset of characters that provides a maximization of among-groups differences.

TABLE 7.—POSTERIOR PROBABILITY CLASSIFICATION MATRICES FOR 4 TAXA. FIRST VALUES ARE FOR MALES FOLLOWED BY VALUES FOR FEMALES IN PARENTHESES.

	davisi	*molaris* + *convexum*	*trinitatum*	*bilobatum*
davisi	16 (9)	0 (0)	0 (0)	0 (1)
molaris + *convexum*	5 (2)	53 (53)	0 (0)	0 (0)
trinitatum	0 (0)	0 (0)	11 (3)	0 (0)
bilobatum	0 (0)	0 (0)	1 (0)	4 (4)

Table 5 gives the ranked order of these variables based on their discriminatory ability.

In the males, the optimal subset of characteristics (Table 8) included characters 1, 4, 5, 15, 16, 21, and 24. The first canonical variate separates *bilobatum* and *trinitatum* from *davisi* and the combined group of *molaris* and *convexum*. The second eigenvector discriminates *davisi* whereas the third separates *trinitatum*.

Separation as seen in the classification matrix is still good with these seven characters. Of the seven specimens of the *molaris-convexum* group that are "misclassified" with *davisi*, six are *convexum*. The *F*-tests for paired means indicate that all pairs are significantly different at the $P < .001$ level except *bilobatum* and *trinitatum*, which differ at the $P < .01$ level.

For the females, the optimal subset of characters includes 1, 2, 5, 15, 17, 21, and 24. The first eigenvector separates *bilobatum* and *trinitatum* from the other groups whereas the second distinguishes *davisi* and *trinitatum*. The final vector discriminates *bilobatum* and *trinitatum*.

The classification matrix again indicates the distinctness of the four groups based

TABLE 8.—CANONICAL VARIATE COEFFICIENTS, GROUP MEANS AND CLASSIFICATION MATRIX BASED ON AN OPTIMAL SUBSET OF MORPHOMETRIC VARIABLES FOR DISCRIMINATION OF 4 *Uroderma* TAXA. VALUES FOR FEMALES ARE GIVEN IN PARENTHESES IN CLASSIFICATION MATRIX.

	♂				♀		
	I	II	III		I	II	III
1.	−7.35	−3.42	3.71	1.	9.14	0.55	−5.24
4.	17.21	−21.78	23.57	2.	−8.55	6.91	0.41
5.	−13.63	17.12	3.39	5.	25.68	2.54	4.56
15.	28.80	14.67	−15.21	15.	−17.47	−10.81	−2.21
16.	−2.44	−14.11	−36.14	17.	−18.03	2.12	−17.41
21.	14.69	-9.36	−10.65	23.	61.84	−33.60	−6.68
24.	−20.29	7.61	−25.70	24.	−68.20	48.40	45.85
% Variance explained	82.3	14.5	3.2		75.0	18.0	7.0
Group Means							
davisi	−0.49	1.67	−0.07		0.93	−1.47	0.58
molaris + *convexum*	−0.94	−0.42	−0.02		0.36	0.34	−0.06
trinitatum	4.54	−0.21	−0.40		−6.08	0.52	1.06
bilobatum	2.59	0.04	1.41		−2.77	−1.47	−1.38

Classification Matrix

	davisi	*molaris* + *convexum*	*trinitatum*	*bilobatum*
davisi	15 (7)	1 (2)	0 (0)	0 (1)
molaris + *convexum*	7 (8)	50 (47)	0 (0)	1 (0)
trinitatum	0 (0)	0 (0)	10 (3)	1 (0)
bilobatum	1 (0)	0 (0)	0 (0)	4 (4)

on only seven variables. Again, of eight specimens in the *molaris–convexum* group that are "misclassified," six are from the *convexum* group. The tests for equality of paired means indicate that all taxa are significantly different at the $P < .001$ level.

DISCUSSION

From the morphometric analysis (Tables 1–8), it is clear that the three cytotypes are morphologically distinct and that the subspecies, recognized by Davis (1968) in the $2N = 38$ (*convexum* and *molaris*) and $2N = 42$ (*bilobatum* and *trinitatum*) cytotypes are also morphologically distinct. The degree of chromosomal divergence within *U. bilobatum* is as great as that reported for any bat species. The other case in which there is considerable divergence in chromosome morphology is in *Macrotus waterhousii* where two races have the same fundamental number (60), but a diploid value of 40 and 46, respectively (Nelson-Rees et al., 1968). In the case of *Macrotus*, it is unclear if the cytotypes represent subspecies or sibling species. In *Uroderma*, although the diploid and fundamental numbers vary between cytotypes, our data do not reveal whether or not the $2N = 38$, 42, and 44 cytotypes represent three specifically distinct taxa.

With the limited data from Colombia, where the $2N = 38$ (*convexum*) geographically approaches the $2N = 42$ (*bilobatum*), no comment can be made except that the two types appear to be separated by the Cordillera Oriental of the Andes. It is clear that if the $2N = 38$ (*convexum*) and the $2N = 44$ (*davisi*) cytotypes interbreed, the zone of intergradation will be between La Herradura, El Salvador, and Chinandega, Nicaragua, a distance of 160 kilometers. It is of interest that the only chromosomal variation we found in Central America was at the two localities adjacent to this zone. However, it is doubtful that karyotypic variation at these localities is a product of interbreeding. The $2N = 43$ karyotype is best explained as a result of centric fusion. The fundamental number remains constant in the sample from La Herradura, not reduced as would be expected, if this variation was a result of interbreeding. The fundamental number of the $2N = 39$ variant is increased by one, and could be the result of back crosses of some hybrid variant to the $2N = 38$ parental stock. On the other hand, the $2N = 39$ karyotype also could be explained as a trisomy for one of the medium-sized acrocentric pairs. If the trisomy hypothesis is correct then none of the chromosomal variation suggests hybridization.

Based on these data the degree of morphological divergence of *davisi* from the $2N = 38$ taxon is no greater than the degree of divergence between the two subspecies having a $2N = 38$ (*convexum* and *molaris*). The smallest *D* value between any pair of taxa for both males (5.399) and females (5.581, see Table 4) is between *convexum* and *davisi* indicating that these two taxa are the least divergent from a morphological standpoint. In light of these data and the lack of data from the zone of contact, it seems best to recognize the *davisi* as a subspecies of *U. bilobatum.*

CHROMOSOMAL DATA

From a chromosomal standpoint *davisi* is more like the South American subspecies (*trinitatum* and *bilobatum*) than like the adjacent Central American taxa (*molaris* and *convexum*). Such evolutionary affinities would be hard to explain on a geographical basis. The autosomal complement of *Uroderma* probably evolved from an ancestor with autosomes like those presently found in *Artibeus*, *Sturnira*, and *Vampyrops* (Baker, 1970), which contains 28 biarmed elements ($FN = 56$). The sex chromosomes of the ancestor were probably like those found in *Vampyrops* and most *Sturnira* and have remained unchanged in gross appearance. To derive the $2N = 42$ karyotype from the *Vampyrops* karyotype would require a minimum of 6 centric fissions and terminalization of centromeres (by pericentric inversions or centric shifts) in three biarmed pairs. The $2N = 44$

karyotype could be derived from the 2*N* = 42 karyotype by a centric fission and a terminalization of the centromere in one pair of biarmed elements.

It is not clear if the 2*N* = 38, FN = 44 karyotype is intermediate between the 2*N* = 30 *Vampyrops* and the 2*N* = 42 karyotype, or, if the series of fundamental numbers reveal intermediate stages of 50 (in the 2*N* = 42), 48 (in the 2*N* = 44) and 44 (in the 2*N* = 38 karyotype), making the 2*N* = 38 karyotype the most derived of the series. Obviously several additional changes will be required to explain the 2*N* = 38 karyotype in relationship to the 2*N* = 42 and 44 karyotypes. To speculate on the details of deriving each karyotype is of little value and the above discussion is presented only to document a most significant point, that is, the degree of chromosomal divergence (even if the most parsimonious routes are taken) in the cytotypes of *U. bilobatum* is outstanding when these data are compared with data from the many other bat species (Baker and Patton, 1967; Baker, 1967, 1970a and 1970b; Hsu et al., 1968; Baker and Hsu, 1970; Kiblisky, 1967; Yonenaga et al., 1969; and Capanna, 1968).

In the light of this information, the question is raised as to what facet of the biology of *Uroderma* has favored the establishment of three chromosomal races. From our field observations and from published data (Davis, 1968) there appears to be nothing unique about the reproductive potential of this species as compared to other fruit eating bats such as *Artibeus*, *Sturnira*, and *Vampyrops*. Collecting records do not suggest that the species occurs in isolated pockets, but that it is found in all lower elevation habitats. Although banding studies are not available to show movements of populations and individuals, there is no reason to believe this species is any less mobile than the many similar species that have not evolved chromosomal races.

However, in one facet *Uroderma* does seem to differ from the other closely related fruit bats of the subfamily Stenoderminae. This genus has a higher percentage of individuals that have survived with spontaneous chromosomal aberrations. A total of 146 specimens of *U. bilobatum* have been karyotyped and four specimens (2.7 per cent) had detectable autosomal variation from the karyotype characteristic of most *U. bilobatum* studied from a specific locality. Thirteen specimens of *U. magnirostrum* have been karyotyped and two of these were abnormal (Baker and Lopez, 1970a), bringing the total number of specimens examined from the genus to 161 with 6 individuals carrying aberrations (3.7 per cent). Compilation of data from the literature and our unpublished data for autosomal aberrations in closely related species reveal the following: *Artibeus* (102 specimens, 1 aberration, a trisomy for a small autosome in a male *A. toltecus*, Baker, unpublished data; Baker, 1967; Baker and Hsu, 1970; Kiblisky, 1969; Yonenaga et al., 1969; Becak, et al., 1968); *Sturnira* (51 specimens, no aberrations, Baker, 1967; Baker and Hsu, 1970; Gardner and O'Neill, 1969; Kiblisky, 1969) and *Vampyrops* (31 specimens, no aberrations Baker, 1967; and Baker and Hsu, 1970). A total of 204 specimens with only one aberration or less than .5 per cent of the specimens.

A test of the hypothesis by the chi-square test that the two values are the result of sampling error is rejected at the $P < .05$ level. The number of spontaneous chromosomal variations in a population is the source of variation available for natural selection to act upon and in this aspect *Uroderma* may vary from the other fruit bats.

The above data for other fruit bats represents samples from nineteen species (8 *Artibeus*, 6 *Sturnira*, and 5 *Vampyrops*) and all have autosomes that are indistinguishable from each other. The fact that *Uroderma bilobatum* has three chromosomal races and *U. magnirostrum* has a karyotype unlike *U. bilobatum* contrast strongly against the above data for 19 species of *Artibeus*, *Sturnira*, and *Vampyrops* where speciation has been unaccompanied by autosomal rearrangements that are detectable with our techniques.

MORPHOMETRIC ANALYSIS

Based on the multivariate estimates of morphological divergence (Table 4 and Fig. 7) *davisi* is more like *convexum* and *molaris* than the subspecies with similar chromosome morphology. From a geographic basis this relationship seems more likely than the affinities suggested by the chromosomal data discussed above. *Uroderma b. davisi* and *U. b. convexum* are the two taxa most similar from a morphological standpoint. Five of the 57 specimens of *convexum* had morphometric characteristics that would classify them as *davisi* (Table 3). The other area in which there was a major breakdown in the posterior probability classification was with specimens of *molaris* (5 of 45) being misclassified as *convexum*.

The degree of accuracy of classification suggests considerable morphometric divergence among taxa (Tables 3, 7, and 8) however, the localities from which misclassified specimens were taken should suggest the degree of genetic interchange. None of the *davisi* specimens were misclassified; however, of the five specimens of *convexum* misclassified as *davisi*, one came from the Chinandega sample (from a sample of thirty but this was not the specimen with the $2N = 39$) that is adjacent to the potential zone of interbreeding, two came from Rivas, Nicaragua, some 160 kilometers from the Chinandega sample and the other two are from central Costa Rica, over 300 kilometers away. Clearly the misclassified specimens are not concentrated at localities adjacent to the potential zone of interbreeding. The one specimen of *molaris* that was misclassified as *davisi* came from Rama, Nicaragua, over 200 kilometers from the nearest sample of *davisi*. The specimen of *convexum* that was misclassified as *molaris* came from Chinandega, Nicaragua. Specimens of *convexum* from Chinandega are probably well isolated from *molaris* by the mountain range that divides the lowland habitats of Central America. Misclassified specimens of *molaris*, referred to *convexum* were one from Dulce Nombre de Culmi, Honduras, two from Juticalpa, Honduras, and two from Puerto Viejo, Costa Rica. The most logical place for the *molaris* and *convexum* populations to interbreed would be across the lowlands in south-central Nicaragua. None of the misclassified specimens from the two subspecies came from this area.

SEXUAL DIMORPHISM

Several features of secondary sexual dimorphism are of interest. In vespertilionid bat species that have been studied, females are larger than males (Findley and Traut, 1970; and Engels, 1936). In two phyllostomatid species, *Stenoderma rufum*, and *Amertrida centurio*, females are larger than males (Jones et al., 1971; Peterson, 1965; Choate and Birney, 1968). However, in four species of mormoopid bats, the males were larger (Smith, 1972). This is the first report of phyllostomatid species in which males are larger than females. It has generally been believed that selection has favored larger females, especially in wing dimension, to compensate for the extra load of pregnancy and carrying young. The two wing measurements taken (1 and 2) revealed no sexual dimorphism.

Another interesting aspect of the secondary sexual dimorphism is that the taxa are sexually dimorphic in varying degrees and in different measurements. In the three taxa in which we have an adequate sample size, *molaris* is sexually dimorphic in eight characters, and *convexum* and *davisi* are dimorphic in two. Both *convexum* and *davisi* are dimorphic in the same measurement at the $P < .01$ but in different characters at the $P < .05$. These differences in the degree of dimorphism probably reflect adaptive differences and further document the distinctness of the three taxa.

Specimens examined: Numbers after sex are those of The Museum Collection of Mammals, Texas Tech University. Numbers in parentheses before localities refer to localities plotted in Fig. 2. Of the 178 listed below 142 were karyotyped (the 4 specimens of Baker, 1967 are not listed). Sixteen karyotyped specimens were not used in the morphometric analysis because they were juveniles. Thirty-six specimens not karyotyped were used in the morphometric analysis.

Uroderma bilobatum davisi. 2N = 44 or 43—EL SALVADOR. (4) Cuscatlan: 1.2 mi W Suchitoto, one male, 12668. (5) La Paz: 3 mi NW La Herradura, nine males, seven females, 12649-64. (3) Liberdad: 8.4 mi NW Colon, one male, 12667. (4) San Salvador: 1 mi W Ilopango Airport, one male, two females, 12669-71. MEXICO. Chiapas: (1) 11.9 mi SE Tres Picos, two males, 10700-01; (2) 6.8 mi N Tapachula, Rancho San Jorge, one male, one female, 10708-09; (2) 3.8 mi SW Tapachula on Mex. 18, three males, one female, 10702-05.

U. b. convexum. 2N = 38—COLOMBIA. (14) Tolima: Melgar, two males, three females, 9322, 9325–26, 9329–30. COSTA RICA. Guanacaste: (11) 5 km SW Finca Taboga, one male, four females 12707–11. Puntarenas: (11) 41.2 mi SE Canas, seven males, five females, 12672–83. San Jose: (13) 12.2 mi SSE San Isidro del General, seven males, six females, 12684–96. NICARAGUA. (7) Chinandega: 1.5 mi S Chinandega, five males, three females, 12712–15, 12749–52; 17 km E, 2 km S Chinandega, thirteen males, nine females, 12716–37. (8) Leon: 25 mi by rd WNW Managua on Nic. 28, one male, three females, 12738–41. (9) Rivas: 4 mi N Rivas, three males, one female, 12745–48.

U. b. molaris. 2N = 38—COSTA RICA. (12) Heredia: 7.3 mi SE Puerto Viejo, one female, 12706; 8 mi SE Puerto Viejo, two males, seven females, 12697–705. HONDURAS. (5) Cortez: 23 mi N San Pedro Sula, one male, 12648. (6) Olancho, 12.1 mi by rd SSW Dulce Nombre de Culmi, two males, one female 12637–39; 10.3 mi by rd SSW Dulce Nombre de Culmi, twelve males, seven females, 12618–33, 12635–36, 13276; 31.3 mi by rd NNE Juticalpa, two males, three females, 12640–44. (4) Santa Barbara: 12 mi N Santa Barbara, one male, one female, 12646–47; 31.8 mi N. Santa Barbara, one male, 12645. MEXICO. Tabasco: (3) 5.4 mi N Teapa, two females, 10678–79; (2) 13.6 mi N Villahermosa, one male, 10822. (1) Veracruz: 4.2 mi N Santiago Tuxtla, one female, 10677. NICARAGUA. (10) Zelaya: vicinity of Rama, five males, three females, 12613–17, 12742–44.

U. b. trinitatum. 2N = 42—TRINIDAD. (4) Caura Valley, two females, 5327, 5485; Guayaguayare, six males 5300–03, 9018–19; Las Cuevas, two males, 5254, 8976; San Rafael, two males, one female, 5390, 5829; 5813; Blanchisseuse, one male, 5808.

U. b. bilobatum. 2N = 42—COLOMBIA. (3) Amazonas, Leticia and vicinity of, three males, two females, 8836, 9044, 9082–84; (1) Meta: Restrepo, 1 female, 9477; Puerto Lopez, 1 km from Meta Bridge between Puerto Lopez and Rio Meta, two males, one female, 9511, 9513–14.

ACKNOWLEDGMENTS

We thank William J. Bleier, Brent L. Davis, Stanley Rouk, and Genaro Lopez for field assistance. Dr. Hugh H. Genoways and Dr. J. Knox Jones, Jr. critically read the manuscript. Dr. Don Wilson, Carl Thomason, and H. R. Winogrond assisted by providing data on collecting localities and facilities. Dr. Bernardo Villa-R granted the permits for the collection of the Mexican material. Supported by National Science Foundation grant numbers GB-29132X and GN-29132X.

REFERENCES

BAKER, R. J. 1967. Karyotypes of bats of the family Phyllostomidae and their taxonomic implications. Southwestern Natur., 12:407–428.

BAKER, R. J. 1970a. Karyotypic trends in bats. Pp. 65–96, *in* Biology of bats. (W. A. Wimsatt, ed.). Academic Press, New York and London, 1: xii + 406 pp.

BAKER, R. J. 1970b. The role of karyotypes in phylogenetic studies of bats. Pp. 303–312, *in* About bats. (B. H. Slaughter and D. W. Walton, eds.) Southern Methodist University Press, Dallas, vii + 339 pp.

BAKER, R. J., AND T. C. HSU. 1970. Further studies on the sex-chromosome systems of the American leaf-nosed bats (Chiroptera, Phyllostomatidae). Cytogenetics, 9:131–138.

BAKER, R. J., AND G. LOPEZ. 1970a. Chromosomal variation in bats of the genus *Uroderma* (Phyllostomatidae). J. Mamm. 51:786–789.

BAKER, R. J., AND G. LOPEZ. 1970b. Karyotypic studies of the insular populations of bats on Puerto Rico. Caryologia, 23:465–472.

BAKER, R. J., AND V. R. MCDANIEL. 1972. A new subspecies of *Uroderma bilobatum* (Chiroptera: Phyllostomatidae) from Middle America. Occas. Papers Mus., Texas Tech University, 7:1–4.

BAKER, R. J., AND J. L. PATTON. 1967. Karyotypes and karyotypic variation of North American vespertilionid bats. J. Mamm. 48:270–286.

BECAK, M. L., R. F. BATISTIC, L. D. VIZOTTO, AND W. BECAK. 1969. Sex determining mechanism XY_1Y_2 in *Artibeus lituratus* (Chiroptera, Phyllostomidae). Experientia 25:81.

Blackith, R. E., and R. A. Reyment. 1971. Multivariate morphometrics. Academic Press, London and New York ix + 412 pp.

Capanna, E. 1968. Some considerations on the evolution of the karyotype of microchiroptera. Experientia 24:624–626.

Choate, J. R., and E. C. Birney. 1968. Subrecent insectivora and Chiroptera from Puerto Rico, with the description of a new bat of the genus *Stenoderma*. J. Mamm. 49:400–412.

Davis, W. B. 1968. Review of the genus *Uroderma* (Chiroptera). J. Mamm., 49:676–698.

Engels, W. L. 1936. Distribution of races of the big brown bat (*Eptesicus*) in western North America. Amer. Midland Nat., 17:653–660.

Findley, J. S., and G. L. Traut. 1970. Geographic variation in *Pipistrellus hesperus*. J. Mamm. 51:741–765.

Gardner, A. L., and J. P. O'Neill. 1969. The taxonomic status of *Sturnira bidens* (Chiroptera: Phyllostomidae) with notes on its karyotype and life history. Occ. Pap. Mus. Zool. Louisiana State University, 38:1–8.

Hsu, T. C., R. J. Baker, and T. Utakohi. 1968. The multiple sex chromosome system of American leaf-nosed bats (Chiroptera, Phyllostomidae). Cytogenetics 7:27–38.

Jones, J. K., Jr., H. H. Genoways, and R. J. Baker. 1971. Morphological variation in *Stenoderma rufum*. J. Mamm. 52:244–247.

Kiblisky, P. 1969. Chromosome patterns of 7 species of leaf-nosed bats of Venezuela (Chiroptera-Phyllostomidae). Experientia, 25:1203–1204.

Morrison, D. F. 1967. Multivariate statistical methods. McGraw-Hill, New York, xiii + 338 pp.

Nelson-Rees, W. A., A. J. Kniazeff, R. J. Baker, and J. L. Patton. 1968. Intraspecific chromosome variation in the bat, *Macrotus waterhousii* Gray. J. Mamm. 49:706–712.

Patton, J. L. 1967. Chromosome studies of certain pocket mice, genus *Perognathus* (Rodentia:Heteromyidae). J. Mamm. 48:27–37.

Peterson, R. L. 1965. A review of the bats of the genus *Ametrida*, family Phyllostomidae. Contrib. Life Sci., Royal Ontario Mus., 65:1–13.

Seal, H. 1964. Multivariate statistical analysis for biologists. Methuen and Co. Ltd., London, xii + 209 pp.

Smith, J. D. 1972. Systematics of the chiropteran family Mormoopidae. Misc. Publ. U. Kan. Mus. Nat. Hist., 56:1–132.

Yonenaga, Y., P. O. Frota, and K. R. Lewis. 1969. Karyotypes of seven species of Brazilian bats. Caryologia 22:63–79.

Received July 26, 1972

15

Reprinted from *Biometrics,* **22**(1), 96–110 (1966)

GROWTH-INVARIANT DISCRIMINANT FUNCTIONS AND GENERALIZED DISTANCES

T. P. BURNABY

Department of Geology, University of Keele, Staffordshire, England.

SUMMARY

A method is described for computing discriminant functions and canonical vectors orthogonal to vectors representing variation which is extraneous to the desired comparisons. The same procedure allows the generalized distance D^2 to be resolved into additive components attributable to taxonomically relevant and non-relevant factors. Applications to biological taxonomic work are discussed and illustrated with a numerical example.

1. INTRODUCTION

The discriminant function of Sir Ronald Fisher, and the related generalized distance D^2 of Mahalanobis, were developed for discriminating between closely similar animal or plant species by means of multiple measurements. The discriminant function is essentially a weighted average of a set of measurements, the weights being computed so as to compensate for redundancy of information due to intercorrelations between measurements. The D^2 statistic is a sum of squares of differences between corresponding mean values of two sets of suitably weighted measurements.

Biological taxonomists however often seem reluctant to employ multivariate methods of this type, even when they are prepared to concede that sets of multiple measurements are of value in taxonomic work. One reason for this seems to be the belief that discriminant functions can give valid results only in the case of organisms in which growth ceases abruptly on the attainment of maturity, such as the higher mammals, man, and most groups of insects. Organisms in which growth continues throughout life, as in most invertebrate groups, are commonly regraded as presenting special difficulties. The procedure usually employed in such cases is to compute ratios of pairs of measurements to serve as measures of shape characterizing particular species or varieties. It is well known that there are objections to this practice: in particular that growth is not necessarily isometric, so that ratios are not always unvarying during the growth of an individual organism; but there being apparently no satisfactory alternative, the use of ratios persists. The use of angular measurements, rather than linear dimensions, is effectively a disguised use of ratios (an angle can be designated by its sine or tangent) and is subject to the same objections.

Again, growth is not necessarily the only nuisance factor in morphometric work. It has been shown for example that the environment exerts an influence on the shape of the shell in several groups of bivalved molluscs: in palaeontological work this may be of some importance. In one recent study of fossil and living primates (Boyce, [1964]) it was found that sexual dimorphism sometimes tended to overshadow the observed differences between related species.

There is thus a need for a general procedure, capable of eliminating either a single growth factor or several nuisance factors from discriminant functions, canonical vectors, or generalized distances between a number of populations.

2. LINEAR GROWTH MODELS

Suppose we have p measurements of an individual organism $(x_1 \cdots x_p)$. The growth of the individual can be thought of as the movement of the point $(x_1 \cdots x_p)$ along a line in a p-dimensional Euclidean space: we assume that this is a straight line, represented by the equations

$$\frac{x_1 - x_{1A}}{\mu_1} = \frac{x_2 - x_{2A}}{\mu_2} = \cdots = \frac{x_p - x_{pA}}{\mu_p}. \tag{1}$$

The vector $(\mu_1 \cdots \mu_p)$ is the set of direction numbers of the line: the set of constants $(x_{1A} \cdots x_{pA})$ represents a fixed point on the line, which might for example denote the adult dimensions of the individual on completion of growth.

Clearly, the usefulness of the model will depend on (a) whether the growth process conforms sufficiently closely to a straight line, and (b) whether the vector $(\mu_1 \cdots \mu_p)$ is not too specific. Thus, if each individual needs a different vector to represent its growth, we may be faced with an intractable problem of estimation. Ideally, we would like to have a constant or nearly constant vector of direction numbers to represent the growth of all individuals of a given species, or of a group of related species if possible.

Experience indicates that these requirements can often be met by a suitable preliminary transformation of the data. The fact that growth is in many cases a multiplicative process makes it appropriate to try taking the logarithms of the measurements, but there need be no rule about this: some other transformation, or none at all, may give better results. The effect of putting $x_i = \log z_i$, where $(z_1 \cdots z_p)$ are now the original measurements, is to convert the model represented by equations (1) to the multivariate form of the Huxley-Teissier allometric growth law (Jolicoeur, [1963]), familiar to most biologists in the bivariate form $z_2 = bz_1^a$ (where $a = \mu_1/\mu_2$). It may be well worth while converting a multiplicative growth law into an additive one in this way, even if it is known or suspected that growth is isometric, corresponding to a value of $a = 1$ in the bivariate case. In the multivariate case this implies that the elements of the vector of direction numbers are all equal, and hence that the directions are necessarily the same for all individuals.

Let us now consider what can be done with the help of the linear model of equations (1). In Fig. 1, a population of n individuals is represented by n points x_A, x_B, $\cdots$, located on n parallel lines, having the same set of direction numbers $\boldsymbol{\mu}$ as a one-dimensional parameter

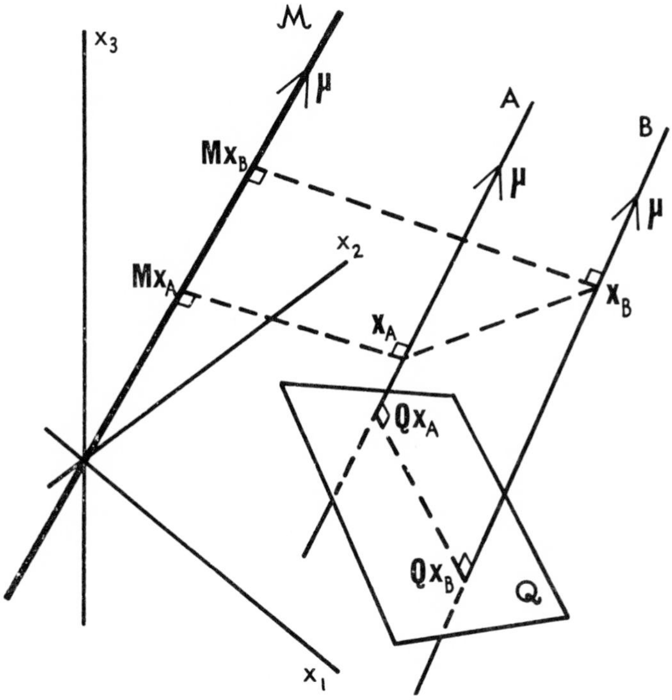

FIGURE 1

Resolution of a Measure of Overall Dissimilarity $|\mathbf{x}_A - \mathbf{x}_B|$ between Two Individual Organisms A and B into a Growth-Invariant Component and a Component Due to Difference in Amount of Individual Growth. The Growth of all Members of the Population is in the Constant Direction $\boldsymbol{\mu}$.

space $\mathfrak{M}$ which is a subspace of the p-dimensional parameter space $\mathfrak{X}$.

Although it will not generally be correct to say that the proportions of the measurements of an organism do not vary during growth, nevertheless the projection of the point $\mathbf{x}$ on to the $(p - 1)$-dimensional subspace $\mathfrak{Q}$ complementary to $\mathfrak{M}$ does not change its position during the growth of the individual organism. The distance between the projections of the points $\mathbf{x}_A$ and $\mathbf{x}_B$ on to $\mathfrak{Q}$ will thus be a growth-invariant measure of the dissimilarity between the organisms A and B.

To find an expression for this distance, note that the matrix $\mathbf{M} = \boldsymbol{\mu}(\boldsymbol{\mu}'\boldsymbol{\mu})^{-1}\boldsymbol{\mu}'$ projects any vector $\mathbf{x}$ on to the one-dimensional subspace $\mathfrak{M}$, since the product vector $\mathbf{Mx}$ is always a scalar multiple of $\boldsymbol{\mu}$. The matrix $\mathbf{Q} = \mathbf{I} - \boldsymbol{\mu}(\boldsymbol{\mu}'\boldsymbol{\mu})^{-1}\boldsymbol{\mu}'$ projects any vector $\mathbf{x}$ on to the space $\mathfrak{Q}$ complementary to $\mathfrak{M}$, since the product vector $\mathbf{Qx}$ is orthogonal to $\boldsymbol{\mu}$. It is easily seen that $\mathbf{MQ} = \mathbf{0}$, and that $\mathbf{M}^2 = \mathbf{MM} = \mathbf{M} = \mathbf{M}'$, $\mathbf{Q}^2 = \mathbf{QQ} = \mathbf{Q} = \mathbf{Q}'$, in other words $\mathbf{M}$ and $\mathbf{Q}$ are idempotent symmetric matrices of order $p \times p$, and of rank 1 and $(p - 1)$ respectively. Hence the required measure of dissimilarity is the distance between the points $\mathbf{Qx}_A$ and $\mathbf{Qx}_B$, which is the length of the vector $\mathbf{Q}(\mathbf{x}_A - \mathbf{x}_B)$.

Suppose now that instead of $\boldsymbol{\mu}$ being constant for the whole group of organisms, we have k different directions $\boldsymbol{\lambda}, \boldsymbol{\mu}, \boldsymbol{\nu}, \cdots$, characterizing the growth of k sub-groups of individuals. Let $\boldsymbol{\Lambda} = (\boldsymbol{\lambda}, \boldsymbol{\mu}, \boldsymbol{\nu}, \cdots)$ be a $p \times k$ matrix whose column vectors span the k-dimensional subspace $\mathfrak{M}$ of the parameter space $\mathfrak{X}$. (Fig. 2 illustrates the case where $p = 3$, $k = 2$, $n = 2$, and $\boldsymbol{\Lambda} = (\boldsymbol{\mu}_A, \boldsymbol{\mu}_B)$.) Let $\mathbf{M} = \boldsymbol{\Lambda}(\boldsymbol{\Lambda}'\boldsymbol{\Lambda})^{-1}\boldsymbol{\Lambda}'$, and let $\mathbf{Q} = \mathbf{I} - \mathbf{M}$. Then $\mathbf{M}$ and $\mathbf{Q}$ are symmetric matrices of order $p \times \mathrm{p}$ and rank k and $(p - k)$ respectively. It is again easily seen that $\mathbf{M}^2 = \mathbf{MM} = \mathbf{M} = \mathbf{M}'$, $\mathbf{Q}^2 = \mathbf{QQ} = \mathbf{Q} = \mathbf{Q}'$, and $\mathbf{MQ} = \mathbf{0}$. The matrix $\mathbf{Q}$ projects all vectors $\mathbf{x}$ on to the $(p - k)$-dimensional subspace $\mathfrak{Q}$ complementary to $\mathfrak{M}$, and any product vector $\mathbf{Qx}$ is orthogonal to the matrix $\boldsymbol{\Lambda}$, implying that the point $\mathbf{x}$ can move in any of the k directions $\boldsymbol{\lambda}, \boldsymbol{\mu}, \boldsymbol{\nu}, \cdots$, without changing the position of its projection on the subspace $\mathfrak{Q}$. Hence the length of the vector $\mathbf{Q}(\mathbf{x}_A - \mathbf{x}_B)$ is a growth-invariant measure of dissimilarity between two individuals A and B.

3. LINEARLY CONSTRAINED DISCRIMINANTS AND CANONICAL VECTORS

Let $\boldsymbol{\delta} = \mathbf{x}_A - \mathbf{x}_B$ be the vector of differences between the means of two p-variate populations A and B. Let $\mathbf{W}$ be the non-singular matrix of within-population variances and covariances. Let $\mathbf{G} = (\mathbf{g}_1, \mathbf{g}_2, \cdots \mathbf{g}_k)$ be a $p \times k$ matrix, $k < p$, whose column vectors span a k-dimensional subspace of the p-dimensional parameter space $\mathfrak{X}$.

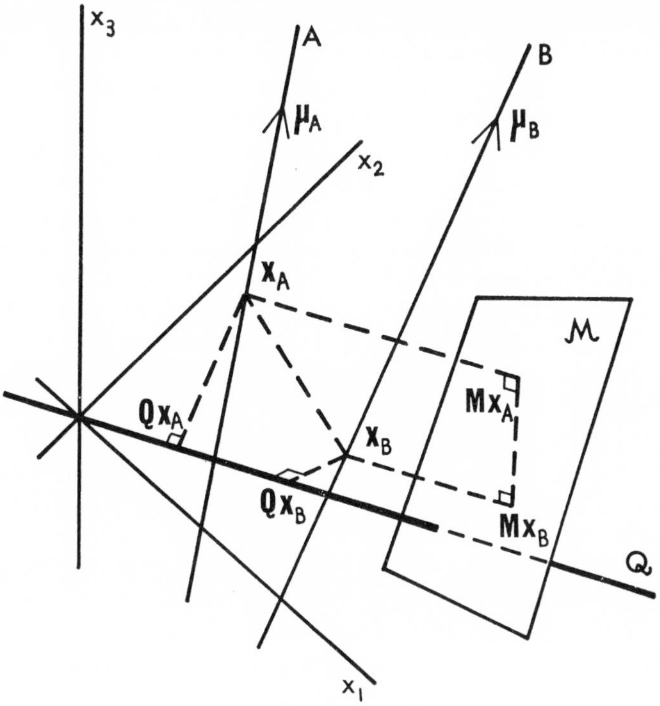

FIGURE 2

As for Figure 1, except that the Direction of Individual Growth May Vary among the Members of the Population, While Remaining Parallel to the Subspace $\mathfrak{M}$. (For Clarity, the Origin of Coordinates is Shown Shifted Away from the Intersection of $\mathfrak{M}$ and $\mathfrak{Q}$.)

Lemma I

If $\mathbf{l}$ is any vector satisfying the constraints $\mathbf{l}'\mathbf{W}\mathbf{l}$ = constant, $\mathbf{l}'\mathbf{G} = 0$, then

$$\frac{(\mathbf{l}'\boldsymbol{\delta})^2}{\mathbf{l}'\mathbf{W}\mathbf{l}} \leq \frac{(\mathbf{l}^{*\prime}\boldsymbol{\delta})^2}{\mathbf{l}^{*\prime}\mathbf{W}\mathbf{l}^*} \tag{2}$$

where $\mathbf{l}^*$ is defined by

$$\mathbf{l}^* = \mathbf{W}^{-1}\{\mathbf{I} - \mathbf{G}(\mathbf{G}'\mathbf{W}^{-1}\mathbf{G})^{-1}\mathbf{G}'\mathbf{W}^{-1}\}\boldsymbol{\delta}. \tag{3}$$

Proof. $\mathbf{l}^*$ as defined by (3) is subject to the same constraints as $\mathbf{l}$, and therefore, premultiplying both sides of (3) by $\mathbf{W}$ and by $\mathbf{l}^{*\prime}$ or $\mathbf{l}$, we find

$$\mathbf{l}^{*\prime}\mathbf{W}\mathbf{l}^* = \mathbf{l}^{*\prime}\boldsymbol{\delta}$$

$$\mathbf{l}'\mathbf{W}\mathbf{l}^* = \mathbf{l}'\boldsymbol{\delta}.$$

Then by Schwarz's inequality we have

$$(\mathbf{l}'\mathbf{W}\mathbf{l}^*)^2 \leq (\mathbf{l}'\mathbf{W}\mathbf{l})(\mathbf{l}^{*\prime}\mathbf{W}\mathbf{l}^*)$$

$$(\mathbf{l}'\boldsymbol{\delta})^2 \leq (\mathbf{l}'\mathbf{W}\mathbf{l})(\mathbf{l}^{*\prime}\boldsymbol{\delta})$$

$$\therefore \frac{(\mathbf{l}'\boldsymbol{\delta})^2}{\mathbf{l}'\mathbf{W}\mathbf{l}} \leq \mathbf{l}^{*\prime}\boldsymbol{\delta} = \frac{(\mathbf{l}^{*\prime}\boldsymbol{\delta})^2}{\mathbf{l}^{*\prime}\mathbf{W}\mathbf{l}^*}.$$

Thus, $\mathbf{l}^{*\prime}\mathbf{x}$ is a discriminant function whose value does not change if the point $\mathbf{x}$ is moved in any of the directions $\mathbf{g}_1 \cdots \mathbf{g}_k$. If the point $\mathbf{x}$ is moved in any other direction, the change in the value of $\mathbf{l}^{*\prime}\mathbf{x}$ is the maximum possible under the stated constraints.

Lemma II

Let B be a $p \times p$ matrix of variances and covariances between the means of a set of populations having a common within-population variance-covariance matrix W. Then the maximum value of the ratio $(\mathbf{l}'\mathbf{B}\mathbf{l})/(\mathbf{l}'\mathbf{W}\mathbf{l})$ is that for which $\mathbf{l}$ is the latent vector corresponding to the largest latent root of the matrix $\{\mathbf{W}^{-1} - \mathbf{W}^{-1}\mathbf{G}(\mathbf{G}'\mathbf{W}^{-1}\mathbf{G})^{-1}\mathbf{G}'\mathbf{W}^{-1}\}\mathbf{B}$, if $\mathbf{l}$ as in Lemma I is required to be orthogonal to a $p \times k$ matrix G.

Proof. We use the method of undetermined multipliers to maximise $\mathbf{l}'\mathbf{B}\mathbf{l}$ subject to the constraints $\mathbf{l}'\mathbf{W}\mathbf{l}$ = constant, $\mathbf{l}'\mathbf{G} = 0$.

Taking undetermined multipliers λ, $(\rho_1 \cdots \rho_k)' = \boldsymbol{\rho}$ the Lagrangian equations are

$$\frac{\partial}{\partial\{l_i\}} \{\mathbf{l}'\mathbf{B}\mathbf{l} - \lambda(\mathbf{l}'\mathbf{W}\mathbf{l} - \text{const.}) - 2\mathbf{l}'\mathbf{G}\boldsymbol{\rho}\} = 0$$

whence we obtain

$$\mathbf{B}\mathbf{l} - \lambda\mathbf{W}\mathbf{l} - \mathbf{G}\boldsymbol{\rho} = 0$$

$$\therefore \mathbf{G}'\mathbf{W}^{-1}\mathbf{B}\mathbf{l} - \mathbf{G}'\mathbf{W}^{-1}\mathbf{G}\boldsymbol{\rho} = 0 \tag{4}$$

$$\boldsymbol{\rho} = (\mathbf{G}'\mathbf{W}^{-1}\mathbf{G})^{-1}\mathbf{G}'\mathbf{W}^{-1}\mathbf{B}\mathbf{l}.$$

Substituting for $\boldsymbol{\rho}$ in (4) gives

$$\{(\mathbf{W}^{-1} - \mathbf{W}^{-1}\mathbf{G}[\mathbf{G}'\mathbf{W}^{-1}\mathbf{G}]^{-1}\mathbf{G}'\mathbf{W}^{-1})\mathbf{B} - \lambda\mathbf{I}\}\mathbf{l} = 0 \tag{5}$$

showing that $\mathbf{l}$ is a latent vector corresponding to one of the latent roots λ of the given matrix (cf. Rao [1964]). Premultiplying (4) by $\mathbf{l}'$ gives $\lambda = (\mathbf{l}'\mathbf{B}\mathbf{l})/(\mathbf{l}'\mathbf{W}\mathbf{l})$, showing that the ratio is maximised by the vector corresponding to the largest root.

If the rank of G is k, and if B and W are both non-singular, there will be $p - k$ non-zero latent roots, and the same number of canonical vectors, which are also solutions of equations (5). This follows from the fact that the number of constraints imposed, including those represented by the latent vectors other than the one being sought, must be less than the rank of **B** or **W**.

4. RESOLVING THE GENERALIZED DISTANCE D^2 INTO ADDITIVE COMPONENTS

The ordinary definition of Mahalanobis' generalized distance is

$$D^2 = \boldsymbol{\delta}'\mathbf{W}^{-1}\boldsymbol{\delta} = \mathbf{l}'(\mathbf{x}_A - \mathbf{x}_B)$$

where $\mathbf{l}'\mathbf{x}$ is the discriminant function for the populations A and B.

It is natural to extend the above definition to include the case where we have imposed a set of k linear constraints $\mathbf{l}^{*\prime}\mathbf{G} = 0$. With $\mathbf{l}^*$ as defined in Lemma I, we define the constrained distance D_Q^2, and the component D_M^2 of the total distance lost through imposing the constraints, by the expressions

$$\begin{aligned} D_Q^2 &= \mathbf{l}^{*\prime}(\mathbf{x}_A - \mathbf{x}_B) \\ &= \mathbf{l}^{*\prime}\boldsymbol{\delta} \\ &= \boldsymbol{\delta}'\{\mathbf{W}^{-1} - \mathbf{W}^{-1}\mathbf{G}(\mathbf{G}'\mathbf{W}^{-1}\mathbf{G})^{-1}\mathbf{G}'\mathbf{W}^{-1}\}\boldsymbol{\delta}, \qquad (6) \end{aligned}$$

$$D_M^2 = \boldsymbol{\delta}'\mathbf{W}^{-1}\mathbf{G}(\mathbf{G}'\mathbf{W}^{-1}\mathbf{G})^{-1}\mathbf{G}'\mathbf{W}^{-1}\boldsymbol{\delta}. \qquad (7)$$

It follows immediately that

$$D^2 = D_Q^2 + D_M^2 . \qquad (8)$$

The notation can be simplified by effecting a change of basis. Let $\mathbf{R}$ be a $p \times p$ non-singular matrix such that $\mathbf{R}'\mathbf{R} = \mathbf{W}^{-1}$. We define

$$\mathbf{y} = \mathbf{R}\mathbf{x}, \qquad \mathbf{d} = \mathbf{y}_A - \mathbf{y}_B = \mathbf{R}\boldsymbol{\delta}, \qquad \boldsymbol{\Lambda} = \mathbf{R}\mathbf{G}.$$

We then have

$$\begin{aligned} D^2 &= \mathbf{d}'\mathbf{d}, \\ D_M^2 &= \mathbf{d}'\boldsymbol{\Lambda}(\boldsymbol{\Lambda}'\boldsymbol{\Lambda})^{-1}\boldsymbol{\Lambda}'\mathbf{d} \\ &= \mathbf{d}'\mathbf{M}\mathbf{d} = (\mathbf{M}\mathbf{d})'\mathbf{M}\mathbf{d}, \end{aligned}$$

where $\mathbf{M} = \boldsymbol{\Lambda}(\boldsymbol{\Lambda}'\boldsymbol{\Lambda})^{-1}\boldsymbol{\Lambda}'$ is a symmetric idempotent matrix of rank k.

$$\begin{aligned} D_Q^2 &= \mathbf{d}'(\mathbf{I} - \mathbf{M})\mathbf{d} \\ &= \mathbf{d}'\mathbf{Q}\mathbf{d} = (\mathbf{Q}\mathbf{d})'\mathbf{Q}\mathbf{d}, \end{aligned}$$

where $\mathbf{Q} = \mathbf{I} - \mathbf{M}$ is a symmetric idempotent matrix of rank $p - k$.

Thus, if $\boldsymbol{\delta}$ is a vector of mean differences relative to the basis $\mathbf{R}$, then $\mathbf{d}$ is the same vector relative to the initial basis $\mathbf{I}$. The generalized distance D is the length of the line segment **d**, which is $(d_1^2 + \cdots + d_p^2)^{1/2}$, and the distances D_M and D_Q are the lengths of the projections of the line segment **d** upon two flats which lie respectively parallel and perpendicular to the set of directions **A**.

5. COMPUTATION OF CANONICAL VECTORS

Ashton, Healy, and Lipton [1957] suggested solving the equations $(\mathbf{B} - \lambda\mathbf{W})\mathbf{l} = \mathbf{0}$ by computing **R** such that $\mathbf{R}'\mathbf{R} = \mathbf{W}^{-1}$ and then solving $(\mathbf{RBR}' - \lambda\mathbf{I})\mathbf{m} = \mathbf{0}$, using a standard computer program for the latent roots and vectors of a symmetric matrix. If **m** is a latent vector corresponding to the latent root λ of the symmetric matrix $\mathbf{RBR}'$, the required canonical vector **l** is given by $\mathbf{l} = \mathbf{R}'\mathbf{m}$.

This technique can be extended to the case where the canonical vectors satisfy the conditions $\mathbf{l}'\mathbf{G} = 0$. We can show that, if m is a latent vector of the symmetric matrix $\mathbf{QRB}(\mathbf{QR})'$, the required canonical vector is given by $\mathbf{l} = (\mathbf{QR})'\mathbf{m}$, where **Q** is the matrix defined in Section 4.

From equations (5) we have

$$(\mathbf{R}'\mathbf{QRB} - \lambda\mathbf{I})\mathbf{l} = \mathbf{0}$$

$$(\mathbf{QRBR}' - \lambda\mathbf{I})(\mathbf{R}')^{-1}\mathbf{l} = \mathbf{0}$$

where **l** is a canonical vector satisfying the conditions $\mathbf{l}'\mathbf{G} = 0$. Define a vector **u** such that $\mathbf{m} = \mathbf{Qu} = (\mathbf{R}')^{-1}\mathbf{l}$. We thus have

$$\begin{aligned}(\mathbf{QRBR}' - \lambda\mathbf{I})\mathbf{Qu} &= \mathbf{0} \\ (\mathbf{QRBR}'\mathbf{Q}^2 - \lambda\mathbf{Q})\mathbf{Qu} &= \mathbf{0} \qquad (\text{since } \mathbf{Q}^2 = \mathbf{Q}) \\ (\mathbf{QRBR}'\mathbf{Q} - \lambda\mathbf{I})\mathbf{m} &= \mathbf{0}\end{aligned} \tag{9}$$

and since $\mathbf{Qu} = \mathbf{m} = \mathbf{Qm}$, the required canonical vector is given by $\mathbf{l} = \mathbf{R}'\mathbf{m} = \mathbf{R}'\mathbf{Qm} = (\mathbf{QR})'\mathbf{m}$, where **m** is a latent vector of the matrix $\mathbf{QRBR}'\mathbf{Q} = \mathbf{QRB}(\mathbf{QR})'$.

6. A RECURSIVE METHOD FOR COMPUTING THE MATRIX Q.

Assuming that we are given the matrix **W** and the set of vectors $\mathbf{G} = (\mathbf{g}_1 \cdots \mathbf{g}_k)$, the square-root inverse matrix **R** can be computed by standard methods (see for example Rao, [1952] pp. 293, 347). We can then compute **M** and **Q** from the formulae given in Section 4.

However, an alternative method which is well adapted to automatic computing is the following. Suppose $k = 1$; we then have

$$\mathbf{Q}_1 = \mathbf{I} - \frac{\mathbf{Rg}_1(\mathbf{Rg}_1)'}{(\mathbf{Rg}_1)'\mathbf{Rg}_1} = \mathbf{I} - \mathbf{M}_1 . \tag{10}$$

The matrix Q_1R will, as we have seen, project any vector in $\mathfrak{X}$, relative to the basis R, on to the $(p - 1)$-dimensional subspace $\mathfrak{Q}_1$, relative to the initial basis I. We now define

$$Y_2 = Q_1Rg_2 \tag{11}$$

$$Q_2 = I - \frac{Y_2Y_2'}{Y_2'Y_2} = I - M_2 . \tag{12}$$

Then the matrix Q_2 will project any vector in the subspace $\mathfrak{Q}_1$, relative to I, on to a $(p - 2)$-dimensional subspace $\mathfrak{Q}_2$, relative to I; and further, the matrix Q_2Q_1R will project any vector in $\mathfrak{X}$, relative to R, on to the subspace $\mathfrak{Q}_2$, relative to I. The process may be continued by defining

$$Y_3 = Q_2Q_1Rg_3 , \qquad Q_3 = I - \frac{Y_3Y_3'}{Y_3'Y_3} = I - M_3$$

so that finally we obtain the required matrix Q as the product[1] of the k matrices

$$Q = Q_kQ_{k-1} \cdots Q_2Q_1 . \tag{13}$$

It will be as well at each stage to test the sum of squares $y_i'y_i$, and if it is very small, to reject the corresponding vector g_i and pass to g_{i+1}. The computation should then be numerically well-conditioned, even if there should be collinearities among the vectors $g_1 \cdots g_k$.

7. THE EFFECT OF HETEROGENEITY IN THE DATA

We have not yet considered whether the within-population matrix of variances and covariances W may itself legitimately contain components of variance in the directions $g_1 \cdots g_k$. It may in practice be impossible to obtain samples which are free from heterogeneity with respect to a growth factor or other factors of variation to be eliminated, so the matter is of some importance.

Accordingly, we require to determine the effect, if any, upon equations (3), (5), or (6), of substituting for W the matrix $A = W + c_1^2g_1g_1' + \cdots + c_k^2g_kg_k'$, where $c_1^2, \cdots c_k^2$ are the unknown components of variance in the known directions $g_1 \cdots g_k$.

Let $C = [c_{ii}]$ be a diagonal matrix of order $k \times k$. We can then write A in the form $W + GC^2G'$. Note however that if we substitute GC for G wherever it occurs in (3), (5), and (6), the C matrices cancel, so that the scaling of the columns of G is arbitrary, and without loss

[1] Since $M_2M_1 = 0$, we have $Q_2 = I - M_2 - M_1$, and by induction $M = M_k + M_{k-1} + \cdots + M_2 + M_1$.

of generality we can write $\mathbf{A} = \mathbf{W} + \mathbf{GG}'$. The inverse of $\mathbf{A}$ can be expressed in the form

$$\mathbf{A}^{-1} = (\mathbf{W} + \mathbf{GG}')^{-1} = \mathbf{W}^{-1} - \mathbf{W}^{-1}\mathbf{G}(\mathbf{I} + \mathbf{G}'\mathbf{W}^{-1}\mathbf{G})^{-1}\mathbf{G}'\mathbf{W}^{-1}. \quad (14)$$

(I am indebted to Dr. C. R. Rao for suggesting the above expansion: compare a slightly less general formula given by Bartlett [1951].) Then by straightforward algebra we obtain

$$\mathbf{A}^{-1}\mathbf{G} = \mathbf{W}^{-1}\mathbf{G}(\mathbf{I} + \mathbf{G}'\mathbf{W}^{-1}\mathbf{G})^{-1}$$

$$(\mathbf{G}'\mathbf{A}^{-1}\mathbf{G})^{-1} = \mathbf{I} + (\mathbf{G}'\mathbf{W}^{-1}\mathbf{G})^{-1}$$

and finally

$$\begin{aligned}\mathbf{A}^{-1} - \mathbf{A}^{-1}\mathbf{G}(\mathbf{G}'\mathbf{A}^{-1}\mathbf{G})^{-1}\mathbf{G}'\mathbf{A}^{-1} \\ &= \mathbf{W}^{-1} - \mathbf{W}^{-1}\mathbf{G}(\mathbf{G}'\mathbf{W}^{-1}\mathbf{G})^{-1}\mathbf{G}'\mathbf{W}^{-1} \\ &= \mathbf{R}'\mathbf{Q}\mathbf{R} \qquad (15)\end{aligned}$$

thus proving that the substitution of the matrix $\mathbf{A}$ for $\mathbf{W}$ in (3), (5), or (6) has no effect whatsoever upon the discriminant function, canonical vectors, or generalized distance.

A method employed by Delaney and Healy [1964], consisting in the elimination of a growth factor from the entire data before commencing a canonical vector analysis, is thus equivalent in its result to the algebraic technique described in this paper for the case where $k = 1$.

8. NUMERICAL EXAMPLE

We consider an imaginary example in which the data consist of sets of three measurements, x_1, x_2, x_3, so that $p = 3$. The matrix of variances and covariances, presumed to have been computed from a large number of observations, is

$$\mathbf{W} = \begin{bmatrix} 1 & 0 & 2 \\ 0 & 4 & 2 \\ 2 & 2 & 6 \end{bmatrix}.$$

We have two nuisance factors to eliminate, of which (1) is a growth factor and (2) is the phenotypic response to an ecological gradient. The vectors of direction numbers are known to be

$$\mathbf{g}_1 = [1 \quad 4 \quad 3]'$$

$$\mathbf{g}_2 = [1 \quad 2 \quad 2]'.$$

k is thus equal to 2. The estimated mean points of two populations in which we are interested are

$$\mathbf{x}_A = [10 \quad 14 \quad 12]'$$

$$\mathbf{x}_B = [\ 7 \quad 10 \quad 10]'.$$

The vector of differences is thus

$$\boldsymbol{\delta} = \mathbf{x}_A - \mathbf{x}_B = [3 \quad 4 \quad 2]'.$$

The matrix **R** may be computed during the process of inverting W (it is not really necessary to complete the inversion).

$$\mathrm{R} = \begin{bmatrix} 1 & 0 & 0 \\ 0 & \frac{1}{2} & 0 \\ -2 & -\frac{1}{2} & 1 \end{bmatrix} \qquad \mathrm{W}^{-1} = \mathrm{R'R} = \begin{bmatrix} 5 & 1 & -2 \\ 1 & \frac{1}{2} & -\frac{1}{2} \\ -2 & -\frac{1}{2} & 1 \end{bmatrix}.$$

To find the ordinary generalized distance D^2, we first compute

$$\mathbf{R}\boldsymbol{\delta} = [3 \quad 2 \quad -6]'$$

and the sum of squares $(\mathbf{R}\boldsymbol{\delta})'\mathbf{R}\boldsymbol{\delta}$ then gives

$$D^2 = 3^2 + 2^2 + (-6)^2 = 49.$$

To find the coefficients of the discriminant function, we compute

$$\boldsymbol{\delta}'\mathrm{W}^{-1} = (\mathrm{R}\boldsymbol{\delta})'\mathrm{R} = [15 \quad 4 \quad -6].$$

In order to compute discriminants and generalized distances which are invariant with respect to the growth factor and to the ecological gradient, we construct the matrix Q as defined in Section 4, as follows:

$$\mathrm{RG} = \begin{bmatrix} 1 & 0 & 0 \\ 0 & \frac{1}{2} & 0 \\ -2 & -\frac{1}{2} & 1 \end{bmatrix} \begin{bmatrix} 1 & 1 \\ 4 & 2 \\ 3 & 2 \end{bmatrix} = \begin{bmatrix} 1 & 1 \\ 2 & 1 \\ -1 & -1 \end{bmatrix}$$

$$(\mathrm{RG})'\mathrm{RG} = \begin{bmatrix} 6 & 4 \\ 4 & 3 \end{bmatrix} = \mathrm{G'W^{-1}G}$$

$$(\mathrm{G'W^{-1}G})^{-1} = \begin{bmatrix} 3/2 & -2 \\ -2 & 3 \end{bmatrix}$$

$$\mathrm{M} = \mathrm{RG(G'W^{-1}G)^{-1}(RG)'} = \begin{bmatrix} \frac{1}{2} & 0 & -\frac{1}{2} \\ 0 & 1 & 0 \\ -\frac{1}{2} & 0 & \frac{1}{2} \end{bmatrix}$$

$$\mathbf{Q} = \mathbf{I} - \mathbf{M} = \begin{bmatrix} \frac{1}{2} & 0 & \frac{1}{2} \\ 0 & 0 & 0 \\ \frac{1}{2} & 0 & \frac{1}{2} \end{bmatrix}.$$

It is easy to verify that the rank of $\mathbf{M}$ is 2 (= k) and the rank of Q is $p - k$, also that $\mathbf{Q}^2 = \mathbf{Q}$, $\mathbf{M}^2 = \mathbf{M}$, and $\mathbf{MQ} = \mathbf{0}$.

To find the generalized distance D_Q^2, we require

$$\mathbf{QR\delta} = [-3/2 \quad 0 \quad -3/2]',$$

and D_Q^2 is then given by the sum of squares $(\mathbf{QR\delta})'\mathbf{QR\delta} = D_Q^2$,

$$D_Q^2 = (-3/2)^2 + 0^2 + (-3/2)^2 = 4\tfrac{1}{2}.$$

Similarly to find D_M^2, we compute

$$\mathbf{MR\delta} = [9/2 \quad 2 \quad -9/2]'$$

which gives $D_M^2 = 44\frac{1}{2} = D^2 - D_Q^2$.

The vector of discriminant function coefficients $\mathbf{l}^*$ is given by

$$\mathbf{l}^{*\prime} = (\mathbf{QR\delta})'\mathbf{R} = [3/2 \quad 3/4 \quad -3/2]$$

and it is easy to confirm that this is orthogonal to both $\mathbf{g}_1$ and $\mathbf{g}_2$, as was intended.

If the number of populations to be discriminated is no more than two, some computational labour may be saved by not actually computing the matrices Q and $\mathbf{M}$. For example, a suitable procedure would be to evaluate successively (1) $\mathbf{R\delta}$, (2) $\mathbf{\Lambda} = \mathbf{RG}$, (3) $\mathbf{a} = \mathbf{\Lambda}'\mathbf{R\delta}$, (4) $\mathbf{U} = (\mathbf{\Lambda}'\mathbf{\Lambda})^{-1}$, (5) $\mathbf{b} = \mathbf{Ua}$, and (6) $\mathbf{l}^* = \mathbf{R}'(\mathbf{R\delta} - \mathbf{\Lambda b})$. I am grateful to Dr. C. R. Rao for suggesting this method.

Using the same imaginary data, the recursive method of Section 6 proceeds as follows. We find

$$\mathbf{Rg}_1 = [1 \quad 2 \quad -1]'$$

$$(\mathbf{Rg}_1)'\mathbf{Rg}_1 = 6$$

$$\mathbf{M}_1 = \frac{\mathbf{Rg}_1(\mathbf{Rg}_1)'}{(\mathbf{Rg}_1)'\mathbf{Rg}_1} = \frac{1}{6}\begin{bmatrix} 1 & 2 & -1 \\ 2 & 4 & -2 \\ -1 & -2 & 1 \end{bmatrix}$$

$$\mathbf{Q}_1 = \mathbf{I} - \mathbf{M}_1 = \frac{1}{6}\begin{bmatrix} 5 & -2 & 1 \\ -2 & 2 & 2 \\ 1 & 2 & 5 \end{bmatrix} \qquad [(10)]$$

$$\mathbf{Rg}_2 = [1 \quad 1 \quad -1]'$$

$$\mathbf{Y}_2 = \mathbf{Q}_1\mathbf{Rg}_2 = [1/3 \quad -1/3 \quad -1/3]' \qquad [(11)]$$

$$\mathbf{Y}_2'\mathbf{Y}_2 = 1/3$$

$$\mathbf{M}_2 = \frac{\mathbf{Y}_2\mathbf{Y}_2'}{\mathbf{Y}_2'\mathbf{Y}_2} = \frac{1}{3}\begin{bmatrix} 1 & -1 & -1 \\ -1 & 1 & 1 \\ -1 & 1 & 1 \end{bmatrix}$$

$$\mathbf{Q}_2 = \mathbf{I} - \mathbf{M}_2 = \frac{1}{3}\begin{bmatrix} 2 & 1 & 1 \\ 1 & 2 & -1 \\ 1 & -1 & 2 \end{bmatrix} \qquad [(12)]$$

$$\mathbf{Q} = \mathbf{Q}_2\mathbf{Q}_1 = \begin{bmatrix} \frac{1}{2} & 0 & \frac{1}{2} \\ 0 & 0 & 0 \\ \frac{1}{2} & 0 & \frac{1}{2} \end{bmatrix}. \qquad [(13)]$$

In an actual computer program, one would compute successively the products **IR**, $\mathbf{Q}_1\mathbf{R}$, $\mathbf{Q}_2\mathbf{Q}_1\mathbf{R}$, etc., and the final output would be the matrix **QR**, ready to take the place of **R** in the computing of canonical vectors and generalized distances.

9. CONCLUDING REMARKS

There remains the question of how the directions $\mathbf{g}_1 \cdots \mathbf{g}_k$ are to be estimated. This is a practical as well as a statistical problem. Two methods of estimating the direction of a growth factor are described in Delaney [1964] and Burnaby [1965]. A third method is to take the first principal components of the within-sample matrices of variances and covariances (Jolicoeur [1963]).

The testing of the significance of an observed D^2 value with linear constraints imposed is one aspect of a wider class of problems considered by Rao [1962;1964]. The large-sample test statistic is $n_A n_B (n_A + n_B)^{-1} D^2$. On the hypothesis of zero population distance, this is distributed as a χ^2 variate with p degrees of freedom when there are no linear constraints. When k independently estimated linear constraints are imposed, the distribution is that of a χ^2 variate with $p - k$ degrees of freedom.

The requirement that the linear constraints be independently estimated means that the first principal component of the matrix W—which has been suggested as an estimator of the direction of a growth factor in relative growth studies (Jolicoeur [1963])—cannot here be used. However, in practice **W** would usually be the pooled matrix for k samples, whose individual **g** vectors could be separately estimated.

The ordinary tests of significance rely upon the assumption that any differences there may be between the separate within-population multivariate distributions are small enough to be ignored. They also depend upon the assumption of multivariate normality. In the writer's experience, the only sure defence against a successful disproof of one or both of these assumptions, is to abstain from collecting, or presenting, too much data! It is true that exact tests are available for the case where the separate **W** matrices are unequal (a useful source is Kullback [1959]), but the question of robustness in the presence of non-normality seems largely unexplored.

It may happen that a linear factor that we wish to eliminate is itself the main source of the differences between the dispersions of the separate populations. Again, the multivariate distributions may be anisotropically non-normal, the main component of non-normal variation having the same direction in each of the sampled populations. In fossil material both these effects are frequently associated with a growth factor. Eliminating the growth factor thus helps to mould the data to fit the distributional model. This is a point which ought to be borne in mind when the data are first inspected with a view to deciding whether to use the logarithmic transformation. It could be worth while to accept some initial worsening of skewness and/or heteroscedasticity, in exchange for an improvement in the linearity and parallelism of the individual growth factors.

For investigating the effect of departures from the distributional model, and of the efficacy or otherwise of action taken to mitigate these effects, use might be made of individual discriminant or canonical variate scores, to provide univariate cross-sections of the transformed multivariate sample distributions.

ACKNOWLEDGMENTS

The author is indebted to Mr. M. J. R. Healy, Dr. H. Liebeck, Dr. G. S. Mudholkar, and to the referees, for their helpful criticisms and suggestions. They are naturally not responsible for any errors or other faults remaining.

REFERENCES

Ashton, E. H., Healy, M. J. R., and Lipton, S. [1957]. The descriptive use of discriminant functions in physical anthropology. *Proc. Roy. Soc. B146*, 552–72.

Bartlett, M. S. [1951]. An inverse matrix adjustment arising in discriminant analysis. *Ann. Math. Statist. 22*, 107–11.

Blackith, R. E. [1960]. A synthesis of multivariate techniques to distinguish patterns of growth in grasshoppers. *Biometrics 16*, 28–40.
Boyce, A. J. [1964]. The value of some methods of numerical taxonomy with reference to hominoid classification. In *Phenetic and Phylogenetic Classification*, Syst. Assoc. Publ. *6*, 47–65.
Burnaby, T. P. [1965]. Reversed coiling trend in *Gryphaea arcuata*. *Geol. Journ. 4*, 257–78.
Delaney, M. J. [1964]. Variation in the long-tailed field-mouse (*Apodemus sylvaticus* (L.)) in north-west Scotland.
I. Comparisons of individual characters. *Prox. Roy. Soc. B161*, 191–9.
Delaney, M. J., and Healy, M. J. R. [1964]. Variation in the long-tailed field-mouse, etc., II. Simultaneous examination of all characters. *Proc. Roy. Soc. B161*, 200–7.
Fisher, R. A. [1938]. The statistical utilization of multiple measurements. *Ann. Eugenics 8*, 376–86. (Revised 1950, in *Contributions to Mathematical Statistics*, John Wiley, N. Y.)
Hotelling, H. [1935]. The most predictable criterion. *J. Educ. Psychol. 26*, 139–42.
Huxley, J. S. [1932]. *Problems of Relative Growth.* Methuen, Lond.
Jolicoeur, P. [1963]. The multivariate generalization of the allometry equation. *Biometrics 19*, 497–9.
Kullback, S. [1959]. *Information Theory and Statistics.* John Wiley, N. Y.
Mahalanobis, P. C. [1936]. On the generalized distance in statistics. *Proc. Nat. Inst. Sci. India 12*, 49–55.
Olson, E. C. [1964]. Morphological integration and the meaning of characters in classification systems. In *Phenetic and Phylogenetic Classification*, Syst. Assoc. Publ. *6*, 123–56.
Penrose, L. S. [1954]. Distance, size and shape. *Ann. Eugenics 18*, 337–43.
Rao, C. R. [1952]. *Advanced Statistical Methods in Biometric Research.* John Wiley, N. Y.
Rao, C. R. [1962]. A note on a generalized inverse of a matrix with applications to problems in mathematical statistics. *J. R. Statist. Soc. B24*, 152–8.
Rao, C. R. [1964]. The use and interpretation of principal component analysis in applied research. *Saukhyā, Series A*, *26*, 329–58.
Tukey, J. W. [1957]. On the comparative anatomy of transformations. *Ann. Math. Statist. 28*, 602–32.

ERRATA

Page 104, lines 4, 5, and 11; and page 108, lines 2, 3, and 4: for $\mathbf{Y}_1$, $\mathbf{Y}_2$ read $\mathbf{h}_1$, $\mathbf{h}_2$ throughout.
Page 104, line 15: "sum of squares $\mathbf{y}_i'\mathbf{y}_i$" should read "sum of squares $\mathbf{h}_i'\mathbf{h}_i$."

Additional References

Cooper, P. W. [1963]. Statistical classification with quadratic forms. *Biometrika 50,* 439–48.
Cooper, P. W. [1965]. Quadratic discriminant functions in pattern recognition. *IEEE Trans. on Information Theory IT 11,* 313–15.
Rempe, U. [1965]. Lassen sich bei Säugetieren Introgressionen mit multivariaten Verfahren nachweisen? *Zeitschrift zool. Syst. Evolutionsforschung 3,* 388–412.

16

Reprinted from *Ecology*, **52**(4), 543–556 (1971)

A MULTIVARIATE STATISTICAL APPROACH TO THE HUTCHINSONIAN NICHE: BIVALVE MOLLUSCS OF CENTRAL CANADA[1]

ROGER H. GREEN

Department of Zoology, University of Manitoba, Winnipeg 19, Manitoba, Canada

Abstract. The use of multiple discriminant analysis to identify the significant and independent ecological factors separating species distributions is proposed and discussed. Such an analysis was performed on 345 samples, containing a total of 10 bivalve mollusc species, from 32 lakes in Manitoba, Ontario, and Saskatchewan. Measurements of nine ecological parameters were associated with each sample. Five discriminant functions account for 95% of the among-species variance, and 4 of the 5 are ecologically interpretable. Three of these, accounting for 80% of the among-species variance, are interpreted as bases of trophic, rather than physical or chemical, separation. There is separation of species on each discriminant function. The use of discriminant scores to classify lakes with maximum relevance to species distributions is demonstrated and discussed. A generally applicable measure of environmental heterogeneity based upon this type of analyisis is proposed. The value of this type of analysis in quantifying ecological concepts derived from the Hutchinsonian niche model is discussed. An example is given of a reduced available niche resulting in the loss of two species, smaller realized niches for the remaining species, and greater niche overlap.

INTRODUCTION

As MacArthur (1968) points out, "The term niche was almost simultaneously defined by Elton and Grinnell to mean two different things." Elton's emphasis was placed on the species' role in the community and Grinnell's on a subdivision of the environment within which the species lived. Hutchinson (1957a) reformulated the niche concept in terms of set theory, proposing that if each measurable feature of the environment is given one coordinate in an N-dimensional space, then the niche might be defined as that region in which the fitness of the individuals is positive.

This concept of the niche as an "N-dimensional hypervolume" has proved useful as a theoretical tool, especially in relation to competition and competitive exclusion. However, when application to field data has been attempted, the model has often proved to be either nonoperational or trivial. That is, if the requirements for a given ecological feature are obviously different for two species, then introduction of the "niche" concept become superfluous.

There are basically three operational problems: (i) There is always a practical limit to the number of environmental parameters which can be measured. No matter how many have been measured without separating two species, the consideration of one more might shrink the niche intersection or eliminate it (MacArthur 1968). Therefore one can in theory demonstrate that two species do not occupy the same niche, but one can never demonstrate that they do occupy the same niche. (ii) A large proportion of the parameters measured are likely to be correlated (redundant), relatively invariant, or irrelevant. Representation of a niche as a region in N space where each of the N dimensions is an ecological parameter and the N axes are orthogonal implies that those parameters act independently, if at all, on the species being considered. If they are in fact correlated, the proper representation would be an acute angle; if they are so highly correlated that the two contribute no more information than one alone, then one less dimension can be used with no loss of information. As Hutchinson (1968) states, "In practice, some provision against using great numbers of highly correlated parameters is probably often needed." (iii) Consideration of each measurable feature of the environment that might be relevant results in a large mass of multidimensional data which is difficult to interpret. One common approach is to graph the data in two dimensions in relation to all possible pairs of axes. However, the number of these goes up very rapidly with increasing N. Furthermore, one or more of the relevant niche dimensions may

[1] Received February 8, 1971; accepted April 20, 1971.

not be one of the measured parameters, but rather a feature of the environment related in a definable way to two or more of the measured parameters. Correct interpretation in such a case is difficult.

These difficulties increase further when one wishes to assess the ecological factors separating the niches of g species from n samples on m measurable parameters, where g, n, and m are large. Maguire (1967) has used such data, collected by Cairns (1964, 1965), to calculate "versatility indices" as a "partial analysis of the niche." He concludes that "it is possible to utilize the concepts of niche hypervolume size and shape in a useful analytical manner," and suggests the possible future extension of using discriminant analysis "to permit assignment of values of relative importance to the contribution of each factor."

The bivalve mollusc species of central Canada are appropriate for determination of factors of niche separation by a multivariate statistical approach such as multiple discriminant analysis. A large number of common species of the families Unionidae and Sphaeriidae occur in lakes of this region, and most of the potentially relevant ecological parameters are measurable, continuous variables. Furthermore, this region contains a wide variety of lake types, ranging from eutrophic and dystrophic to oligotrophic. It was decided that the environments to be considered would be limited to lacustrine benthos at depths greater than the lower margin of rooted aquatic vegetation. This frame of reference eliminates the difficulties of quantifying measures of possible mollusc-plant associations, and of microcurrent effects in rivers and streams.

METHOD: THE MULTIPLE DISCRIMINANT MODEL

Multiple discriminant analysis (Rao 1952, Cooley and Lohnes 1962) starts with a data set consisting of n measurements on m parameters, each of the n measurements being associated with an individual belonging to one of g groups. The data may be visualized as g sets of points in an m-dimensional space. The analysis reduces the data set to n measurements on k new parameters which are linearly independent (orthogonal), additive functions ("discriminant functions") of the original m parameters. The data may now be visualized as g sets of points in a $k \leq m$-dimensional space, and the standardized coefficients of the discriminant functions indicate the relative contributions of the original m parameters to each of the k discriminant functions. There is an approximate test of significance of separation of groups (Rao 1952), which unfortunately requires an assumption not likely to be satisfied by most ecological data (see next section).

Multiple discriminant analysis provides an appropriate statistical model for the Hutchinsonian niche, stochastically defined. Any statistical model of the Hutchinsonian niche must necessarily be a multivariate model, but there are many other multivariate statistical models such as factor analysis, canonical correlation analysis, and multivariate analysis of variance. However, none is appropriate to this problem except the last, which is the basis of the significance test for separation of groups in the multiple discriminant analysis.

Species abundance and species presence are usually not related to environmental parameters in a linear additive manner. In fact, the curve of species abundance, and therefore of probability of occurrence, along any environmental gradient is usually unimodal, with the maximum at some ecological optimum. Slobodkin (1968) and others have pointed out some of the inadequacies of predictive models of species abundance and distribution which are based on multiple linear regression equations with species as dependent variables. In multiple discriminant analysis, as used here, species are groups separated in ecological space by linear additive functions of ecological parameters, rather than dependent variables supposedly related to the ecological parameters in a linear additive manner. That is, the linear additive relationship is among the ecological parameters, and not between the species and the ecological parameters.

This analysis, if the necessary statistical assumptions can be satisfied, provides a solution to two of the operational problems in niche analysis mentioned previously. First, any of the m ecological parameters that are highly correlated, invariant, or irrelevant to group (species) separation will be eliminated or combined in the reduction from m to k dimensions. Second, the ecological dimensionality is reduced before interpretation of species separation is necessary; any unmeasured ecological parameter which is related in a known manner to any measured ecological parameters should be detectable by examination of the standardized discriminant function coefficients. The third operational problem remains: species can be shown to have different ecological requirements in terms of $k \leq m$ parameters, but if there are no differences, any unmeasured parameter might have provided a basis for discrimination.

Because each of the n samples consists of a battery of m ecological measurements associated with the presence of an individual of one of g species, it follows that the analysis is based on species presence, but not on species absence. Hutchinson (1968) has also used this redefinition of niche, where presence, implying ability to live at a point in ecological space, replaces positive fitness at that point. As Hutchinson reasons, "Since the specimens were living at the recorded temperatures, these temperatures must be within the tolerances of the organisms."

Failure to consider the absence of species in a sample undoubtedly results in the loss of information. However, consideration of absence, besides being inappropriate for multiple discriminant analysis as used here, also creates ambiguity. If a species is present, one can conclude that the species can live there, and that the total niche for the species must include that point in ecological space. If the species is absent, there are three possible interpretations: (i) The species cannot live there; that is, its niche does not include that point. (ii) The species can live there, but never had the opportunity for zoogeographic reasons. (iii) The species can and does live there, but the sample failed, by chance, to include a representative of that species.

Assumptions

The statistical assumptions for the valid use of multiple discriminant analysis are:

1) The "groups" can be defined a priori. This means that species must be "good" species and not ecotypes. All specimens must therefore be identified and verified without reference to the site of collection and doubtful identifications eliminated. The circularity of using ecological criteria, in whole or in part, to assign specimens to groups (species) for ecological separation of such groups should be obvious.

2) The values assumed by the m ecological parameters are samples from an m-dimensional multivariate normal distribution. This assumption of normality is as likely to be satisfied here as in any set of ecological data; however, the k discriminant functions are more likely to be normally distributed than are the m original parameters, as a consequence of the Central Limit Theorem (Cooley and Lohnes 1962).

3) The postulated orthogonal discriminant functions are, in fact, linear functions of the original correlated parameters. The observations on the m original ecological parameters may be transformed, taking into account known or presumed relationships among parameters, in order to minimize nonlinearity and to improve normality.

4) If the pooled within-groups $m \times m$ variance-covariance matrix is estimated on a large number of degrees of freedom, the total variation is approximately distributed as chi-square (Rao 1952); it may serve as a test criterion for overall group differences, if the variance-covariance matrices for the groups are estimates of a common variance-covariance matrix. That this assumption of homogeneous within-group matrices is unlikely to be satisfied with ecological data is indicated by the common use by ecologists of the prefixes "steno-" and "eury-," and by the emphasis on the importance of "niche size" in many ecological and evolutionary studies. However, significance of group separation may also be judged by examining separation of groups in $k \times k$ discriminant space. If the overall chi-square test (biased or not) is highly significant, the discriminant function coefficients are ecologically interpretable, and there are obvious species separations on each discriminant function, then it would, in my opinion, be reasonable to conclude that the species differences are greater than would be produced by drawing random samples from a multivariate swarm.

Relevance of parameters

The choice of each of the m environmental parameters was based upon four criteria: (i) The parameter should have a reasonable theoretical possibility of effect on molluscs, should have been described in the literature as having such an effect, or should be correlated with a parameter which satisfies one of these requirements. (ii) It must be possible to easily and rapidly determine the parameter in the field, or to collect a sample in the field and determine the parameter at a later date in the laboratory without the delay causing error. (iii) The *in situ* seasonal variation of the parameter should be insignificant relative to the variation of the parameter among all samples collected. (iv) The parameter should describe the environment in direct contact with the organism sampled, not a geographic area in which the organism lives. Lake area would be an example of the latter type of parameter.

The third criterion eliminates a number of otherwise appropriate parameters, such as dissolved oxygen concentration, nutrient levels (phosphorus and nitrogen), primary production rate, and biological oxygen demand. All of these parameters are related to productivity or "trophic state." However, I decided that it was better to omit such parameters than to introduce an artifact caused by differences in sampling times over a 5-month period. Furthermore, several of the parameters which meet all four criteria, and are included, have been shown to be correlated with production and eutrophic state. Parameters such as lake area, which are eliminated by the fourth criterion, can be considered when interpretations of the discriminant functions are attempted.

The $m = 9$ parameters which were chosen and used are:

1) pH, which influences the deposition of calcium and the availability of dissolved CO_2 for conversion to $CO_3^{=}$ to form the $CaCO_3$ component of the shell (Wilbur 1964). Hunter (1964), for example, states that in Britain few molluscs are found at $pH < 6.0$.

2) Total alkalinity, which, along with total dissolved solids, is correlated with lake productivity or trophic state (Moyle 1956, Northcote and Larkin 1956, and Rawson 1958). Since I found a 98%

correlation between reported total alkalinity and TDS values over a wide range of lake types, I chose the former as easier to measure. Clarke and Berg (1959) and Hunter (1964) state that mollusc distribution is strongly influenced by total alkalinity.

3) Calcium, which is necessary for shell formation, and is mentioned as an important limiting factor by many authors. Calcium concentration can vary widely in shallow waters over dense rooted vegetation, but such areas were not included in this study.

4) Total hardness less calcium hardness, which is a measure of divalent and trivalent cations other than calcium—primarily magnesium in fresh water. Wilbur (1964) cites the work of several authors who have shown that *in vitro* $CaCO_3$ crystal formation can be affected by concentration of magnesium and other ions.

5) Sodium chloride. Cvancara (1967) lists chloride as a significant limiting factor affecting the distribution of unionid clams in the Red River Valley, North Dakota.

6) Depth, an obvious factor separating species distributions within lakes.

7) Mean sediment particle size, which "depends largely on the current strength of the local environment" (Folk 1968). Marzolf (1965) also discusses the relationship between sediment composition and current velocity, the latter of which would determine rate of turnover of food for filter-feeders.

8) Percentage loss on ignition, as a measure of percentage organic matter of sediment, which would be related to food available for detritus-feeders.

9) Sorting coefficient of sediment. According to Folk (1968), "For best sorting—currents must be of intermediate strength and also of constant strength."

The importance of sediment type in determining mollusc distributions is emphasized by many authors, including Eggleton (1952), Clarke and Berg (1959), Murray and Leonard (1962), Khozov (1963), and Parmalee (1967).

Limitations of the method

This analysis can only detect factors of niche separation which influence the distribution of the species in space. Species which occupy different niches because they feed on different types of food in the same place will not be separated, although there is no reason for parameters related to food type not to be used in such an analysis.

I have frankly chosen a species assemblage and a set of environmental parameters which appear to be appropriate to the analysis, but modifications of the analysis could no doubt be devised which would allow the use of qualitative parameters, or of parameters which vary widely throughout the year. The latter would obviously be essential for use of this type of analysis to separate species by ecological criteria in many species assemblages.

Sampling procedure

It was planned that during the sampling period, May–September 1969, as many lakes as possible would be sampled. It was assumed that lakes ranging from the Precambrian Shield to the limestone Lake Agassiz escarpment would provide a wide variety of lake types. The lakes sampled, their locations, and a brief description of the geological substrates are given in Table 1, and their geographical distribution is shown in Fig. 1.

On small lakes, or near-shore portions of large lakes, one or more rough transects of benthic samples were taken from deep water to the shore, using a variety of sampling gear. Supposedly "quantitative" sampling devices were not required because estimates of abundance or density were not necessary. The number of samples taken from a lake was roughly proportional to the lake area. Biological sampling methods included the Ekman grab, a small bottom dredge with a fine net and an adjustable cutting edge, oyster tongs with 5 m handles, and diving. Near-shore sampling on Lake Winnipeg was supplemented by samples collected during a Fisheries Research Board of Canada limnological research cruise during July 1969, which covered the entire lake.

On collection, all live molluscs were placed on ice, and preserved within a few hours. In small lakes, a vertical series of water samples was collected with a PVC Van Dorn bottle. If temperature stratification was absent, one water sample was analyzed for pH, total alkalinity, calcium hardness, total hardness, and sodium chloride, using a Hach Chemical kit, and the values were applied to all samples from that lake (or to that location in a large lake). If temperature stratification was present, the depth of the thermocline was noted; two such analyses were carried out, one for samples from less than and one for samples from more than the thermocline depth.

A sediment sample was collected for each biological sample. If the Ekman grab was used for the biological sample, a sample was taken from the undisturbed sediment surface in the grab. If it was not, an Ekman sample was taken at the spot where the biological samples had been taken. When the biological sample was collected while diving, a sediment sample was collected with a jar. In some cases, particularly on hard substrates in deep water, sediment samples were difficult to obtain, and in a few cases the sample was no doubt inadequate and somewhat biased. However, the comparability of samples from similar adjacent substrates was good.

Sediment samples were sealed in Whirl-Pak bags, placed on ice, and frozen within a few hours. Freez-

TABLE 1. Index numbers, names, locations, and brief geological descriptions of lakes sampled during the period May–September 1969

Lake No.	Name	Latitude	Longitude	Description
1	West Hawk L.	49°46′N	95°12′W	Shield (meteor crater?)
2	Falcon L.	49°43′N	95°15′W	East end shield, west end glacial drift
3–4	Norris L.	50°27′N	97°23′W	Limestone
5	North Shoal L.	50°28′N	97°38′W	Limestone
6	Gull L.	50°25′N	96°32′W	Limestone
7	"Mussel L."	49°41′N	93°57′W	Shield
8	FRB 239	49°39′N	93°47′W	Shield
9	Mad Dog L.	49°36′N	93°48′W	Shield
10	Hillock L.	49°39′N	93°53′W	Shield
11	L. Winnipeg	52° N	97° W	West shore limestone, east shore shield
12	Whitemouth L.	49°15′N	95°41′W	Limestone glacial drift
13	Crescent L.	49°58′N	98°17′W	Oxbow lake on river plain
14	Swan L.	49°22′N	98°55′W	River valley, galcial drift
15–16	St. Malo L.	49°19′N	96°56′W	Artificial, on glacial drift
17	L. Athapapuskow	54°31′N	101°40′W	North shore shield, south shore limestone
18	Neso L.	54°40′N	101°32′W	Shield
19	First Cranberry L.	54°35′N	101°16′W	Shield and limestone
20	Payuk L.	54°38′N	101°32′W	Shield
21	L. St. Martin	51°42′N	98°24′W	Limestone
22	L. Manitoba	51° N	98°30′W	Limestone
23	Clear L.	50°42′N	100° 3′W	Escarpment shale and limestone
24	Sandy L.	50°34′N	100°10′W	Escarpment shale and limestone
25	Minnedosa L.	50°16′N	99°48′W	Artificial, on shale and limestone
26	Whirlpool L.	50°44′N	99°48′W	Escarpment shale and limestone
27	Roddy L.	49°41′N	93°47′W	Shield
28	Fox L.	49°55′N	94°44′W	Shield
29	Dauphin L.	51°15′N	99°45′W	Limestone
30	L. Winnipegosis	52°20′N	100°10′W	Limestone
31	Steeprock L.	52°37′N	101°25′ W	Escarpment shale and limestone
32	Whitefish L.	52°20′N	101°35′W	Escarpment shale and limestone
33	L. Madge	51°40′N	101°35′W	Escarpment shale and limestone
34	Childs L.	51°37′N	101°10′W	Escarpment shale and limestone
35	West Blue L.	51°37′N	100°50′W	Escarpment shale and limestone
36	East Blue L.	51°37′N	100°50′W	Escarpment shale and limestone
37	Singoosh L.	51°37′N	100°50′W	Escarpment shale and limestone

ing was necessary to keep the organic content unchanged, but it may have slightly affected the particle size-frequency distribution. Just before the analysis the sample was thawed, thoroughly mixed, and two subsamples were taken for particle size-frequency analysis and for organic content analysis. Size-fractions down to 0.0625 mm were determined by wet sieving, drying, and weighing. The hydrometer method (Bouyoucos 1927) was used for the fraction less than 0.0625 mm. The mean particle size in phi units (Krumbein 1936) and the sorting coefficent (the standard deviation of particle size in phi units) were calculated by the method of moments (Folk 1968). Organic content was estimated by loss on ignition, after drying, in a muffle furnace for 1 hour at 500–550°C.

Preliminary taxonomic identifications were verified by H. B. Herrington (sphaeriids) and A. H. Clarke (unionids).

Data compilation

It seemed reasonable to assume that the relationships among the chemical and sediment parameters would be more multiplicative than additive. The values for the chemical parameters were transformed logarithmically, with the exception of pH, which is already in logarithmic form. The arcsine transformation seemed most appropriate for percentage organic content of sediment (Steel and Torrie 1960).

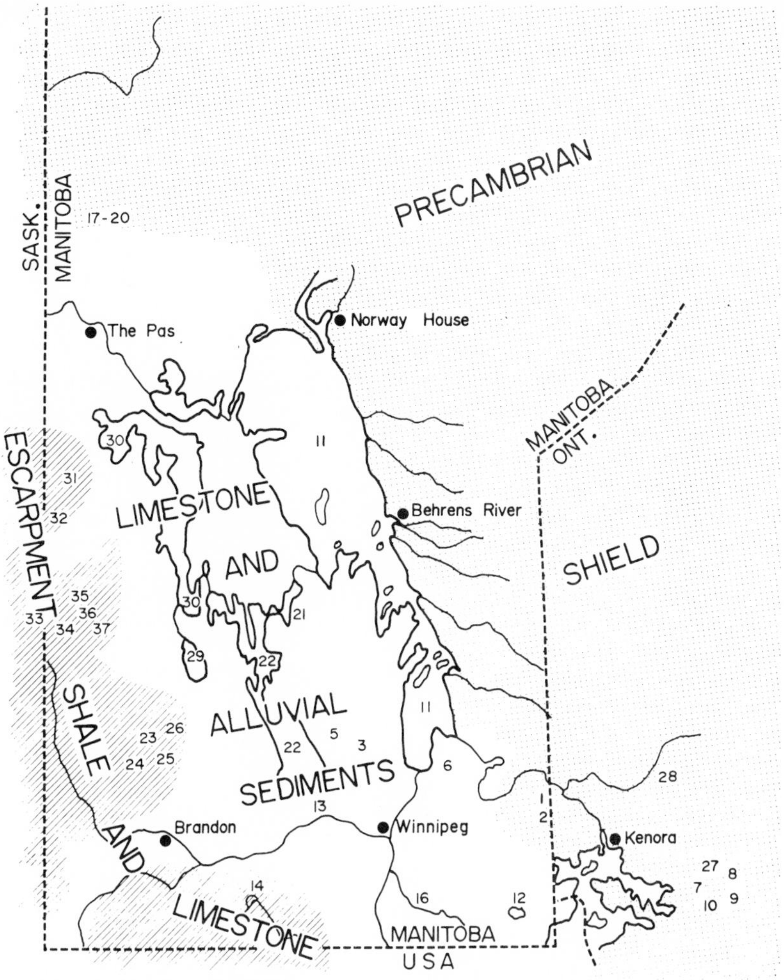

FIG. 1. The geographic distribution of the lakes sampled, in relation to large-scale geological features. Lake numbers refer to Table 1.

Spencer (1966[2] and personal communication) has found that in factor analysis logarithmically transformed chemical and sediment data from marine sediments yield more interpretable results than such data untransformed. The use of logarithmic "phi units" to describe mean sediment particle size and sorting coeffiicent is standard practice, since particle size is usually more normally distributed in these units. Therefore, these two parameters did not require further transformation.

It was assumed that some species would be represented by a few samples from one or two lakes, and that some a priori criteria for a cut-off point should be established. Therefore, before sampling began, it was decided that no species would be included in the analysis if it was represented by less than nine samples or was collected from fewer than three lakes. It was thought that these criteria would lessen the likelihood that the ecological position of a species would be biased by collection in a few samples from one or two unusual lakes.

[2] Spencer, D. W. 1966. Factor analysis. Ref. No. 66-39, Woods Hole Oceanographic Institution, Woods Hole, Mass. Unpublished manuscript.

Calculations

Following Rao (1952) and Cooley and Lohnes (1962), the $m \times m$ deviation scores cross products matrices $W_{i=1...g}$ were calculated for each of the g species. The pooled within-groups mean squares and cross products matrix is then

$$W\ (n-g\ \text{df}) = \sum_{1}^{g} W_i\,,$$

which leads to the estimate of the within-groups variance-covariance matrix $D = W/(n-g)$, assuming homogeneity of dispersion among groups. The influence of each species on W and D is therefore proportional to the number of samples n_i in which the species occurs ($n = \Sigma\, n_i$).

The $m \times m$ total sample mean squares and cross products matrix T was then calculated for all n samples without regard for group (species) membership; the $m \times m$ among-groups deviation scores cross products matrix A was calculated from the relationship $A\ (g-1\ \text{df}) = T - W$. The discriminant function vectors and roots are associated with the determinantal equation $|D^{-1}A - \lambda I| = 0$. The 9×9 matrix inversion, multiplication, and roots and vector solution for this equation were performed on an IBM 360 computer at the University of Manitoba Computer Centre.

The vectors associated with each root provide the discriminant function coefficients, but the position of the k new orthogonal axes in relation to the original m axes is quite arbitrary. Therefore, a varimax rotation (Kaiser 1958), which rotates the axes so as to maximize the number of high and low coefficients and to minimize the number of intermediate ones, was performed. The discriminant function axes remain orthogonal, and the origin and the relationships among the group centroids in the reduced (k-dimensional) space remain unchanged.

RESULTS

The primary data

A total of 382 samples, each containing one of 22 live mollusc species, were collected from 37 lakes. The associated measurements on the 9 environmental parameters did, as expected, vary over a wide range (Table 2). Low values for the water chemistry parameters fall below the sensitivity limits of the methods used, as indicated. The raw data are summarized in Table 3, which shows that 10 of the 22 species met the criteria of at least nine samples from at least three lakes. For these 10 species, the untransformed mean values for each of the nine ecological parameters are given.

The multiple discriminant analysis

The overall test of significance of among-species differences yielded $\chi^2[m(g-1) = 81\ \text{df}] = 1{,}359$, which is highly significant. As previously mentioned, this chi-square is biased due to the heterogeneity of the within-groups variance-covariance matrices $D_{i=1...9}$, which is apparent in Fig. 2.

TABLE 2. Ranges of ecological parameters measured (Lake numbers refer to Table 1.)

Parameter	Low	Lake no (s)	High	Lake no (s)
pH	6.5	8	9.5	13
Tot. alkalinity (ppm, as $CaCO_3$)	<14	Several shield lakes	752	5
Calcium (ppm, as $CaCO_3$)	<17	Several shield lakes	188	14, 16, 25
Tot. hardness – Ca (ppm, as $CaCO_3$)	<17	Several shield lakes	256	24
Sodium chloride (ppm)	<12	Many lakes	850	21
Depth (m)	—	—	87	1
Mean sediment particle size (mm)	0.019	11	2.7	30
% organic matter	0	Several lakes	48	4
Sorting coefficient ($\sigma\phi$)	0.26	30	3.45	11

The chi-square tests of the significance of each extracted discriminant function (hereafter abbreviated DF) would be even more biased; therefore, the common procedure in factor analysis was adopted whereby factors (or DF's) are extracted until a certain percentage of the variance (among-species variance, in this case) is accounted for. Table 4 summarizes the results for the minimum number of DF's needed to account for 95% of the among-species variance. The coefficients have been standard-

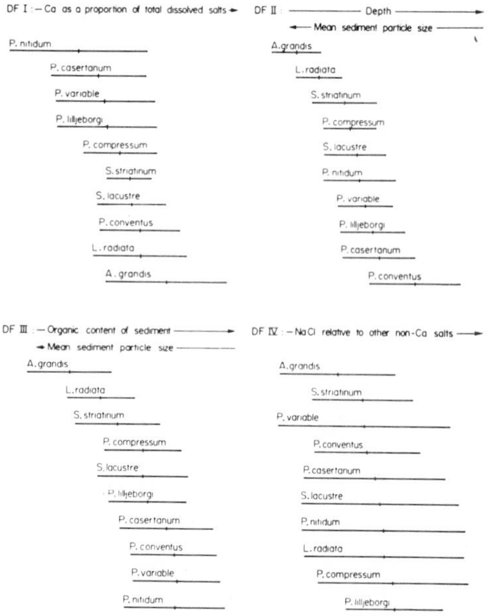

FIG. 2. The mean plus and minus one standard deviation for each species, in terms of the discriminant scores for DF I, DF II, DF III, and DF IV. The length of each horizontal line segment is the range for two-thirds of the individuals of that species.

TABLE 3. Summary of biological and ecological data used for the multiple discriminant analysis. Species below the line were not used (see text)

Species	No. samples	No. lakes	pH	Tot. alk. (ppm)	Ca (ppm)	Total hardness less Ca (ppm)	NaCl (ppm)	Depth (m)	Part. size $\bar{x}(\phi)$	% org.	Sorting coeff. ($\sigma\phi$)
			Untransformed species means on original parameters						Sed. parameters		
Anodonta grandis	62	18	7.9	78	63	57	29	1 4	1.02	1.7	1.46
Pisidium casertanum	52	12	8.2	156	99	114	51	6.8	1.56	6.9	1.89
Lampsilis radiata	52	9	8.2	145	98	115	90	2.0	1.05	1.5	1.71
Sphaerium striatinum	49	4	8.0	109	81	67	34	3.9	1.69	3.2	2.11
Pisidium compressum	36	7	8.4	189	122	133	93	3.7	2.31	3.2	1.74
Pisidium nitidum	32	11	8.6	198	97	168	93	3.8	2.06	10.1	1.91
Pisidium lilljeborgi	19	3	8.2	125	88	71	50	6.7	1.72	4.6	1.90
Pisidium variable	18	10	8.4	184	114	127	50	4.7	2.39	10.8	1.88
Pisidium conventus	14	8	7.4	41	42	38	24	16.6	1.32	10.2	2.25
Sphaerium lacustre	11	5	8.3	151	111	109	63	4.6	1.68	4.3	2.08
Subtotal	345	32 diff. lks.									
P. idahoense	9	1									
P. ferrugineum	7	4									
P. ventricosum	6	4									
S. transversum	2	1									
Lampsilis ovata ventricosa	2	1									
P. punctatum	2	1									
Lasmigona complanata	2	1									
S. rhomboideum	1	1									
P. walkeri	1	1									
Amblema costata	1	1									
P. subtruncatum	1	1									
Total	380	37 diff. lks.									

TABLE 4. Standardized and normalized discriminant function coefficients, showing the relative contribution of the original parameters to each DF. Coefficients greater than $\pm$ 0.4 are italicized

	Discriminant functions I	II	III	IV	V
Percentage of among-species variance	45.0	23.8	11.2	9.4	6.5
Cumulative percentage	45.0	68.8	80.0	89.4	95.9
Parameters:					
pH	0.006	−0.039	0.138	−0.385	−0.200
Tot. alkalinity	*−0.771*	0.207	0.200	0.280	*0.677*
Calcium	*0.613*	−0.124	0.071	−0.062	0.378
Tot. hardness — Ca	0.119	0.064	0.069	*−0.451*	0.069
Sodium chloride	0.024	0.142	0.292	*0.724*	−0.145
Depth	−0.095	0.822	0.369	0.092	*−0.549*
Sed. part. size	−0.069	*0.404*	*0.517*	−0.156	0.121
Sed. % org. matter	0.007	0.159	*0.663*	−0.004	−0.042
Sed. sorting coef.	−0.025	0.228	0.058	0.091	0.127

Interpretations of discriminant functions:

- I Concentration of calcium relative to total alkalinity.
- II Depth and depth-related sediment mean particle size.
- III Organic content of sediment, related to sediment particle size.
- IV Concentration of sodium chloride relative to concentration of non-calcium dissolved salts.
- V Apparently interaction between alkalinity and depth, possibly influence of wave action in shallows of large alkaline lakes.

ized and may therefore be interpreted as relative contributions by the nine original ecological parameters to the DF's.

Interpretation of the discriminant functions

Empirical interpretations of the DF's are given at the bottom of Table 4. The discriminant scores were calculated for each sample and for each DF, and the mean and standard deviation for each species were calculated. The means, plus and minus one standard deviation (the range for approximately two-thirds of the individuals of each species), are shown in Fig. 2. The scales for the discriminant scores are omitted because the absolute values are arbitrary and are of no significance other than on a comparative basis.

The last discriminant function (DF V) accounts for only 6.5% of the among-species variance and is difficult to interpret. The strong relationships of opposite sign with total alkalinity and depth suggest an interaction between lake type and species depth distributions; one possible interpretation would be the influence of wave action, which is greatest in the shallows of the larger and more alkaline lakes, such as Lake Winnipeg and Lake Manitoba. Because of the low contribution to the separation of species, DF V will not be considered further.

DF I, which accounts for 45% of the among-species variance, is largely a function of calcium and total alkalinity. The signs are opposite and, since the analysis was performed on log-transformed data, the empirical interpretation is concentration of calcium relative to total alkalinity. The basis of separation on DF I (Fig. 2) is that *Pisidium nitidum, P. casertanum, P. lilljeborgi,* and *P. variable* are the only species found in small, shallow, eutrophic lakes with little or no outflow, such as lakes 4, 6, 13, 24, and 26 (see Table 1). Where only one species is present (lakes 4, 13, and 26), it is *P. nitidum.* Hutchinson (1957b) states that the proportion of calcium in the dissolved salts typically rises in lakes with increasing salinity; but, in closed basins where salt concentration reaches very high values through evaporation, calcium carbonate precipitates out first, resulting in a lower proportion of calcium. Large filter-feeders such as the unionids *Anodonta grandis* and *Lampsilis radiata* never occur in such small, closed lakes, but can occur in high salinity lakes which are large enough so that there is wind-generated current and wave action. Lakes 11, 21, 22, 30, and 32 are such lakes, and they are the only lakes in which *A. grandis* and *P. nitidum* co-occur.

In order to verify this hypothesis, the mean discriminant score for DF I was calculated for each of 14 lakes whose drainage areas do not include Precambrian Shield. If DF I is an index of rate of water turnover, the means of these discriminant scores should show a clear relationship to an independent function of rate of water turnover. The function chosen was

$$f(\mathrm{DA,LA}) = \log(\mathrm{DA/LA}) + b\log(\mathrm{LA})^{\frac{1}{2}},$$

where DA and LA are lake drainage area and lake surface area, respectively. The function is in logarithmic form because DF I is primarily a function of two log-transformed parameters. The ratio of DA to LA should be proportional to turnover rate of the lake volume, if lake morphometries, precipitation, and evaporation are similar for the lakes considered. The square root of LA should be proportional to average wind fetch and therefore to wind-generated water movement. The DA and LA values were determined from topographic sheets, using a planimeter. Only lakes with off-Shield drainage areas were used for the following reasons: (i) It is difficult to accurately measure the complicated Shield drainage areas. (ii) The off-Shield lake morphometries tend to be similar, whereas Shield lake morphometries do not. (iii) Precipitation is higher and evaporation lower in the area of most of the Shield lakes studied than elsewhere in the area covered by the study.

The relationship between DF I scores and the function of DA and LA is shown in Fig. 3. The higher variability on the left-hand side of the graph is probably due to the greater error in calculating lake areas and drainage areas for very small lakes on flat, marshy terrain, such as lakes 3 and 6. The hypothesis that DF I is a measure of rate of water turnover is accepted.

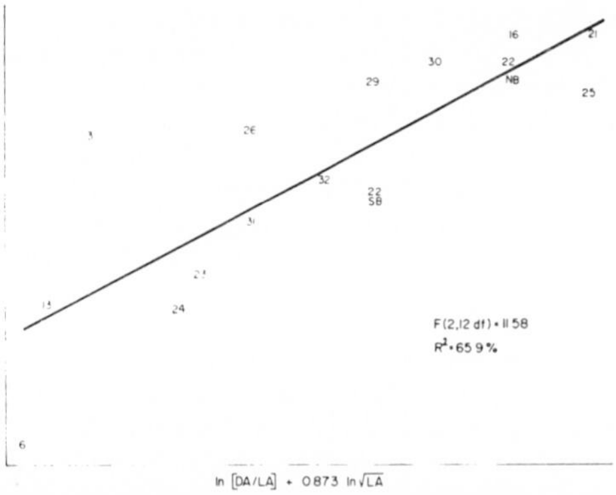

FIG. 3. The relationship between mean DF I scores for 14 lakes and the function of lake drainage area (DA) and lake surface area (LA): ln (DA/LA) + b ln (LA)$^{\frac{1}{2}}$. Lake numbers refer to Table 1.

DF II, which accounts for an additional 24% of the among-species variance, is a function of depth and sediment particle size. Particle size decreases (increases in phi units) with increasing depth, as would be expected. It is not possible to identify the actual ecological factor represented by DF II with certainty, but it is certain that DF II is not a function of depth per se. Discriminant scores for DF II are consistently higher or lower for given lakes at comparable depths, presumably because of differences among the lakes in factors affecting sediment particle size at given depths.

DF III, which accounts for an additional 11% of the among-species variance, is also related to sediment characteristics—but to particle size and organic content rather than to depth.

DF II and DF III separate different species, as can be seen in Fig. 2. In the large lakes of the Lake Agassiz basin (lakes 11, 22, and 30) the common unionid species is *Lampsilis radiata*, and the most common sphaeriid species are *Sphaerium striatinum, Pisidium compressum, P. nitidum, P. lilljeborgi,* and *P. casertanum*. These six species form a depth-sediment series on DF II, with the poorest separation being that among the second, third, and fourth of these species, and between the last two. However, the species within these two "depth" assemblages show better separation on DF III. *P. nitidum, P. lilljeborgi, P. casertanum,* and *P. variable* are found in high organic sediments characteristic of the small, eutrophic, closed basin lakes in which they are the only species. Indeed, DF III is in part a measure of lake productivity or "trophic state" and, as such, varies from lake to lake, as well as within lakes.

It is tempting to hypothesize that DF II is a measure of available food for filter-feeders utilizing plankton or detritus suspended just above the surface of the sediment; this would be affected by primary production and oscillating water turbulence, both of which are depth-dependent. Such water turbulence at the sediment surface would also affect sediment particle size. DF III would then be a measure of food available for deposit-feeders, which would be consistent with the high contribution by sediment organic content and the reduced dependence on depth. The situation as presented in Fig. 2 may be confused by the fact that some species may feed both ways to some extent.

DF IV provides the least separation of species, accounting for approximately 9% of the among-species variance, and only about one-fifth of the discriminating power of DF I. The two species which occur commonly on the Precambrian Shield, *A. grandis* and *P. conventus,* rarely occur in high chloride waters. *S. striatinum,* which is abundant in the relatively high alkalinity waters of Lake Winnipeg, does not often occur in such waters either.

Discussion

The trophic basis of niche separation

On the basis of the above results, I conclude that there is real separation of these 10 species in the defined ecological space, and that four ecological factors are sufficient to account for almost 90% of the variance attributable to differences among species. With the exception of percentage organic matter, as estimated by percentage loss on ignition, all of the *m* original parameters are physical or chemical rather than biotic. The three most important discriminant functions, accounting for 80% of the among-species variance, are however, most easily interpreted on a trophic rather than a physical or chemical basis. This trophic basis of separation of niches, which I frankly did not anticipate at the outset of this study, is all the more impressive when one considers the range of physical and chemical parameters in the environments studied (Table 2). Such results as this, and that of Paine (1966) for the rocky intertidal, suggest that Elton's (1950) concept of the niche as trophically defined tends to hold for even what would seem, a priori, to be physically rather than biologically accommodated communities (Sanders 1968).

Apparently the species diversity that we observe can often be explained by the distribution of species on variables which are a small subset of possible variables. Moreover, these subsets tend to be of a trophic rather than an abiotic nature.

Generation of new hypotheses

For many of the 10 species included in this analysis the basis of ecological separation is reasonably clear. For some species these four factors do not appear to provide significant ecological separation. However, new hypotheses and experiments to test them are suggested.

This approach provides a useful analytical tool for generation of new ecological hypotheses, if only by reduction of the number of potential ecological interactions in a species assemblage to a statistical minimum. For example, is *P. nitidum* barred from low salinity lakes by osmotic problems, unsuitable substrates, or competition for limited food? Is *P. conventus* limited to deep water by competition or by low tolerance to high temperature? What is the physiological basis of NaCl as a limiting factor? What biological processes are first affected? Which ion, Na^{+} or Cl^{-}, or both, are acting? Is DF IV a function of NaCl relative to Mg^{++} and other divalent and trivalent cations because such ions have a buffering effect?

The separation between *A. grandis* and *L. radiata* is reasonably good, but in eight of the lakes sampled the two species co-occur. Are they feeding on the same food material in these lakes, or do they reduce competition by filtering different particle size distributions or by employing different digestive enzyme complexes? Or is neither species food-limited? The seperation between *P. nitidum* and *P. variable* is poor, both species occurring in high organic-content sediments of high-salinity lakes at about the same depth. Are both species detritus-feeders? If so, are they selective detritus-feeders, separated trophically?

Classification of environments

If one wishes to classify environments in a manner that is both as efficient and as relevant to species distribution as possible, then a useful procedure is to base the classification on discriminant scores from an analysis such as this. Considering lakes as environments, they may be positioned in a two-dimensional space defined by axes DF I and DF IV, the two DF's which mainly describe "among-lakes" variation. In Fig. 4, each lake is represented by a smooth, closed curve of minimum area, containing all points defined by DF I and DF IV scores for all samples from that lake.

Such a representation is useful in several ways. The values of two parameters define the relationship of a lake to other lakes with respect to species distributions. The "area" of each lake is proportional to the heterogeneity of the "bivalve mollusc environment" as described by DF I and DF IV. The similar negative slopes of many of the elliptical lake "areas" reflect similar differences between epilimnetic and hypolimnetic water chemistry in lakes that were stratified at the time of sampling.

It is obvious that the heterogeneity of the water mass of Lake Winnipeg (lake 11) is far greater than that of any other lake studied. Although Lake Winnipeg is shallow and well mixed, and generally unstratified except to a limited extent in the north basin, its large size (surface area = 24,530 km^2), irregular shoreline (shoreline development = 3.4), and geologically varied drainages result in marked heterogeneity of its waters. Rivers entering from the east shore, such as the Winnipeg, Berens, and Poplar Rivers, drain the Precambrian Shield and contribute water characteristic of oligotrophic lakes. Alkaline waters enter from the west by the Red River in the south basin and by the Dauphin and Saskatchewan Rivers in the north basin. The Dauphin River, draining Lakes Manitoba and Winnipegosis by way of Lake St. Martin, contributes water of high NaCl content.

Environmental heterogeneity is often used by ecologists as a criterion to explain, in part, variations in species diversity among environments. A generally applicable measure of environmental heterogeneity has not been available; I therefore suggest as such a measure the total area, volume or hypervolume (depending on the number of DF's deemed significant), of the environment in the sense of Fig. 4, with each DF scaled so that the variation in that dimension is proportional to the contribution to the separation of the species.

A representation such as Fig. 4 may also be used to describe the distributions of individual species as bivariate normal probability distributions in relation to the environments. As an example, the bivariate probability ellipses containing 50 and 90% of the samples of *Anodonta grandis* are shown in Fig. 4. Such ellipses are easily calculated from the reduced ($k \times k$) within-group variance-covariance matrix for the k discriminant functions. In four of the six lakes which lie wholly or more than half outside the 90% ellipse, *A. grandis* was not recorded. In the other two lakes it was very uncommon relative to other species, and *Lampsilis radiata* was the characteristic unionid. *A. grandis* was recorded from 10 of the 16

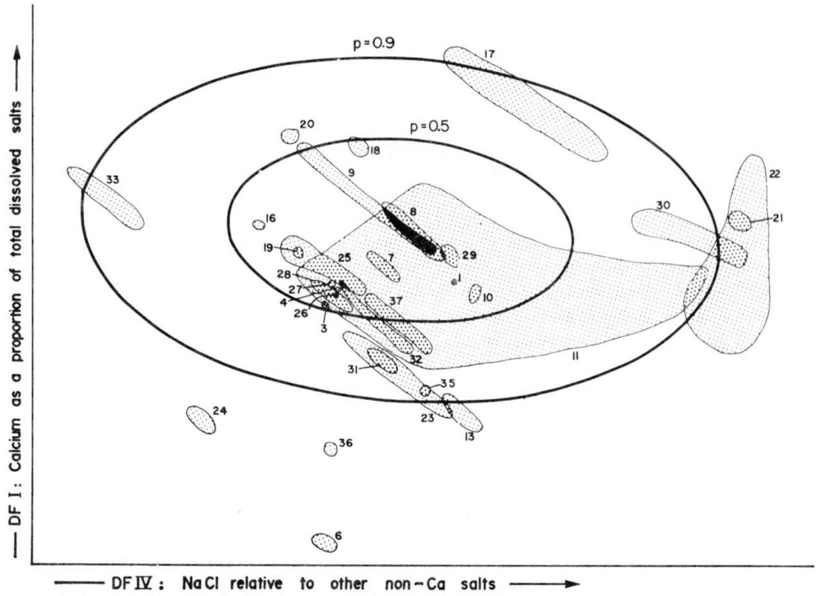

FIG. 4. Each shaded area represents a lake for which all points, defined by DF I and DF IV discriminant scores for all samples from that lake, fall within the area. Lake numbers refer to Table 1. The two concentric ellipses contain 50 and 90% of all samples of *Anodonta grandis,* and are calculated from the means, variances, and covariance in DF I–DF II space for the samples containing *A. grandis.*

lakes which lie wholly or more than half within the 50% ellipse, and was almost certainly present (empty shells observed or collected by other persons) in two of the others.

Such a "map" of environments in relation to species distributions would be useful for the ecologist evaluating potential stocking success in areas where possibly viable species do not naturally occur. It must be remembered, however, that the DF's are determined by maximal separation of all of the species considered, with the contribution by each species being proportional to the number of samples in which it was recorded. Therefore, if the primary objective is to choose between species A and species B for stocking in each of a series of lakes or other habitats, then the discriminant analysis should be based upon occurrences of these two species only (i.e., $g = 2$). If one is not concerned with a choice among g species for stocking given environments, but rather with a choice among given environments of those suitable for one species chosen a priori, then a format such as Fig. 4 might be used with actual environmental parameters instead of DF's as axes.

It must also be remembered that a high probability of survival is only in relation to the parameters, and functions of the parameters, used in the analysis. Parameters not considered might place a given habitat outside the niche of a species, even if the habitat is optimum for all of a large number of parameters that were considered.

Available and realized niches

The available niche for bivalve species in Lake Winnipeg may be defined as the space, relative to the DF axes, within which all samples recorded from Lake Winnipeg lie. Even in this lake, the most chemically heterogeneous of the study, most of the within-lake variation is attributable to DF II and DF III, which are primarily functions of depth and sediment type. The available niche, in DF II–DF III space, is shown on the left in Fig. 5. The white space contains all the DF II and DF III discriminant scores for all Lake Winnipeg samples.

Within this space are drawn the 50% contour ellipses for the four most frequently collected species from Lake Winnipeg: *Sphaerium striatinum*, *Pisidium casertanum*, *Lampsilis radiata*, and *P. lilljeborgi*. These species, accounting for 71% of the recorded specimens from the lake, subdivide and almost fill the available niche. The near-coincidence of the ellipses for *P. lilljeborgi* and *P. casertanum* suggests at least partial trophic separation of these two species.

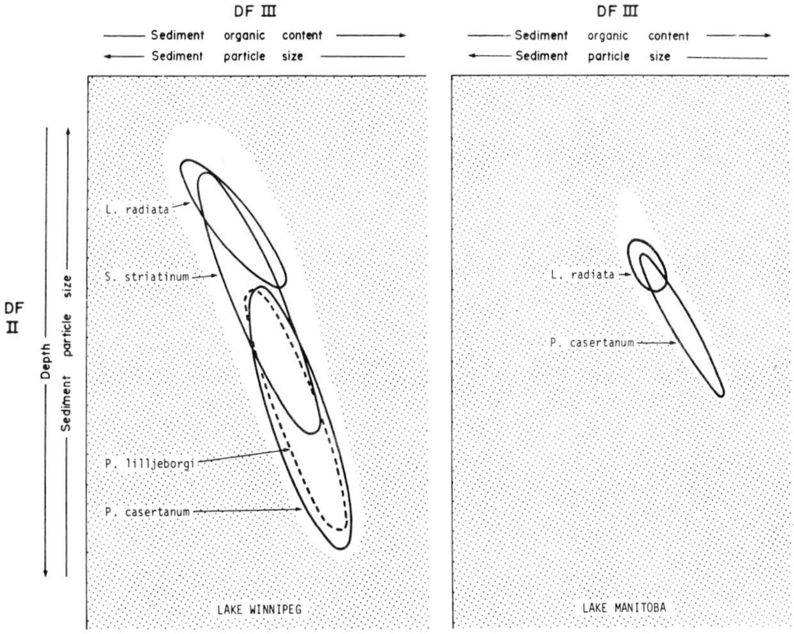

FIG. 5. All DF II and DF III discriminant scores for all Lake Winnipeg samples fall within the white area on the left. All scores for all Lake Manitoba samples fall within the white area on the right. The scales of the two ordinates and the two abscissas are identical. The 50% contour ellipses for each of the four species most frequently collected from Lake Winnipeg are shown for the two lakes. Two of the species were not collected from Lake Manitoba.

The available niche for bivalve species in Lake Manitoba, as shown on the right in Fig. 5, is much smaller than in Lake Winnipeg; it is shifted in position toward decreased average depth and toward increased average organic content of sediment. Lake Manitoba is a large but extremely shallow and windswept lake. The maximum depth of the large south basin is about 6 m, whereas the average depth of Lake Winnipeg is 13 m. The shoreline tends to be uniform, unprotected, and of gradual slope. The decreased depth range reduces the available niche in the DF II dimension, and the uniform sediment composition at a given depth reduces it in the DF III dimension.

Two of the four most common species in Lake Winnipeg, *S. striatinum* and *P. lilljeborgi*, were not collected in Lake Manitoba. The 50% contour ellipses for the other two species, *L. radiata* and *P. casertanum*, are smaller than in Lake Winnipeg. They overlap in Lake Manitoba, but not in Lake Winnipeg. It therefore seems reasonable to conclude that fewer bivalve species can occupy the smaller available niche in Lake Manitoba, and that the Lake Winnipeg species that are not eliminated are forced into smaller niches, packed more tightly together. Another niche dimension, sodium chloride concentration (DF IV, Fig. 2), may also play a role in the absence of *S. striatinum* from Lake Manitoba.

In the format of Fig. 5, such poorly defined ecological concepts as niche size, niche overlap, open or available niches, and subdivision of niches by competitive exclusion, may be objectively defined and quantitatively measured, at least on a comparative basis. There is no formal problem in going to more than two independent niche axes, although the representation becomes more difficult. In any case, this type of analysis does enable the ecologist to reduce the niche dimensionality to a minimum, with known contributions by each dimension to the separation of species in ecological space. This having been done, interpretation of the dimensions and consideration of concepts such as the above can proceed even in studies where the original data set includes large numbers of samples, species, and ecological parameters.

Acknowledgments

I would like to thank the following persons for their contributions to this study. Peter Lemon and Gian Vascotto assisted in the field, and the latter carried out the routine sediment analyses. The Fisheries Research Board of Canada provided the opportunity to participate in a Lake Winnipeg research cruise and to collect in their experimental lakes area in northwestern Ontario. H. B. Herrington and A. H. Clarke verified preliminary taxonomic identifications and identified doubtful cases. I would in particular like to thank Joyce Cook, who sorted most of the biological material and made preliminary identification of the sphaeriids. Grant support was provided by the National Research Council, the Aquatic Biology Research Unit, the Northern Studies Committee, and the National Museum of Canada. N. Arnison, B. G. E. Asmuss, G. E. Hutchinson, K. Patalas, N. Snow, and G. L. Vascotto critically reviewed an early draft of this manuscript.

Any errors are my own.

Literature Cited

Bouyoucos, G. J. 1927. The hydrometer as a new method for the mechanical analysis of soils. Soil Sci. **23**: 343–354.

Cairns, John, Jr. 1964. The chemical environment of common freshwater protozoa. Notulae naturae of the Academy of Natural Sciences of Philadelphia, No. 365, p. 1–6.

———. 1965. The environmental requirements of freshwater protozoa, p. 48–52. *In* Biological problems in water pollution, seminar. Third Public Health Service Publication No. 999 WP 25.

Clarke, A. H., and C. O. Berg. 1959. The freshwater mussels of central New York. Cornell Univ. Agr. Exp. Sta. Mem. 367. Ithaca, New York. — p.

Cooley, W. W., and P. R. Lohnes. 1962. Multivariate procedures for the behavioral sciences. Wiley, New York. — p.

Cvancara, A. M. 1967. Mussels in the Red River Valley in North Dakota and Minnesota and their use in deciphering drainage history, p. 187–196. *In* W. J. Mayer-Oakes [ed.] Life, land, and water (Proc. 1966 Conf. on Envir. Stud. of the Glacial Lake Agassiz Region). Univ. Manitoba Press, Winnipeg.

Eggleton, F. E. 1952. Dynamics of interdepression benthic communities. Amer. Microscop Soc., Trans. **71**: 189–228.

Elton, Charles. 1950. The ecology of animals. Methuen, London. 97 p.

Folk, R. L. 1968. Petrology of sedimentary rocks. Hemphill's, Austin, Texas. 170 p.

Hunter, W. R. 1964. Physiological aspects of ecology in nonmarine molluscs, p. 59–126. *In* K. M. Wilbur and C. M. Younge [ed.] Physiology of mollusca. Academic Press, New York.

Hutchinson, G. E. 1957a. Concluding remarks. Cold Spring Harbor Symp. on Quant. Biol. **22**: 415–427.

———. 1957b. A treatise on limnology. Vol. 1. Wiley, New York, 1015 p.

———. 1968. When are species necessary? p. 177–186. *In* R. C. Lewontin [ed.] Population biology and evolution. Syracuse Univ. Press, Syracuse, New York.

Kaiser, H. F. 1958. The varimax criterion for analytical rotation in factor analysis. Psychometrika **23**: 187–200.

Khozov, M. M. 1963. Lake Baikal and its life. W. Junk, The Hague. 344 p.

Krumbein, W. C. 1936. Application of logarithmic moments to size frequency distributions of sediments. J. Sediment. Petrol. **6**: 35–47.

MacArthur, R. H. 1968. The theory of the niche, p. 159–176. *In* R. C. Lewontin [ed.] Population biology and evolution. Syracuse Univ. Press, Syracuse, New York.

Maguire, B., Jr. 1967. A partial analysis of the niche. Amer. Natur. **101**: 515–523.

Marzolf, G. R. 1965. Substrate relations of the burrowing amphipod *Pontoporeia affinis* in Lake Michigan. Ecology **46**: 579–592.

Moyle, J. B. 1956. Relationships between the chemistry of Minnesota surface waters and wildlife management. J. Wildl. Manage. **20**: 303–320.

Murray, H. D., and A. B. Leonard. 1962. Handbook of

unionid mussels in Kansas. Mus. Nat. Hist. Misc. Publ. No. 28. Lawrence, Kansas.

Northcote, T. G., and P. A. Larkin. 1956. Indices of productivity in British Columbia lakes. J. Fish. Res. Bd. Can. **13**: 515–540.

Paine, R. T. 1966. Food web complexity and species diversity. Amer. Natur. **100**: 65–76.

Parmalee, P. W. 1967. The fresh-water mussels of Illinois. Ill. State Mus. Pop. Sci. Ser., Vol. 8. — p.

Rao, C. R. 1952. Advanced statistical methods in biometrical research. Wiley, New York. — p.

Rawson, D. S. 1958. Indices to lake productivity and their significance in predicting conditions in reservoirs and lakes with disturbed water levels, p. 27–42. *In* P. A. Larkin [ed.] H. R. Macmillan lectures in fisheries. Univ. British Columbia.

Slobodkin, L. B. 1968. Aspects of the future of ecology. Bioscience **18**: 16–23.

Steel, R. G. D., and J. H. Torrie. 1960. Principles and procedures of statistics. McGraw-Hill, New York. — p.

Wilbur, K. M. 1964. Shell formation and regeneration, p. 243–282. *In* K. M. Wilbur and C. M. Yonge [ed.] Physiology of mollusca. Academic Press, New York.

Part III

CLUSTER ANALYSIS

Editors' Comments on Papers 17 Through 25

The various methodologies of cluster analysis attempt to sort a heterogeneous set of previously unpartitioned objects into groups that adequately reflect the original interobject relationships. Unlike other fields of multivariate analysis clustering procedures have been developed largely by applied data analysts, emphasizing solutions to specific problems, rather than by theoreticians. As a result, a general theoretical framework of cluster analysis is lacking and one is faced with a great diversity in clustering procedures. For example, the specific sorting algorithm may be divisive or agglomerative (i.e., dividing the collection into successively smaller sets or sequentially fusing previous groups), and the clustering method may produce mutually exclusive or overlapping clusters, each defined monothetically or polythetically (i.e., by partitioning objects on a few characters or on overall similarity). The final result of these various procedures is usually a graphic display, such as the dendrogram of hierarchical schemes, and one measures the fidelity with which the model represents the data by comparing original object interrelationships with those reproduced from the cluster diagram (e.g., the cophenetic correlation coefficient of Sokal and Rohlf, Paper 23).

Two clustering procedures are commonly encountered in biological literature. Sokal and Michener in 1958 (Paper 17) proposed a polythetic agglomerative method for displaying systematic relationships that has been widely adopted by numerical taxonomists. On the other hand, plant ecologists have utilized a monothetic divisive procedure introduced into phytosociology by Williams and Lambert (Paper 18) under the name of *association analysis*. Hence, even though modern cluster analysis can be traced to such early workers as Zubin (1938) and Tryon (1939), and many other alternative algorithms and clustering methods have been proposed in a variety of fields (e.g., Rogers and Tanimoto, 1960; Ward, 1963; Edwards and Cavalli-Sforza, 1965; Rubin, 1966; Orloci, 1967), these two papers introduced the methodology currently popular among biologists. For a more comprehensive treatment of clustering procedures the reader may consult Sokal and Sneath (1963) (or the revised version by Sneath and Sokal in 1973).

In a field as diverse as cluster analysis, reviews and comparative assessments of different procedures are always welcome, for these often represent important contributions in their own right. One such important article, by Gower (Paper 19), provides a comparative assessment of three common clustering procedures. Another, by Rohlf (Paper 20), gives an excellent review and discussion of hierarchical clustering schemes. In addition, Williams (1971) has recently reviewed the various clustering procedures utilized in biology. Although we do not reprint it here, the latter paper can also serve as a general introduction to cluster analysis for the uninitiated reader.

The logistics of cluster analysis can be divided into a choice of a similarity coefficient for evaluating interobject relationships, and a clustering algorithm that subsequently defines the *a posteriori* grouping of these objects. Both decisions will alter the outcome of the analysis, and comparative assessments of the effect of these alternatives on the resultant clusters are important, especially to the practitioner. Rohlf and Sokal (1965) and Eades (1965) have discussed the relative merits of utilizing correlation or distance for displaying taxonomic relationships, and Orloci (1972) has recently provided a comprehensive review of the various measures of similarity used in phytosociology. In addition to these direct measures of similarity, other measures have also been suggested, including information indices (e.g., Orloci, 1968, 1969, 1972) and generalized distance. Moss (1968) has discussed the effect of various alterations in clustering technique, particularly with respect to character standardization, on a numerical classification in systematics. One of the more thorough comparative discussions of the effect of both similarity coefficient and sorting strategy on cluster results, however, is given by Williams, Lambert, and Lance (Paper 21). Utilizing data from plant communities, they recommend centroid clustering on an information statistic rather than the more popular correlation or Euclidean distance coefficients. Finally, in a recent paper, Goodman (Paper 22) provided a much needed discussion on the relative merits of utilizing direct or generalized distance for cluster analysis in systematics.

An important need in cluster analysis is to be able to evaluate objectively how well a clustering result reflects the original interobject relationships, as well as to be able to compare alternative cluster results. An early development in this area is the *cophenetic correlation coefficient* of Sokal and Rohlf (Paper 23). Even though this procedure has been widely adopted in biology, Farris (Paper 24) was the first to attempt a statistical evaluation of the cophenetic correlation as an optimal measure of goodness of fit. While other statistical approaches to goodness of fit have been proposed (e.g., Hartigan, 1967), clearly this area of cluster analysis deserves further attention.

Since cluster analysis developed largely as an applied methodology, theoretical developments have lagged behind a profusion of empirical assessments of extant clustering procedures. While a generally recognized theory does not yet exist, two early papers by Lance and Williams (1967) and Jardine and Sibson (Paper 25) represent important steps in this direction. We have reprinted the latter paper here because it appears to be the first real attempt to define a rigorous theoretical framework for a general clustering system (the interested reader may wish to consult their more recent text on mathematical taxonomy for additional discussions on this subject). There is little doubt that cluster analysis will continue to be expanded and quantified, especially in the direction of graph theory (e.g., Rohlf, 1973) and we alert the reader to these future developments.

BIBLIOGRAPHY

Carmichael, J., J. A. George, and R. S. Julius. 1968. Finding natural clusters. *Syst. Zool. 17:*144–150.

Crovello, T. J. 1968. Key communality cluster analysis as a taxonomic tool. *Taxon 17:*241–258.

Eades, P. C. 1965. The inappropriateness of the correlation coefficient as a measure of taxonomic resemblance. *Syst. Zool. 14:*98–100.

Edwards, A. W. F., and L. L. Cavalli-Sforza. 1965. A method for cluster analysis. *Biometrics 21:*363–375.

Fleiss, J. L., and J. Zubin. 1969. On the methods and theory of clustering. *Multivariate Behav. Res. 4:*235–250.

Hartigan, J. A. 1967. Representation of similarity matrices by trees. *J. Amer. Stat. Assoc. 62:*1140–1158.

Jardine, N., and R. Sibson. 1971. *Mathematical Taxonomy.* John Wiley & Sons, Inc. New York. 286 pp.

Johnson, S. C. 1967. Hierarchical clustering schemes. *Psychometrika 32:*241–254.

Lance, G. N., and W. T. Williams. 1967. A general theory of classificatory sorting strategies: I. Hierarchical systems. *Computer J. 9:*373–380.

Moss, W. W. 1968. Experiments with various techniques of numerical taxonomy. *Syst. Zool. 17:*31–47.

Orloci, L. 1967. An agglomerative method for classification of plant communities. *J. Ecol. 55:*193–206.

———. 1968. Information analysis in phytosociology: partition, classification, and prediction. *J. Theoret. Biol. 20:*271–284.

———. 1969. Information theory models for hierarchic and nonhierarchic classifications, in *Numerical Taxonomy* (A. J. Cole, ed.). Academic Press, New York. 324 pp.

———. 1972. On objective functions of phytosociological resemblance. *Amer. Midland Naturalist 88:*28–55.

Rogers, D., and T. Tanimoto. 1960. A computer program for classifying plants. *Science 132:*1115–1118.

Rohlf, F. J. 1973. Algorithm 76 hierarchical clustering using the minimum spanning tree. *Computer J. 16:*93–95.

———, and R. R. Sokal. 1965. Coefficients of correlation and distance in numerical taxonomy. *Univ. Kansas Sci. Bull. 45:*3–27.

Rubin, J. 1966. An approach to organizing data into homogeneous groups. *Syst. Zool. 15:*169–182.

Sneath, P. H. A. 1969. Evaluation of clustering methods, in *Numerical Taxonomy* (A. J. Cole, ed.). Academic Press, New York. 324 pp.

Sokal, R. R., and P. H. A. Sneath. 1963. *Numerical Taxonomy.* W. H. Freeman, San Francisco, 354 pp. Revised 1973 by Sneath and Sokal.

Tryon, R. C. 1939. *Cluster Analysis.* Edwards Bros., Ann Arbor, Mich.

———. 1958. General dimensions of individual differences: cluster analysis vs. multiple factor analysis. *Educ. Psychol. Measurement 18:*477–495.

———, and D. G. Bailey. 1970. *Cluster Analysis.* McGraw-Hill Book Company, New York. 347 pp.

Ward, J. 1963. Hierarchical grouping to optimize an objective function. *J. Amer. Stat. Assoc. 58:*236–244.

Williams, W. T. 1971. Principles of clustering. *Ann. Rev. Ecol. Syst. 2:*303–326.

Zubin, J. 1938. A technique for measuring likemindedness. *J. Abnormal Social Psychol. 33:*508–516.

17

Reprinted from *Univ. Kansas Sci. Bull.*, **38**, p. 2, (22), 1409–1438 (1958)

A Statistical Method for Evaluating Systematic Relationships [1]

BY

ROBERT R. SOKAL and CHARLES D. MICHENER [2]

Department of Entomology

University of Kansas, Lawrence

ABSTRACT. Starting with correlation coefficients (based on numerous characters) among species of a systematic unit, the authors developed a method for grouping species, and regrouping the resultant assemblages, to form a classificatory hierarchy most easily expressed as a treelike diagram of relationships. The details of the method are described, using as an example a group of bees. The resulting classification was similar to that previously established by classical systematic methods, although some taxonomic changes were made in view of the new light thrown on relationships. The method is time consuming, although practical in isolated cases, with punched-card machines such as were used; it becomes generally practical with increasingly widely available digital computers.

INTRODUCTION

The purpose of the study reported here was to develop a quantitative index of relationship between any two species of a higher systematic unit, as well as to exploit such indices of association in the establishment of a satisfactory hierarchy. The authors became interested in the development of such a method when they attempted to find a technique for classifying organisms that was free from the subjectivity inherent in customary taxonomic procedure.

---

1. Contribution number 945 from the Department of Entomology, University of Kansas.

2. We wish to acknowledge the constructive criticism received in connection with this and related work from the following individuals who kindly gave their time to read and comment upon the manuscript: Paul R. Ehrlich, University of Kansas; Raymond B. Cattell, University of Illinois; Alfred E. Emerson, University of Chicago; Warwick E. Kerr, Universidade de São Paulo; Ernst Mayr, Harvard University; Louis L. McQuitty, Michigan State University; G. G. Simpson, American Museum of Natural History; Peter C. Silvester-Bradley, University of Kansas and University of Sheffield; and Paulo E. Vanzolini, Departmento de Zoologia, Secretaria de Agricultura, São Paulo. These persons, however, are not responsible for the opinions which we have expressed.

Acknowledgment is also due to the University of Kansas General Research Fund for assistance.

The systematic group chosen as a test of the feasibility of this undertaking was one consisting of 97 species of solitary bees in the family Megachilidae. This choice was made because one of us (C. D. M.) has made recent systematic studies of these insects, so that conclusions as to the relationships obtained by the usual systematic procedure could be compared with the results of the new method.

The findings of our study as well as the philosophical bases of our attempts at quantifying systematic relationships have been reported elsewhere (Michener and Sokal, 1957). In this paper we propose to describe in some detail the actual method employed, as well as our reasons for adopting it and for rejecting several alternate procedures. It is our intention to illustrate the procedures in sufficient detail so that persons with a limited knowledge of statistical methods will be able to follow our method. We expect our system to be applicable to most organisms, provided they exhibit a variety of characters, and the account to follow is consequently phrased in general terms. However, our practical illustrations are based on the bee group cited above in order to provide the reader with concrete examples.

A quantitative method of finding the relationship between two species must be based on a number of taxonomic characters in a manner similar to the traditional systematic approach. However, whereas the latter technique generally uses few characters and weights these quite unequally and subjectively, the former method employs numerous but unweighted characters. Our reasons for not weighting characters have been detailed in the companion paper (Michener and Sokal, 1957). In the absence of an objective criterion of character weight it seems best to rely on a large number of equally weighted characters. In our bee study we employed 122 characters per species; however, we feel significant results may be obtained from as few as 60 characters.

Our use of the word "character" will require some elaboration. In its commonest taxonomic usage, a character is any feature of one kind of organism that differentiates it from another kind. Thus the red abdomen of one bee is a character distinguishing it from another bee with the abdomen black. In this paper we use the word in a second connotation only; that is, as a feature which varies from one kind of organism to another. Now, to use the above example, abdominal color is the character, which occurs in two "states" or alternatives, red and black.

For each character the states were coded: 1, 2, 3, etc. In the bee study the number of states per character ranged from two to eight. Much variation in the number of states is undesirable from the point of view of the methods discussed below. In the study we undertook most characters had either three or four states. However, when variation exceeds desirable bounds it might be preferable to divide the character state codes by a common denominator or to normalize them.

The kinds of characters used in the bee study and the manner in which they were coded are discussed at length by Michener and Sokal (1957). The possible effect of parallelism is also treated in the same article. For purposes of the present paper the available data might be summarized as follows: we have records of a given number (n) of species. For each species we have k records, k being the number of characters considered in the study. The coded values for any character may range from 1 to 9 depending on the number of states in which this character occurs in the group under consideration. As was mentioned previously it is desirable to have the number of states not differ too widely for the various characters. While it is not necessary to limit the number of possible character states to nine, our particular computational setup was greatly facilitated by the use of a single digit code.

PROCEDURES

Character correlations and species correlations

Two obvious ways suggested themselves to the authors regarding a procedure for deducing relationships from the character states of a group of species. We could either correlate characters with each other or species with each other. Since both of these methods would lead to interpretable, although differing, results a brief discussion of the implications of the two approaches follows.

Sturtevant (1942) undertook a study of the genus *Drosophila* with objectives and procedures somewhat similar to ours. He recorded 33 morphological, cytological and life history characters for each of 56 species of *Drosophila* and two species of the genus *Scaptomyza*. In his aim to develop a classification "as free from personal bias as I could make it," Sturtevant set up two tables. The first was a table of the total number of differences with respect to the 33 chosen characters between any two of the 58 species. These give the degree of difference between the species concerned and are analogous to the complemental values of the "matching coefficient" discussed in the section on Choice of a Correlation Coefficient below.

A second table showed correlations between characters, expressed as two-way frequency distributions. By examining the three highest character correlations Sturtevant found that six species consistently fell into the exceptional classes of the two-way frequency distributions. They were the two *Scaptomyza* species and four species of *Drosophila* which he thereupon placed in separate subgenera. On the basis of the number of character differences between and within subgenera Sturtevant was able to confirm this classification and arrive at some ideas on the relationships and origins of the various groups. He also performed a similar analysis on 29 characters of 40 genera of flies (*Scatophaga, Conops* and 38 assorted Acalypterae) to establish the relations of the family Drosophilidae. Unfortunately the paper cited lists only summaries of the above tables and it is therefore difficult to compare Sturtevant's findings with ours.

Correlation between characters (R-technique in the idiom of the factor analysts) is the customary technique in biological and psychological studies involving correlational analysis. In character correlation matrices involving studies within one species each correlation represents the sum total of the common forces acting on any pair of characters. When analyzed by some method of factor analysis, the matrix customarily yields a so-called general size factor, a series of group factors affecting various groups of characters, and residual specific factors affecting single characters only. The foregoing is an example of a factor constellation involving morphological characters and is not necessarily the only possible constellation. As a matter of fact much psychometric work and the biometric papers by Howells (1951) and Stroud (1953) use the method of "simple structure" which *a priori* rejects solutions involving general factors.

Regardless of the constellation preferred, the factors common to two characters and causing them to be correlated could be visualized as developmental forces, genetic or environmental in the final analysis. The range of these genetic or environmental forces is dependent on the causes of variation within the sample of individuals studied. Thus a sample of individuals from an inbred, isogenic, line of animals would yield character correlations reflecting common nongenetic, physiological (*i. e.*, caused by microecological differences) factors only. Another sample comprising individuals from various races or subspecies would provide correlations based on common factors representing (1) genetic differences between individuals; (2) genetic differences between races; (3) nongenetic physiological differences between individuals and (4) nongenetic ecological differences between races. One of the authors (R. R. S.) has been able to accumulate a series of character correlation matrices from various organisms representing these levels of variation. Matrices on correlation of aphid characters within galls (clones) and between galls have been published (Sokal, 1952) while similar matrices on aphid correlations between localities and morphological correlations within and between strains of houseflies and *Drosophila* await suitable analysis and publication.

When the sample transcends the bounds of the species the factors behind a character correlation matrix take on new meaning: They now represent genetic divergence or the results of evolutionary processes. In the one case they were ontogenetic forces, in the other they are phylogenetic forces. This type of analysis was pioneered by Stroud (1953) who analyzed correlations of 14 characters for soldiers of 48 species and imagines of 43 species of the termite genus *Kalotermes.* He was able to interpret some factors extracted from his correlation matrices as recognizable evolutionary trends.

Another method of correlational analysis is called the transposed matrix method or the Q-technique (as compared with the R-technique of character correlations, discussed above).[3] It consists of correlations between individuals based on measurements of characters which they have in common. In psychology this involves correlations between persons based on scores for common tests which these persons have taken. In the Q-technique we are in effect dealing with the same kind of raw data as in the R-technique, but we compute the correlation coefficients by summing squares and products at right angles to the direction previously taken (or we transpose the matrix before computation which amounts to the same thing).

A Q-technique correlation coefficient in a study correlating individuals of one species represents common forces or factors acting on the two individuals concerned. In this case we cannot speak of the "sum total of common forces" as we could in the case of the

3. In a recent paper Cattell (1954) has suggested restricting the Q and R symbolism to studies involving factor analysis and proposed Q′ and R′ for studies, such as the present one, employing more superficial methods.

R-technique. Insofar as the characters used are indicative of the entire spectrum of potential variation of the individuals we can say that the resulting correlation coefficient is representative of the real affinity between two individuals. When scanned for clusters of high correlation coefficients the Q-type matrix reveals types of individuals which are similar. It is thus especially suited to classificatory problems. When subjected to factor analysis the resulting factors are now of a different nature. The general size factor has been lost and in its place we find a general taxonomic group factor which accounts for the overall correlations of all the individuals in the study.

When, as in the present study, the correlation is between species of a taxonomic unit the general factor is a general systematic factor denoting overall relationship within the systematic group. The species having the highest factor loading would be most representative of the group. Other factors would describe subgroups within the systematic unit and describe the relationships of these subgroups with each other and of the species to the subgroups. It should be clear from the above that for purposes of biological classification the relationships represented by a Q-technique matrix are more meaningful by far than are those of a R-technique matrix. Except for the above-mentioned work of Sturtevant (1942) which involved not correlations but character differences, the only Q-type study in systematics of which the authors are aware is in a publication by one of them (Sokal, 1958) containing factor analyses of selected portions of the present data. A number of the phytosociological coefficients of association and similarity can be considered as of the Q-type.

Psychologists have used Q-technique repeatedly (*e. g.*, Burt 1937, Stephenson 1936), although R-technique is still preferred in most studies. Cattell (1952) has listed 5 points of criticism of the Q-technique. It is appropriate that we discuss briefly their relation to the problems under study here. The first objection is that Q-technique loses the general size factor, yielding in its place a common species factor. This latter is claimed to be trivial by Cattell, and correctly so, for psychological work. However, in a matrix of correlations between species such a general systematic factor delineates the relation of individual species to the taxonomic group and indicates the proportion of the variance of each species explained by the general systematic factor.

Cattell's second objection to Q-technique is that it is unreasonable to assume simple structure in the factorization of a Q-matrix. The authors agree with this argument, but for the purposes of the present paper it is not important since they are not here undertaking a factor analysis. Furthermore, they feel that simple structure (*i. e.*, each factor having essentially zero effects on certain species) is not necessarily a very suitable constellation for many biological factorizations.

The third objection refers to a customary shortcoming of Q-matrices. They are based on few individuals and generalizations about the entire population are drawn from them. In this study, the matrix is of course of more than adequate size. Furthermore our conclusions are not intended to extend to species not included in our study.

It is true that the species recorded are an eclectic sample from those extant in the world today. On the other hand we are of course dealing with a sample obtained by natural selection from the multitude of species or specieslike entities that have existed since the origin of the four genera of this study. Hypotheses regarding these extinct species will be valid only insofar as recent species reflect the course of evolutionary history.

Another point in connection with the third objection is the number of characters employed. True relationships will become apparent only insofar as the characters adequately represent the sources of variation within the species.

A fourth objection relates to the lack of equivalence in recording and interpreting the factors from the Q- and R-matrices. It compares the relative permanence of psychological tests with the relative impermanence of persons. In this study we are confronted with characters and species varying in their relative permanence, but both equally permanent when based on the time scale of the scientist investigating them.

The fifth criticism, labelling the Q-technique as descriptive rather than predictive, again is invalid when applied to the present data. Since the purpose of the study is historically descriptive and one of our aims is to divide the population of species into categories, the technique's fault for psychological research becomes a virtue in our field of investigation.

There are two evolutionary situations under which it is important to examine the two types of matrices. The first might be referred to as breakage of correlation. It occurs when in a certain evolutionary line two characters that were correlated in ancestral lines and are still correlated in related lines become independent of each

13—8050

other. Under such conditions the R-matrix is a poor representation of the true relation between the two characters. There is no good way of representing such a correlation, close in one line, absent in the other. On the other hand a Q-matrix is not affected by such data.

Convergence of species for a number of characters is a second disturbing phenomenon. Here the R-matrix is not affected while the Q-matrix is affected if the convergent characters outweigh the nonconvergent ones in numbers.

We do not believe this is likely if an adequate number of characters is studied. In case of a preponderance of convergent characters and in the absence of paleontological data it is doubtful whether the systematists would be able to distinguish convergence from relationship by descent.

From a consideration of the above arguments it follows that given the objectives and material of the present study the Q-technique is to be preferred to the R-technique and the objections made by Cattell to the former method do not apply to our case. However, besides the theoretical reason for adopting the Q-technique as reflecting relationships between species there were several practical reasons for so doing. The problem of finding a suitable type of correlation coefficient between characters would have been formidable in view of the coding system adopted. Since some of the characters were present in two states only while others were present in as many as eight states, there would probably not have been any one type of correlation coefficient for all possible character combinations. A matrix based on correlation coefficients of different types would be far from desirable. Furthermore, uniformity of computational procedure was essential to efficient handling of the data by International Business Machines (IBM) equipment.

Not to be underestimated is the saving in computation resulting from adoption of a 97 x 97 species correlation matrix vs. a 122 x 122 character correlation matrix. The former requires the computation of only 4656 correlation coefficients while the latter would necessitate 7381 such coefficients.

The choice of a correlation coefficient

As a next step a suitable correlation coefficient had to be chosen to represent the correlations between species. There were serious considerations against the use of the product-moment correlation coefficient since the variables (species) are anything but normally distributed. Table 1 presents frequency distributions of state codes

TABLE 1

Frequency distributions of state codes for the characters of species 19, 56, 83 and 84.

State code	Sp. 19 f	Sp. 56 f	Sp. 83 f	Sp. 84 f
1	54	56	48	46
2	31	40	42	41
3	31	14	23	26
4	3	11	7	6
5	2	1	2	2
6				1
7	1			
Σ f	122	122	122	122

for four representative species. The distributions are highly asymmetrical. Those for species 19 and 56 approach Poisson distributions for their means when the class codes are reduced by one. Any interpretation of this agreement is dubious, however, in view of the variable number of states possible per character.

Other correlation coefficients were considered and rejected. The correlation ratio, η, is unsuitable since ${}_x\eta_y$ does not necessarily equal ${}_y\eta_x$. Tetrachoric r would have lost some of the information available because it would necessitate reducing all characters to two states. Furthermore the theoretical assumptions of underlying normality essential to correct application of the tetrachoric correlation coefficient cannot be defended for all characters.

Another method of demonstrating an association between species would be the very simple one of counting the numbers of matches in states for the 122 characters of any pair of species of bees and then dividing this number by 122, the highest possible number of such matches. The results for species 19, 56, 83, and 84 are shown on table 2 where these "matching coefficients" are compared with product-moment correlation coefficients. The "matching coefficients" are somewhat higher than the correlation coefficients but resemble them in relative magnitude. In spite of this fact, "match-

TABLE 2

"Matching coefficients" (below diagonal) and product-moment correlation coefficients (above diagonal) between species 19, 56, 83 and 84.

	19	56	83	84
19	X	.40	.37	.37
56	.52	X	.47	.38
83	.53	.61	X	.93
84	.50	.54	.87	X

ing coefficients" were not used since they have an unknown sampling distribution, they distort resemblances by counting a 3 to 4 mismatch the equal of a 1 to 7 mismatch, and finally they would have been harder to handle by the IBM equipment available to us.

Lacking a more suitable means of correlation we adopted the product-moment *r*, in spite of nonnormal distribution of variates and possible heteroscedasticity. Various ways of improving the distributions by means of transformations were tried. Table 3 shows the same correlation coefficients as the upper half of the matrix of table 2, but based on $\sqrt{X}$ and $\sqrt{X+.5}$ transformations. The slight differences obtained do not justify the extra computational labor involved.

We have already briefly touched on the desirability of coding the data in such a way as to put all character states on the same scale. In a character with two states the code 2 indicates a situation differing greatly from one given by code 2 in a character with 7 states. This problem is also encountered in Q-matrices in psychology where the scores for different tests are often not in comparable units. This situation is usually met by normalizing the rows (tests, or in our case characters) of the raw score matrix. The authors did not per-

TABLE 3

Product-moment correlation coefficients between species 19, 56, 83 and 84 based on variates coded as $\sqrt{X}$ (below diagonal) and as $\sqrt{X+.5}$ (above diagonal). Compare with uncoded product-moment correlation coefficients in table 2.

	19	56	83	84
19	X	.42	.36	.37
56	.42	X	.50	.41
83	.36	.51	X	.93
84	.37	.41	.93	X

form this transformation since (1) it would have removed the common systematic factor from the matrix of correlations and would thus have lowered the correlation coefficients considerably; (2) application of the character state codes does standardize the data to a certain extent because 76 percent of the characters have either three or four states and only 3 percent have six or more states; (3) although the additional labor of normalizing the variates would not have been excessive the amount of IBM work involved in computing correlation coefficients would have been prohibitive, since a one-digit code would not have sufficed for normalized data.

The authors are well aware that their methodology of coding and correlation could profit by refinement. It is, however, our point of view that in a pilot study of this nature such refinements are premature. Should the general method prove of value, significant results will surely emerge in spite of minor imperfections in technique.

Computation

The computation of a large matrix of correlation coefficients such as the 97 x 97 bee matrix presents serious technical difficulties. Only high speed electronic computing machines are able to perform this operation with real dispatch. At the time our bee data were being processed we had only punched-card tabulating machines at our disposal. It might be noted here that a computational operation of this magnitude cannot reasonably be undertaken without some automatic computing facilities. The equipment used by the authors is that available in the University of Kansas IBM laboratory: a card punch (type 26), a verifier (type 56), an accounting machine (type 402) and a reproducing machine (type 514).

The computational problem was simplified somewhat by the fact that the variates consisted of single digits only. This increased the number of variables that the machine could process simultaneously. Each IBM card represented a character with the state code of each species for the particular character listed in separate columns. Since there are only 80 columns per card, it was impossible to record all species on any one card. A different approach was therefore adopted and the card divided as follows:

Column 1—Project code
Columns 2-4—Character code number
Column 5—Deck code (explained below)
Columns 6-8—Left blank for possible subsequent use
Columns 9-44—Multiplier columns for 36 species
Columns 45-80—Multiplicand columns for 36 species.

The 97 species were divided into group I for species 1-36, group II for species 37-72 and group III for species 73-97. Since group III used only 25 columns another 5 columns were taken up by a repetition of data on species 1 through 5, which we used as a check on computational procedure. Six decks of 122 cards each, one card per character, were then prepared. The decks were constituted as follows:

Deck	Multiplier	Multiplicand
1	Group I	Group I
2	Group II	Group II
3	Group III	Group III
4	Group I	Group II
5	Group II	Group III
6	Group I	Group III

Different card colors besides a punched code were used to distinguish the decks.

By running these decks in succession through the tabulator we were able to reduce rewiring of the board to one half of what it would have been with the minimum number of decks (3).

The method of arriving at the Σx^2 and Σxy was the customary one of progressive digiting with interspersed "X-cards." Running time on the 402 tabulator was some 24 hours. Punching and verifying of the cards had taken a similar amount of time. Thus the preparation of the Σx^2 and Σxy for the entire matrix took about a week. These values were computed for a half-matrix only. However, a test deck and five test variables detected wiring errors and machine malfunction with a reasonable limit of safety.

The next step was the computation of the correlation coefficients. This was done by computers using desk calculators.[4] The matrix of squares and products was subdivided into manageable sections, 30 variables (species) square. All computations were checked by a different computer and, where possible, by different steps. The computational procedure employed was the customary L method.[5] It does not seem necessary to elaborate on the details of this method. Any good textbook of statistics will contain a section on the computation of a product-moment correlation coefficient. Furthermore, each computation center has its own setup for correlation coefficients depending on the capabilities of the machines and thus no general account need be presented here.

The correlation coefficients were calculated to four significant decimal places and entered on a matrix. Three decimal places would have been quite sufficient for this study; however four were computed in case later statistical work required greater refinement. Total computation time for this phase of the work was 160 man-hours. It should be emphasized that the time estimates given above refer to the relatively simple equipment available to us. Digital computers are now available which would handle the entire computation, from raw data to completed correlation matrix without human intervention in less than an hour. This would be only one-two hundredths of the time it took us to compute the same informa-

4. The writers at this point wish to express their appreciation to Misses Betty Becker, Marion Clyma, Jacqueline Johnson, Normandie Morrison, and Messrs. D. A. Crossley, Jr., Ralph Jones and Roger Price for their conscientious assistance with IBM work and desk computation.

5. $r_{xy} = L_{xy} / \sqrt{L_x}\sqrt{L_y}$, where $L_{xy} = N\Sigma XY - \Sigma X \Sigma Y$ and $L_x = N\Sigma X^2 - (\Sigma X)^2$, $L_y = N\Sigma Y^2 - (\Sigma Y)^2$.

tion! With every passing year electronic computers are becoming more efficient and more widely distributed. Thus the computational aspects of our method will become a progressively less important impediment.

Since the matrix of correlation coefficients was unwieldy (it also had to be subdivided into sections) and since further work with the correlation coefficients was contemplated, the latter were punched on 4656 IBM cards, one to a card. These cards were duplicated by means of the reproducing punch in order to obtain cards for a complete matrix of 9312 correlation coefficients. Information on these cards included matrix row and column numbers for the particular correlation, the coefficient with sign, and a class code for the coefficient. These class code numbers (1-22) represented 22 classes of a frequency distribution of the correlation coefficients arrayed in ascending order of magnitude with class intervals of .05. In addition, the cards contained codes for the relationship between the two species involved as evaluated by conventional systematic methods (by C. D. M.).

The correlation coefficients on punched cards have so far been put to the following uses: We have compiled a printed tape record of the full matrix, column by column, which has been very useful for reference and further computation. Another tape has been compiled giving a listing and frequency distribution of the correlation coefficients grouped in the 22 size classes. This tape has been of great value in various approaches to a classification of the relationships demonstrated by the matrix. A third tape lists the sums of the correlation coefficients, column by column. This has been necessary for the B-coefficient method briefly described below. A fourth tape presents a two-way frequency distribution showing the relation between correlation coefficients and the relationship code developed by conventional systematic methods. These tapes were prepared in a few hours running time from the correlation coefficient cards, which we still expect to use in a variety of ways.

The matrix of correlation coefficients

In the bee study the 4656 correlation coefficients computed in the above manner ranged in magnitude from —.0626 for the correlation between species 26 and 92, to .9747 for the correlation between species 43 and 44.[6] As was mentioned previously, a fre-

6. For lack of space the matrix cannot be reproduced here. Microfilm or IBM-tape or card copies can be obtained through the Secretary, Department of Entomology, University of Kansas. Lawrence.

quency distribution of these coefficients, grouped into 22 classes with class intervals of .05 was set up. The modal class showed a class mark of .38; this represents the most frequent class of correlation coefficients found between species in this study. However, a second mode was located at .78. This bimodality would indicate that we are dealing with two populations of correlation coefficients: those indicating close, possibly intrageneric relations and others representing more distant relations. Codes representing Michener's previous views on the relationships among the species were correlated with the above coefficients. The single correlation coefficient between the correlation matrix and Michener's codes was .80. It was encouraging to find that magnitude of the correlation coefficients in our matrix was apparently an estimate of systematic relationship as indicated by the previous classification.

Another way of examining these correlation coefficients is to study frequency distributions of the coefficients for any single species against all other species. By this means we were able to distinguish members of closely related groups of species from isolated species within a genus and these in turn from very isolated species representing monotypic genera or subgenera. For a detailed discussion and illustrations of this procedure, the reader is referred to Michener and Sokal (1957).

The absence of significant negative correlations from our matrix requires some discussion. Q-technique matrices of correlations between people (based on psychological tests) are quite likely to yield such correlations. If there are distinct, antithetical types of persons represented in the matrix, such as extroverts and introverts, it is likely that a high score for one type will be a low score for the other and vice versa. In our case evolutionary progress may be represented by either an increase or a decrease in state codes. In the majority of characters the supposedly primitive situation is an intermediate state code with two diverging evolutionary trends represented by the lower and higher code numbers. Furthermore, characters representing correlated trends were not necessarily coded along the same scale or in the same direction. It is clear that under such circumstances distantly related forms are likely to be uncorrelated rather than negatively correlated.

The search for group structure

The matrix of correlation coefficients between species can be put to a variety of uses and the analysis reported below represents merely an initial effort at an exploitation of the data. The correlation coefficients serve as an absolute measure of relationship between any two species in our study, limited only insofar as the characters chosen do not represent the total correlated variation of the two species.

The search for structure among the correlation coefficients of the matrix is of course no different in aim from the search by the systematist for a natural system in an array of species. Such a system consists of a hierarchy of groups. Various methods can be used for discovering a hierarchy in data such as ours. A customary, rather simple device of the psychometrician is so-called "cluster analysis," developed to a fine art by Tryon (1939).

A concise description of the procedure (the ramifying linkage method) is given in Cattell (1944) and Thomson (1951). Because of the simplicity of the procedure, cluster methods are used extensively, although Cattell (1944, 1952) and others have pointed out that cluster analysis cannot be considered a substitute for the more involved factor analytic methods. Attempts to employ cluster analysis for finding structure in our matrix were only partially successful, since the resulting clusters were partly overlapping, *i. e.*, a given species might be a simultaneous member of two clusters. This makes good sense for intermediate forms in an abstract scheme of relationships. In a systematic hierarchic classification, however, groups at the same level have to be mutually exclusive for practical as well as for theoretical reasons, except for low level groups exhibiting reticulate evolutionary pattern (rare above the species level in animals). A further reason for the unsuitability of cluster analysis is the complexity of the clusters as more species are added to them. Although clusters are therefore not convenient in an initial search for structure, the diagram of relationships established by methods to be described below could be easily recognized in the clusters outlined by cluster analysis. A method essentially similar to cluster analysis is the ρ- group and ρF-group method of Olson and Miller (1951) applied to three paleontological R-technique matrices. It suffers from the same drawbacks as cluster analysis.[7]

7. After the present research and manuscript had been completed one of us (R. R. S.) became acquainted with the psychometric work of Professor Louis L. McQuitty of the Michigan State University, who in recent years has developed a whole battery of refined cluster methods (McQuitty, 1955, and a series of papers in press in The British Journal of Statistical Psychology, Educational and Psychological Measurement, and Psychological Monographs). Several of these papers deal with psychological problems which are closely related to those of biological classification. One of the methods invented by McQuitty bears a close resemblance to our variable group method developed below. It is interesting (as well as reassuring to us) that workers in different fields had unknown to each other developed some of the same formulations. We hope to try some of McQuitty's other methods on our material. They have the advantage of simplicity and can be programmed for electronic computation without much difficulty. Indeed the time may not be far off when computation for a study such as our bee work will be a minor matter routinely handled by a computing center in a very few hours and the remaining problem will be the collection of data for the machine and the interpretation of the voluminous answers that are produced.

As a technique for grouping the species we experimented extensively with the coefficient of belonging (B-coefficient) of Holzinger and Harman (1941). It is the sum of the correlations among the members of a group divided by the sum of the correlations of these group members with the other variables (species) of the study.

Results of our B-coefficient analyses for the bees were reasonably good, as judged by the previous classification and by our subsequent investigations. There was one main drawback, however. Large species groups showed a lack of structure and relatively low B-coefficients which would make the species in these groups appear a good deal less related to one another than members of groups of two or three species. The cause of this phenomenon is not hard to find. In large species groups the denominator of the B-coefficient would include high correlations due to correlations of group members with numerous other prospective members not yet included in the group. This would tend to depress the B-coefficient values. By the time all such members have been admitted to the group, it has become so large that even the admission of a relatively unrelated variable will effect the B-coefficient only slightly.

In view of the disadvantages of the B-coefficient we developed our procedure which is presented below in a general manner together with some of the reasons for its adoption. This presentation is followed by a detailed step-by-step account of the computational procedure for readers who wish to become more familiar with it.

A nucleus of a group was established, using the two species having the highest coefficient of correlation. Then species would be added to this nucleus, one at a time, always adding first the species having the highest average correlation with members of the group. The limit of the groups could be found by decreases ($\overline{L}_{n+1} - \overline{L}_n$) in the level of the average correlation $\overline{L}_n$, where the subscript refers to the number of members in the group. As in the B-coefficient a significant drop is empirically determined since sampling distributions of average correlations, such as $\overline{L}_n$, are unknown. By developing first lower groups (species groups), then by the same method grouping these into larger groups (sometimes subgenera), and these into still larger or higher groups, etc., it has been possible to develop a hierarchy of groups for which the diagram of relationships (figure 1) can serve as a representative. Each number in this figure represents a different species; for a list of the species concerned see Michener and Sokal (1957).

Since $\overline{L}_n$ is not amenable to rigorous statistical treatment it was decided to recompute correlation coefficients (using Spearman's sum

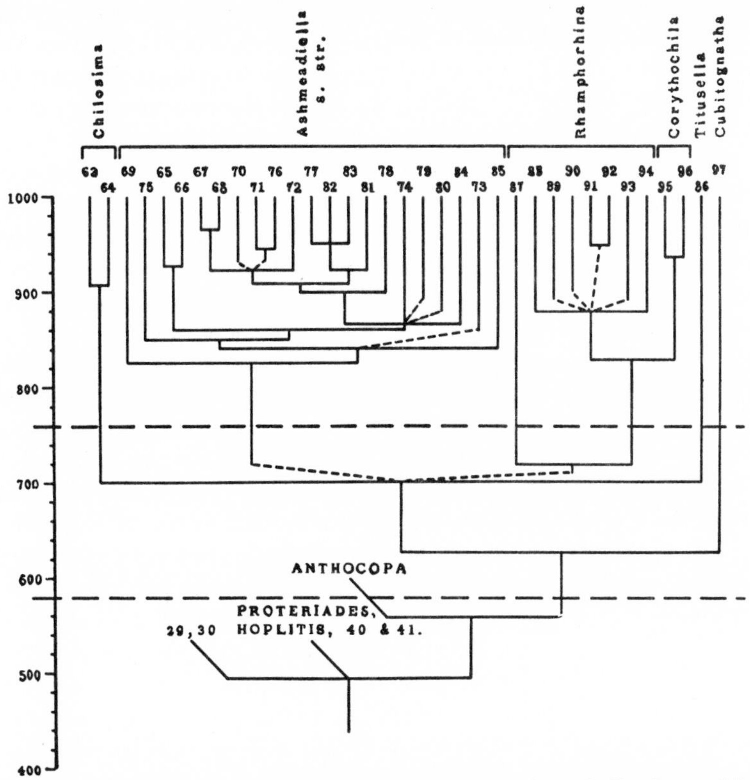

FIG. 1. Diagram of relationships for the genus *Ashmeadiella* obtained by the weighted variable group method. Ordinate: magniture of correlation coefficient multiplied by 1000. Exact correlations between any two joining stems can be found by reading the value on the ordinate corresponding to the horizontal line connecting the stems. This value becomes approximate and maximal in cases of multifid furcations. Broken lines used where more than three stems join are for convenience only; the horizontal connecting line has the same significance as elsewhere. "Roofs" over species numbers at the summits of the lines delimit subgenera containing more than one species, as based on C. D. M.'s previous findings and not on this study. Generic names are in small capitals. The horizontal broken lines are not relevant to the present account; they are explained by Michener and Sokal (1957).

of variables method) after the group limits at each hierarchic level had been reached. Thus we returned at the end of each grouping procedure to a new matrix of correlation coefficients about which confidence statements might be made. Two further considerations in the final choice of a method for grouping remain to be mentioned:

We might have admitted only one new member for each group at a given hierarchic level, thus obtaining a diagram of relationships consisting of bifurcations only. We have called this method the pair-group method as contrasted with the variable-group method, where any number of new members can be admitted to the group at any one hierarchic level, the limit of the group being determined by a significant drop in $\bar{L}_n$. The pair-group method has some theoretical justification in that much evolutionary ramification is believed based on speciational processes involving the splitting of one species into two. However, there must also occur some speciation as a result of the splitting of a species into more than two isolates and, on the assumption of equal evolutionary rates for these new lines, the pair-groups method would fail to represent the true situation. Moreover, many of the groups must be markedly different, not merely because of divergence, but because of extinctions of intermediates. A group might be broken into any number of different subgroups by different extinctions. Furthermore an empirical study of this method (see fig. 2 for an analysis of relationships in the subgenera *Chilosima* and *Ashmeadiella* by the pair-group method and compare with the left side of fig. 1 for the variable-group method) demonstrates that in spite of the pair-group device we are forced into multified furcations by drops in $\bar{L}_n$ too small to plot or by temporary

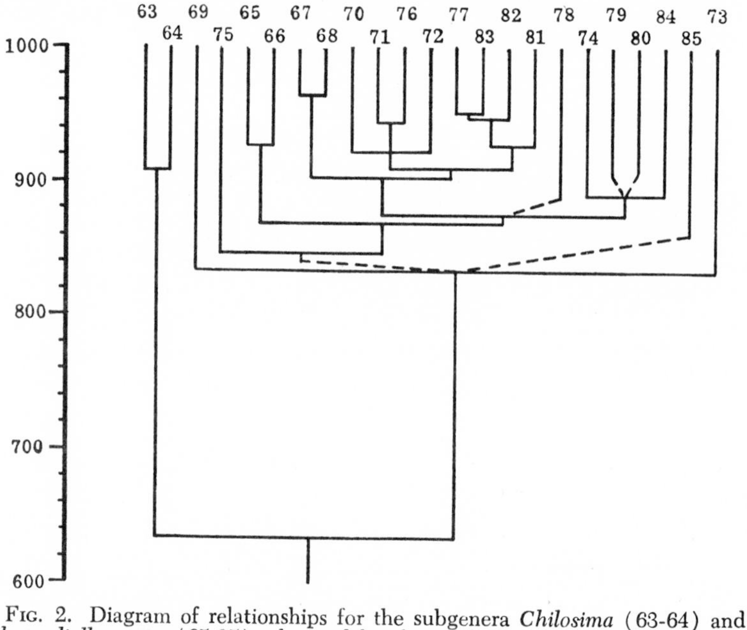

FIG. 2. Diagram of relationships for the subgenera *Chilosima* (63-64) and *Ashmeadiella s. str.* (65-85), obtained by the method of pair-groups, *i. e.* diagrams would ideally consist of bifurcations only. Stems have been weighted. Explanatory comments as for figure 1.

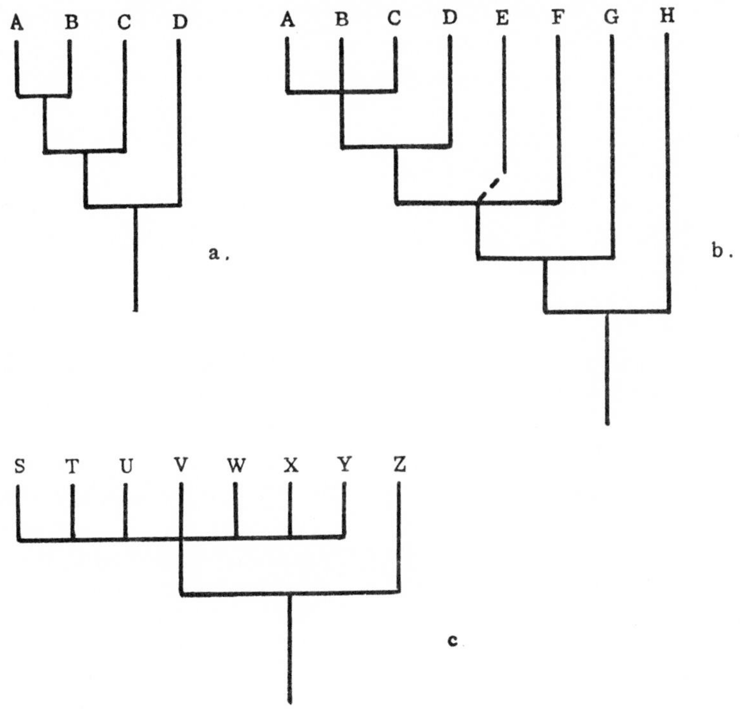

FIG. 3. Hypothetical diagrams of relationships to illustrate effects of different methods of weighting stems. For explanation see text.

reversals of $\overline{L}_n$ values, discussed below. Thus the variable-group method was adopted as the more reasonable and flexible of the two.

A second consideration is how to weight the variables during the recalculation of the correlation matrix after each grouping procedure. A simple diagram (fig. 3a) will make this issue clear. A and B represent the two species with the highest correlation coefficient. The $\overline{L}_n$ for C against A and B is significantly below r_{ab}, so that A and B are represented as being closer to each other than they are to C. When studying the relation of a fourth species D with group ABC we face the following problem: Should we calculate the correlation of ABC against D with A, B and C equally weighted or should we weight A = B and AB = C? Rephrased biologically, the problem is whether to relate species D with the homogeneous group ABC, or with the stem AB-C, where C carries as much weight in determining the relation with D as do A and B together. Although in a simple case, such as the one described above the two alternatives may not produce very different results, in a situation such as depicted in fig. 3b species H might be weighted as $1/8$ of the group A-H, or $1/2$, depending on the system adopted. Similarly species B would be weighted $1/8$ in the former case but only $1/72$ in the latter case. When dealing with fairly large groups the second method would therefore reduce the weight of the early admitted members and increase the weight of those species admitted later.

The same problem is found in a situation such as shown in fig. 3c. By the first method species T is weighted $1/8$, by the second method it is weighted only $1/14$. Neither of the two methods is entirely satisfactory. By method one we are reducing the importance of species H and Z in representing groups A-H and S-Z respectively. If the relationship diagrams of figures 3b and 3c depict true phylogenetic relationships, then H and Z should represent half of their respective lines regardless of subsequent diversification in the other halves. On the other hand giving relatively greater weight to single late arrivals also gives heavier weight to specialized features of such species and thus would tend to distort the relational pattern, while specializations in the diversified branch of the stem tend to cancel each other, permitting a better average picture of the groups to emerge. The optimal system of weighting would be one between these two extremes, weighting each species according to its number of generalized and specialized features. This is clearly impossible without renewed introduction of a subjective element into our procedure. We therefore adopted the second method, *i. e.*, the weighting of new members as equal to the sum total of all old group members, thinking it to be the less objectionable of the two. We feel that this method will represent stems more correctly and that bias introduced by specializations of late joiners will be kept down by the large number of characters considered in our study.

We are reassured in our decision by the results of a comparative study on the subgenera *Chilosima* and *Ashmeadiella*. Figure 4 shows the results of a variable-group analysis of these subgenera by weighting method one, while results by method two can be seen in the left side of the diagram of fig. 1. General agreement as to relationships and level of furcations is very good. The main difference between the two diagrams is that in method one group 77-81 [8] first receives 79 before receiving group 67-72, 78 and 84, while in method two it first receives group 67-72, then 78 and then 79 among others.

8. In the interest of brevity groups will be identified by their leftmost (in the diagram) and rightmost members with a dash separating the two. Thus 77-81 means group 77, 82, 83, 81. It clearly does *not* include all species ranging in number from 77 through 81, and includes some beyond that numerical range.

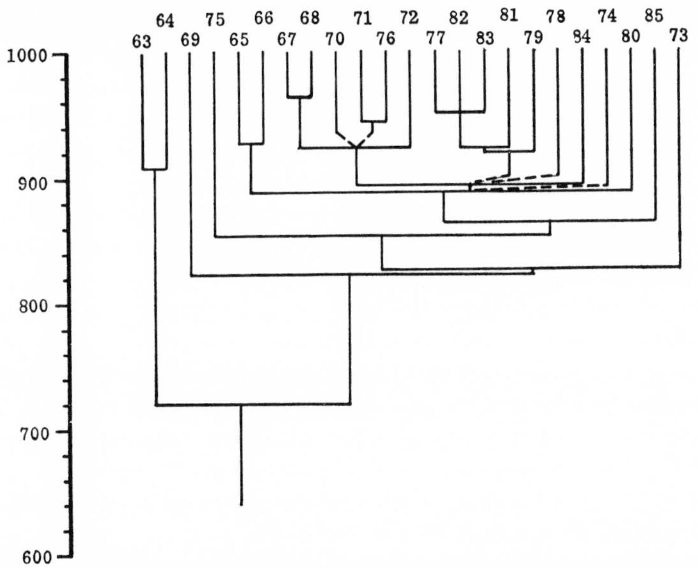

FIG. 4. Diagram of relationships for the subgenera *Chilosima* (63-64) and *Ashmeadiella s. str.* (65-85), obtained by the variable group procedure under weighting method one, *i. e.*, equal weights for all stems. Explanatory comments as for figure 1.

Careful examination of the original correlation coefficients makes the reasons for these differences clear. Group 77-81 is closer to 79 than to 78 except that 81 is closer to 78 than to 79. Also 81 is closer to 67-72 than are 77, 82, and 83. Therefore in method two, where 81 receives as much weight as 77, 82 and 83 together, 67-72 joins the nuclear group first. This is also partly due to the fact that unequal weighting of the species in the 67-72 group favors those close to the 77-81 group. Since 78 is closer to 67-72 than 79, the latter, while originally quite close to 77-81 is now temporarily delayed and 78 joins the combined group 67-81 before 79 does. These relations are at too low a phyletic level to be included in the original diagram of relationships drawn by C. D. M. who feels that there is little that can be obtained from classical systematic studies of these species to suggest whether method one or method two is preferable. In view of the small over-all differences between the two methods and especially in view of the fact that the lines concerned all join by either method with a difference in correlation coefficients of less than .06, it may well be that we have made too great an issue of the matter. In any case we feel confident that weighting method two will present us with a reasonably bias-free picture.

The Weighted Variable Group Method

It was thought advisable to give a detailed account of our method in order to enable readers to repeat the operations should they so desire. The subgenera *Chilosima* and *Ashmeadiella,* which have been used as a testing group before, will serve as an illustrative example. These subgenera include species 63 through 85 (see figure 1).

Correlation coefficients among these 23 species are shown in table 4. All values are significant with probability values of less than one percent. The highest correlation coefficient among these species is .965 for 67 x 68.[9] This is also the highest correlation involving either of these two species. The next to enter group 67-68 is species 70 which has the greatest average correlation ($\bar{L}_n$ = .892) with 67 and 68 since 67 x 70 = .896 and 68 x 70 = .889. No other species in the study has as high an average correlation with 67-68, as can be learned from a few trials. We established empirically, as a result of numerous trials, that a drop in $\bar{L}_n$ of .030 gave a satisfactory limit for groups; therefore 70 is not to be admitted to group 67-68 at this particular time. Another high correlation involving species other than 67 and 68 is 77 x 83 = .951. This is also the highest correlation for the two species concerned. Next to join this nucleus is species 82 with an $\bar{L}_n$ value of .936. The drop is less than .030; therefore 82 is admitted. Next to join is species 79 with an $\bar{L}_n$ value against 77, 82 and 83 of .905. There is now a significant drop from the previous $\bar{L}_n$ value and 79 is excluded for the time being. Drops in $\bar{L}_n$ are always measured from the previous $\bar{L}_n$, not from the initial $\bar{L}_n$. Our second group is therefore 77-83. In a similar manner we established groups 63-64, 65-66 and 71-76, each consisting of only two species.

So far only 11 species out of the 23 of the study have been placed into groups. A systematic survey was then made of the remaining 12 species to see if any group had been missed. For example, examination of species 69 revealed that its highest correlation was with species 72 (69 x 72 = .820). However, this latter value was not the highest correlation for 72, since 72 x 76 = .904. Thus 72 might eventually join the group containing 76, and 69 might join the group containing 72, both of which events came to pass at a later stage of the analysis. At the present time, however, species 69 and 72 are left unattached to any group. Similarly the remaining ten species in the study were shown not to belong to any nuclear group. To

9. We shall use this symbolism in place of the more formal $r_{67\text{-}68}$

TABLE 4—Matrix of Correlation Coefficients Among Species 63 to 85
(Extracted from larger matrix of 97 species)

	63	64	65	66	67	68	69	70	71	72	73	74	75	76	77	78	79	80	81	82	83	84
63.....	XXX																					
64.....	.908	XXX																				
65.....	.652	.653	XXX																			
66.....	.624	.623	.927	XXX																		
67.....	.692	.681	.806	.771	XXX																	
68.....	.682	.672	.812	.767	.965	XXX																
69.....	.523	.511	.725	.676	.775	.817	XXX															
70.....	.752	.718	.860	.832	.896	.889	.772	XXX														
71.....	.685	.725	.831	.793	.864	.858	.728	.897	XXX													
72.....	.669	.710	.859	.843	.849	.870	.820	.889	.902	XXX												
73.....	.560	.611	.699	.620	.739	.719	.706	.727	.779	.789	XXX											
74.....	.582	.637	.803	.731	.770	.758	.697	.748	.805	.779	.823	XXX										
75.....	.569	.611	.786	.777	.737	.754	.771	.712	.771	.847	.683	.796	XXX									
76.....	.697	.741	.837	.809	.883	.868	.734	.922	.945	.904	.751	.791	.752	XXX								
77.....	.566	.608	.826	.784	.804	.833	.764	.780	.809	.861	.767	.862	.838	.798	XXX							
78.....	.677	.719	.762	.718	.814	.792	.704	.848	.893	.811	.772	.762	.693	.895	.802	XXX						
79.....	.567	.610	.830	.763	.748	.763	.768	.728	.764	.817	.772	.832	.826	.769	.926	.800	XXX					
80.....	.643	.605	.822	.769	.787	.757	.716	.783	.828	.786	.789	.865	.802	.800	.863	.822	.852	XXX				
81.....	.635	.679	.822	.788	.827	.847	.776	.842	.873	.866	.805	.834	.780	.872	.885	.883	.857	.816	XXX			
82.....	.617	.661	.853	.783	.793	.816	.787	.821	.859	.888	.809	.857	.827	.850	.925	.844	.897	.855	.923	XXX		
83.....	.589	.631	.866	.821	.815	.819	.738	.823	.861	.855	.762	.887	.821	.843	.951	.838	.891	.894	.907	.948	XXX	
84.....	.542	.584	.794	.747	.725	.738	.659	.705	.740	.741	.667	.845	.766	.737	.886	.725	.852	.849	.788	.851	.926	XXX
85.....	.555	.605	.759	.695	.790	.819	.800	.774	.787	.854	.832	.843	.763	.784	.829	.723	.778	.762	.859	.849	.827	.744

set up one of the latter we required a correlation coefficient which was the highest one for both participating species (*i. e.*, the reciprocally highest correlation).

After the groups had been delimited a new correlation matrix was computed considering the newly formed groups as single variables, *i. e.*, the previous matrix of 23 variables (matrix 1) was reduced to one of 17 variables (matrix 2). It is self-evident that the only correlation coefficients in need of recomputation were those involving new groups. Correlations involving only species that had remained single were not altered in any way. As a matter of fact, a procedure was devised by means of which the correlations were not even recopied, but variables joined into groups were crossed out and the new group variables were entered along the margins of the old matrix. The actual computational procedure is quite simple and considerably less complicated than the computations for finding the original correlation coefficients. It is described in the paper by its originator (Spearman, 1913) and also by Holzinger and Harman (1941). Let us illustrate this method by computing the correlation between groups $(63)_1$ and $(67)_1$[10]. The general formula for this computation is

$$r_{q.Q} = \frac{\Box qQ}{\sqrt{q + 2\Delta q}\ \sqrt{Q + 2\Delta Q}}$$

where $\Box qQ$ is the sum of all correlations between members of one group with the other group, Δq is the sum of all correlations between members of the first group, ΔQ is a similar sum between members of the second group, q is the number of species in group one and Q the number of species in group 2. Thus in this particular case $\Box qQ$ equals $(\overline{63 \times 67}) + (\overline{63 \times 68}) + (\overline{64 \times 67}) + (\overline{64 \times 68}) = .692 + .682 + .681 + .672 = 2.727$; Δq in this case equals only $(\overline{63 \times 64}) = .908$ while ΔQ equals $\overline{67 \times 68} = .965$, since each of these groups consists of two species only. In cases where a group consists of 3 species, for example, the Δ term consists of the sum of $(1 \times 2) + (1 \times 3) + (2 \times 3)$. In the present case $q = Q = 2$ species. Substituting into the formula given above:

$$\overline{(63)_1 \times (67)_1} = \frac{2.727}{\sqrt{2 + 2(.908)}\ \sqrt{2 + 2(.965)}} = .704$$

10. The notation $(63)_1$, refers to the group of species formed in matrix 1, the lowest numbered member of which is species 63, *i. e.*, to group 63-64. Similarly $(67)_1$, refers to 67-68, and $(77)_1$, to 77-82-83.

These computations can be set up in a systematic manner and are then neither particularly complicated nor time consuming. In the special case where we wish to calculate the correlation coefficient between a single species (x) and a new group (q), the formula is amended as follows:

$$r_{x \cdot q} = \frac{\Sigma r_{x \cdot q}}{\sqrt{q + 2\Delta q}}$$

An illustration is the correlation of species 69 with group $(77)_1$. $\Sigma r_{x \cdot q}$ equals $(69 \times 77) + (69 \times 82) + (69 \times 83) = .764 + .788 + .738 = 2.290$, while $\Delta q = (77 \times 82) + (77 \times 83) + (82 \times 83) = .925 + .951 + .948 = 2.824$.

$$\text{Then } 69 \times (77)_1 = \frac{2.290}{\sqrt{3 + 2(2.824)}} = .779$$

In such a manner a new 17 x 17 correlation matrix (matrix 2) was constituted. From this point on the species groups $[(63)_1, (65)_1, (67)_1, (71)_1,$ and $(77)_1]$ were tested as though they were single species.

Once matrix 2 had been computed the identical grouping procedure was followed. Group $(71)_1$ had a mutually highest correlation with species 70 at .923. They were then joined by 72 and group $(67)_1$ at $\bar{L}_n$ levels of .903 and .885 respectively. These affiliations of 72 and $(67)_1$ were also their highest correlations. The next prospective joiner was species 81 at $\bar{L}_n = .859$, *i. e.* not quite the established drop of .030. However species 81 had highest relations not with the previous species but with group $(77)_1$ with a correlation of .923. Therefore it was excluded from consideration as a candidate for the earlier group and the runner up, species 78, used instead. The latter gave an $\bar{L}_n$ value of .844, clearly a significant drop from .885. Species 81 meanwhile was used in a nucleus of a new group 77-81$[(77)_2]$. Situations such as the above were the exception. In general the relations and choices were entirely straightforward and could be left to the discretion of the computing assistants.

At the end of each grouping procedure the remaining single variables were checked to avoid missing groups with low correlations between members. With each grouping procedure the matrix of correlation coefficients became smaller and the job of recomputation less. The weighting procedure adopted by us was automatic in that all correlation coefficients used were from the previous matrix and not the initial one. It took eleven matrices to obtain a single group out of the 23 species of the two subgenera. This amount of work could have been reduced by raising the minimum recognized difference in $\bar{L}_n$ level above .030 but there would have been a resulting loss of detail in the diagram of relationships. Conversely, however, reducing the recognized difference below .030 would not have increased the meaningful detail, since even .030 was too small to prevent the occasional reversal of *r* values discussed below.

Once computed, the relations were represented as diagrams of relationships as in figure 1. The ordinate at the left of each diagram is graduated in units of 1000 x *r*. The correlations between any joining stems in the diagram can be read by measuring the level along the ordinate of the horizontal line connecting the stems. Thus species 63 and 64 are correlated at a level of .908, while group 63-64 is related to group 67-72 at .702. Furcations involving more than three lines are shown by broken lines converging on the midpoint of the horizontal line as in group 88-96 of the above figure. The tops of the figures are at a level of 1000 (correlation of 1) since obviously each species is perfectly correlated with itself.

In cases of groups of only two stems the $\bar{L}_n$ level corresponds to the correlation coefficient of the two stems. When more than two stems join to form a group the highest $\bar{L}_n$ level was graphed for all group members. Thus while 77 x 83 equals .951, $\bar{L}_n$ for 82 against 77 and 83 equals .936. The group of these three species (77-83) is shown related at levels .951. Occasionally the correlation coefficients for the same group in successive matrices will rise a little. Thus in this same figure groups 63-64, 69-85, 87-96, and species 86 are shown joining at .702. The first three groups actually joined at .671, but species 86 which joined their group at the next matrix did so at level .702. This type of situation, which occurred infrequently, might lead one to express concern about the validity of the method, since regular decreases in levels of correlation coefficients and $\bar{L}_n$ values are expected. However, it can be shown from Spearman's formula for the correlation of sums of variables that slight increases in the levels of correlation coefficients of the sums of variables above the correlations of their component variables are possible. For example, if A and B have formed the nucleus of a group at $r_{a \cdot b} = .9$ and C is about to join them, then by the rules of the variable group method both $r_{a \cdot c}$ and $r_{b \cdot c}$ must $< r_{a \cdot b} = .9$. It can then be shown that $r_{(ab) \cdot c}$ must be $< .925$. Thus $r_{ab \cdot c}$, while it will usually be $< r_{a \cdot b}$, could be slightly more than .9. Similar situations can be shown to exist with larger-sized groups. The increases found by us were well below the mathematically possible limits. In all such cases the relations were represented as multifid furcations of all the stems involved in the reversal and at the highest of the several $\overline{L}_n$ levels considered.

In a successful method of studying relationships, the results of the analysis should be relatively independent of the number of species in the correlation matrix. If at least one species per species-group is included in a matrix, the ideal method of analysis should reproduce the diagram of relationships based on an earlier study of a larger matrix. If the method can be shown to produce similar results, the fact that our matrix contains only a sample from the population of species can be ignored with greater assurance.

We tested this question by subjecting the odd-numbered species in the entire genus *Ashmeadiella* to a weighted variable group analysis. This should give an adequate cross-section of relationships in that genus. Since some trends might be lost by exclusion of the

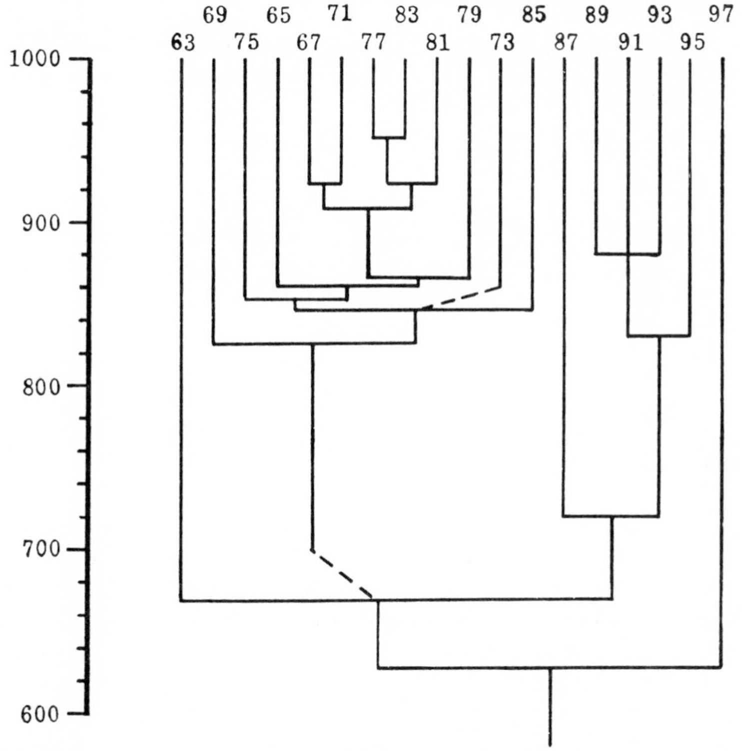

FIG. 5. Diagram of relationships predicted for odd numbered species of the genus *Ashmeadiella* on the basis of the relationship diagram of figure 1. Explanatory comments as for figure 1.

even-numbered species a special diagram of relationships was prepared from figure 1 by using only odd-numbered species. Figure 5 shows this predicted diagram of relationships.

Figure 6 shows the results of the weighted variable group analysis on the odd-numbered species. There is less structure in this dia-

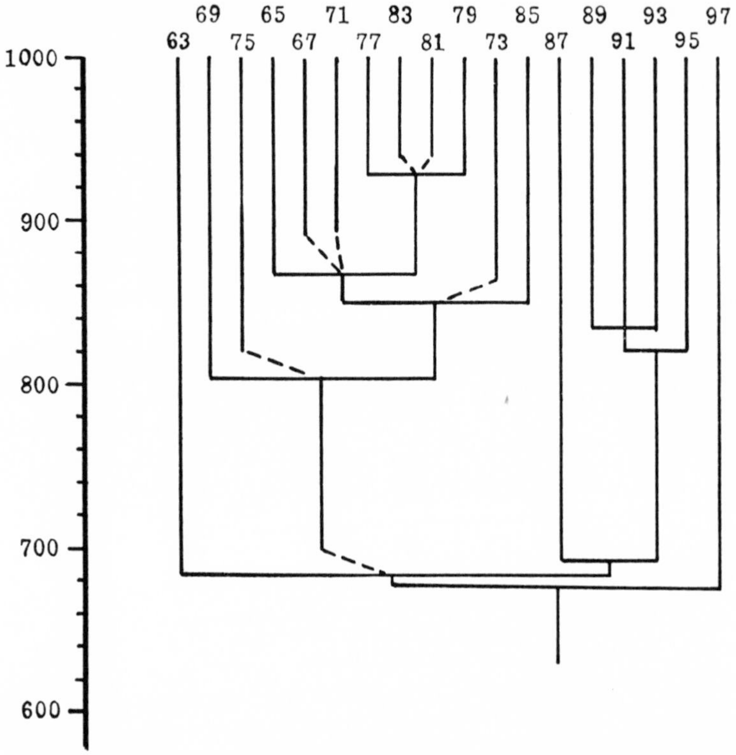

FIG. 6. Diagram of relationships obtained by an independent weighted variable group analysis of the odd-numbered species of the genus *Ashmeadiella*. Explanatory comments as for figure 1.

gram as compared with the predicted diagram of Figure 5. This was to be expected since some of the structure was based on relations involving the missing even-numbered species. In general, agreement between the two diagrams is very good, however, and we therefore feel reassured that the species left out in our study would not have changed greatly our diagrams of relationships.

CONCLUDING COMMENTS

A detailed discussion of the comparisons of our findings with the previous classifications of the four genera of bees is given by Michener and Sokal (1957). It will suffice here to state that general agreement was good but that a number of taxonomic re-evaluations seemed necessary as an outcome of these analyses. It should be remembered that while diagrams such as figure 1 may suggest phylogenies, in reality they only indicate static relationships. As indicated in the paper referred to, additional refinements were devised to give diagrams of relationships which we believe more nearly approach phylogenetic trees.

In view of these results we are encouraged to believe that, since the methods we have described are increasingly practical with the growing availability of high speed computers, this or similar schemes will be more widely utilized with different groups of organisms. Although the method we have described is a first attempt and would profit by either simplification or refinement, we believe it is a step toward reducing the subjectivity of systematic work, and therefore a step in the right direction.

LITERATURE CITED

BURT, C. L.
1937. Correlations between persons. Brit. Journ. Psychol., vol. 28, pp. 59-96.

CATTELL, R. B.
1944. A note on correlation clusters and cluster search methods. Psychometrika, vol. 9, pp. 169-184.
1952. Factor analysis. Harper & Brothers, New York, pp. xiii + 462.
1954. Growing points in factor analysis. Austral. Journ. Psychol., vol. 6, pp. 105-140.

HOLZINGER, K. U., and H. H. HARMAN.
1941. Factor analysis. Univ. of Chicago Press, Chicago, pp xii + 417.

HOWELLS, W. W.
1951. Factors of human physique. Amer. Journ. Phys. Anthrop., n. s., vol. 9, pp. 159-191.

MCQUITTY, L. L.
1955. A method of pattern analysis for isolating topological and dimensional constructs, Research Report AFPTRC-TN-55-62, Air Force Personnel and Training Center, Lackland Air Force Base, San Antonio, Texas, v + 38 pp.

MICHENER, C. D., and R. R. SOKAL.
1957. A quantitative approach to a problem in classification. Evolution, vol. 11, pp. 130-162.

OLSON, E. C., and R. L. MILLER.
1951. A mathematical model applied to a study of the evolution of species. Evolution, vol. 5, pp. 325-338.

SOKAL, R. R.
1952. Variation in a local population of *Pemphigus*. Evolution, vol. 6, pp. 296-315.
1958. Quantification of systematic relationships and of phylogenetic trends. Proc. Tenth Internat. Congress Entomology, Montreal, (in press).

SPEARMAN, C.
1913. Correlations of sums or differences. British Journ. Psychology, vol. 26, pp. 344-361.

STEPHENSON, W.
1936. The inverted factor technique. Brit. Journ. Psychol., vol. 26, pp. 344-361.

STROUD, C. P.
1953. An application of factor analysis to the systematics of *Kalotermes*. Systematic Zool., vol. 2, pp. 76-92.

STURTEVANT, A. H.
1942. The classification of the genus *Drosophila* with descriptions of nine new species. Univ. Texas Publ., no. 4213, pp. 1-51.

THOMSON, G. H.
1951. The factorial analysis of human ability. 5th ed. Houghton Mifflin Company, New York, pp. xv + 383.

TRYON, R. C.
1939. Cluster analysis. Edwards Bros., Ann Arbor, Mich., pp. 1-122.

18

Reprinted from *J. Ecol.* **47,** 83–101 (Mar. 1959)

MULTIVARIATE METHODS IN PLANT ECOLOGY

I. ASSOCIATION-ANALYSIS IN PLANT COMMUNITIES

BY W. T. WILLIAMS AND J. M. LAMBERT
Botany Department, University of Southampton

(*With two Figures in the Text*)

CONTENTS

I. INTRODUCTION

A. *General*

The valuable pioneer work by Goodall (1953) on the use of interspecific associations for sorting quadrats into groups is based on his definition of a homogeneous unit of vegetation as one in which all species-associations are indeterminate or non-significant. The method he eventually recommends, from empirical studies on data from the Australian Mallee, is to sort on the most abundant species involved in positive associations, pooling the *residuum* at each stage. Since statistical methods of this kind, however, inevitably require much large-scale computation, it is necessary both to examine the statistical foundations of any method proposed and to assess whether the ecological information obtained in fact justifies the time and labour involved. Furthermore, the analysis of complex communities can normally only be brought within the reach of the practising ecologist if the method is programmed for a digital computer, and this requirement must be borne continuously in mind if prohibitive expense in computing time is to be avoided.

It is implicit in the use of any sorting method that it is expected to reveal an underlying structure simpler than the raw matrix of associations. This is equivalent to assuming that the associations are not all independent; but if this is the case, the system is multivariate, and must be studied as such. With this in mind, we shall show, on theoretical grounds and by reference to two selected heathland communities, that a method of subdivision more effective than Goodall's can be devised: and we shall then attempt to assess the type of ecological information thereby obtained.

B. *Terminology, symbols and definitions*

(i) *Association* will be used throughout in its statistical sense. It is the traditional term where attributes rather than quantitative measurements are concerned, and we prefer to reserve the term 'correlation' (employed by Goodall) for use in its normally accepted contexts. The risk of confusion with phytosociological 'associations' is slight in the present paper.

(ii) *Community* is used as a convenient neutral term to denote any set of species growing together, without implying a particular statistical or ecological status.

(iii) Individual species will be represented by letters, capitals for presence and lower case for absence.

(iv) Class-symbols will be traditional—*e.g.* (abD) is the class defined by the absence of A and B, the presence of D, and irrespective of the presence or absence of C, E, . . . etc. The complete population is denoted by (N).

(v) Classes defined by one—(B), two—(Ce), . . . etc., letters will be said to be of the first, second, . . . etc., order.

(vi) We shall use $\Sigma\chi^2$ to denote the sum of all the significant χ^2 values associated with a particular species in the class under study, both positive and negative associations being included.

II. Theoretical considerations

A. *The homogeneous grouping*

Goodall's definition of a 'homogeneous grouping' rests on two quite independent concepts: associations may be absent either by being indeterminate—either or both species being present or absent in all quadrats—or by falling below significance. Such groupings can be sought only in presence-or-absence data; it is easily shown theoretically—and Goodall has shown empirically—that a true correlation, based on the *quantities* of the species concerned, cannot in general be removed by subdivision.

Three aspects of the basic definition require examination:

(i) *The use of negative associations.* Goodall rejects negative associations. Admittedly they can be treacherous; if the quadrat size is too small they are apt to appear simply by virtue of the fact that space occupied by one plant cannot be occupied by another. Nevertheless, their presence may greatly strengthen the analysis, for in a large and obviously heterogeneous area—a dry heath bordering a bog, or a wood bordering a pasture—the main subdivision is likely to be reinforced by strong negative associations. We therefore recommend reverting to the original suggestion of Tuomikoski (1942) and regarding positive and negative associations as of equal importance.

(ii) *Homogeneity of a group of associations.* Both Goodall (1953) and Greig-Smith (1957) suggest that at the $P = 0{\cdot}05$ level of significance 1 in 20 species-pairs may be expected to reach significance, and that a population can be regarded as homogeneous if this proportion is not exceeded. However, we have already pointed out (Sect. 1 A) that the method tacitly assumes that the associations are not independent; and in this case Goodall's criterion is equivalent to a test of the significance of an entire correlation matrix. Although many—mostly empirical—tests have been proposed for this purpose, no simple valid test concerning the number of significant entries is in fact known. We therefore prefer at this stage to remove *all* significant associations.

(iii) *Relative importance of the basic criteria.* The alternative criteria—indeterminacy or non-significance—by which a Goodall grouping is defined are not of equal value. A non-significant association has been tested and found wanting; an indeterminate association has not been tested, the circumstances being such that no association can be manifested. We shall therefore take the attitude that, if a choice has to be made, it is more important to reduce the level of associations than to render them indeterminate.

B. *The problem of subdivision*

(i) *The requirement.* The basic problem is to subdivide a population so that all associations disappear; but there will in general be a large number of alternative subdivisions fulfilling this requirement. We therefore propose the concept of *efficient* subdivision, by which we intend subdivision on that species which, in the two subclasses resulting, produces the smallest total number of residual significant associations. This species can, of course, always be identified by examining every possible means of subdivision at each stage; but, even with a computer, this would be an unacceptably lengthy operation were more than a few species involved. Our immediate problem, therefore, is to find a parameter which, *when applied to the class to be subdivided,* is most likely to bring about efficient subdivision.

(ii) *The available parameters.* The definition of the final groupings sought is such that a test of significance is required. The obvious test for small-scale computation would be that which Goodall uses: χ^2 (with Yates's correction) where the cell-frequencies are large enough, and Fisher's 'accurate method' where they are small, or where the χ^2 is marginally significant. However, if the method of subdivision is to be capable of translation into a relatively simple programme for a digital computer, a single test is desirable; and since when the cell-frequencies are large Fisher's method is extremely lengthy, and moreover involves high factorials which would greatly complicate the programme, we propose the use of χ^2 with Yates's correction for *all* tests. Fortunately, the Yates's correction is known to over-correct, so that it is most unlikely that we shall obtain spurious associations; and in this case we are prepared to sacrifice marginally significant associations to the overriding need for simplicity of analysis.

If we wish to reduce the level of association an actual *measure* of association is required; and the parameter selected for subdivision is inherently likely to be related to the parameter selected for measurement. (It is for this reason that the *abundance* of a single species (cf. Goodall), which is not in itself an associative property, cannot fulfil our requirements for subdivision.) Consider, therefore, the $s \times s$ matrix whose entries are our selected association-index, say I, taking the values in the principal diagonal to be invariant. Consider, too, the column sums, which (neglecting the principal diagonal) will represent, for each species, its ΣI value. Now if, in a search for a postulated simpler structure, we subjected this matrix to a factorial analysis, the species with the highest ΣI would necessarily have the highest loading on the first centroid axis, which is itself an approximation to the first principal axis. If, therefore, the associations are due to underlying factors, division on the species with the largest ΣI is likely to produce the greatest possible discontinuity in the first common factor; it will tend to reduce the residual ΣI—*i.e.* the residual general level of association—to a minimum. Since this is precisely what we require to do, ΣI may be expected to be the parameter we need.

There remains the choice of the index I. Since we have in any case decided to calculate χ^2 as a test of significance, it is natural to inquire whether it cannot also serve as an index of association. It is not normally used as such, since the degree of association (but not the significance) which it represents depends on the size of the population in which it is measured; but this is not important in our case, since the values are always compared within a single population or class. The fact that it can serve this double purpose gives it an overwhelming advantage for hand computation over the several alternative indices available. However, Yates's correction is not applicable to χ^2 when this is used as an index of association, and it would be theoretically preferable to construct $\Sigma\chi^2$ from uncorrected values, using the corrected values solely as a test of homogeneity. This refinement would, however, effectively put the problem out of reach of hand computation for any but the simplest communities; and we shall in this paper necessarily concentrate on the results obtainable from corrected $\Sigma\chi^2$ values. It would, of course, be simpler still to select the single greatest χ^2 value, but this would tend to emphasize trivial features at the expense of the population as a whole. Similarly, it would be simpler to select the species with the greatest *number* of significant associations, which would be equivalent to rendering the greatest possible number of associations indeterminate; but we have already pointed out (Sect. II A (iii)) that this is less fundamental than reducing the level of association.

(iii) *Ambiguity.* Whatever parameter is in use, indeterminate values must be taken as zeros, the species concerned being excluded from the analysis; and if corrected χ^2 is used, non-significant values must also be taken as zeros. Any path of subdivision may thus terminate in a choice between two or more species of equal importance—for instance, in a class with a single non-zero χ^2. Such ambiguities will be rare with uncorrected χ^2, since they are then only likely to occur in classes with a high level of indeterminacy. We propose here to resolve all ambiguities by reference to the next highest class in which discrimination is possible.

(iv) *Application of the parameter.* Imagine a population divided on species X into (X) and (x). (X) is still found to be divisible on species Y into (XY) and (Xy). Goodall would 'pool' (Xy) with (x) to form a new population. Our chief objection to this system is that information relating to the discontinuity (X)/(x) is being discarded, and on these grounds we would prefer a 'hierarchical' system—a division, once made, would remain inviolate throughout the analysis. It is conceivable that, from the point of view of the investigation as a whole, the information thus discarded would be better discarded, and an empirical test on actual populations is desirable.

There are, however, other objections to 'pooling'. First, the final groups produced are not necessarily capable of simple statistical definition in terms of presence or absence of key species; secondly, the route by which these groups are obtained is not in itself meaningful, so that *only* the final groups, and not the successive subdivisions which have produced them, are available for examination; and thirdly, the computation is inevitably much longer, since large 'pool' populations are examined at intervals throughout the analysis. With a powerful associative criterion such as $\Sigma\chi^2$, it is unlikely that the matter will be important; for pooling is inherently likely to recreate the intense associations which have previously been removed, so that the pooled group is statistically rejected and the hierarchical pattern retained.

A disadvantage of the use of corrected $\Sigma\chi^2$ values for hierarchical subdivision is that the path of subdivision may change with significance level; with uncorrected values only the degree of subdivision will change. We shall accept this inconvenience for the purpose of the present paper, and must therefore select a 'standard' level of significance for subdivision. We have chosen to work at $P = 0{\cdot}05$; only experience on actual populations can show whether this results in an excessive fragmentation of the population.

(v) *Recombination.* Goodall tests his final groupings for their ability to combine without recreating significant associations; but, as with pooling, combination of groups across a hierarchical division is inherently likely to recreate such associations when an associative parameter is used. Recombination might, however, occur in two rather special cases. First, if the subdivision has produced any very small groups—of, say, less than about 8 quadrats—these may well recombine indiscriminately, since the total number of quadrats resulting is too small for any but the most intense associations to be detected by the crude significance test we are using. Such recombinations—and, for that matter, the individual groups themselves—are statistically meaningless, and are merely an indication that the significance level selected is too low. Secondly, recombination may occur if many of the possible associations are indeterminate. Such cases are likely to be rare, but would be statistically meaningful; they represent cases in which the criterion of indeterminacy can override that of non-significance. Nevertheless, since (Sect. II A (iii)) we believe indeterminacy to be of less importance than lack of significance, the final groups so obtained are likely to be less informative than the primary divisions. In any case, a simple recombination test provides little new information; failure to recombine tells us only that two classes are different, and nothing of the extent to which they are related. Methods can be devised for the quantitative comparison of such communities, and we shall return to these in a later paper. Although we shall test our final groupings for recombination here, it would in any case be quite impracticable to incorporate a general test in a computer programme; for r final groupings the scanning of all $\frac{1}{2}r(r-1)$ possibilities would lengthen the process to an intolerable extent if r were large.

III. COMMUNITY ANALYSES

A. *Introduction*

In the experimental section which follows we are testing two independent hypotheses. The first of these is purely statistical: it is that division on the species with the highest $\Sigma\chi^2$ will subdivide the population in the most efficient manner possible. Theoretical considerations, it will be recalled, tell us that it is *likely* to do so, not that it necessarily will.

The second hypothesis is that the resulting subdivision will provide information of ecological value. It is impossible to examine all possible subdivisions and compare them with a known ecological situation: first, because it is difficult to conceive of a *completely* known ecological situation; and secondly, because the total number of such subdivisions is prohibitively great.* We can only divide on a selected statistical basis and ask, 'what ecological information has this division provided?' and thereby try to assess the efficacy of the method.

B. *Methods*

(i) *Selection of areas.* The general problem can most easily be investigated in a community with sufficiently few species to allow of a full analysis of all associations in all possible classes within the population.† The test-communities chosen were two areas of New Forest heathland in the proposed Denny Reserve. The first, containing a number of well-defined burnt strips of different ages, not only provided a very simple population for complete analysis, but also promised a reasonably clear-cut ecological comparison of results from different sorting methods; while the second, more complex yet sufficiently similar to permit cross-comparison, gave greater opportunities for investigating the potentialities of the new method.

(ii) *Sampling methods.* For a discussion of random *versus* systematic sampling, *see* Greig-Smith (1957). Since we were concerned only with presence-or-absence data and therefore required no estimates of density, and since we were concerned with the pattern of an area as a whole, we chose to sample systematically by means of a rectangular grid. The unlikely possibility that our grid might 'resonate' with the vegetation is one we were prepared to accept, in view of its immense practical advantages in setting out, mapping and later revisiting the sample sites. The actual sampling was carried out by M. J. Hudson and P. F. Hunt, both students of the Southampton Botany Department.

(iii) *Species considered.* To reduce the species to manageable numbers for hand computation, only vascular plants occurring in more than 2% of the quadrats in either or both of the communities are included in the analyses. They are represented by the following letters:

A, *Calluna vulgaris* (L.) Hull; B, *Molinia caerulea* (L.) Moench; C, *Erica tetralix* L.; D, *E. cinerea* L.; E, *Pteridium aquilinum* (L.) Kuhn; F, *Trichophorum caespitosum* (L.) Hartman; G, *Ulex europaeus* L.; H, *U. minor* Roth; J, *Festuca ovina* L.; K, *Potentilla erecta* (L.) Rausch.; L, *Polygala serpyllifolia* Hose; M, *Carex panicea* L.; P, *C. binervis* Sm.

C. *Results*

1. *A 5-species community ('Beaulieu Road')*

(i) *The area.* A Callunetum in the New Forest, E. of Beaulieu Road station and N. of Bishop's Purlieu (map ref. SU/349054). It lies for the most part directly over Barton Sand, though towards the S. end a slight ridge is covered with a cap of Plateau Gravel, tailing out northwards with some downwash. The soil on both sand and gravel is a well-developed podsol. The area is subjected to an approximately 7-year rotational burning, and is further frequently affected by casual burning through sparks from passing trains; local report and counts from *Calluna*

* The number of ways in which a population containing p final classes can be divided into groups is a well-known problem in combinatorial analysis, and no explicit general solution is known. For $p = 4$ (2 spp.) and $p = 8$ (3 spp.) it is easily shown by enumeration that the answers are 15 and 4140 respectively.

† For s species this involves the examination of $\frac{1}{2}s(s-1)$ associations in 3^{s-2} classes selected from the $3^s - 2^{s-1}(s+2)$ classes of order $(s-2)$.

Table 1. χ^2 *matrix for Beaulieu Road 5-species community*

Significant associations are entered as their corrected χ^2 values, in normal type if the association is positive, in italics if it is negative. Non-significant values are denoted by x, and are treated as zeros for the computation of $\Sigma\chi^2$.

	A	B	C	D	E
A	...	51·31	45·66	x	x
B	51·31	...	93·76	*12·62*	*68·64*
C	45·66	93·76	...	*4·84*	*14·08*
D	x	*12·62*	*4·84*	...	6·92
E	x	*68·64*	*14·08*	6·92	...
$\Sigma\chi^2$	96·97	226·33	158·34	24·38	89·64

rings suggested that the last deliberate burning of the area as a whole had taken place $4\frac{1}{2}$ years before the survey, but that certain strips had later been accidentally burnt between 1 and 2 years ago and others during the spring of the current year.

The area was sampled in July 1957 by a rectangular grid of 44×14 1-metre quadrats spaced 5 m apart in both directions. One record was lost in transferring the data to cards, and (N) is thus effectively 615.

(ii) *The population.* Six vascular species were actually present, A to F in our key. However, F was represented in less than 2% of the quadrats and has been ignored. Of the 32 possible final classes, only the following 12 exist:

(ABCDe) = 6	(AbCde) = 9	(Abcde) = 69
(ABCde) = 244	(AbcDE) = 4	(aBcde) = 23
(ABcDe) = 8	(AbcDe) = 11	(abcdE) = 5
(ABcde) = 193	(AbcdE) = 12	(abcde) = 31

Of the 131 possible classes of order 3 in which associations are to be sought, 26 are empty; as a result, a further 47 vanish, having become identical with classes of higher order; 3 more are excluded as being below the minimum size (8) in which, using Yates's correction, significant associations can be manifested; and 55 remain.

(iii) *Association-analysis.* Although we have examined all possible associations in all possible classes in this community, we here present only (in Table 1) the χ^2 matrix for (N), and (in Table 2) the χ^2 values within the first-order classes. The

Table 2. *First-order classes in Beaulieu Road 5-species community*

The first column gives the class under examination, the second the number of quadrats in that class. In the association-columns, – denotes that the association is *necessarily* indeterminate by virtue of the species whose presence or absence defines the class; o denotes that the association is indeterminate *in this population*; and x denotes that the association fails to reach significance. Significant associations are entered as their corrected χ^2 values, in normal type if the association is positive, in italics if it is negative.

		{A, B}	{A, C}	{A, D}	{A, E}	{B, C}	{B, D}	{B, E}	{C, D}	{C, E}	{D, E}
(A)	556	–	–	–	–	73·29	*19·34*	*65·41*	*7·18*	*12·50*	9·25
(a)	59	–	–	–	–	o	o	x	o	o	o
(B)	474	–	24·8	x	o	–	–	–	x	o	o
(b)	141	–	x	4·34	x	–	–	–	x	x	x
(C)	259	o	–	o	o	–	x	o	–	–	o
(c)	356	16·16	–	x	x	–	*25·63*	*35·06*	–	–	3·85
(D)	29	o	o	–	o	5·63	–	x	–	x	–
(d)	586	58·12	47·91	–	*5·20*	83·08	–	*59·22*	–	*11·55*	–
(E)	21	o	o	x	–	o	o	–	o	–	–
(e)	594	48·50	43·99	x	–	77·88	*7·69*	–	x	–	–

associations in (N) show that the species can be roughly divided into two groups—A, B and C on the one hand, D and E on the other; there are positive associations within each group, negative associations between them. Some of these associations are very strong; $\{A, B\}$, $\{B, C\}$, $\{B, D\}$ and $\{B, E\}$ exist at almost every class-level, and we note that all four involve B. Furthermore, in the second- and third-order classes, the existence of the powerful association $\{A, C\}$ is manifested only in the presence of B. This suggests that the first division should be on B, which is strikingly confirmed by an examination of the first-order classes. Division on any other species produces two classes, one of which is almost entirely free from associations (though mainly by indeterminacy), the other retaining most or all associations unimpaired. If the population is divided on B, however, all negative associations disappear, and we are left with a single positive association in each class, one of which ($\{A, D\}$ in (b)) is only marginally significant.

B is therefore the species which our decision-parameter is required to select. Reference to Table 1 shows that B is the species with the highest $\Sigma\chi^2$; this parameter has therefore here proved successful. On the other hand, Goodall's criterion—the most abundant species—would have required the population to be divided on A.

We may now continue the subdivision. An ambiguity (A or C) arises in (B); reference to the $\Sigma\chi^2$ values in the next highest class—(N)—requires division on C. The ambiguity (A or D) in (b) is similarly resolved by division on A. All classes are now homogeneous, and we find that they cannot in fact be recombined without recreating significant associations. The complete hierarchy is therefore as follows:

(N)

(B) (b)

(BC)	(Bc)	(bA)	(ba)	
250	**224**	**105**	**36**	= No. of quadrats

Division on *uncorrected* $\Sigma\chi^2$ would have given the following:

(N)

(B) (b)

(BA)	(Ba)	(bA)	(ba)
451	**23**	**105**	**36**

(Ba) and (ba) could now be recombined across the hierarchy to produce the homogeneous first-order class (a). Table 1 shows that division of the original population into (A) and (a) is one of the three possible ways of making the greatest possible number of associations indeterminate at the first stage; and that this is a case of true indeterminacy-recombination (Sect. II B (v)) is shown by considering the effect of raising the significance level to $P = 0{\cdot}01$. (b) now fails to subdivide, and the three remaining groups—(BA), (Ba) and (b)—cannot be recombined; recombination is possible only if we can recreate the almost completely indeterminate group (a).

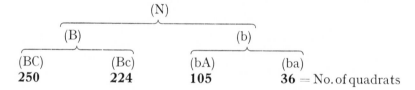

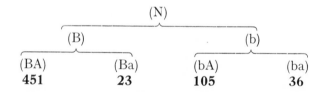

Goodall's method—abundance and pooling—would in this simple case also give the three groupings (AB), (Ab) and (a), though without, of course, any information as to their relative importance.

(iv) *Ecological significance.* Fig. 1 shows the distribution of the corrected $\Sigma\chi^2$ groupings in relation to the burning lines; Table 3 gives their full species-composition. A valuable feature of a hierarchical system is that its successive subdivisions can be identified on a map at successive levels of importance. Here, the first-order (B)/(b) division forms a well-marked '*Molinia* extinction line' which divides the area cleanly into two; but although it coincides almost exactly with a burning line, the presence of both young and old communities on either side suggests that the burning here has merely sharpened the effect of a more powerful underlying factor. Contouring of the site (Fig. 1—inset) and examination of a limited number of soil pits later established that the *Molinia* boundary is roughly associated with the beginning of the Plateau Gravel ridge. In essence, therefore, this primary division of the population reflects the basic features of topography and soil, rather than the more conspicuous but superficial burning regime.

By contrast, the second-order (final) divisions directly follow the burning pattern. Within the *Molinia* zone, (BC) covers all areas which have not recently been burnt;

Table 3. *Species composition of final groupings at Beaulieu Road*

Ref. No. on map	1	2	3	4
Grouping	(BC)	(Bc)	(bA)	(ba)
No. of quadrats	250	224	105	36
A = Calluna vulgaris	250	201	105	0
B = Molinia caerulea	250	224	0	0
C = Erica tetralix	250	0	9	0
D = E. cinerea	6	8	15	0
E = Pteridium aquilinum	0	0	16	5

it is a fairly closed community consisting almost entirely of *Calluna, Molinia* and *Erica tetralix.* Areas accidentally fired during the current year bear only (Bc); *Molinia* has regenerated throughout and *Calluna* is rapidly following—so rapidly that, even during the time that this area was being recorded, a number of quadrats changed from (aBc) to (ABc). *Erica* clearly takes longer to re-establish itself, though the mixture of (BC) and (Bc) over the 1½-year-old areas suggests that it follows *Calluna* fairly soon; it had in fact begun to regenerate from stools even in the youngest areas when the site was revisited 3 months after the survey. South of the *Molinia* boundary, the division between (Ab) and (ab) is again related to a burning line. (Ab) is a 4½-year-old *Calluna* strip with a little *Pteridium, Erica cinerea* and *E. tetralix* admixed; (ab) is recently burnt bare ground, with sparse *Pteridium* at one end, but with *Calluna* only regenerating after the survey was complete. The division between these two is clearly, from its significance level, of very minor importance, and relates to the poor degree of association within the parent population (b). In fact, their floristic separation at the time of the survey is entirely dependent on the delayed regeneration of *Calluna* in (ab) compared with its performance in the other recently burnt areas; and this is probably correlated with the position of the (ab) strip on the dry south slope of the gravel ridge.

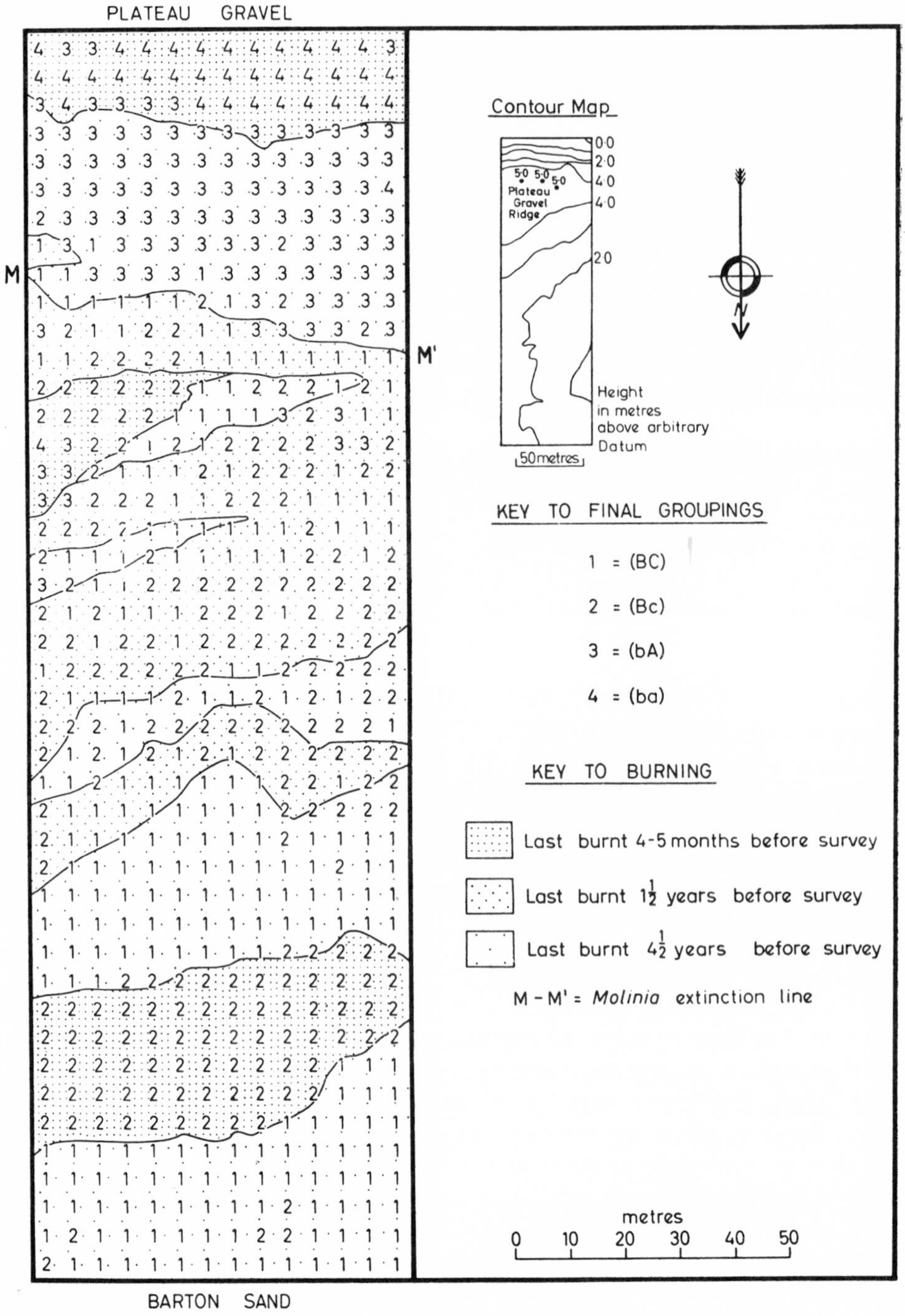

Fig. 1. Map of final groupings at Beaulieu Road.

The four final groupings (BA), (Ba), (bA) and (ba) obtained by the hierarchical use of uncorrected $\Sigma\chi^2$ without recombination differ only from those set out above by the division of (B) at an earlier phase of the burn succession, *i.e.* the re-establishment of *Calluna* rather than *Erica tetralix* becomes the discriminating feature. Since the time lag between the entry of these two species is very short, the exact course of this secondary division relates only to very minor differences in phase, and there is little to mark either method of division as more important than the other. Subsequent recombination of (aB) and (ab) into the highly indeterminate (a) combines 23 scattered quadrats still lacking *Calluna* from the most recently burnt strips on the Barton Sand with those of the bare area on the south side of the Plateau Gravel ridge. Since the immediate effect of burning is simply to remove species without replacing them by others, the actual number of species is reduced in areas where this second (burning) factor is strong; the sub-population consequently exhibits a level of indeterminacy sufficiently high to override the significance barrier of the first (soil) factor, though the latter is still evident at the earlier stage of the analysis.

A classification directly into (AB), (Ab) and (a) on Goodall's abundance/pooling method is ecologically less convincing and revealing: a major division is made in relation to a markedly evanescent phase of the burn succession and the underlying soil relationships are largely obscured.

2. *A 10-species community ('Matley Ridge')*

(i) *The area.* A *Calluna* heath essentially similar to the foregoing, about 2 miles (3·2 km) S.E. of Lyndhurst on the N. slope of Matley Ridge (map ref. SU/325075). The subsoil is also similar, with Plateau Gravel at the top (S.) end, some gravel downwash, and Barton Sand at the bottom. An old cart-track, intercepting three other minor tracks at right-angles, runs down a central gully from S. to N., and the whole area is much crossed by ruts made by military vehicles in World War II. The slope is also longitudinally divided by a well-marked burning line: the west side bears a fairly open community, known to have been burnt 2½ years before the survey; the east a conspicuously older one which, though much disturbed during the war and not immune from accidental local fires, had not been rotationally burnt for many years.

The area was sampled in November 1957, using a more widely spaced grid of 36×11 1-metre quadrats 10 m apart, and (N) is thus 396.

(ii) *The population.* Of the 17 vascular plants which were actually recorded, 4 were present in too few quadrats for study. Of the remainder (A-D and F-P in our key), A and B are statistically inactive in that they each occur in every quadrat but two; they are therefore omitted from the analysis, and the final population to be considered thus consists of 10 statistically effective species with a background of A and B.

Only 63 of the 1024 possible final classes actually exist; and, of these, many are represented by a single quadrat.

(iii) *Association-analysis.* We present the χ^2 matrix for (N) in Table 4. Once more, the species tend to divide into two groups: C and F, though not mutually associated, are separated from the remaining 8 species by negative associations. It is clear, however, that the main interest in the subdivision of this population will lie in the extent to which it reveals any general structure underlying the positive

Table 4. χ^2 *matrix and residual* 1*st-order associations for Matley Ridge* 10-*species community*

Significant associations are entered as their corrected χ^2 values, in normal type if the association is positive, in italics if it is negative. Non-significant values are denoted by x, and are treated as zeros for the computation of $\Sigma\chi^2$.

The last three rows of the table represent the *number* of residual associations at the significance level indicated in both 1st-order classes (taken together) obtained by division on the species heading each column.

	C	D	F	G	H	J	K	L	M	P
C	...	x	x	*40·37*	x	*13·52*	*23·84*	x	x	*34·83*
D	x	...	*18·43*	x	39·37	x	x	8·48	x	x
F	x	*18·43*	...	x	x	x	x	x	x	x
G	*40·37*	x	x	...	x	36·23	48·62	x	4·23	13·99
H	x	39·37	x	x	...	51·94	x	11·70	x	5·23
J	*13·52*	x	x	36·23	51·94	...	53·84	4·61	31·82	43·07
K	*23·84*	x	x	48·62	x	53·84	..	x	33·21	23·84
L	x	8·48	x	x	11·70	4·61	x	...	x	x
M	x	x	x	4·23	x	31·82	33·21	x	...	15·15
P	*34·83*	x	x	13·99	5·23	43·07	23·84	x	15·15	...
$\Sigma\chi^2$	112·56	66·28	18·43	143·44	108·24	235·03	183·35	24·79	84·41	136·11
				Number of residual associations						
P = 0·05	18	22	25	16	21	15	16	30	18	19
P = 0·01	13	20	17	13	19	8	8	26	14	11
P = 0·001	11	18	13	9	13	4	5	24	10	9

associations within the 8-species group. To avoid presenting a 45×20 table, we have included in Table 4 information as to the total *number* of significant associations remaining in both first-order classes when the population is subdivided on each species in turn. At the 0·05 level, 3 species (G, J and K) are almost equal with 16, 15 and 16 residual associations each respectively; at the 0·01 level the choice is reduced to J or K, with 8 each; and at the 0·001 level, J has 4 to K's 5. The level of the residual associations after division on J, moreover, lies slightly below that of those from K. There is no doubt, therefore, that the required first division is on J; and we note with satisfaction that J has the highest $\Sigma\chi^2$ value. (The most abundant species, C, would clearly produce a relatively inefficient division.) Continued hierarchical division using corrected $\Sigma\chi^2$ at the P = 0·05 level produces the following final subdivision:

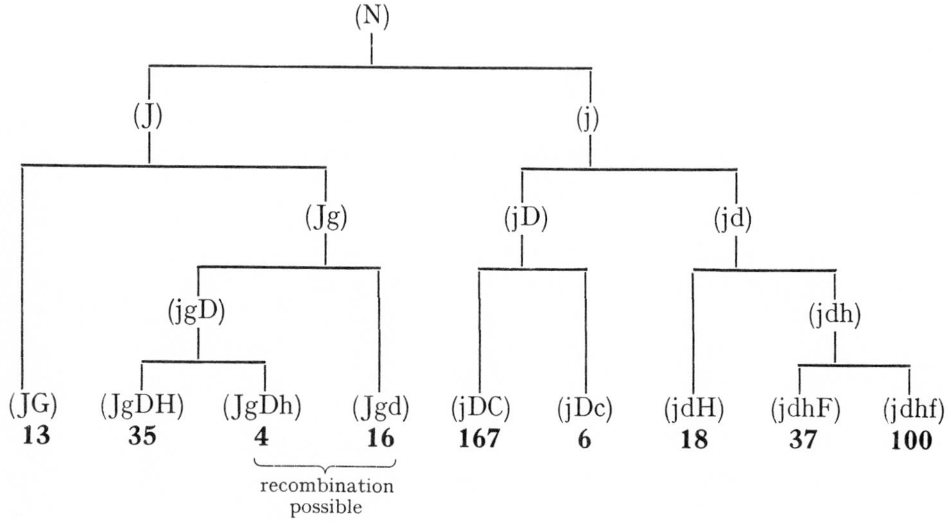

An ambiguity in (Jg)—D or H—is resolved by reference to (J); a second in (Jgd)—H or L—by reference to (Jg); a third in (jd)—H or L—by reference to (j); and a fourth in (jdh)—F or M—remains unresolved in (jd) and is resolved in favour of F by reference to (j). Of all the final groupings, only two—(JgDh) and (Jgd)—can be recombined; and, since (JgDh) is represented by only 4 quadrats, this recombination is without statistical meaning (Sect. II B (v)). It will be noted that two of the groupings contain less than 8 quadrats, and three more are close to the size (15) at which not even perfect associations could be detected at the P = 0·001 level of significance; this fineness of subdivision suggests that the significance level chosen was too low.

This population proves, incidentally, to be one in which the hierarchy changes with significance level; at the 0·01 level (J) divides on C, though division on G is re-established (via a C/G ambiguity) at the 0·001 level. We shall confine our attention to the 0·05 level result given above.

Table 5. *Species composition of final groupings at Matley Ridge*

Ref. No. on map	1′	2′	3′	4′	5′	6′	7′	8′
Grouping	(JG)	(JgDH)	(Jgd) + (JgDh)	(jDC)	(jDc)	(jdH)	(jdhF)	(jdhf)
No. of quadrats	13	35	20	167	6	18	37	100
A = Calluna vulgaris	11	35	20	167	6	18	37	100
B = Molinia caerulea	13	35	20	165	6	18	37	100
C = Erica tetralix	7	33	20	167	0	18	37	100
D = E. cinerea	4	35	4	167	6	0	0	0
F = Trichophorum caespitosum	2	4	3	17	0	6	37	0
G = Ulex europaeus	13	0	0	3	0	0	0	2
H = U. minor	5	35	8	58	5	18	0	0
J = Festuca ovina	13	35	20	0	0	0	0	0
K = Potentilla erecta	6	4	2	0	0	0	0	0
L = Polygala serpyllifolia	8	25	8	89	0	14	11	34
M = Carex panicea	3	3	6	1	0	0	4	0
P = C. binervis	4	7	1	2	0	0	0	0

The greater number of species involved at Matley compared with Beaulieu Road made it impracticable to analyse this community by more than one method without the aid of a computer. The following assessment of ecological results from the analysis above must therefore omit any immediate comparison with results from the use of uncorrected $\Sigma\chi^2$.

(iv) *Ecological significance.* The map of the final groupings and their species lists are given in Fig. 2 and Table 5 respectively. The mosaic of communities on the map partly reflects the patchy disturbance of the ground, and its ecological interpretation is clearly less easy than in the almost diagrammatic Beaulieu Road community. Nevertheless, an underlying pattern can be seen, which was by no means obvious when the area was surveyed.

The 68 quadrats separated as (J) (*Festuca ovina*) at the important first dichotomy are ringed on the map for clarity. They are obviously concentrated into two main clusters, both lying athwart the 2½-year burning line. One occurs near the top of the slope, with extensions along the line of the old cart-track to end in a

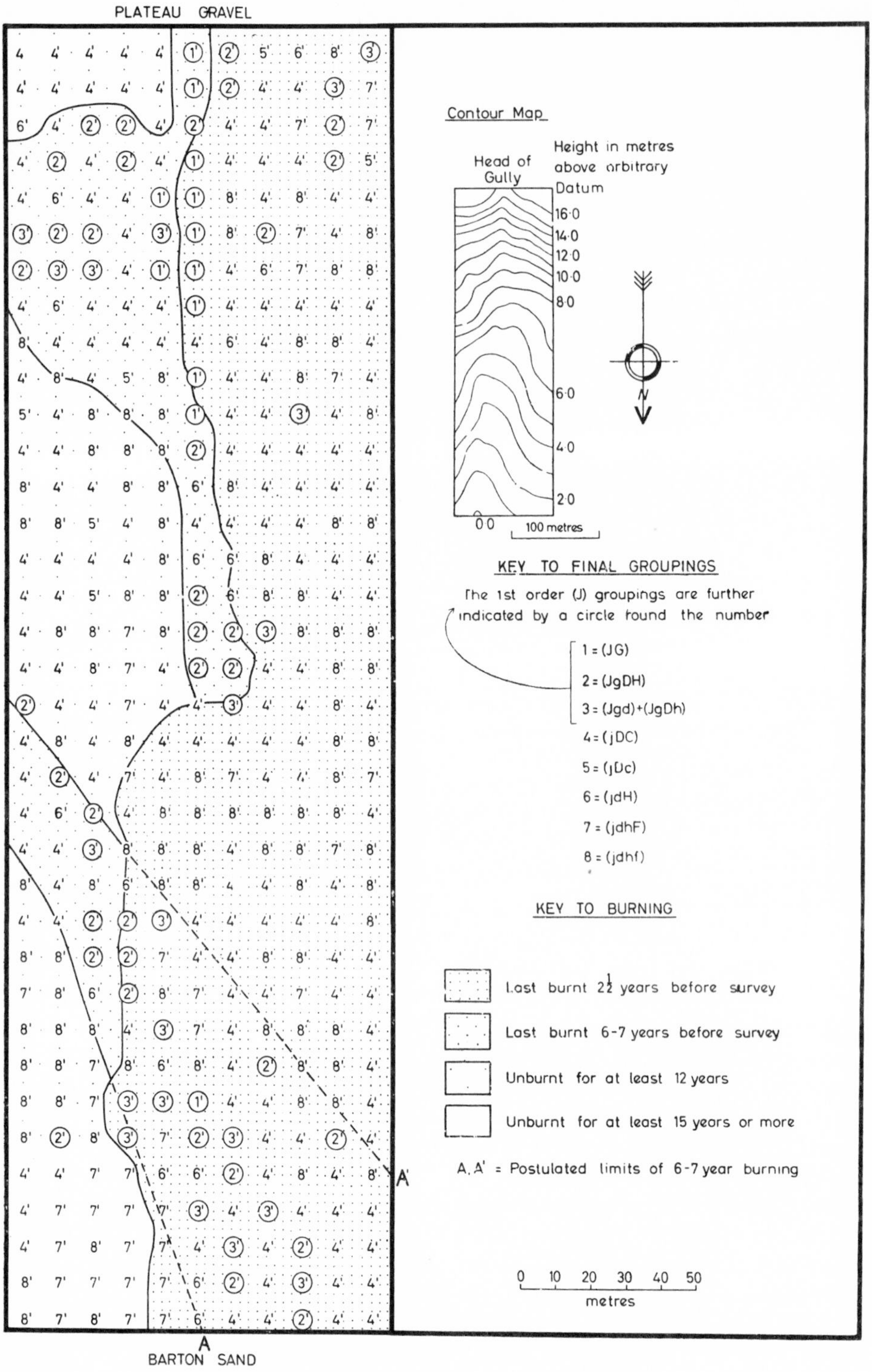

Fig. 2. Map of final groupings at Matley Ridge.

small outlier further down the gully; the other crosses the lower part of the area obliquely in a widening S.E./N.W. strip. Contouring of the site (Fig. 2—inset) and a few exploratory soil pits showed that neither bears any consistent relationship with an obvious soil or topographical feature, apart from the local focus on the gully; and the real key to the distribution of (J) was only revealed when the burning history of the area was later reconstructed in detail from counts of *Calluna* rings. Except for a very old pocket of 15-20 (-25)-year-old plants at the wettest end, most of the Callunetum on the east side of the area was roughly 12-13 years old, and clearly indicated recolonization after the war-time disturbance ceased. But two fairly large patches of 6/7-year-old *Calluna* also occurred within this zone, presumably as the result of later localized fires; they were not markedly different from the rest in either cover or size of plants, and had not been specifically noted in the original survey. On the west side of the $2\frac{1}{2}$-year burning line, any direct indication of these accidental fires had naturally been lost. Nevertheless, it is possible to suggest the path of at least one of them by extrapolation; and it will be noted from the map that the boundaries thus defined strikingly enclose the whole of the oblique strip containing (J). Moreover, the eastern part of the other (J) cluster similarly coincides with the position of the other 6/7-year-old patch; and there is no reason to suppose that, despite its N./S. deflection along the gully, this fire did not also cross to the western side to affect that part of the area now bearing the rest of (J).

Again, as at Beaulieu Road, we find that the first stage of the analysis has ignored the most conspicuous feature of the area—in this case the $2\frac{1}{2}$-year burning line—in favour of a more subtle but ecologically legitimate division. It is outside the scope of this paper to inquire closely into the specific connection between (J) and the 6/7-year burning. *Festuca* is notoriously resistant to trampling and similar disturbance; the most economical hypothesis on present information is that it spread over the area during the military occupation, that it was later ousted by taller competitors except where the intermediate burning reopened the community sufficiently for *Festuca* to persist, and that the surrounding parts completed their reversion to more typical heath before the later $2\frac{1}{2}$-year burning could act in the same way. Even so, there is still some circumstantial evidence, both from inspection of the site and from examination of air photographs showing the direction of former tank tracks, that a more complex story may be partly involved. The fact that a simple causal explanation for the distribution of *Festuca* and its associates is not possible on the available data by no means diminishes the value of the statistical analysis; on the contrary, it has directed attention to a phase of the burning pattern which might otherwise have been missed, and has succeeded in extricating a concrete ecological problem from an otherwise incoherent community mosaic.

It will be noted, incidentally, that, although the Matley and Beaulieu areas both lie across a boundary between Plateau Gravel and Barton Sand, the striking primary vegetation/soil relationships exposed at Beaulieu are not reiterated by the first division here: this was already foreshadowed by the presence of *Molinia*—the operative species at Beaulieu—throughout the Matley area. The explanation probably lies in the fact that the range of drainage conditions may be less extreme at Matley; it lies entirely on a damp north slope, and has no topographical counterpart to the dry, raised Beaulieu ridge. Moreover, the geological boundary at

Matley is very patchy and indistinct, with correspondingly less clear-cut subsoil differences.

To return to the map, we find that, within (J), the small (JG) final grouping is very sharply circumscribed; apart from one isolated quadrat, it is restricted to the top part of the disused cart-track down the gully, which has become filled with *Ulex europaeus*. The division between (JgDH) and the recombined group (Jgd) and (JgDh) is unimportant; it is based primarily on a small positive association between *U. minor* and *Erica cinerea*, and the 35 quadrats separated as (JgDH) have no easily discernible ecological focus.

It is the negative (j) side of the original dichotomy which provides the main bulk of the Matley population. Here the first division is on *E. cinerea*, mainly by virtue of a fairly strong positive association with *Ulex minor* and a slightly weaker negative one with *Trichophorum*. All but 6 of the (jD) quadrats emerge at the next division as the final grouping (jDC). Remembering that these 167 quadrats also contain the universal A and B, we note that (jDC) is defined by the presence of four species—*Calluna, Molinia, Erica cinerea* and *E. tetralix*—very widespread in New Forest heaths in general, so we are not surprised to find it abundant and widely distributed here. It is most poorly represented in the oldest and densest vegetation at the N.E. corner, and this would accord with the observations of Fritsch & Parker (1913) at Hindhead that *E. cinerea* is gradually eliminated in the late phases of a burn succession. The separation of (jDc) is based only on two very small associations; nevertheless, we find that its 6 quadrats are confined to the top half of the slope, suggesting here, at least, the possible operation of a soil or drainage factor.

Within (jd), the 18 quadrats of (jdH) have particular interest in that they nearly all lie within the postulated limits of the 6/7-year burning and partly fill the spaces between the *Festuca* quadrats. But since there is a strong positive association between *Festuca* and *Ulex minor* in the population as a whole, it is reasonable to find (jdH) sharing the same habitat zones which favour (J).

Lastly, we have a minor division of (jdh) on *Trichophorum*. Although separated on a single small positive association, the 37 quadrats of (jdhF) have a recognizable ecological niche in that they lie preponderantly in the lower and wetter part of the area, with nearly half of them clustered in the pocket of rank vegetation at the N.E. corner. The big group (jdhf) thus remains to fill the gaps left after all the disturbing elements in the original community have been allotted their places. All its 100 quadrats contain *Erica tetralix* as well as *Calluna* and *Molinia*; we recognize it in fact as virtually identical in composition with the (BC) community of Beaulieu Road, but this time appearing at the extreme negative end of the hierarchy.

The emergence of this same triple community from the independent Beaulieu and Matley analyses is perhaps one of the more encouraging results of the method. In each case, the *Calluna-Molinia-Erica tetralix* grouping forms an appreciable part of the total population; and in certain respects it may be regarded as the basic heathland community of the region. It has become further impoverished in parts of Beaulieu Road by unfavourable soil conditions or by very recent burning: it has become locally enriched or modified at Matley by the entry of other disturbing species to give the more complex Matley mosaic.

IV. DISCUSSION

It is arguable that vegetation as an organic whole consists, not of individual species, but of relationships between them. A painting is *made* of pigments, oil and canvas, but it may reasonably be said that it *is* a set of relationships between forms and colours; similarly, vegetation is *made* of plants, but it *is* a matrix of correlation coefficients. It is for this reason that we agree with Greig-Smith (1957, p. 163) as to the importance of factor analysis, although in the only published example known to us—Goodall (1954)—we believe that the method adopted is inappropriate to the study of vegetation (we shall support this view in a later communication). We have already pointed out that association-analysis is an approximation to factor analysis, since it seeks to divide the population by causing the greatest possible discontinuity in a postulated first factor. (A closer approximation would be obtained by summing, not χ^2 values, but the corresponding correlation coefficients $\left(\sqrt{\frac{\chi^2}{N}}\right)$ taken regardless of sign; but for hand computation this would be extremely laborious.) Of the various forms of factorial analysis, we have invoked the very flexible, and now conventional, multifactorial analysis; the methods of Goodall (1953) and Hopkins (1957) are, however, approximations to the older and rather more crude method of group factor analysis. Both attempt to extricate groups of species internally linked only by positive associations. Goodall seeks these directly, but by a method we have shown to be inefficient; Hopkins seeks them by the active elimination of negative associations. Goodall's quadrats, like ours, become members of discrete groups. Hopkins, on the other hand, 'scores' his quadrats according to the groups represented within them. This we believe to be well intentioned but ill advised; the problem of specification of individuals when qualitative data have been factorized is controversial and difficult, and is not to be solved by a simple arbitrary scoring system.

In short, therefore, we believe that the foregoing pages have shown that our method fulfils its statistical function of dividing a population as efficiently as possible, and that it does this by selecting the greatest possible discontinuity in the first centroid axis of the association-matrix under scrutiny. However, even a complete multifactorial analysis does not *necessarily* provide any useful ecological information; the 'factors' it discloses are mathematical postulates, arising from a transformation designed to express the information contained in the matrix in the most economical form possible. The solution is not unique, nor do the 'factors' necessarily relate to any real entities in the outside world. Whether such corresponding entities can be recognized, what their nature may be, and whether the information they provide is useful for other purposes, can only be decided by appeal to the area under investigation. In the two test-communities studied here, we have found that our major groupings do in fact correspond with recognizable habitat factors, and that the analysis moreover directs attention to important points which—even in these very restricted communities—might otherwise have remained obscured by more conspicuous but superficial features.

But, although a sorting system may be only a crude approximation to a factor analysis, it provides, in compensation, additional information of a different kind; for the homogeneous groups which result may be used in their own right for phytosociological study. The groups, considered as subclasses of the original population, are likely to be most meaningful if defined in their lowest possible order; for example, at Beaulieu Road, (BC) happens also to be (ABCe), and at Matley (jdhf) is equivalent to (ABCdfhjkmp). The positively defining species of a class may together be regarded broadly as its 'characteristic species'. They provide its 'constants'; and in our hierarchical system they also carry a strong element of the 'fidelity' concept. It is tempting to try to identify the various classes further in terms of their total species. Percentage composition will not do; the fiducial limits of a percentage depend markedly both on its value and on the size of the parent population, and the danger of comparative lists of percentage compositions is that they suggest a precision which they do not in fact possess. We have preferred in this paper to publish only the figures of actual occurrences in quadrats; an approximate comparison between the abundance of a species in different communities can then be made by the χ^2 test.

It must be emphasized that any one set of final groupings relates only to the particular population which has been examined. The findings from one area thus cannot be used directly to provide the basis for a more generally applicable phytosociological classification. Nevertheless, where certain species are consistently associated in the vegetation of a region, the group containing them is likely to reappear in a number of independent analyses from separate areas. The recognition of the substantially similar (ABC) grouping at both Beaulieu Road and Matley is a case in point; it provides a common focus to which specific local variations can be related.

The number of final groupings to be considered will necessarily change with the precise definition of homogeneity adopted. If non-significance is, as we recommend, to be the major criterion, then our method may be expected to give the smallest possible number of final groups; indeterminacy may, if a population contains several abundant and highly associative species, reduce the number slightly—as it does at Beaulieu Road—but is objectionable on other grounds. In any case, the number of groupings by any system is a direct reflection of the heterogeneity of the vegetation; and if in a very complex area the complete separation becomes too unwieldy to be appraised, it is possible either to raise the significance level or to examine only the major subdivisions. At Matley, for instance, the first-order division makes a useful distinction between the groupings which approach a poor 'grass-heath' in composition and those of more typical heath.

The use of association-analysis in ecological research is most likely to be in connection with the primary survey of an area rather than in its later more detailed investigation. It will serve to *expose* the problems, not to *solve* them. For this purpose a comparatively crude but easily applicable method is likely to be of more value than a more precise but less straightforward technique; and this is an additional reason for our advocation of statistical approximations where feasible. The method is at the moment being programmed for a 'Pegasus' digital computer; when the programme is fully tested we propose to carry out a comparative investigation of the Matley data with a view to making definite recommendations as to the basic parameter to be used (corrected χ^2, uncorrected χ^2, or $\sqrt{\frac{\chi^2}{N}}$), the path of subdivision (hierarchical or by pooling), and the value of the different significance levels. We shall also examine populations involving larger numbers of species. The results presented in this paper already suggest that association-analysis of this type will prove a powerful ecological tool; and it seems likely, even at the present stage of this investigation, that we shall not require to change the basic principles of the method.

SUMMARY

1. Goodall's method of subdividing a set of sample quadrats into homogeneous groups, in which all species-associations have been made non-significant or indeterminate, is subjected to a theoretical analysis.

2. A new sorting method is described, consisting of hierarchical division on the species with the highest aggregated value of the chosen association-index in the class under study. The properties of suitable indices are briefly considered, but in order to bring this exploratory study within the reach of hand-computation, the index used ($\Sigma\chi^2$) was constructed from corrected χ^2 values, non-significant and indeterminate values being taken as zero, and ambiguities being resolved in the next highest class in which discrimination is possible. Equal weight is given to positive and negative associations.

3. The statistical efficiency of the method is confirmed by its application to populations from two heathland communities.

4. The nature of the ecological information thus obtained is assessed, and it is concluded that a method of this type is likely to prove a very useful tool in primary survey.

ACKNOWLEDGMENTS

It is with pleasure that we record our indebtedness to the following: To Mr M. J. Hudson and Mr P. F. Hunt, for the entire collection of the quadrat data, and for help in the preliminary sorting of the record cards; to Mr J. A. Bailey, for other assistance in the field; to Mr A. C. Clark, of the Geography Department, for his careful drawing of the maps; to Mr E. Wynne Jones, the Deputy-Surveyor of the New Forest, for permission to work on the areas; to Dr J. G. Manners, for placing his extensive knowledge of New Forest heaths at our disposal; and to Dr G. N. Lance, of the Mathematics Sub-Department of Computation, for advice on the application of a digital computer to this type of work.

REFERENCES

FRITSCH, F. E. (1913). The heath association on Hindhead Common. *New Phytol.*, **12**, 149-63.

GOODALL, D. W. (1953). Objective methods for the classification of vegetation. I. The use of positive interspecific correlation. *Aust. J. Bot.*, **1**, 39-63.

GOODALL, D. W. (1954). Objective methods for the classification of vegetation. III. An essay in the use of factor analysis. *Aust. J. Bot.*, **2**, 304-24.

GREIG-SMITH, P. (1957). *Quantitative Plant Ecology*. Butterworths, London.

HOPKINS, B. (1957). Pattern in the plant community. *J. Ecol.*, **45**, 451-63.

TUOMIKOSKI, R. (1942). Untersuchungen über die Untervegetation der Bruchmoore in Ost-finnland. I. Zur Methodik der pflanzensoziologischen Systematik. *Ann. Soc. Zool.-Bot. Fenn. Vanamo*, **17**, 1-203. (*ex* Goodall 1953).

(*Received* 8 *March* 1958)

19

Reprinted from *Biometrics,* **23**(4), 623–637 (1967)

A COMPARISON OF SOME METHODS OF CLUSTER ANALYSIS

J. C. GOWER

Rothamsted Experimental Station, Harpenden, Herts., England

SUMMARY

Each individual of a multivariate sample may be represented by a point in a multidimensional Euclidean space. Cluster analysis attempts to group these points into disjoint sets which it is hoped will correspond to marked features of the sample. Different methods of cluster analysis of the same sample may assume different geometrical distributions of the points or may employ different clustering criteria or may differ in both respects. Three superficially different methods of cluster analysis are examined. It is shown that the clustering criteria of all these methods, and several new ones derived from or suggested by these methods, can be interpreted in terms of the distances between the centroids of the clusters; the geometrical point distribution is found in most instances. The methods are compared, suggestions made for their improvement, and some of their properties are established.

1. INTRODUCTION

Cluster analysis has become popular amongst some systematists engaged in devising classifications. There are many techniques of cluster analysis and it is difficult to judge their relative merits and demerits, because a cluster is not a well defined concept. It may well be that several different types of cluster need defining and in this case any investigator should consider carefully which definition fulfills his requirements. With any definition, attempts could be made to devise algorithms, to operate on data to determine the clusters so defined. More than one suitable algorithm might be found, but they should all give the same result except perhaps when the data are pathological, admitting more than one set of clusters consistent with the definition. It might sometimes be more convenient to use simple algorithms giving only approximate results; the efficiency of these approximate methods could be studied. In practice, the procedure is commonly reversed, the algorithm implicitly defining a cluster. Some algorithms conceal the form of the clusters they find and it is our aim here to expose the underlying cluster structure for a few methods. This amounts to trying to determine what the definition of a cluster must be, for a given cluster-finding algorithm.

Three algorithms (those described by Sokal and Michener [1958], Edwards and Cavalli-Sforza [1965], and Williams and Lambert [1959]) referred to as Methods I, II and III respectively, are considered in this paper. These algorithms have been selected because they are superficially dissimilar but it is shown that they have much in common.

New algorithms which approximate, or are suggested by, the primary ones have also been constructed and discussed. This procedure is helpful because the new algorithms are often more amenable to analysis, and also because the study of parallels reveals properties of the cluster definitions which might otherwise go unnoticed.

2. GEOMETRICAL REPRESENTATION OF THE MULTIVARIATE SAMPLE

Most methods of cluster analysis operate on a multivariate sample of N individuals with observations on the same v variates for each individual. Some methods require that all the variates are dichotomous with scores 0 to 1 to indicate presence and absence of characters; others require all variates to be quantitative (i.e., measured on a scale). In practice, combinations of such variates occur and multi-valued qualitative variates may also be included.

It is convenient to imagine each individual as represented by a point $P_r(r = 1, 2, \cdots, N)$ in a multidimensional Euclidean space. There are many ways of getting such a representation. For example, when all the variates are quantitative (and possibly when they are dichotomous) their values can be taken as the coordinates of P referred to a set of rectangular axes—this is the model assumed as a basis for principal components analysis. Alternatively, Gower [1966] has shown how the coordinates of a set of points referred to their principal axes can be calculated given any set of distances between every point pair: distances might, for example, be chosen as $1 - S_{ij}$ when S_{ij} is a similarity coefficient comparing the ith and jth individuals and distances between populations might be given by Mahalanobis' generalized distance (D^2). With this representation there is no need for all (or any) of the variates to be quantitative.

Although we shall constantly have in mind this geometrical representation of the multivariate sample we shall not explicitly require the coordinate values.

3. METHOD I (SOKAL AND MICHENER)

The weighted mean-pair method of Sokal and Michener was originally applied to an entomological problem [1958]. A more recent description has been given by Sokal and Sneath [1963] who recommended it as the best of a class of commonly used methods of cluster analysis.

The method operates on a $N \times N$ similarity matrix, but originally formal product-moment correlations between pairs of individuals were used. The algorithm pairs those two individuals, i and j say, which have the highest similarity and replaces columns (and rows) i and j by a single column with suitably chosen 'average' similarity coefficients. To replace correlations, the Spearman formulae giving the correlations between sums of standardized variables are used, and for similarities, the means of the elements in rows (and columns) i and j are taken. The process is then repeated on the new $(N - 1)$ order matrix, when either two new individuals have the highest similarity and form a new pair, or the existing pair combines with a further individual to make a cluster of three. The process continues by pairing individuals and clusters of previously combined individuals.

3.1. *Comments on Method I*

This is an agglomerative method, building up clusters from the individuals.

The Spearman formulae correlate two sets of sums of variates where the constituent variates are all standardized to have unit variance. In Method I, when correlations are used, the original matrix is a matrix of correlations between individuals. To make correlations between individuals independent of the scales of measurements of the variates, each true variate is first standardized by dividing by its standard error. The process of correlating individuals effectively destroys this standardization because the scores for each individual are implicitly restandardized by dividing by the standard error of all its standardized scores and then subtracting the mean restandardized score. Such correlation coefficients between individuals are best regarded as just another association coefficient; it is difficult to justify using the Spearman formulae on them and even more difficult to justify using Fisher's z transformation, as has been recommended, for this is based on the sampling distribution of correlations between variates drawn from normal distributions. We give a simple alternative averaging procedure in the next section, equally valid for correlations and similarities.

3.2 *The evaluation of similarities between clusters*

In the geometrical interpretation we assume that if S_{ij} is a similarity between individuals i and j, then the distance between their point representations P_i and P_j is $[2(1 - S_{ij})]^{\frac{1}{2}}$. Gower [1966] has shown that the latent vectors of the similarity matrix, scaled so that the sum of squares of the elements of the rth vector is equal to the rth latent root, gives directly a set of coordinates with this distance property.

Other monotonic transformations of dis-similarity might have been selected but this choice simplifies the algebra and leads to a direct comparison with the procedure of averaging similarities. The replacement of similarities by correlations creates no problem. The algorithm of Method I replaces the point pair P_i and P_j which are closest together by a new point; with the Spearman formulae or average similarity method the position of this new point relative to P_i and P_j is unknown (in our method we ensure that the new point lies on the line P_iP_j).

Suppose a stage in the operation of Method I has been reached when there are k clusters (some, or all, of which may still represent individuals) and that these are represented by points $P_1, P_2, \cdots, P_k$. Suppose also that P_n and P_m are the closest neighbours whose distance apart is $[2(1 - S_{nm})]^{\frac{1}{2}}$. Now P_n and P_m are to be replaced by a single new point P' and we require the similarity $S_{r,nm}$ between P' and any individual or cluster r, represented by P_r. The geometrical situation is illustrated in Figure 1. We also suppose that P_n has weight n and P_m weight m (the weight of P_r is immaterial); these weights might, for example, be the number of individuals comprising the clusters represented by P_n and P_m, but other types of weighting will also be considered.

We place P' at the centroid of P_n and P_m so that P' has weight $n + m$. In Figure 1 we have given the square of each side. We define $S_{r,nm}$ by equating P_rP' with $[2(1 - S_{r,nm})]^{\frac{1}{2}}$. Therefore

$$1 - S_{r,nm} = \frac{n}{n+m}(1 - S_{rn}) + \frac{m}{n+m}(1 - S_{rm}) - \frac{nm}{(n+m)^2}(1 - S_{nm}), \qquad (1)$$

or

$$S_{r,nm} = \frac{n}{n+m} S_{rn} + \frac{m}{n+m} S_{rm} + \frac{nm}{(n+m)^2}(1 - S_{nm}). \qquad (2)$$

Formula (2) is very easy to use and ensures that as new individuals or clusters are added to existing individuals or clusters, the new cluster so formed is always represented by its centroid; this procedure is not always desirable. We might, for example, wish to give each cluster unit weight, regardless of the number of individuals in it; this would place P' at the midpoint of P_nP_m and give

$$S_{r,nm} = \tfrac{1}{2}(S_{rm} + S_{rn}) + \tfrac{1}{4}(1 - S_{nm}). \qquad (3)$$

At this point we must discuss briefly the terminology referring to weighted and unweighted analyses. Normally we would call formula 2 weighted and formula 3 unweighted, but Sokal and Sneath [1963],

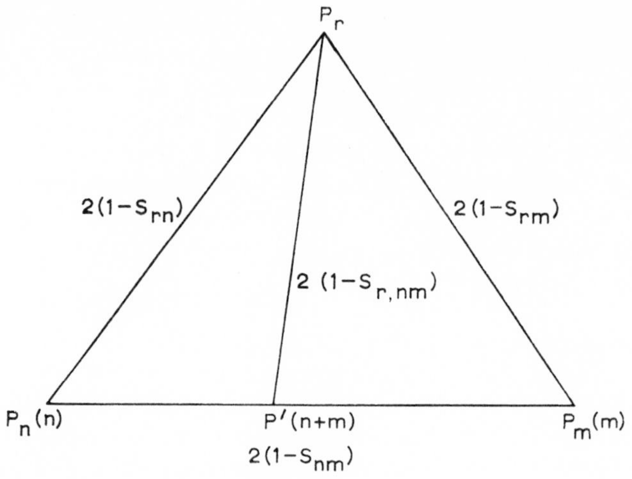

FIGURE 1

MEAN PAIR AGGLOMERATIVE CLUSTERING. WHEN P_n AND P_m COMBINE THEY ARE REPRESENTED BY P' (DIVIDING P_n, P_m IN THE RATIO $m : n$). IF P_r REPRESENTS ANY OTHER INDIVIDUAL (OR CLUSTER) IT IS NECESSARY TO CALCULATE P_rP'.

following Sokal and Michener [1958] reverse this nomenclature; we shall adhere to their terminology to avoid confusion. This nomenclature can be defended if we fix attention on the original individuals and ignore clusters which may have been derived from them; in this case we see that P' is the unweighted centroid of these original units when formula 2 is used, whilst 3 implies a rather complicated weighting system.

Formula 3 would be useful when, in an extreme case, several identical individuals had been included in the sample. These would be represented by identical points and we would not want them to bias, by sheer weight of numbers, the cluster they would inevitably form. A similar point is made by Sokal and Sneath ([1963], pp. 190–1) when discussing the relative merits of weighting or not weighting:

> 'Sokal and Michener [1958], adopted the method of weighting each new member as equal to the sum total of all old group members. They thought such a procedure to be the less objectionable method of the two, in view of the underlying assumed phylogenetic causes of the phenetic relationships under study. Thus it was felt that clusters of several OTU's [individuals] or stems bearing a single OTU [individual] represent independent evolutionary lines.'

Accordingly, we would often favor the unequal weighting system and use equation (3). This system is also easier to compute.

Note that, when the groups are close (i.e., S_{nm} is near to unity) equation (3) gives a method similar to that of averaging similarities, but in the later stages of the computations when dis-similar groups are being joined it may lead to different results. This equivalence suggests that Method I embodies a metric which equates distances between the original individuals approximately proportional to $(1 - S_{ij})^{\frac{1}{2}}$.

4. METHOD II (EDWARDS AND CAVALLI-SFORZA)

Edwards and Cavalli–Sforza [1965] suggest dividing the points into two sets such that the sum of squares of distances between sets is a maximum. This defines what they mean by a cluster and they only require an algorithm to find the two sets with the desired property. Because the total sum of squares is a constant for a given sample, maximizing the between-set sums of squares is equivalent to minimizing the within-set sum of squares. Their algorithm is to examine all $2^{N-1} - 1$ partitions of the N points and select the one which gives the minimum within-set sum of squares. The sum of squares within a set of p points is $1/p$ times the sums of the $\binom{p}{2}$ squared interpoint distances; this property is helpful when examining all partitions generated by successively exchanging one point between the two current sets.

4.1 *Comments on Method II*

The computational labour of examining all $2^{N-1} - 1$ splits is enormous and the authors give $(N - 1)^2 2^{N-11}$ seconds as the time required on a computer with a 5μ-second access time. Thus with N as low as 21, over 100 hours are required and with $N = 41$ the time is over 54,000 years. Even with the fastest of projected computers these times could only be decreased by a factor of about 100, so the method is impracticable even for small values of N, and an alternative, possibly approximate, algorithm must be devised. This the authors appear to have done because they write '(with more than 16 points) a sequential method is necessary to avoid taking all the splits, and this has also been developed,' but give no details.

The sum of squares of distances between two sets X and Y with n and m members respectively $(n + m = N)$ is

$$(nm/N) \sum_{i=1}^{v} (\bar{x}_i - \bar{y}_i)^2, \qquad (4)$$

where $(\bar{x}_1, \bar{x}_2, \cdots, \bar{x}_v)$ is the centroid of set X and $(\bar{y}_1, \bar{y}_2, \cdots, \bar{y}_v)$

the centroid of set Y. Thus for given values of n and m, maximizing the between-sets sum of squares is equivalent to dividing the N points into two sets whose centroids are maximum distances apart. In section 3.2 reasons have been given for sometimes avoiding methods which depend on sample sizes when determining a cluster. The present method has this property. For example, if we have three positions A, B, and C on a straight line with co-ordinates 0, $\frac{5}{7}$, 1, then on a nearest neighbour basis any individuals at B and C would be grouped together irrespective of sample sizes. If there is one individual at A, 49 at B, and 50 at C, the best sum of squares split puts the individuals at A and B into one group and those at C into the other. When there are only five individuals at B, then A forms one group and B and C the other. Accordingly, we also investigate below the simpler grouping criterion $\sum_{i=1}^{r} (\bar{x}_i - \bar{y}_i)^2$ which maximizes the distance between centroids. This will group B with C in both the above cases but is still not independent of the values of m and n although we might expect it to be less sensitive to them. For example, if there are 20 individuals at A, 7 at B, and 3 at C, centroid splitting groups A and B; incidentally, in this case sum of squares splitting groups B and C. Thus in both sum of squares and centroid splitting, nearest neighbouring individuals need not be grouped together (see Method I for clustering on nearest neighbours). It may even be possible that the biggest distance or sum of squares can be obtained by assigning some of the individuals at B to be grouped with those at A and the remainder to be grouped with those at C.

4.2 *Properties of Method II*

Suppose that the $N(= n + m)$ points have been divided into two sets X of n points and Y of m points. The situation is set out geometrically in Figure 2 where X is the centroid of the set X and Y the centroid of the set Y. For convenience we can take XY as one coordinate axis with x as the coordinate of X and y that of Y. XY must pass through G, the centroid of the whole system. Suppose P is one of the members of set Y, then we can take another coordinate axis in the plane PXY. P can then be defined by two coordinate values (u, v) in this plane irrespective of the total number of dimensions used in the representation. Now if P is assigned to the other set X, then the centroid X will move to X' on PX and the centroid Y to Y' on PY, where $X'Y'$ must also pass through G. The coordinates of X' and Y' are

$$X'\left(\frac{nx + u}{n + 1}, \frac{v}{n + 1}\right); \qquad Y'\left(\frac{my - u}{m - 1}, \frac{-v}{m - 1}\right).$$

Thus

$$(n + 1)^2(m - 1)^2X'Y'^2 = [nm(y - x) - Nu + (nx + my)]^2 + N^2v^2.$$

When X and Y are the two sets which maximize the between set sums of squares, moving P from Y to X must reduce this sum of squares. By formula (4) this means that:

$$nmXY^2 \geq (n + 1)(m - 1)X'Y'^2;$$

i.e.,

$$nm(n + 1)(m - 1)(y - x)^2 \geq [nm(y - x) - Nu + (nx + my)]^2 + N^2v^2. \quad (5)$$

Thus P, and hence all members of Y, must lie within a sphere with centre at $G + nm(y - x)/N$ on XY and radius r_y given by

$$N^2r_y^2 = nm(n + 1)(m - 1)(y - x)^2.$$

Similarly, the members of X lie within a sphere with centre at $G - nm(y - x)/N$ on XY and radius given by

$$N^2r_x^2 = nm(n - 1)(m + 1)(y - x)^2.$$

These two spheres intersect for the imaginary value of v given by

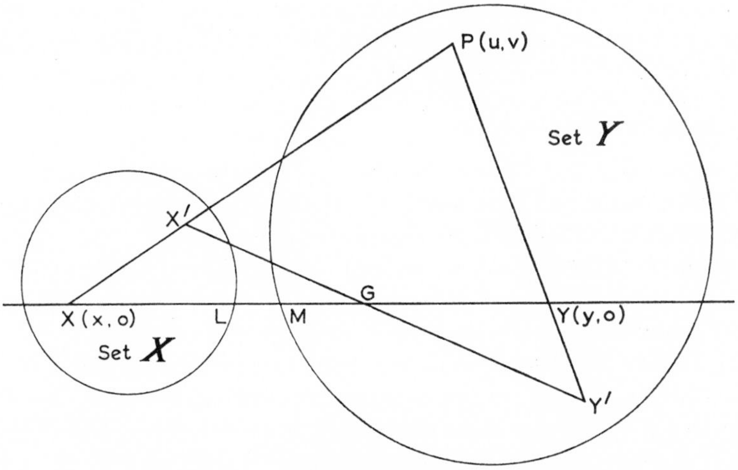

FIGURE 2

DIVISIVE CENTROID CLUSTERING. WHEN P IS INCLUDED IN X INSTEAD OF Y, THE CENTROID X MOVES TO X' AND THE CENTROID Y TO Y'. WHEN **X, Y** ARE OPTIMUM CENTROID CLUSTERS THE DISTANCE XY BETWEEN THE CENTROIDS CANNOT BE LESS THAN THE NEW DISTANCE $X'Y'$.

$v^2 = -\frac{1}{4}(y - x)^2$, so that the sets X and Y are disjoint. A series of multidimensional flats can be drawn perpendicular to XY which separate the two sets; one of these flats is given by

$$u = G - (m - n)(y - x)/2N.$$

When we define optimum grouping as that which maximizes the distance between centroids, we have

$$XY \geq X'Y'$$

i.e.,

$$(n + 1)^2(m - 1)^2(y - x)^2 \geq [nm(y - x) - Nu + (nx + ny)]^2 + N^2v^2. \quad (6)$$

In this case all members of Y again lie within a sphere with the same centre as before but with radius $N\rho_y = (n + 1)(m - 1)(y - x)$. The members of X lie within a sphere with radius $N\rho_x = (n - 1)(m + 1)(y - x)$. When n and m are large, ρ_x and ρ_y are nearly equal to r_x and r_y. These two spheres are also disjoint and the two flats perpendicular to XY given by

$$NU_x = (n - 1)y + (m + 1)x \quad \text{and} \quad NU_y = (n + 1)y + (m - 1)x$$

bound a region containing no members of either set.

Suppose the spheres meet XY in L and M as shown in Figure 2 (for centroid splitting L and M coincide with U_x and U_y); then a member of Y can be closer to X than Y if $XM < MY$. It is easy to show that this is impossible for sum of squares splitting but *may* occur with centroid splitting when $m - n > 2$. The possibility of the existence of points nearer to the other centroid than their own seems to be a disadvantage of the centroid method. This is unlikely to arise when the groups are well separated, when one would expect the sum of squares and centroid methods to give the same result.

The properties established above do not help provide a simple, efficient, and exact algorithm for either sum of squares splitting or centroid splitting. The centroid interpretation of both methods suggests that it might be simpler to work with the coordinates of points rather than with the distances between them. We know the radii and the positions of the centres of the hyperspheres giving optimum division, so any partition giving one or more points inconsistent with formulae (5) and (6) (as appropriate) can be rejected. The disjoint properties of the hyperspheres mean that partitions giving overlapping hyperspheres can be ignored. Unfortunately even if these properties could be conveniently used to find a partition consistent with (5) or (6), there would still be no guarantee that the optimum had been found, for it has not been shown that such partitions are unique—we have proved necessity but not sufficiency. It is also worth noting that when the N points occupy $(N - 1)$ dimensions, it is always possible to construct a flat which separates any subset of these points from the remainder. For example, when three points are at the vertices of a triangle, a straight line can always be drawn so that any selected point lies on a different side to the other two, but this is not possible if the three points are co-linear.

It is not difficult to devise approximate algorithms for establishing this type of division; for example, we could start with two sets X, Y containing those points X_0, Y_0 which are farthest apart. Find the point P nearest to X_0 and the point Q nearest to Y_0. Suppose $PX_0 < QY_0$; then P is added to the set X and X_0 replaced by X_1, the centroid of X_0 and X_1. Points are thus added successively to either X or Y, and at any stage each set is represented by the centroid of the points currently assigned to it. This process is a divisive method similar to the agglomerative Method I, discussed above.

5. METHOD III (WILLIAMS AND LAMBERT)

Williams and Lambert [1959] consider the case when the data matrix **X** consists entirely of presence and absence data (denoted by 1 and 0, respectively). Their original applications were all ecological, where the 'variates' were the different plant species present (or absent) in N quadrats (corresponding to the individuals). The N quadrats are to be divided into two subsets on the basis of the species k which best separates them (in a defined sense); one set being all those quadrats containing species k and the other all quadrats without species k. The matrix **R** is calculated whose elements r_{ij}^2 are proportional to the values of χ^2 derived from the 2×2 contingency table of species i versus species j. The species used for primary division is that for which $\sum_{i=1}^{v} \chi_{ik}^2$ is a maximum: i.e., k is the species which has maximum row-total in **R**. After primary division the algorithm is repeated on each subset until some stopping rule terminates the process. As Lance and Williams [1965] point out, the value of χ_{ij}^2 is N times the square of the correlation r_{ij} between species i and j. These authors give as a rationale of the method (here slightly modified) that if we divide on the basis of variate x_k, then the between-groups sum of squares for any other single variate x_i is given by the usual 'sum of squares due to regression' formula $r_{ik}^2 \sum_j (x_{ij} - \bar{x}_i)^2$, so that the total over all variates is

$$\sum_{i=1}^{v} r_{ik}^2 \sum_{j=1}^{N} (x_{ij} - \bar{x}_i)^2.$$

If now every variate is standardized to have unit variance this value becomes $\sum_{i=1}^{v} r_{ik}^2$.

5.1 *Comments on Method III*

The rationale given above shows that as in Method II the sum of squares between the two groups formed is a maximum, but here only v of the $2^{N-1} - 1$ splits are to be examined. The coordinates of the points representing the quadrats are the scores of **X** standardized to have unit variance. The squared distance between two points P_i and P_j measured on this scale is $\sum 1/s_r^2$, where s_r^2 is the variance of species r and summation takes place over all species not both present or both absent in quadrats i and j. Now s_r is smallest for very rare or very common species and is largest for species which occur in half the quadrats; thus the distance between a pair of quadrats, one of which contains a rare species, or does not contain a common species, will be exaggerated compared with distances between 'typical' quadrats. This may not always, or even usually, be a desirable property. If two quadrats differ in only one species, are they more different when this is a rare species than when it is a common species? The rare species may occur merely by chance, and hence be unimportant, whereas the lack of a common species may well be due to a real difference in ecological conditions. A distance such as $(1 - S_{ij})^{\frac{1}{2}}$ would normally be preferable, where S_{ij} is a simple matching coefficient between quadrats i and j (i.e., S_{ij} is the proportion of species either both present or both absent). This is equivalent to regarding the original 0/1 data as coordinate points without standardization.

Lance and Williams]1965] have found that the $\sum \chi^2$ criterion does indeed tend to 'fragment the analysis by initially splitting off outliers from the population', and say that the maximum value of $\sum_{i=1}^{v} |r_{ik}|$ provides a better general purpose solution. This is equivalent to defining the distance between the centroids of the two sets into which the sample is split by $\sum_{i=1}^{v} |m_{1r} - m_{2r}|/s_r$, where m_{pr} is the mean of variate r in group p, and is an example of the city block metric (see Kruskal [1964]). In this system of geometry a centroid is defined by the usual formula, but the distance between two points $P_i(x_{i1}, x_{i2}, \cdots, x_{iv})$, $P_j(x_{j1}, x_{j2}, \cdots, x_{jv})$ is defined as $\sum_{r=1}^{v} |x_{ir} - x_{jr}|$. The distance between two quadrats is now $\sum 1/s_r$ where, as before, summation is over those species not both present or both absent in the two quadrats. Thus there is an improvement in the treatment of aberrant quadrats because species representation is now weighted by $1/s_r$ instead of $1/s_r^2$.

The city block metric is not invariant under rotation of axes. Multivariate statistical methods commonly transform quantitative variates to new sets of axes and this suggests that the $\sum |r|$ criterion would be inappropriate for any extension of the method to include quantitative variates.

Both the $\sum r^2$ and $\sum |r|$ criteria are based on sums of squares between groups, so the factor nm of equation 4 is relevant. To get centroid division these criteria should be replaced by $\sum_{i=1}^{v} r_{ik}^2/n_k m_k$ and $\sum_{i=1}^{v} |r_{ik}|/(n_k m_k)^{\frac{1}{2}}$, where n_k and m_k are the sizes of the two sets with and without species k.

5.2 *Maximizing the multiple correlation*

An alternative, but related, procedure to Method III is to maximize the multiple correlation R_k^2 instead of maximizing the sum of simple correlations. Such a procedure is equivalent to finding the best linear discriminator (the multiple regression function) which separates one species on the basis of the other $v - 1$ species.

We can select each of the v variates in turn and calculate the multiple correlation with the remaining $v - 1$ variates. Division is made on the variate x_k for which R_k^2 is a maximum. To avoid inverting v different matrices each of order $v - 1$ we can use the following easily verified result from regression theory which needs the inverse of only one matrix.

Let $\mathbf{X}$ be the original data matrix of N rows and v columns; i.e., v is the number of species and N the number of quadrats. We shall write $\mathbf{X} = (\mathbf{x}_1, \mathbf{x}_2, \cdots, \mathbf{x}_v)$, where $\mathbf{x}_i$ is the column vector giving the values for the ith species (assumed, for convenience, measured as deviations from their mean) for all N quadrats. Suppose species t is taken as the dependent variate, then $(\mathbf{X}'\mathbf{X})_{t,t}^{-1}$, the tth diagonal element of $(\mathbf{X}'\mathbf{X})^{-1}$ is given by

$$(\mathbf{X}'\mathbf{X})_{t,t}^{-1} = [(\mathbf{x}_t'\mathbf{x}_t)(1 - R_t^2)]^{-1}.$$

Thus to maximize R_t^2 we have only to find the value of t for which

$$(\mathbf{x}_t'\mathbf{x}_t)(\mathbf{X}'\mathbf{X})_{t,t}^{-1} \tag{7}$$

is a maximum. Formula (7) is trivial to compute for all values of t once $(\mathbf{X}'\mathbf{X})^{-1}$ has been computed. If variate $t = k$ is the one for which R_t^2 is a maximum and $\mathbf{x}_k$ consists entirely of 0's and 1's, then the two groups are the quadrats with 0 and 1, respectively, in $\mathbf{x}_k$; when $\mathbf{x}_k$ is quantitative we might split the individuals (quadrats) into two groups by selecting a value σ such that one group consists of those individuals for which $x_k < \sigma$, and the other of those for which $x_k > \sigma$, a possible choice of σ being the value which maximizes the difference between the mean values of x_k in the two groups. This device could also be used if Method III were used with quantitative variates.

6. A COMPARISON OF THE THREE METHODS

Method II chooses, out of $2^{N-1} - 1$ possible divisions, that split which maximizes the between set sum of squares. Method III does much the same, except that only v possible splits are examined and the split is made on the single species (variate) which maximizes the sum of squares. We have given reasons for replacing the sum of squares criterion by the closely related but simpler concept of distance between centroids but neither method is entirely satisfactory. Method III embodies a particular distance function which may often have undesirable properties; in Method II any distance function can be chosen. Method II, however, becomes computationally impracticable for quite low values of N; it is hoped that some of the properties given in section 5.2 might help in devising a more convenient algorithm for achieving the purpose of Method II but this would only be useful when it is agreed that the centroid is a desirable measure of the centre of location of a cluster; in section 3.2, we have discussed how the use of the centroid can sometimes give misleading results. Many of the difficulties outlined in the examples at the end of section 4.1 arise from the use of the centroid.

Cluster analyses have been classified into agglomerative versus subdivisive, and polythetic versus monothetic methods. Method I is an agglomerative method because clusters are gradually extended by adding further individuals or previously defined clusters; Methods II and III are divisive. Method III is monothetic because division is made on a single character whilst Methods I and II are polythetic because the division is into groups which may, and usually will, contain no constant characters. It is held that divisive methods will not lead to any spurious groupings and although this is probably mostly true there appears to have been no formal investigation. For example, suppose we have three well defined groups; then no harm is done if division is made as in Figure IIIa, but can we guarantee that it will not occur as in Figure IIIb? We would, however, probably be happier if divisions were made as in Figure III(c), which is the type of clustering found by agglomerative methods such as Method I. A further disadvantage of divisive methods is that there appears to be no simple way of adjusting for the difficulties discussed in section 4.1 which may arise from using the centroid; it might be possible by putting into the analysis the right proportion of each type of individual to get any desired type of nesting.

The main criticism of monothetic methods is the obvious one that

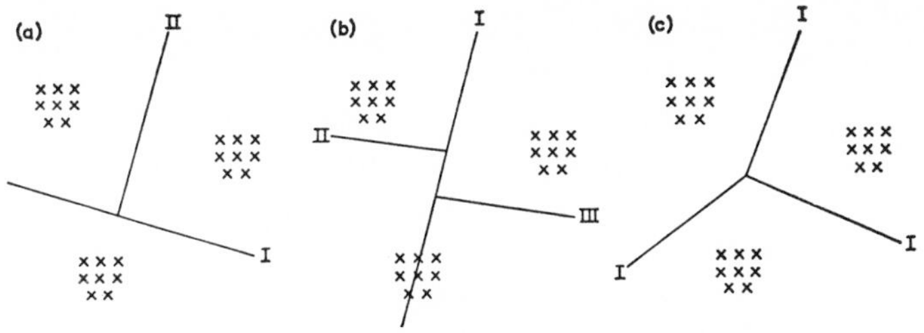

FIGURE 3

POSSIBLE DIVISIVE AND AGGLOMERATIVE METHODS OF CLUSTERING. THE ROMAN FIGURES GIVE THE SUCCESSIVE LINES OF DIVISION.

individuals (quadrats) which just happen to lack the attribute (species) on which division is made, but are otherwise very similar to other individuals, will be wrongly classified. It also seems that monothetic methods commit one to a divisive form of clustering; the converse is not true. Consequently we do not recommend Method III for most purposes but it might be useful for devising a key. However, even here we suggest that the method (outlined in section 5.2) which maximizes the multiple correlation, might be more efficient because this method is related to discriminant analysis with its properties of minimum errors of mis-classification.

For general purpose classification work we recommend Sokal and Michener's 'Weighted Mean-Pair Group' method (i.e., Method I) using equation 5 to calculate the similarities (or correlations) between old clusters and new clusters. We would expect Method I using the centroid formula 2 to be at least an approximate algorithm to achieve the aims of Method II; the clusters would, of course, be found in the reverse order to those of Method II.

6.1 *Warning*

A cluster analysis imposes a hierarchical structure on the sample. Although this is probably a legitimate requirement for taxonomic purposes (at least in differentiating the higher taxonomic orders) and when a key is being constructed, it may be quite inappropriate in other situations. When cross-classifications exist, for example, the more traditional forms of multivariate statistical analysis will be more useful (Gower [1966]).

7. CONCLUSION

There appears to be general agreement between the three methods that clusters are defined by selecting those sets of individuals which maximize some measure of interset distance. Because distance itself can be defined in many ways, such a cluster definition must be relative to the particular measure of distance chosen. The lack of a precise definition of a cluster will not stop people from using methods of cluster analysis; so it is as well to understand just what the various methods set out to do and what assumptions are made implicitly and explicitly. We have found the geometrical approach very revealing and hope that some of our results will help to clarify the subject.

ACKNOWLEDGEMENTS

I thank Mr. A. J. B. Anderson and Mr. C. K. Kirton for their comments on an earlier draft of this paper.

UNE COMPARAISON DE QUELQUES METHODES D'ANALYSES DE GRAPPES

RESUME

Chaque élément d'un échantillon multivarié peut être représenté par un point dans un espace Euclidien multidimensionnel. L'analyse de grappes s'efforce de grouper ces points enensembles disjoints dont on espère qu'ils correspondent à des traits marquants de l'échantillon. Différentes méthodes d'analyses de grappes pour un même échantillon peuvent s'appuyer sur des hypothèses différentes concernant la distribution géométrique des points ou peuvent employer des critères différents de réunion en grappes ou peuvent différer à la fois par ces deux aspects. Trois méthodes apparemment différentes d'analyses en grappes sont examinées. On montre que les critères de mise en grappes de toutes ces méthodes, et plusieurs nouveaux critères dérivés de ceux-ci, ou suggérés par ces méthodes, peuvent être interprétés en termes de distances entre les centroïdes des grappes. Dans la plupart des cas on trouve la distribution géometrique. Les méthodes sont comparées et des suggestions sont faites pour leur amélioration; quelques-unes de leurs propriétés sont établies.

REFERENCES

Edwards, A. W. F. and Cavalli-Sforza, L. L. [1965]. A method for cluster analysis. *Biometrics 21*, 362–75.

Gower, J. C. [1966]. Some distance properties of latent roots and vector methods used in multivariate analysis. *Biometrika 53*, 325–38.

Kruskal, J. B. [1964]. Multidimensional scaling by optimizing goodness of fit to a non-metric hypothesis. *Psychometrika 29*, 1–27.

Lance, G. N. and Williams, W. T. [1965]. Computer program for monothetic classification ('Association analysis'). *Comp. J., 8*, 246–9.

Sokal, R. R. and Michener, C. D. [1958]. A statistical method for evaluating systematic relationships. *Univ. Kansas Sci. Bull. 38*, 1409–38.

Sokal, R. R. and Sneath, P. H. [1963]. *Principles of Numerical Taxonomy. W. H.* Freeman, San Francisco and London.

Williams, W. T. and Lambert, J. M. [1959]. Multivariate methods in plant ecology, I. *J. Ecol., 47*, 83.

20

Reprinted with permission from *Syst. Zool.*, **19**(1), 58–82 (1970)

ADAPTIVE HIERARCHICAL CLUSTERING SCHEMES

F. JAMES ROHLF

Abstract

Rohlf, F. J. (Biological Sciences, State Univ., Stony Brook, N. Y. 11790) 1970. Adaptive hierarchical clustering schemes. Syst. Zool., 18:58–82.—Various methods of summarizing phenetic relationships are briefly reviewed (including a comparison of principal components analysis and non-metric scaling). Sequential agglomerative hierarchical clustering schemes are considered in particular detail, and several new methods are proposed. The new algorithms are characterized by their ability to "adapt" to the possible trends of variation found within clusters as they are being formed. A nonlinear version allows the isolation and description of clusters which are parabolic, ring-shaped, etc., by the introduction of appropriate dummy variables. Procedures for computing the best fitting trend line through the cluster are also presented, and problems in measuring the amount of information lost by clustering are discussed. [Phenetics; cluster analysis; numerical taxonomy.]

This paper is concerned with a brief review of some of the techniques of summarizing phenetic similarities that have been proposed for use in numerical taxonomy. One class of methods (sequential agglomerative) is considered in detail and new procedures which allow for elongated and curvilinear clusters are proposed.

The "taxonomy problem" in biology can be described as follows: Given a set of specimens ("operational taxonomic units" or OTU's, Sokal and Sneath, 1963, which may represent taxa of any rank) known only by a list of their properties or characters, we wish to find the "best" way of describing their often complex patterns of mutual similarities (*phenetic* relationships). Such relationships do not necessarily imply evolutionary (cladistic) relationships (for a discussion of these approaches, see Sokal and Camin, 1965).

The methods that have been developed appear to have a more general application than just in biological taxonomy, but there are certain facts and assumptions that can be made in biology which influence our choice of methods. As a result, the techniques may or may not be completely valid in other fields. Some of the considerations which influence the development of cluster analyses in biological taxonomy are the following: (1) "All things being equal" we would hope that a system of nested clusters would be found. This is due to the fact that evolution is believed usually to be a divergent process and the distribution of OTU's in a phenetic space should to some extent reflect this. There are, of course, exceptions to this overall rule which are very important, such as those provided by hybridization and clinal variation. (2) Another consideration is the nature of the character set representing each OTU. We would like to use a "random sampling of characters" or at least a "representative" sampling of characters. But since different sets of characters seem to yield slightly different systems of relationships (Rohlf, 1963; Ehrlich and Ehrlich, 1967; Michener and Sokal, 1966), biologists may have to get used to the idea of using different classifications, based upon different sets of characters, each best for its own special purpose, with overall similarities based on the total character set available at any one time. (3) The selection of OTU's is also not random. Since we cannot study all organisms, we must select those which are of immediate interest. But even with a specified group of organisms, we usually cannot sample at random. This is so because the distributions of recent (and even fossil) organisms are clumped in a phenetic hyperspace. One needs to pass up many very similar, common specimens to obtain a more interesting sampling of different kinds of organisms. Thus, a preliminary screening of individuals according to their apparent similarities must be made before one can make detailed measurements to analyze their phenetic relationships quantitatively. The problem here is mainly the great amount of effort required to record large numbers of characteristics for large numbers of individuals. It is hoped that with the development of automatic data recording devices, it will be possible to gather large amounts of data easily so that random sampling will become practical. As a consequence of points (2) and (3) given above, cluster analytic procedures are only quasi-statistical (see also Sneath, 1967, for further discussion).

A major problem in cluster analysis is that there is no universal agreement on what constitutes a cluster; and, most investigators think that clustering methods depend on the clusters to be found. This may be the reason why almost everyone in this field has his own version of cluster analysis. Moreover, it is important not to cluster one's data blindly, but to develop techniques of clustering which allow one to check the reasonableness of the clusters for the data to which it is applied.

An additional problem which must be faced is the fact that present applications in biological taxonomy require the use of cluster analysis on data with relatively few OTU's. Thus, one does not have a high density of points within clusters (if they exist), but, rather, only vague outlines and suggestions of clusters.

SUMMARY OF NUMERICAL TAXONOMIC METHODS

The standard computational steps performed in a "typical" numerical taxonomic study are outlined below for reference purposes later in the paper.

The following *preliminary steps* are usually carried out.

A. Data are gathered on p characters (variables) for t individuals (operational taxonomic units or OTU's). The data are usually quantitative measurements of some sort and often are coded into a few class intervals. The result is a p by t matrix $\mathbf{Y}$. Frequently there are missing observations in the matrix.

B. The characters are standardized over the OTU's to place them in equivalent units. Usually they are transformed so as to have a mean of zero and a unit standard deviation (or else the ranges may be equalized). It should be pointed out that some procedures are unaffected by such scaling (see below and Friedman and Rubin, 1966, for examples).

C. A matrix of coefficients is computed which measures the relative degree of similarity between all pairs of OTU's. Some coefficients which have been used are: the product-moment correlation coefficient, r; average distance, d (Sokal, 1961); generalized distance, D^2 and various association coefficients (see Sokal and Sneath, 1963; Rohlf and Sokal, 1965; Huizinga, 1962).

The matrices generated in step C are usually quite large ($t(t-1)/2$ elements). It is therefore necessary to summarize the information on the phenetic relationships implied by them so that the results can be easily comprehended and communicated. Since the results must be simplified, there will almost always be some loss of information and consequent distortion in the final results. The problem is to select a method which gives the proper balance between preservation of information (good fit) and simplicity.

Three classes of methods may be distinguished:

1.—*Multi-dimensional scaling* (ordination). These are techniques in which one obtains the coordinate axes of each OTU in the smallest dimensional space that still preserves sufficient information about the interpoint distances (Kruskal, 1964). This can be accomplished in various ways depending on the restrictions placed upon the system. A special case (in which it is required that the new coordinated axes be linear combinations of the original variables and that the maximum percent of the original variance of the original variables be preserved) is called "principal components analysis" (Rao, 1952; Anderson, 1958; for a taxonomic example see Rohlf, 1967, 1968). Alternatively one has nonmetric scaling (Kruskal, 1964) and parametric mapping

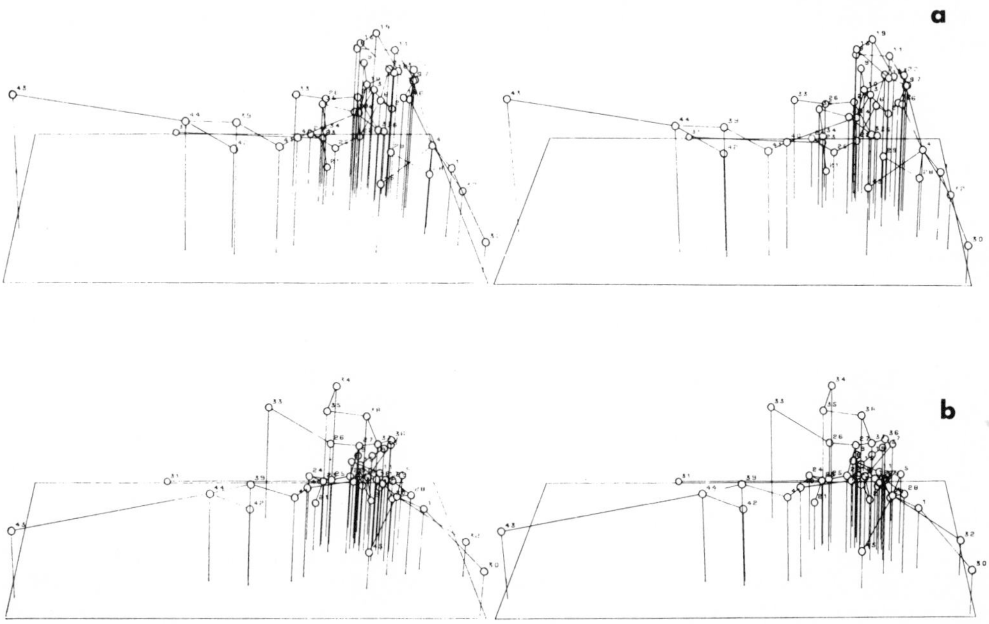

FIG. 1.—Stereo pairs of a perspective view of a three-dimensional model of 45 species of mosquito pupae (Rohlf, 1967). a. Projections onto the first three principal components computed from a matrix of correlations among 74 characters. Horizontal lines between points show the connections for the shortest simply connected network in the original space projected onto the three-dimensional model. These links connect closest neighbors in the 74-dimensional space. The three factors accounted for 59.36% of the variation. The cophenetic correlation equals 0.941. b. Projections computed by non-metric scaling analysis. Stress = 0.120 cophenetic correlation = 0.985.

(Shepard and Carroll, 1966). The advantage of these techniques is that they allow one to examine a scatter diagram displaying a summary of the structure of the data without having to first assume that clusters are present. A possible disadvantage for taxonomy is that they do not yield a hierarchical classification. I know of no published applications of the nonmetric scaling techniques to biological taxonomy but the few such analyses I have tried have given very satisfactory results (good [low stress] compression of the data into few dimensions). Fig. 1 shows stereo pairs based on a principal components analysis (Fig. 1a) and nonmetric scaling (Fig. 1b).

2.—*Network Analyses.* In these procedures one attempts to find networks (with varying degrees of connectedness) which will reflect the underlying structure of the data. Examples are the use of shortest simply connected networks (Kruskal, 1956; called "Prim" networks by Cavalli-Sforza and Edwards, 1967, who cited the work of Prim, 1957). Jardine and Sibson (1968a and b) and Neely (1969) have suggested the use of networks that are more than simply connected and thus capable of more adequately describing the data structure, but these are more complex to represent. Jardine and Sibson (1968a and b) also discuss methods for constructing a classification from their networks. An example of a shortest connection network is given in Fig. 1.

3.—*Cluster Analysis.* In these procedures OTU's are allocated to classes. Some of the classes may be nested within other classes if a hierarchical clustering scheme (HCS, Johnson, 1967) is used. If compact clusters are actually present with large gaps between clusters, then these procedures yield very adequate summaries. A large number of clustering algorithms which have been proposed. Ball (1965), Lance and Williams (1966), Gower (1967), and Bonner (1964) present accounts which compare some of the properties of various methods. Different methods are most suitable for data with certain structure, but relatively little is known about the type of data structure for which each method is most suitable. This point has been considered by Rohlf (1963 and 1967) and Sokal (1966).

Regardless of which of the above types of summarization techniques are used, it is important that they be accompanied by measures of the degree of fit (lack of distortion) between the original matrix of similarity coefficients and the pattern of similarity values implied by the summary. A number of such techniques have been proposed. For example, in principal components analysis one commonly quotes the percentage of the total variance which is contained in the first k dimensions. A useful approach to this problem of fit was proposed by Hartigan (1967) and Jardine et al. (1967). They suggested that one consider the process of constructing a hierarchical clustering scheme as one of fitting a hierarchical tree structure (which has the properties of an ultrametric; Hartigan, 1967) to the original distances (which usually are metric). If the distance between two OTU's in a phenogram is the level of the lowest cluster which contains both OTU's, then such distances satisfy the ultrametric inequality ($d_{ij} \leq \max\{d_{ik}, d_{jk}\}$, for all OTU's in the set). This approach implies the restriction that the level of each cluster ("cophenetic value" of Sokal and Rohlf, 1962) is a monotonically increasing function. Thus, "reversals" found in certain clustering methods (Michener and Sokal, 1957; Farris, 1969) are not permitted. Sokal and Rohlf (1962) proposed the cophenetic correlation (correlation between the elements of the original similarity matrix and the similarity values implied by a scheme of representation such as a cluster analysis, phenogram, principal components analysis, etc., see Fig. 9 below). Hartigan (1967) proposed a weighted sum of squared differences between the original euclidean distances and distances given by the phenogram (the fitted ultrametric). Jardine et al. (1967) proposed a class of measures of distortion, $\Delta\mu$, which involve sums of the absolute values of the differences between the original and implied distances raised to various powers ($1/\mu$). Jardine and Sibson (1968a) give a normalized version $\hat{\Delta}\mu$. By varying μ ($0 \leqslant \mu \leqslant 1$) one places greater or less emphasis on the larger or smaller differences. These are equivalent to the Minkowski metrics used by Kruskal (1964) to measure the average magnitude of the differences. Work such as that of Farris (1969) is needed to study the consequences of using these new coefficients.

As shown by Rubin (1966), Friedman and Rubin (1966), and Hartigan (1967), if one can define a measure of optimality, one can attempt to find the "optimal" classification by trial and error even if a direct algorithm cannot be devised.

It would seem desirable to use measures such as the cophenetic correlation or $\hat{\Delta}\mu$ which can be applied to the results of many types of analyses, so that one can decide which method gives the best fit to a given set of data. For example, using the cophenetic correlation coefficient as a measure of fit, a principal components analysis using only the first three components gives a better fit to data in which there is little tendency for OTU's to occur in clusters (e.g., Rohlf, 1967), whereas a cluster analysis can give a better fit if distinct, tight clusters actually exist (e.g., Rohlf, 1968). Since different methods of cluster analysis can yield rather different results when applied to the same data, an objective procedure is to select the method which provides the best fit (if the cophenetic correlation coefficient is used, then one is searching for the least squares best fitting summary of the phenetic relationships).

Principal components analysis is a useful complement to cluster and network analyses. A two- or three-dimensional scatter diagram shows the longer distances fairly well but it distorts the smaller

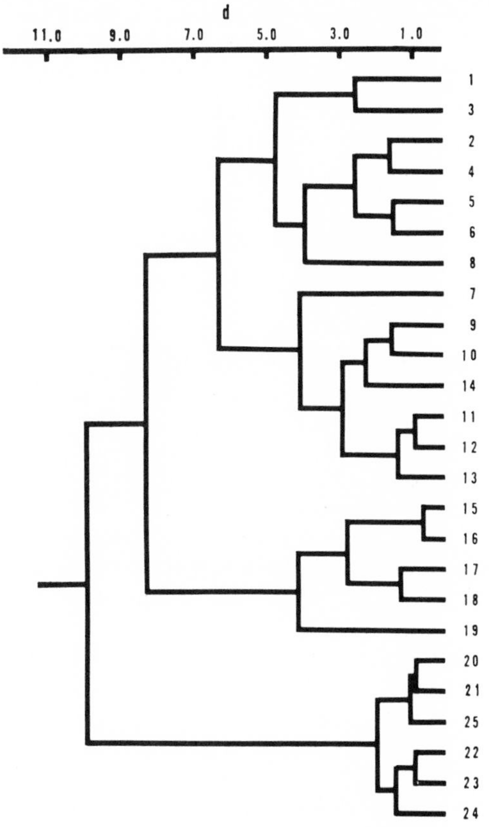

FIG. 2.—Phenogram constructed using UPGMA cluster analysis applied to artificial data in Fig. 4a. Cophenetic correlation, r = 0.781.

distances (Rohlf, 1968). The opposite seems to be true of most forms of cluster and network analysis. Thus, a principal components analysis is useful for assessing the overall gross phenetic structure of one's data. By projecting simple networks onto the principal components space one can display both types of relationships in a single figure. Fig. 1a is an example using the mosquito pupae data from Rohlf (1967). One will note that the genera *Aedes, Culex, Culisetea,* and *Orthopodomyia* each form a connected graph. *Anopheles* does not since species 45 is connected to an *Aedes* before it connects to an *Anopheles.* Fig. 2b is an example of nonmetric multi-dimensional scaling using the same data. Nonmetric scaling tends to evenly distribute the distortions among both the large and small distances (see Fig. 9 below).

HIERARCHICAL CLUSTERING SCHEMES (HCS)

In the present study, we shall consider only the class of clustering algorithms which can be characterized by being sequential and agglomerative.

A *sequential* clustering process forms clusters in a regular stepwise manner rather than "simultaneously" as in procedures such as those proposed by Rohlf (1966) and Hartigan (1967) which attempt to optimize the entire tree. Sequential procedures are usually much faster than simultaneous ones. This is important if large numbers of OTU's are to be considered.

Agglomerative clustering procedures begin with pairs of similar OTU's and build up clusters by contrast with divisive methods which start with the entire set of OTU's and divide it into subsets and these further into sub-subsets, etc. (see for example the method of Williams and Lambert, 1959). Gower (1967) points out some problems which arise in divisive-type cluster analyses.

Examples of cluster analytic techniques that satisfy the above restrictions are single linkage, complete linkage, weighted and unweighted pair group methods (see Sokal and Sneath, 1963; and Sokal, 1966).

The computational algorithm for all of these methods can be outlined as follows (the only differences between the methods being in the details of step 3 below):

1. Read in the matrix to be analyzed [in the account given below we will assume, for notational simplicity, that an average distance matrix (Sokal, 1961) is used].

2. Find all reciprocally closest pairs of OTU's (i.e., OTU's I and J are mutually closest if the smallest d in column I is in row J and the smallest d in column J is in row I). If the matrix is symmetric at least one such pair must exist. In the case of ties an arbitrary tie-breaking rule must be used.

3. For each of these pairs of OTU's add a new row and column to the matrix. The

coefficients, which give the distance between each of these new clusters and the previous OTU's are then computed. For example, if column $t+1$ corresponds to the cluster of OTU's I and J, then

$$d_{k,\,t+1} = d_{k,\,I+J} = f(d_{kI}, d_{kJ}; d_{IJ})$$

for $k = 1, t\ (\neq I, J)$. Also distances of the type

$$d_{t+1,\,t+2} = d_{I+J,\,L+M} = f(d_{IL}, d_{IM}, d_{JL}, d_{JM}; d_{IJ}; d_{LM}).$$

Column $t+2$ corresponds to a cluster consisting of the pair of OTU's L and M. A variety of functions, $f(\ldots;\cdot;\cdot)$, have been used (such as the minimum, maximum, or average of the distances listed before the first semicolon and correspond to what are called single linkage, complete linkage, and average linkage cluster analysis, respectively, by Sokal and Sneath, 1963). Proctor (1966) and Gower (1967) have proposed functions (see below) which result in a "centroid" clustering. Lance and Williams (1966) also give several such functions.

4. Rows and columns corresponding to the OTU's which have clustered are now deleted. The size of the matrix is reduced by the number of pairs found in step 2.

5. Steps 2 through 4 are repeated until there is only a single row and column left in the matrix.

After steps 3 and 4 have been taken for the first time, some of the columns in the matrix correspond to single OTU's while other columns may correspond to a cluster containing a variable number of OTU's. This makes it possible to make the new distances computed in the second and subsequent passes through step 3 a function of the number of OTU's within each cluster as well. For example, in the unweighted pair group method using arithmetic averages (UPGMA) one calculates a weighted average distance:

$$d_{k;I+J} = (n_I d_{KI} + n_J d_{KJ})/(n_I + n_J).$$

In centroid clustering the following functions are used:

$$d_{I+J,K} = f(d_{KI}, d_{KJ}; d_{IJ}) = \frac{n_I d_{KI} + n_J d_{KJ} - n_I n_J d_{IJ}/(n_I + n_J)}{n_I + n_J}$$

$$\begin{aligned} d_{I+J,L+M} &= f(d_{IL}, d_{JL}, d_{IM}, d_{JM}; d_{IJ}; d_{LM}) \\ &= (n_I n_L d_{IL} + n_J n_L d_{JL} + n_I n_M d_{IM} \\ &\quad + n_J n_M d_{JM})/(n_I + n_J)(n_L + n_M) \\ &\quad - n_I n_J d_{IJ}/(n_I + n_J)^2 \\ &\quad - n_L n_M d_{LM}/(n_L + n_M)^2, \end{aligned}$$

where n_I, n_J, etc., are the number of OTU's in cluster I, J, etc.

It is possible to diagram the results of a cluster analysis in the form of a tree-like structure called a "phenogram" by Camin and Sokal (1965). An example is given in Fig. 2. A computer program "TAXON" (available from the author) has been prepared which performs the analysis described above with built-in options for a variety of functions and special provisions for large matrices which cannot fit in the main core storage. The program also depicts a phenogram.

A simpler program is available in Bonham-Carter (1967) which also has provision for drawing a phenogram on an automatic plotter.

ANALYSIS OF THE RESULTS OF A HCS

The statistical reliability of a phenogram is not known. However, the effectiveness (as measured by the cophenetic correlation coefficient) of the phenogram in summarizing the original distance matrix indicates that, of the various functions (to be used in step 3 of the HCS) which have been proposed so far, a *weighted* average has always yielded the best fit (Sokal and Rohlf, 1962, Farris, 1969). This procedure has been called the "*unweighted* pair group method using arithmetic averages" or UPGMA (Sokal and Sneath, 1963). The question arises, however, whether a simultaneous clustering procedure might produce a phenogram with less distortion. Rohlf (1966) and Hartigan (1967) have shown this to be possible. It would not be computationally practical to evaluate the criterion

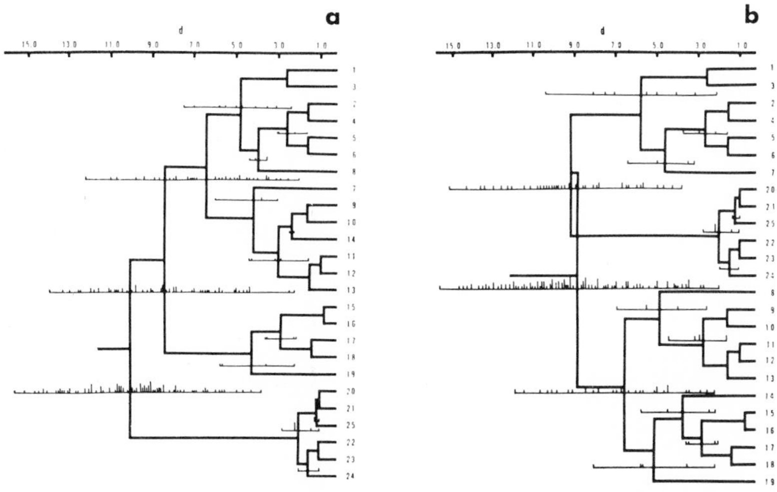

FIG. 3.—Phenograms based on the artificial data given in Fig. 4a. Superimposed upon the branches of the phenograms are frequency distributions of the distance coefficients between all OTU's in one branch against all OTU's in the other branch. a. Phenogram constructed using UPGMA cluster analysis (as in Fig. 1). b. Phenogram constructed using the AHCS discussed in text, artificial variance equals 0.65. See text for further explanation.

for all possible phenograms, nor does it seem possible to develop an algorithm for finding the "optimum" least squares best-fitting phenogram. Therefore a trial-and-error approach must be taken in which an initial phenogram is adjusted by moving OTU's from one branch to another branch of the phenogram in an attempt to decrease the measure of distortion. The existing algorithms require large amounts of computer time. Thus this approach is not yet practical for large scale use. Farris (1969) gives an algorithm which yields phenograms with higher cophenetic correlations than those given by UPGMA but his technique permits reversals.

When comparing different sets of data using the same clustering procedure, one may also use these measures of distortion as an index to measure the extent to which the phenetic relationships are hierarchic (Rohlf and Fisher, 1968). If, for example, the cophenetic correlation is very high, say above 0.95, one may feel satisfied that imposing a system of nested clusters has not caused undue distortion. If the correlation is much lower, say about 0.6 or 0.7, one may well question the assumption that one has a system of nested clusters. In such cases it has been found very useful to investigate whether the lack of fit pertains to the phenogram as a whole, or if, perhaps, there is distortion only in certain regions.

A helpful technique is to plot frequency distributions of the distance coefficients on each of the branches of the phenogram (Fig. 3). Since each branch unites two clusters, the coefficients in the frequency distribution are the distances between all of the OTU's in one branch and the OTU's in the other.

A detailed study of these "residuals" indicated that the deviations are often not "random." Typically most of these extreme residual distances involved only one or two of the OTU's in one cluster and most of the OTU's in another cluster. Further study revealed that this was caused by an OTU, say *I* (a member of a fairly tight cluster) being relatively close to an OTU, *J*. OTU *J*, however, was not particularly close to the OTU's to which OTU *I* was close. This indicated, geometrically, that OTU *J* was on the "outer edge" of a cluster. If the cluster

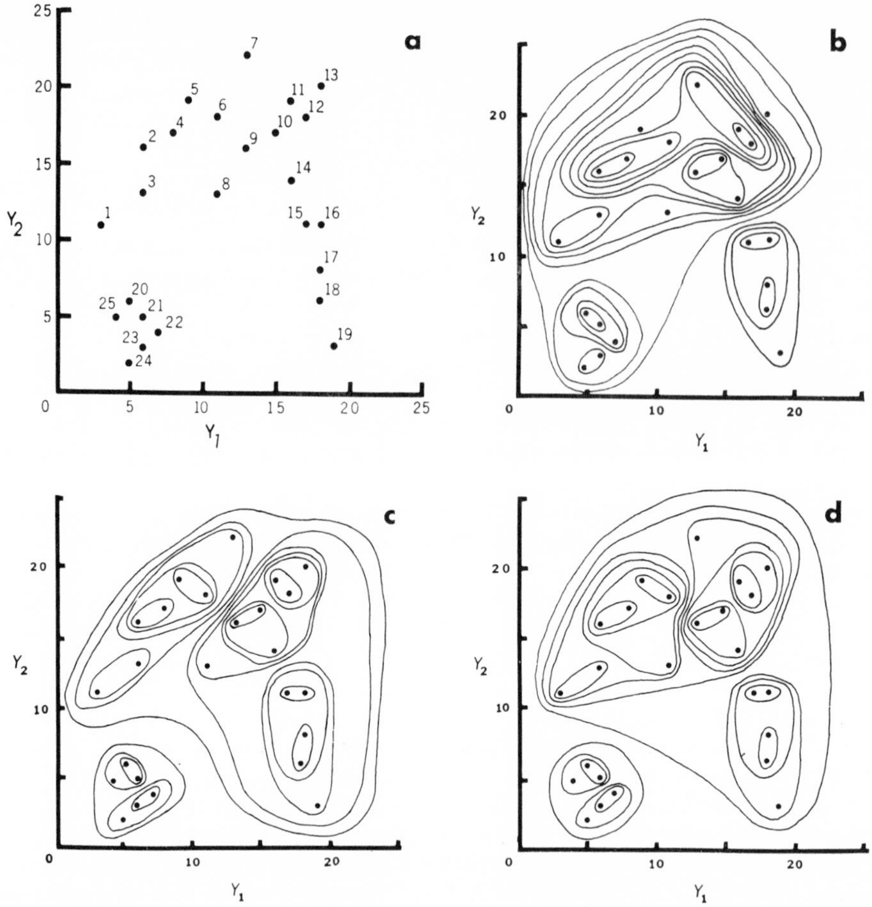

FIG. 4.—Example of effect of varying the value of the artificial variance in the linear AHCS discussed in text. a. Two-way scatter diagram of artificial data. b, c, and d. Results of clustering with variance equal to 100, 1.0, and 0.01, respectively. Lines indicate points included in clusters at each successive step in the clustering process. The exact shape of the lines has no meaning.

were elongated into a hyper-ellipsoid, the effect would be even stronger. For example, compare the placement of OTU's 8 and 14 in Fig. 4 (a bivariate scatter diagram of artificial data) with their placement in the phenogram in Fig. 2 constructed from the same data (Fig. 3 is also of this same data). R-type principal components analysis with computation of the projections of OTU's onto the principal axes and the construction of three-dimensional models enables a visual confirmation of this type of relationship in actual multivariate data (Rohlf, 1967, 1968).

ADAPTIVE HIERARCHICAL CLUSTERING SCHEMES (AHCS)

One can envision many different ways in which the basic HCS could be extended to allow for clusters which are ball shaped, elongated, or, perhaps, even curved in various ways. One could try reclustering the same data using a variety of techniques in order to see which method gives the best results. In practice one would probably find, however, that one method worked well for a few of the clusters but another method worked better for some of the other clusters in the same set of data. It would seem desirable to develop a method sensitive to trends in the shape of the clusters as they are formed, automatically adjusting the algorithm to be more efficient for the types of clusters which seem to be appearing. It would also seem to be important that the clustering procedure not only indicate the cluster to which each OTU belongs but also yield information about the structure of the cluster itself.

The paper by Carmichael, George, and Julius (1968) is one attempt in this direction: their algorithm employs three different criteria for deciding when one should stop adding OTU's to a cluster. Their program points out various statistics including flags indicating the *reason* why clustering was stopped at a specified level for a particular cluster. This yields some indication as to the nature of the cluster.

I propose to use a measure of distance between clusters that takes into account current knowledge of the size and shape trends found within each cluster. By this method, as successive OTU's added to the cluster slightly change its shape, the equation used to compute the distance between a cluster and all new candidates for inclusion will be adjusted successively so that new OTU's consistent with the extrapolated trends in the cluster are considered to be closer to the cluster than OTU's which depart appreciably from this trend. The ultimate success of this procedure is obviously limited by the extent to which OTU's are actually grouped into distinct clusters which follow relatively simple and consistent shape trends. A cluster with a very complex shape (for example, a helix) represented by only a few OTU's and with a large amount of random noise superimposed will be very difficult to resolve and recognize for what it is. If such points are relatively distant from other clusters, then clusters can be recognized but their underlying shapes will not be apparent.

The procedure developed below has the limitation that one must specify beforehand the class of cluster shapes considered permissible. Within that class the algorithm can automatically select the proper shape. I shall first describe the logic of the algorithm in terms of clustering hyperellipsoidal clusters and extend the procedure in the following section to clusters of more general shapes.

LINEAR AHCS

This allows for the possibility that at least some of the clusters may have the form of hyperellipsoids, rather than be limited to hyperspheroid.

The basic modification required is that the functions used in step 3 of the algorithm given above must be changed so that if (at a given state in the clustering procedure) there is evidence that the cluster is elongated, OTU's close to the ends of an elongated cluster should be considered "closer" than OTU's at the same euclidian distance from the mean of the cluster but along the sides of the cluster. Single linkage cluster analysis tends to do this but it does not take into consideration the fact that the density of points at the ends of the clusters is usually quite low, so that the inter-point distances are greater there.

Let us consider a simple example for the case in which one has only two variables. If the two variables are uncorrelated within the cluster (Fig. 5a), then two points such as a_1 and a_2 will be considered equally distant from the mean of the cluster. However, if the two variables are correlated within a cluster (see Figs. 5b and 5c), then points such as b_1 and c_1 should be considered to be closer to the centroid of their clusters than points b_2 and c_2. The concentric ellipses represent the loci of points which are equally distant from the centroids of the clusters. The higher the correlation between the variables, the more marked will be the discrepancy between ordinary euclidian distances between points and the centroid of their cluster and the distance function employed in this algorithm.

The equation for the distance function employed in this study is as follows:

$$\mathbf{D}_{I\rightarrow J} = \sqrt{(\bar{Y}_J - \bar{Y}_I)^t \mathbf{S}_I^{-1} (\bar{Y}_J - \bar{Y}_I)\,|\,\mathbf{S}_I\,|}$$

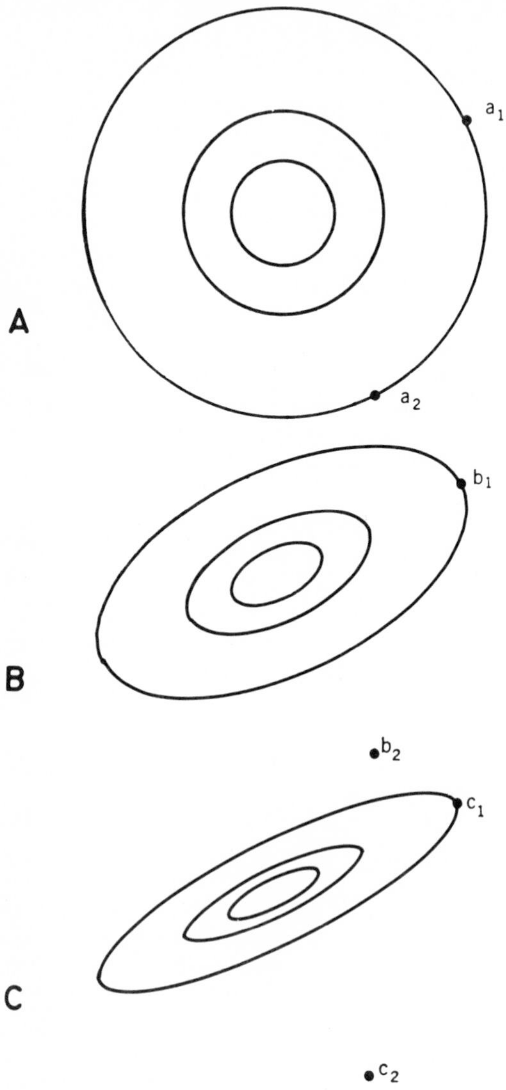

FIG. 5.—Effect of various magnitudes of the correlation coefficient upon the shape of an equal frequency ellipse. See text for discussion.

where $\mathbf{D}_{I \to J}$ is the distance from OTU_I to OTU_J. The reason for the arrow in the subscript is to indicate that it is the distance from I to J, which is not necessarily the same as the distance from J to I. S_I is the within group, variance-covariance matrix, V-CV, for OTU_I for the p variables, $|S_I|$ is the determinant of S_I, and $\bar{Y}_I$ and $\bar{Y}_J$ are column vectors containing the means for OTU's I and J, for the p variables (the superscript "t" indicates matrix transportation).

This distance function is very similar to Mahalanobis's generalized distance, D (see Rao, 1952; Seal, 1964; or Anderson, 1958, for accounts), but it differs from it in two important ways. First, the within group V-CV matrix of just one group is used rather than a pooled V-CV matrix (the reason for not pooling is that we have no reason to expect the different V-CV matrices to be homogeneous). Secondly, we have included the generalized variance $|S_I|$ as a factor. This was found necessary since we wished to take the *shape* of a cluster into consideration and not its size or hypervolume. If this is not done then one finds that all points are close to large clusters and distant from small ones.

It is important that the properties of $\mathbf{D}_{I \to J}$ be understood so that one will know what to expect of this type of cluster analysis. The data plotted in Fig. 6a have the following V-CV matrix:

$$S = \begin{bmatrix} S_1^2 & S_{12} \\ S_{21} & S_2^2 \end{bmatrix} = \begin{bmatrix} .0012388 & .0003788 \\ .0003788 & .0002936 \end{bmatrix}.$$

The eigenvalues are $\lambda_1 = .0013718$ and $\lambda_2 = .0001606$, which are the diagonal elements of the diagonal matrix Λ. The matrix of column eigenvectors (normalized) is

$$F = \begin{bmatrix} C_{11} & C_{12} \\ C_{21} & C_{22} \end{bmatrix} = \begin{bmatrix} 0.9435 & -0.3314 \\ 0.3314 & 0.9435 \end{bmatrix}.$$

The first eigenvector (corresponding to largest eigenvalue) points in the direction of the principal axis of the equal frequency ellipsoid which could be fitted to the data (see Sokal and Rohlf, 1969, for an example of the computation for the bivariate case). The equation of the line representing the major trend of variation is given by the equation

$$C_{12} Y_{1i} + C_{22} Y_{2i} = K_i$$

constructed using the eigenvector corresponding to the *smallest* eigenvalue. Note that if there is relatively little scatter of points (deviations) from the principal axis,

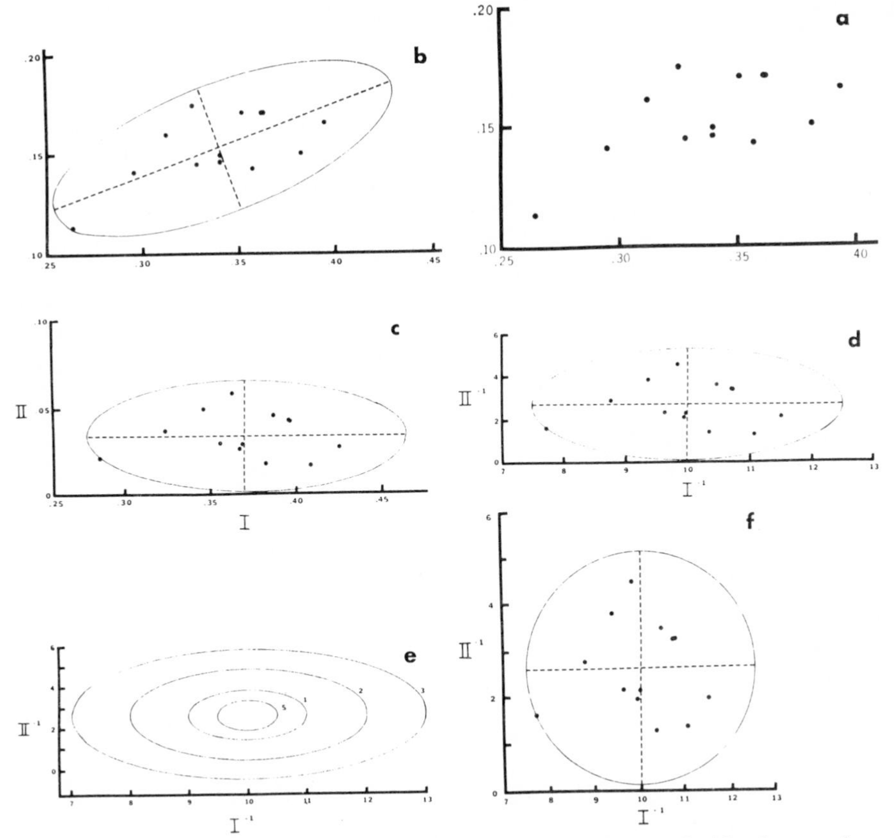

FIG. 6.—Diagram showing geometric interpretation of transformations implied by the use of generalized distance coefficients. a. Two-way scatter diagram of actual data for species number 4 in Fig. 7a. b. Major and minor axes and 90% equal frequency ellipse fitted to the data of Fig. 6a. c. Ellipse in b rotated so that major and minor axes are parallel to the coordinate axes. d. Scale changed by multiplying scale on each axes by the reciprocal of its eigenvalue (variance). e. Loci of points which are of distance .5, 1, 2, and 3 units from the mean of the cluster. f. Ellipse of Fig. 6d replotted so that abscissa and ordinate are drawn to a similar scale demonstrating the fact that transformations have caused the equal frequency ellipse to become circular in outline.

the values of K_i (which would be obtained if the t data points were substituted into this equation) will almost be constant. If we take $\bar{K}$ as the mean value which we would obtain, this gives us the equation of the best fitting trend line, i.e., the line with a minimum squared deviations from it (Fig. 6b):

$$0.3314 Y_{1i} + (-0.9435) Y_{2i} = -0.0330.$$

Analogously with p greater than 2, the vector corresponding to the smallest root gives us the equation of the best fitting line. If the last two eigenvectors are very small, relative to the rest, then the last two vectors yield the equation of the best fitting plane, etc.

This approach appears contrary to that usually taken in a principal components analysis where one usually computes only those eigenvectors corresponding to the largest eigenvalues. In the present case, one is interested in just the opposite, since the inverse of the within group variance-covariance matrix is employed in the equation for $\mathbf{D}_{I \to J}$. As pointed out by Sebestyen (1962) the eigenvectors corresponding to the *smallest* eigenvalues expresses the "property" most invariant within the cluster.

Since $$S = F\Lambda^{-1}F^t,$$ $$S^{-1} = F\Lambda^{-1}F^t.$$

Thus it is possible to write:

$$\mathbf{D}^2_{I\to J} = (\bar{Y}_J - \bar{Y}_I)^t F\Lambda^{-\frac{1}{2}} F^t(\bar{Y}_J - \bar{Y}_I)v,$$

where v equals the generalized variance (a constant in the present context). If $P = F^t Y$, the projection of the OTU's onto the eigenvectors, then

$$\mathbf{D}^2_{I\to J} = v \sum_{i=1}^{p} \lambda_i^{-1} (P_{iJ} - P_{iI})^2,$$

where the index i refers to the p eigenvectors. Thus, the squared differences in projections of each OTU onto each axis is weighted according to the reciprocal of the eigenvalue corresponding to that axis. The largest weight is given to the eigenvector with the smallest λ.

This weighting procedure has some similarities to that used by Pearson (1926) for the coefficient of racial likeness (which ignores the correlations among the characters) and to the weighting scheme of Farris (1966) (which ignores the correlations among the characters and also implies that the within group variation is the same in all groups).

These equations can, perhaps, be better understood if we re-express them in terms of their corresponding geometric operations.

Fig. 6a shows some artificial data representing a single cluster for which two variables Y_1 and Y_2 are correlated within the cluster (they may be uncorrelated in other clusters). In Fig. 6b are shown the major and minor axes for the 90% equal frequency ellipsoid fitted to this data. In Fig. 6c, the data have been rotated and translated so the origin of the graph is now at the mean and the major and minor axes are used as the coordinate axes (a result of the operation $P = F^tY$). In Fig. 6d, the coordinate axes have been rescaled by multiplying them by $\lambda^{-\frac{1}{2}}$. In this new space, the distance of a point at coordinates (A, B) to the mean ($\bar{Y}_1$, $\bar{Y}_2$) can be computed using the standard Pythagorean equation. Fig. 6e shows contours of points which have an equal distance from the mean of the cluster. In Fig. 6f both axes have been drawn to the same scale, causing the contours to become circular in outline. Consequently, in Fig. 6f, ordinary euclidian distances are equal to the generalized distances of Fig. 6e. Thus, OTU's near the ends of an elongated cluster would in fact be nearer to the center of the cluster in the conventional sense and would be added to a cluster in preference to OTU's located along the sides of the cluster.

If this distance function is used in step (3) of the algorithm, step (2) must also be modified, since there may not be any reciprocally close pairs of OTU's. This is due to the fact that in general $\mathbf{D}_{I\to J} \neq \mathbf{D}_{J\to I}$. When such cases arise, we can simply find the smallest $\mathbf{D}_{I\to J}$ value and consider the two OTU's involved as a mutually close pair.

Even when reciprocally close pairs are found, other OTU's may "interfere" in the sense that if OTU's I and J are mutually closest $\mathbf{D}_{K\to I}$ may be less than either $\mathbf{D}_{I\to J}$ or $\mathbf{D}_{J\to I}$. In such cases, OTU's I and J are not treated as a mutually closest pair of OTU's.

Fig. 7 shows the results of this algorithm applied to a set of simple data. The data represent seven species of limpets (marine gastropods) plotted with respect to two variables describing shell shape (ratio of the height of the shell to its length and the ratio of the distance between the apex of the shell, see diagram in Fig. 7a). With only two variables used one should not place much taxonomic importance on the clusters obtained, but the clusters do seem reasonable for the given set of data.

It should be pointed out that this type of clustering is not affected by linear transformations of the original data. Thus, standardization of the characters has no effect and need not be performed (as is also the case in the techniques proposed by Friedman and Rubin, 1966). The effect of the size of the organisms is also minimized if one has samples of both large and small specimens *within* each OTU (thus, the

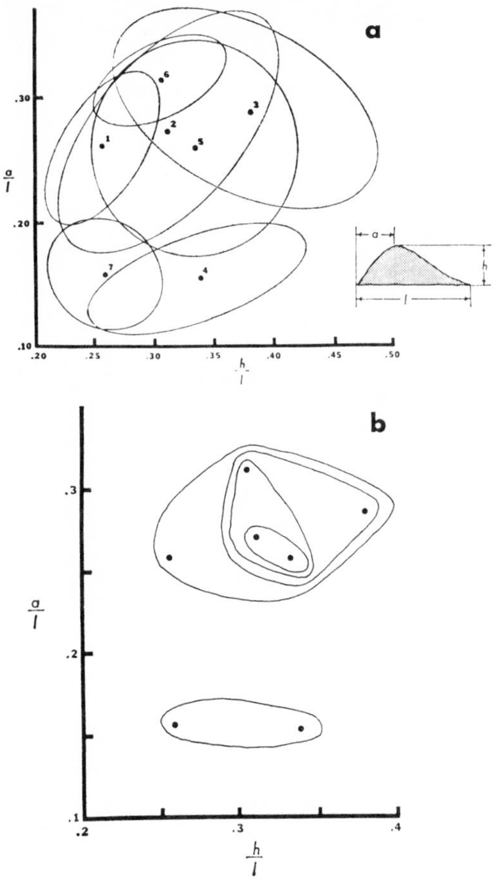

FIG. 7.—Two-way scatter diagram showing the means of 7 species of limpets and 90% equal frequency ellipses fitted to the individual observations for each group. Insert (left side view of a limpet) defines measurements used in the ratios for the abscissa and ordinate of the figure. b. Results of clustering using the linear AHCS. Details are discussed in text.

technique proposed by Burnaby, 1966, is not necessary).

For most numerical taxonomic studies, this procedure would be prohibitive since it would require at least $n+1$ individuals to be measured on all p characters for each of the t OTU's, since $2t-1$ principal components analyses would need to be performed. Computer time would also be a problem if p (the number of characters) were large.

Several shortcuts are possible. It would be most efficient to carry out a principal components analysis on the among OTU's V-CV matrix and use only the first few, k, components as variables in order to lighten the computational load. The V-CV matrices for each OTU must also be transformed so that they are compatible with these principal components (i.e., they are transformed using the same eigenvectors, new $S = F^tS\ F$). If there is no information about the covariation of characters within the OTU's, as when individual specimens are used as OTU's, one can do one of two things: (1) One can enter the OTU's into the analysis and cluster on the basis of ordinary euclidean distances until a cluster has enough OTU's in it so that its variance-covariance matrix is of full rank ($n > p$). A problem with this is that one might run out of OTU's before this happens. This could be avoided by first performing a principal components analysis, using only the first few components as variables. (2) Another procedure would be to generate artificial V-CV matrices for each OTU. In the experiments carried out so far, we have used V-CV matrices which imply that the concentration ellipsoids around each OTU are actually hyperspheroids. The hypervolume of these hyperspheroids was left as a parameter which could be adjusted. The most desirable value is not known. If the volume is relatively large, then the clustering will differ little from centroid clustering described before and hyperspheroidal clusters will tend to be formed. If, however, the value is very small, then, once a cluster composed of two or more OTU's is formed, the method will be *very sensitive to the shapes of the clusters.* The precise value to be used must depend upon several factors: 1, The scale of measurements used for each variable (thus standardization is recommended); 2, The amount by which the OTU's deviate from tightly packed spheroidal or ellipsoidal clusters (since this need not be the same for all clusters no one value may be optimal for all of the clusters); and 3, The true dimensionality of the phenetic space.

Experience so far has not yet enabled the development of optimal values for the variances in the artificial variance-covariance matrices. All one can do at present is to try several values and observe the resulting phenograms and cophenetic correlations. However, it is now less obvious how the cophenetic values should be computed. If the method of Sokal and Rohlf (1962) is used, the cophenetic correlations steadily decrease as the variances decrease even though the phenograms seem subjectively to improve. For example, the phenograms in Figs. 3a, 4b, 4c, and 4d have cophenetic correlations equal to 0.781, 0.733, 0.723, and 0.643, respectively.

There are two ways in which the cophenetic correlation can be raised without changing the branching patterns of the phenogram. First, one can adjust the level of each branch so that it is at a level corresponding to the average of the distances between the OTU's in the two clusters being united (rather than the logarithm of the hypervolume of the resulting cluster). When this is done to yield Fig. 3b we note that such an adjustment can yield reversals which must be eliminated by changing the two bifurcations to a trifurcation at a level corresponding to their weighted average. This adjustment raised the cophenetic correlation for the phenogram in Fig. 3b from 0.656 to 0.689. The small change reflects the fact that the logarithm of the hypervolume is not a bad scale to start with. Secondly, the cophenetic correlation may be raised by adding additional existing information to the phenogram. For example, one could somehow show that three of the clusters were elongated and oriented in particular directions. If this were done the cophenetic values would have to be computed taking this into consideration and the resulting cophenetic correlation for Fig. 3b would undoubtedly be higher. Algorithms and computer programs are being developed to perform such computations.

Fig. 8 shows an example of this method applied to the mosquito pupae data of Rohlf (1967). The projections of the 45 species onto the first nine principal components were used as input. The spacing of

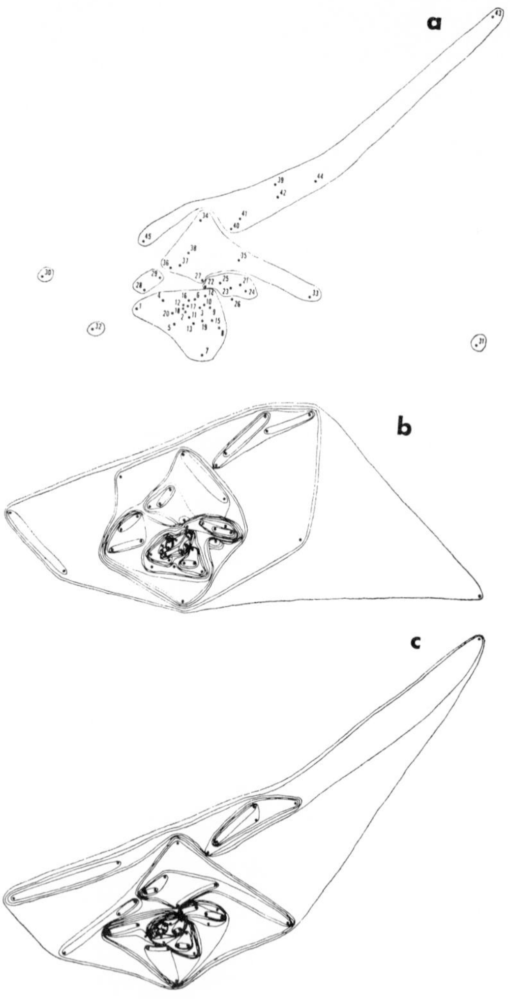

FIG. 8.—Results of linear AHCS applied to the mosquito pupae data of Rohlf, 1967 (see also Fig. 1). a. Results of non-metric multidimensional scaling for two dimensions. Stress = 0.170, cophenetic correlation equals 0.967. Species code numbers and generic groupings are also indicated: 1–20 *Aedes*; 21–25 *Culex*; 26, 27, and 33–38 *Culiseta*; 28 and 29 *Orthopodomyia*; 30 *Mansonia*; 31 *Uranotaenia*; 32 *Wyeomyia*; and 39–45 *Anopheles*. Results of UPGMA cluster analysis. Cophenetic correlation is 0.913. Results of AHCS (presented in text) with artificial variance equal to 0.001 and using projections of the 45 species onto a 9-dimensional space as input data. Cophenetic correlation is 0.643.

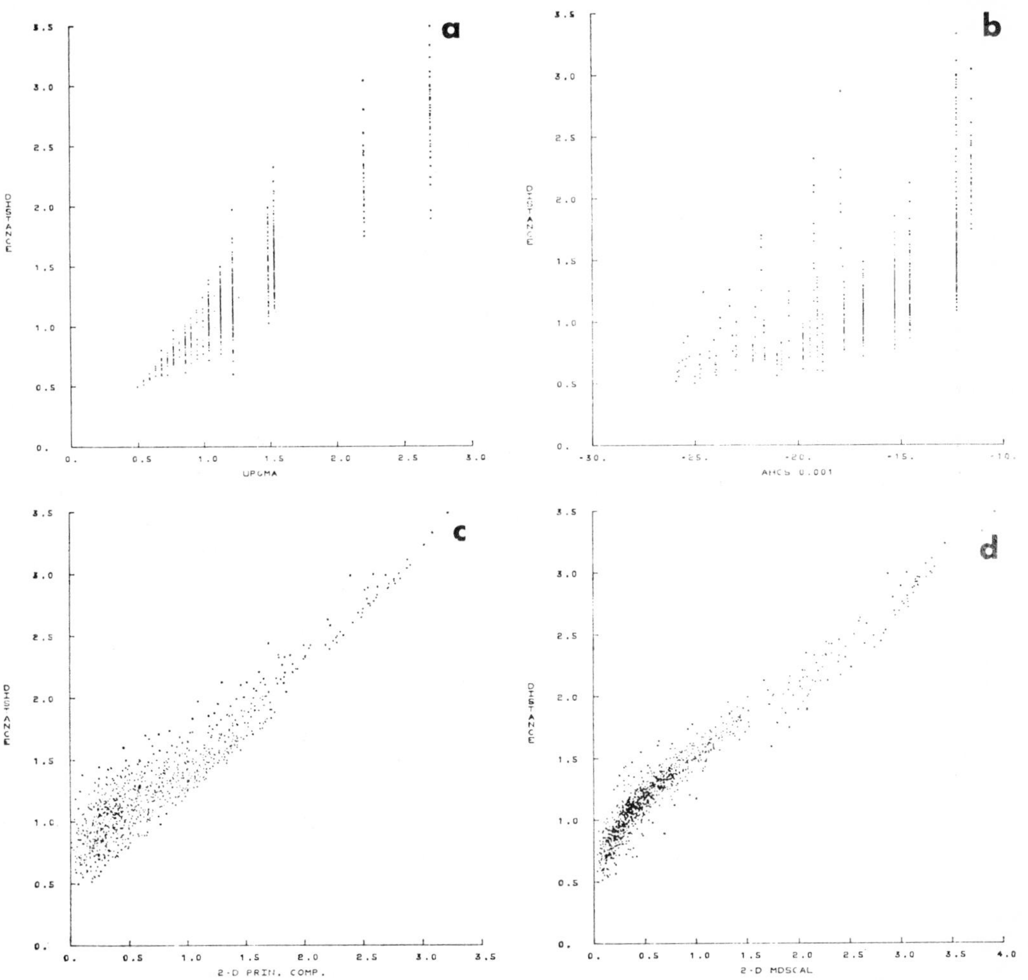

FIG. 9.—Comparisons of amount and patterns of distortion introduced by various methods of summarization of phenetic relationships. On all figures the ordinates are the average taxonomic distance between all pairs of species and the abscissas are the estimated distances or cophenetic values from various methods. The following correlation coefficients refer to cophenetic correlation coefficients. a. UPGMA, r = 0.913. b. AHCS as is Fig. 8c, r = 0.643. c. Principal components analysis with projections based on only two dimensions, r = 0.938. d. Non-metric multidimensional scaling analysis in two dimensions, r = 0.967.

points in Fig. 8 was determined by a multidimensional scaling analysis (cf. Fig. 5 in Rohlf, 1967). Fig. 8a gives the species code numbers and the generic groupings. Fig. 8b indicates the results of a UPGMA cluster analysis applied to the original distance matrix. Similar results were obtained using AHCS with a very large variance. Note the tendency for "round" clusters to be formed. Fig. 8c shows the results of AHCS with a variance of 0.001. The results are very good by the standards of conventional taxonomy. All of the species of *Aedes*, *Culex*, and *Orthopodomyia* form their own clusters. There are only two "misplaced" species. Species 30 joins 45 before it clusters with the rest of *Anopheles* and species 26 is placed with the *Culex*.

Fig. 9 compares the amounts and patterns of distortion introduced to the pupae data by these various techniques. The original distances are plotted along the ordinates of

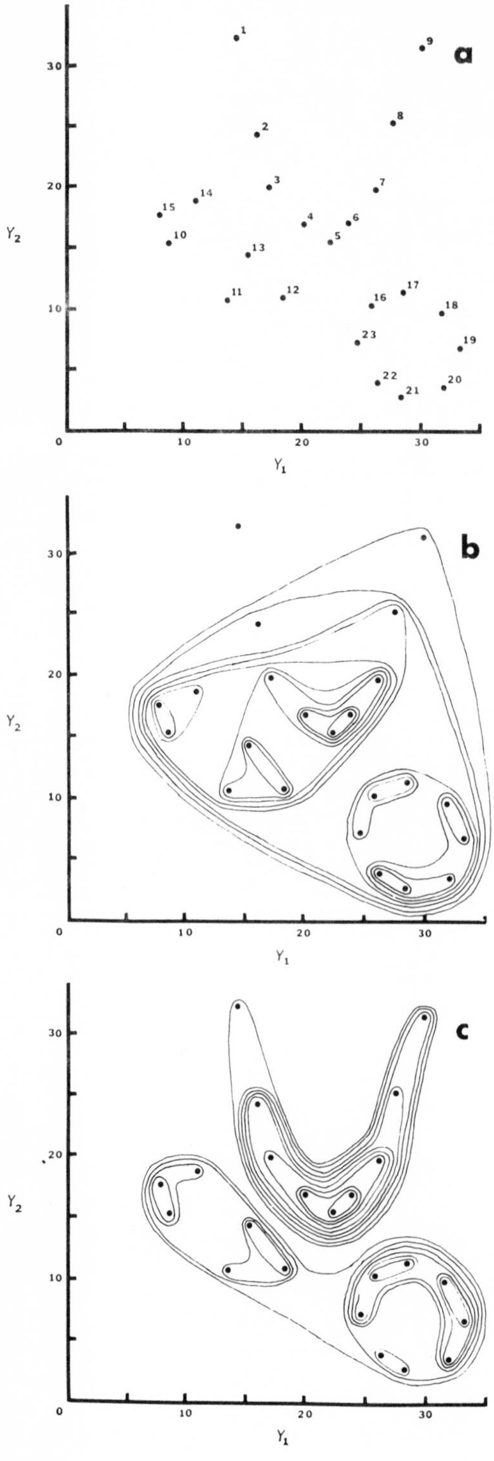

the figures and estimated distances (cophenetic values) are plotted along the abscissas.

NONLINEAR ADAPTIVE HCS

If one suspects that the clusters to be found in a given set of data are neither hyperspheroidal nor elongated into hyperellipsoid but, rather, that they may take the form of clouds of points around some curvilinear function (see Fig. 10 for examples), then the techniques in the preceding section will not adequately isolate and describe the clusters. There are (at least) two main approaches that we may take. First, we could in some way determine the regression function (or at least a numerical approximation; Sneath, 1966) which fits the shape of the cluster and then compute the distance between an OTU and the curve representing the cluster. Sebestyen (1962, section 5.1) points out the problems involved with this method. The second procedure would be a simple extension of the method given in the preceding section. We would wish to compute the distance between an OTU and a cluster using a distance function which takes into consideration the apparent shape of the cluster, as well as the extent to which the OTU's fit the underlying curve (if it exists).

This latter procedure can be easily accomplished by simply adding new dummy variables to the original data matrices. The precise nature of the new dummy variables will depend upon the class of underlying curves that one believes might be appropriate. For example, if we wish to allow the possibility of points being clustered around points (hyperspheroidal clusters), lines (hyperellipsoidal clusters), parabolas, circles (ring-shaped clusters), ellipses, or hyperbolas, then we would simply add the squares of the original variables as well as their cross-products of the variables taken

←

Fig. 10.—Examples of AHCS applied to nonlinear clusters. a. Artificial data with three clusters in the shape of a parabola, an ellipse, and a circle. b and c. Results of clustering, using artificial variances equalling 0.001 and 0.0001, respectively.

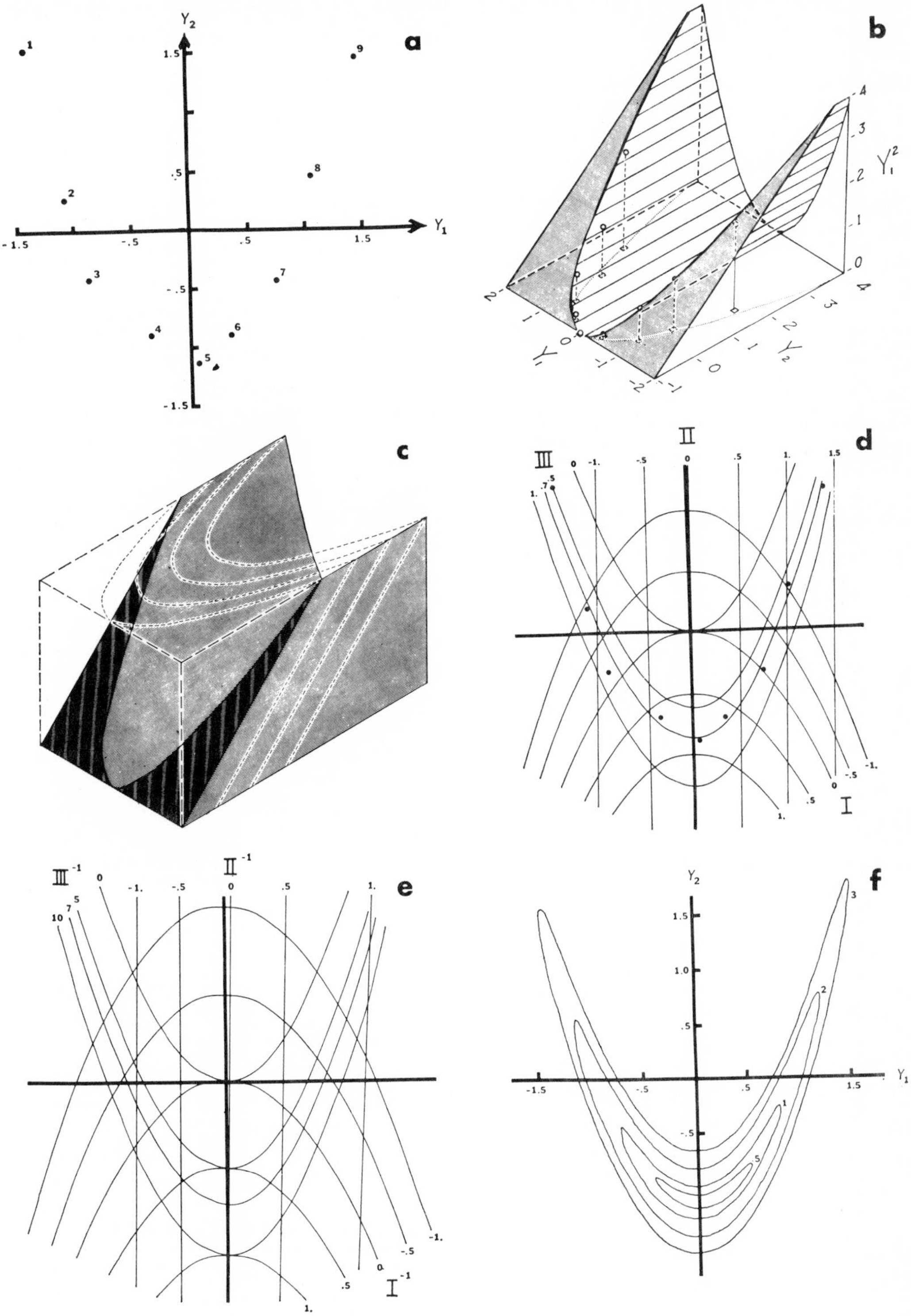
a
Y_2
Y_1
1.5
.5
-.5
-1.5
1
2
3
4
5
6
7
8
9
b
Y_1^2
Y_1
Y_2
c
d
I
II
III
e
I^{-1}
II^{-1}
III^{-1}
f
Y_2
Y_1
1.5
1.0
.5
-.5
-1.5

two at a time. Thus, if we have only two variables, say Y_1 and Y_2 we would now have five variables (in general the total number of variables would be $p + p(p+1)/2$), Y_1, Y_2, Y_1^2, Y_2^2, Y_1Y_2. One could then input into the computer program, described above, the means for each OTU on these variables and the variance-covariance matrices for each group for these variables. In order to avoid very large numbers it is desirable to standardize the original variables before computing their powers.

If, at a given stage in the clustering process, the data points describe a ring, the generalized distance between the OTU's and this cluster would be small for OTU's close to this ring and would be relatively large for OTU's away from the ring. It should be pointed out that OTU's near the mean of the cluster (in the center of the ring) would be considered relatively far away from the cluster, as one would wish for such a procedure. In general, OTU's along the direction of the main trend of the cluster would be considered relatively close and OTU's not in the direction of the general trend of the cluster would be considered distant. The amount of within-group variation, i.e., the extent to which the points closely follow a trend, affects the degree to which the distance function reflects the trends. If there is much scatter and the data only vaguely follow a parabola, for example, then the distance will differ little from what would be obtained if the nonlinear dummy variables had not been included. Since the properties of the nonlinear generalized distances employed in this method are not well known, an account is given below which describes their more important properties.

The mathematical basis for this technique is a simple generalization of the procedure described in the preceding section (Sebestyen, 1962). Let us consider an example such as the parabolic shaped cluster extracted from Fig. 10 and displayed in Figure 11a. For purposes of illustration, let us add only the one dummy variable, $Y_3 = Y_1^2$, which we know we will need (Figure 11b). The same results would be obtained if we included Y_2^2 *and* Y_1Y_2, but that would require one to visualize a five-dimensional space, which is undesirable for illustrative purposes.

Note that since the ordinate in Figure 11b is Y_1^2 the data points must all lie on the surface of a parabolic cylinder as shown in the figure. The mean of the sample of points will lie at the centroid of the distribution and not on the surface of the parabolic cylinder.

The mean, $\bar{Y}$, and the V-CV matrix for the data plotted in Figure 11b are (the p variables are Y_1, Y_2, $Y_3 = Y_1^2$):

$$\bar{Y}^t = (0.0,\ 0.0,\ 0.885314)$$

$$S = \begin{bmatrix} 1.000000 & 0.008470 & 0.012959 \\ 0.008470 & 1.000000 & 0.774072 \\ 0.012959 & 0.774072 & -0.614745 \end{bmatrix}.$$

The eigenvalues of S are $\lambda, = 1.60541$, $\lambda_2 = 0.9967$, and $\lambda_3 = 0.00967$. The matrix of column eigenvectors (normalized) is

$$F = \begin{bmatrix} -0.024199 & 0.995462 & -0.005044 \\ -0.787594 & -0.022171 & -0.615802 \\ -0.615718 & -0.019029 & 0.787900 \end{bmatrix}.$$

←

FIG. 11.—Diagram showing geometric interpretation of generalized distances computed using non-linear data. a. Artificial data from Fig. 10a. b. Data of a with a new dummy variable Y^2_1 added so that the points are now plotted in a three-dimensional space. Since the ordinate is Y^2_1 all points must lie in a parabolic cylinder. The cylinder is cut by the plane determined by the first two principal components of the variance-covariance matrix among the three variables. c. Parabolic cylinder as in b cut by a series of parallel planes. The intersections of these planes with the cylinder are shown as a family of parabolas at the top of the figure. These parabolas represent one set of curvilinear coordinate axes as explained in the text. d. Artificial data in a with the curvilinear coordinate axes generated by principal components I, II, and III. e. Curvilinear coordinate axes as in Fig. d with scales weighted according to the reciprocal of the eigenvalues of the principal components. f. Loci of points of distance .5, 1, 2, and 3 from the mean of the cluster. This figure is analogous to Fig. e.

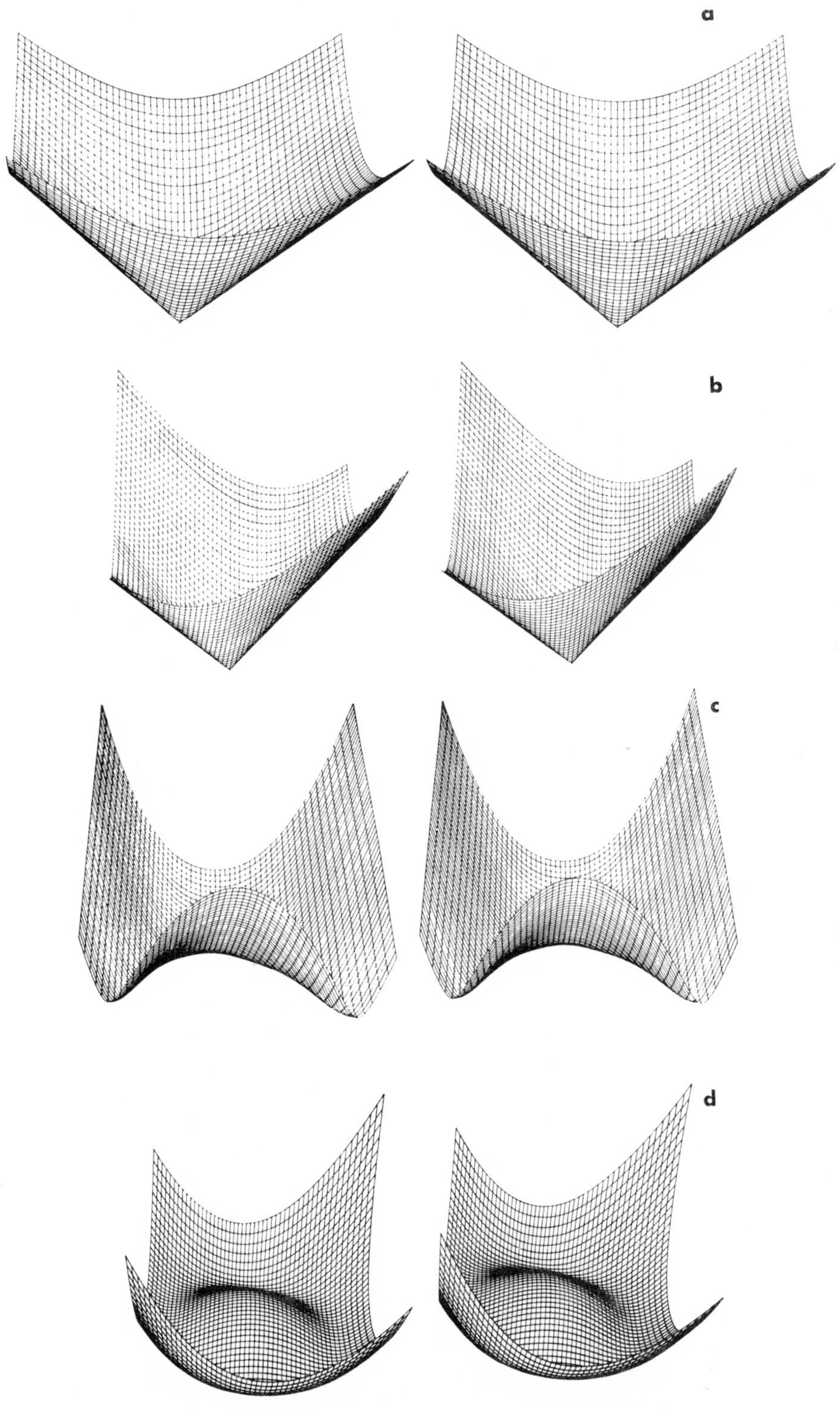
a
b
c
d

As before, the eigenvectors point in directions of maximum variance. In a cartesian space one can interpret the coordinate of a point (a, b) on the Y_1 axis as being the value of K which makes the line $Y_1 = K$ intersect with the point (a, b). K obviously equals a in this simple case. Analogously, the coordinate of a point (Y_1, Y_2, Y_3) on axis I, for example, is the value of P_I which satisfies the equation: $-0.024199\ Y_1 - 0.787594\ Y_2 - 0.615718\ Y_3 = P_I$ (the coefficients are taken from the first eigenvector). The locus of all points which yield a particular value of P_I, say zero, is a plane cutting through the 3-dimensional space as shown in Fig. 11c. If we take into consideration the fact that $Y_3 = Y_1^2$ and consider only points on the parabolic cylinder we note that the projection of the intersection of this plane and the cylinder onto the Y_1, Y_2 plane is a parabola. By varying the value of P_I we get a family of parabolas. All points on the same parabolar have equal projections onto axis I. In a similar manner (using the second and third eigenvectors) we can have planes for axes II and III as shown in Fig. 11c. The projections of these intersections are a series of curves almost parallel to the Y_2 axis for axis II and a series of parabolas for axis III.

Fig. 11d shows these three families of curves simultaneously. This figure is analogous to Fig. 6c in that we can now express the coordinate of each point on the cylinder in terms of its projection (coordinate) on the three *curvilinear* coordinate axes corresponding to the three eigenvectors. For example, the curvilinear coordinates of the second point ($Y_1 = -1.07557$, $Y_2 - 0.28687$) is ($P_I = -0.9122$ $P_{II} = -1.0942$, $P_{III} = 0.7402$) computed by multiplication by the eigenvectors as before ($P = F^t Y$). These coordinates can be confirmed by an examination of Fig. 11d.

A general discussion of curvilinear coordinate systems is given in Hildebrand (1952).

The smaller eigenvalue and its associated eigenvector indicate the general trend of the data and the closeness of the points to this curve. In the present example we see that the best fitting curve is the parabola $-0.005\ Y_1 - 0.616\ Y_2 + 0.788\ Y_1^2 = 0.700$ which can be rewritten as

$$0.616\left[Y_2 + \frac{0.700}{0.616} + \frac{(0.005)^2}{4(0.788)(0.616)}\right] = 0.788\left(Y_1 - \tfrac{1}{2}\frac{0.005}{0.788}\right)^2,$$

or simply $Y'_2 = (0.788/0.616)(Y'_1)^2$ after a simple translation of the axes.

If we had included the additional dummy variables Y_2^2 and Y_1Y_2, the trend equation would be of the general form

$$AY_1^2 + BY_1Y_2 + CY_2^2 + DY_1 + EY_2 = \overline{P}$$

By a rotation of the axes one can eliminate the Y_1Y_2 term. By completing the square and translating the rotated axes one can arrive at a simplified equation which makes the form of the curve more obvious (see standard analytical geometry texts for details of the methods for carrying out these operations). If a miscellaneous collection of dummy variables are employed simplification of the trend equation may be more difficult. The magnitude of the smallest eigenvalue measures the fit of the data to this trend curve. It is the variance of the deviations of the n data points measured perpendicularly from the curve.

Analogously to Figure 6d, we can construct Figure 11e in which the coordinate axes have been weighted by $\lambda^{-\frac{1}{2}}$. The

←

FIG. 12.—Generalized distances of points in the Y_1, Y_2 plane to the mean plotted as a surface in three dimensions and shown as stereo pairs. The height of the surface at a point Y_1, Y_2 is the distance of that point to the mean of the cluster. a. Ordinary euclidean distance, the surface is in the shape of a cone. b. Cluster is in the form of an elliptical scatter of points such as found in Fig. 6a. c. Generalized distances of points to a parabolic shaped cluster (such as that in Fig. 11a). d. Generalized distance to a ring-shaped cluster (such as that in the lower righthand corner of Fig. 10a).

shapes of the axes are unaffected by this operation, only the scales are changed. We cannot measure distance directly in this space. As above, we must compute distances using the standard formula:

$$D^2_{I \to J} = \sum_{i=1}^{p} \lambda_i^{-1} (P_{iJ} - P_{iI})^2,$$

where P_{iI} are the means projections of the sample points onto the curvilinear axes, P_{iJ} are the curvilinear coordinates of a point in the space depicted in Fig. 11b, and the λ_i are the eigenvalues.

Contours of points on the parabolic cylinder, which are equidistant from the sample mean, are shown in Fig. 11f. One will note that, as in Fig. 6e, points which deviate from the mean in the direction of the principal trend of the data (in the present case, parabolic) are indicated to be relatively closer to the mean than are points which deviate in other directions.

It should be pointed out that, unless the points have identical values of Y_1, the mean of the sample will not lie in the surface of the parabolic cylinder shown in Fig. 11b. Thus the mean cannot be plotted in Figs. 11d or 11e and a distance equal to zero cannot be shown in Fig. 11f (of course, the mean of another sample could be identical to the present sample mean, the distance between these two means would be correctly indicated to be zero). It is difficult (but not impossible) to construct a plot analogous to Figure 6f. Considering only data points we can study $\boldsymbol{D}_{I \to J}$ as a function of only two variables Y_1 and Y_2.

Fig. 12 shows the distance functions we have considered as surfaces in three dimensions. Figure 12a represents the ordinary cartesian distance, the height $(\bar{Z})$ of the surface at each point (Y_1, Y_2) is equal to the distance of the point (Y_1, Y_2) from the mean of the sample at $(\bar{Y}_1, \bar{Y}_2)$. Figs. 12b, c, and d are similar plots where the cluster is an ellipsoid, parabola, or a ring, respectively. Figs. 6e and 11f represent some of the contours of Figs. 12b and c respectively. Fig. 12d clearly shows that points at the centroid of a ring shaped cluster are further away from the cluster than are points along the ring itself.

Non-linear AHCS is not limited to clusters which take on the shapes of conic sections (as is the case in the simple example described above). By introducing the appropriate dummy variables a variety of cluster shapes can be handled. The principal limitations of this approach is that one must decide beforehand which class of functions are appropriate for a given set of data. But note that the amount of computation necessary increases very rapidly as additional dummy variables are added.

Figs. 10b and c show the results of the proposed clustering procedure (with dummy variables Y_1^2, Y_2^2 and Y_1Y_2 added) for two values of the diagonal elements of the artificial within group V-CV matrices (see above). A small value clearly results in isolating the clusters properly. An examination of the eigenvectors and eigenvalues at each step in the clustering also correctly indicate the shape of each cluster. In contrast to earlier examples, very small variances still resulted in essentially the same clusters.

As in the case of other clustering procedures, AHCS's success is limited by the closeness and compactness of clusters, but this technique has the ability to follow clear nonlinear trends. AHCS also has the advantage that it yields the *equation* of the best fitting curve through the cluster as well as its degree of fit (in contrast to the techniques described by Sneath, 1966; Carmichael et al., 1968; and Ball, 1965).

As was the case for the linear HCS the cophenetic correlations (using the conventional cophenetic values) were rather low (0.691 and 0.573 for Figs. 10b and c). A general technique for computing cophenetic values for nonlinear clusters is more complex and is being developed.

It is also quite clear that the concept of weighted curvilinear coordinates can be applied to other related statistical methods. For example, Sebestyen (1962) developed nonlinear generalized discriminant functions (his examples were in terms of discrimination between classes made up of disjoint subclasses). Techniques of nonlinear factor analysis are given by MacDonald (1962, 1967). Gnanadesikan and Wilk (1969) discuss and give an example of nonlinear principal components analysis.

DISCUSSION

In order to be worth interpreting, a phenogram must impose a relatively small amount of distortion of the pattern of phenetic relationships to be found in the original distance matrix. Thus, it is very important that the distortion between the phenogram and the original distance matrix be measured by some stated criterion. Without such a measure one has no basis for interpreting a phenogram seriously. Many indices are possible (see above). If (in an actual study) the cophenetic correlation coefficient is quite high one should be encouraged to try to interpret the clusters which emerge. However, this only sets a limit to when one should start interpreting. As can be seen from the examples presented above, even a phenogram that fits the phenetic relationships in a way that intuitively appears quite poor will yield cophenetic correlation coefficients well above the range of values expected, due to chance alone (Rohlf and Fisher, 1968). Thus, even a coefficient near 0.9 does not guarantee that the phenogram serves as a sufficiently good summary of the phenetic relationships. The situation is somewhat analogous to problems in regression where one cannot tell from the percentage of variation explained by linear regression whether or not the deviations from regression represent simple random scatter or are systematic (indicating that, perhaps, a quadratic regression would have been more appropriate). The only way of determining this is to make some sort of plot where one can visually attempt to see patterns in the residuals; or curvilinear regression may be done to see whether a significantly better fit is obtained.

Likewise in cluster analyses, one could either examine the "residuals" or deviations between the estimated and actual distances (as in Figs. 3 and 9) or try a variety of cluster analyses to see which one gives the highest cophenetic correlation. As Sokal and Rohlf (1962) suggested, it has been found that, of the sequential agglomerative HCS's which use distances in a euclidean space and place similar OTU's together, the UPGMA procedure always yields the highest cophenetic correlation regardless of the structure of the data. Thus, for this class of techniques, the UPGMA procedure is clearly to be preferred. However, we have seen above how poorly UPGMA clusters certain types of data structures. Thus we should not be satisfied with results given by UPGMA, but should investigate whether it is possible to get better degree of fit. It is also important that techniques be developed which enable more information to be included in a phenogram. The frequency distributions on the branches of the phenogram as shown in Figure 3 show information about the relative degree of fit of different parts of the phenogram to the data. In addition, it would be desirable to annotate the phenogram with information concerning the nature of the structure of the clusters. For example, it is important whether the points in a cluster are distributed in the form of an ellipsoid, paraboloid, ring, etc. In the case of clusters with a linear order, it would also seem appropriate that the OTU's be listed in the phenogram in some natural sequence reflecting their position in the cluster (rather than simply in the order in which they are added to the cluster, which means that OTU's toward the extreme ends of an elongated cluster will often be placed side by side in the phenogram).

These additional considerations add much complexity to the results of a cluster analysis, but I believe they are necessary in order to achieve an acceptable level of summarization of one's data, so that one may safely use the results in biology. Some of the criticisms by conventional taxonomists of the *results* of numerical taxonomic studies have been justifiable in that *some* of the taxonomic conclusions which they questioned were artifacts of the clustering procedure—more direct (but more laborious) examination of the original distance matrix indicated relationships more consistent with conventional classifications than with the phenograms (see for example Barr and Chapman, 1964, and Rohlf, 1964).

Comprehension of multivariate relationships is difficult. The best advice that can be given to someone who needs practical results from numerical taxonomy is to try a variety of techniques for summarizing his data. Different methods will expose somewhat different aspects of the phenetic relationships. If in doubt about conflicting results one should not hesitate to directly examine the original data. If a single unique "result," rather than a general understanding, is required then that summary with the lowest measure of distortion should be used.

ACKNOWLEDGMENTS

Contribution number 1427 from the Department of Entomology, The University of Kansas and number 7 from the program in Ecology and Evolution, State University of New York, Stony Brook. This work was supported, in part, by a grant (GB-4927) from the National Science Foundation. Computer time was made available on the GE 635 by the University of Kansas Computation Center.

The paper has benefitted from my discussions with a number of people. Particularly helpful were discussions at the Second Numerical Taxonomy Symposium (Sokal, 1969), the Numerical Taxonomy Discussion Group at The University of Kansas, and with Dr. Joseph B. Kruskal and the Mathematics Group at Bell Telephone Laboratories at Murray Hill, N.J. E. J. Carmone furnished me a copy of Dr. Kruskal's MDSCAL-IV program. Drs. R. R. Sokal and J. S. Farris read the manuscript and made many useful suggestions.

Patricia R. Rohlf assisted in the preparation of the illustrations and in the typing of the manuscript. Dr. A. J. Rowell generously prepared Figs. 11b and 11c.

REFERENCES

Anderson, T. W. 1958. An introduction to multivariate statistical analysis. Wiley: New York, 374 pp.

Ball, Geoffrey H. 1965. Data analysis in the social sciences: what about the details? Proc. Fall Joint Computer Conf., 27:533–559.

Birch, L. C. and P. R. Ehrlich. 1967. Evolutionary history and population biology. Nature, 214(5086):349–352.

Bonham-Carter, G. F. 1967. Fortran IV program for Q-mode cluster analysis of non-quantitative data using IBM 7090–7094 computers. Kansas State Geological Survey, Computer Contribution No. 17, 78 pp.

Bonner, R. E. 1964. On some clustering techniques. IBM Journal of Research and Development, 8(1):22–32.

Bromley, D. B. 1966. Rank order cluster analysis. British Jour. Mathematical and Stat. Psych., 19:105–123.

Burnaby, T. P. 1966. Growth-invariant discriminant functions and generalized distances. Biometrics, 22:96–110.

Camin, Joseph H. and Robert R. Sokal. 1965. A method for deducing branching sequences in phylogeny. Evolution, 19:311–326.

Carmichael, J. W., J. Allen George, and R. S. Julius. 1968. Finding natural clusters. Syst. Zool., 17:144–150.

Cavalli-Sforza, L. L. and A. W. F. Edwards. 1967. Phylogenetic analysis: models and estimation procedures. Evolution, 21:550–570.

Constantinescu, Paul. 1967. A method of cluster analysis. Brit. J. of Math. and Stat. Psych., 20:93–106.

DuPraw, E. J. 1964. Non-Linnean taxonomy. Nature, 202(4935):849–852.

Ehrlich, Paul R. and Anne H. Ehrlich. 1967. The phenetic relationships of the butterflies I. Adult taxonomy and the nonspecificity hypothesis. Syst. Zool., 16:301–317.

Estabrook, G. F. 1966. A mathematical model in graph theory for biological classification. J. Theoret. Biol., 12:297–310.

Farris, J. S. 1966. Estimation of conservatism of characters by constancy within biological populations. Evolution 20:587–591.

Farris, J. S. 1969. On the cophenetic correlation coefficient. Syst. Zool. 18:279–285.

Friedman, H. P. and J. Rubin. 1966. On some invariant criteria for grouping data. IBM New York Scientific Center Technical Report:1–49.

Goodall, David W. 1968. Affinity between an

individual and a cluster in numerical taxonomy. Biometrie-Praximetrie, 9:52–55.

GOWER, J. C. 1967. A comparison of some methods of cluster analysis. Biometrics, 23(4): 623–637.

GNANADESIKAN, R. AND M. B. WILK. 1969. Data analytic methods in multivariate statistical analysis. (In preparation).

HARTIGAN, J. A. 1967. Representation of similarity matrices by trees. J. Amer. Stat. Assn., 62:1140–1158.

HILDEBRAND, F. B. 1962. Advanced calculus for applications. Prentice-Hall: New Jersey, 646 pp.

HUIZINGA, J. 1962. From DD to D^2 and back. The quantitative expression of resemblance. Koninkl. Nederl. Akademie van Wetenschappen-Amsterdam, Proc., Series C, 65:1–12.

JACKSON, R. A. AND T. CROVELLO. 1969. (In preparation).

JARDINE, C. J., N. JARDINE, AND R. SIBSON. 1967. The structure and construction of taxonomic hierarchies. Math. Biosciences, 1:173–179.

JARDINE, N. AND R. SIBSON. 1968a. The con struction of hierarchic and non-hierarchic classifications. Computer Journal, 11:177–184.

JARDINE, N. AND R. SIBSON. 1968b. A model for taxonomy. Math. Biosciences 2:465–482.

JOHNSON, STEPHEN C. 1967. Hierarchical clustering schemes. Psychometrika, 32:241–254.

KRUSKAL, J. B. 1956. On the shortest spanning subtree of a graph and the traveling salesman problem. Proc. Amer. Math. Soc., 7:48–50.

KRUSKAL, J. B. 1964. Nonmetric multidimensional scaling. Psychometrika, 29:1–27.

LANCE, G. N. AND W. T. WILLIAMS. 1966. A general theory of classificatory sorting strategies. I. Hierarchical systems. Computer Journal, 9: 373–380.

MCDONALD, R. P. 1962. A general approach to nonlinear factor analysis. Psychometrika, 27: 397–415.

MCDONALD, R. P. 1967. Numerical methods for polynomial models in nonlinear factor analysis. Psychometrika, 32:77–112.

MICHENER, C. D. AND ROBERT R. SOKAL. 1957. A quantitative approach to a problem in classification. Evolution, 11:130–162.

MICHENER, CHARLES D. AND ROBERT R. SOKAL. 1966. Two tests of the hypothesis of nonspecificity in the Hoplitis complex (Hymenoptera: Megachilidae). Ann. Ent. Soc. of America, 59 (6):1211–1217.

MOSS, W. WAYNE. 1967. Some new analytic and graphic approaches to numerical taxonomy, with an example from the Dermanyssidae (Acari). Syst. Zool., 16(3):177–207.

NEELY, PETER M. 1969. Towards a theory of classification. Neighborhood limited classification. (In preparation.)

ORE, O. 1962. Theory of graphs. American Mathematical Society, Providence, Rhode Island.

PEARSON, K. 1926. On the coefficient of racial likeness. Biometrika, 18:105–117.

PRIM, R. C. 1957. Shortest connection networks and some generalizations. Bell System Techn. Journal, 36:1389–1401.

PROCTOR, JEAN R. 1966. Some processes of numerical taxonomy in terms of distance. Syst. Zool., 15:131–140.

RAO, C. R. 1952. Advanced statistical methods in biometric research. John Wiley & Sons Inc.: New York.

REITER, S. AND G. SHERMAN. 1965. Discrete optimizing. J. Soc. Indust. Appl. Math., 13:864–889

ROGERS, D. J. AND T. T. TANIMOTO. 1960. A computer program for classifying plants. Science, 132:1115–1118.

ROHLF, F. JAMES. 1963. Congruence of larval and adult classifications in Aedes (Diptera: Culicidae). Syst. Zool., 12:97–117.

ROHLF, F. J. 1964. Methods for checking the results of a numerical taxonomic study. Syst. Zool., 13:102–104.

ROHLF, F. JAMES. 1966. Stepwise clustering procedures and ellipsoidal clusters. Mimeo; paper presented at Brookhaven meetings of Biometric Society, April, 1966.

ROHLF, F. JAMES. 1968. Stereograms in numerical taxonomy. Syst. Zool., 17:246–255.

ROHLF, F. JAMES AND DAVID L. FISHER. 1968. Tests for hierarchical structure in random data sets. Syst. Zool., 17:407–412.

ROHLF, F. J. AND R. R. SOKAL. 1965. Coefficients of correlation and distance in numerical taxonomy. Univ. Kansas Sci. Bull., 45:3–27.

RUBIN, GERROLD, 1966. An approach to organizing data into homogeneous groups. Syst. Zool., 15:169–182.

SEAL, H. L. 1964. Multivariate statistical analysis for biologists. Wiley: New York. xi+207pp.

SEBESTYEN, G. S. 1962. Decision-making processes in pattern recognition. Macmillan: New York. viii+162pp.

SHEPARD, R. N. AND J. D. CARROLL. 1966. Parametric representation of nonlinear data structures. *In*: Multivariate analysis; Ed. P. Krishnaih. Academic Press: New York. Pp. 561–592.

SNEATH, P. H. A. 1966. A method for curve seeking from scattered points. The Computer Journal, 8:383–391.

SNEATH, P. H. A. 1967. Some statistical problems in numerical taxonomy. The Statistician, 17:1–12.

SOKAL, ROBERT R. 1961. Distance as a measure of taxonomic similarity. Syst. Zool., 10:70–79.

SOKAL, R. R. 1966. Numerical taxonomy. Sci. Amer., 215(6):106–116.

SOKAL, R. R. 1969. The second annual conference on numerical taxonomy. Syst. Zool., 18: 103–104.

SOKAL, ROBERT R. AND JOSEPH H. CAMIN. 1965.

The two taxonomies: areas of agreement and conflict. Syst. Zool., 14:176–195.

SOKAL, R. R. AND F. JAMES ROHLF. 1962. The comparison of dendrograms by objective methods. Taxon, 11:33–40.

SOKAL, R. R. AND F. JAMES ROHLF. 1969. Biometry. Freeman: San Francisco.

SOKAL, R. R. AND P. H. A. SNEATH. 1963. The principles of numerical taxonomy. Freeman: San Francisco. xvi+359pp.

WILLIAMS, W. T. AND J. M. LAMBERT. 1959. Multivariate methods in plant ecology I. Association-analysis in plant communities. J. Ecol., 47:83–101.

Department of Biology, State University of New York, Stony Brook, New York 11790.

21

Reprinted from *J. Ecol.*, **54**, 427–445 (July 1966)

MULTIVARIATE METHODS IN PLANT ECOLOGY

V. SIMILARITY ANALYSES AND INFORMATION-ANALYSIS

BY W. T. WILLIAMS* AND J. M. LAMBERT
Botany Department, University of Southampton

G. N. LANCE
C.S.I.R.O. Computing Research Section, Canberra, Australia

CONTENTS

INTRODUCTION

In the current controversy over the relative merits of classification and ordination in vegetational analysis, it has been argued elsewhere (Lambert & Dale 1964) that the initial choice rests more on the convenience of the user than on preconceptions as to the continuity or discontinuity of the vegetation: if the prime requirement is to produce vegetational units which can be used for mapping or description, then classificatory methods are more applicable. In circumstances where classification is desired, there is a further user choice in the overall type of classification to be adopted, namely between hierarchical and non-hierarchical (i.e. reticulate) systems. It has been pointed out (Lance & Williams 1966) that hierarchical methods seek to subdivide the population progressively by the most efficient steps, while non-hierarchical methods—such as the many variants of cluster analysis—are aimed at the erection of efficient groupings irrespective of the route by which they are obtained; and, since no method is yet available which simultaneously maximizes hierarchical efficiency and group homogeneity, the user must decide whether to optimize the groupings or the route. In general, hierarchical methods are better known, less cumbersome, and more widely used in ecological work than direct clustering techniques, and—without prejudice as to the possible value of the latter under certain requirements—we have so far concentrated on the hierarchical

* Now at the C.S.I.R.O. Computing Research Section, Canberra, Australia.

approach. The first four papers of this series* (Williams & Lambert 1959, 1960, 1961; Lambert & Williams 1962) were concerned with a particular hierarchical method known as association-analysis, with the ecological results assessed over a number of test-communities. The present paper and subsequent communication will examine critically an alternative set of of hierarchical methods known collectively as 'similarity analyses', and compare the strategy and results of the most effective of these methods with those of association-analysis.

II. GENERAL CONSIDERATIONS

Hierarchical methods of classifying elements into sets are subject to two independent choices. First, the strategy may be *divisive*, in that the population is progressively subdivided into groups of diminishing size, or *agglomerative*, in that individuals are progressively fused into groups of increasing size until the entire population is synthesized. Secondly, the strategy may be *monothetic*, every group at every stage (except the entire population) being definable by the presence or lack of specified attributes, or *polythetic*, the groups being defined by their general overall similarity of attribute structure. Of the four systems so generated, agglomerative monothetic methods cannot exist, except in a trivial sense. Furthermore, of the two existing divisive polythetic methods, one (Edwards & Cavalli-Sforza 1965) is computationally out of reach for all except very small populations, and the other (Macnaughton-Smith *et al.* 1964) is not yet sufficiently developed for application to ecological problems. In practice, therefore, the choice at present lies between divisive-monothetic and agglomerative-polythetic.

Agglomerative-polythetic methods (i.e. similarity methods) are historically the older, deriving at least from the work of Kulczynski (1927); in their less-developed forms they are also simpler, and even amenable to hand-calculation. It is not, therefore, surprising that a number of variants, many of them frankly inefficient, already exist in the literature; a good general review is presented by Sokal & Sneath (1963). We can conveniently think of the overall strategy as determined by two choices. The first of these concerns the measure of similarity to be employed. The number of coefficients which at one time or another have been suggested is legion; it would be completely impracticable, and almost certainly unprofitable, to attempt to compare them all. Criteria have been suggested (Williams & Dale 1965) which may be used as a guide in the selection of coefficients for further study, and our present selection has been based on these recommendations. The second choice concerns the precise strategy to be used in making the successive fusions. Here again the alternatives have been discussed elsewhere (Williams & Dale 1965) and two have been selected for comparison.

A difficulty in the past has been that, owing to the restricted computer facilities hitherto available, any one worker has commonly had access to a single sorting method and often a single coefficient; there has therefore been virtually no opportunity for the comparative assessment of different coefficients and different fusion strategies on the basis of the results obtained from a single set of data. With improved computing facilities it is now easier, and more urgently necessary, to undertake comparative studies. Our results have been obtained by the use of the 'flexible similarity programmes' QUALNEAR and CENTROID on the Control Data 3600 computer in the C.S.I.R.O. Computing Research Section at Canberra; these programmes have already been briefly announced in a communication (Williams & Lance 1965) primarily concerned with the inferential problems presented by all intrinsic numerical classifications.

* For brevity, we shall subsequently cite these papers as Papers I–IV.

III. COMPARISON OF SIMILARITY METHODS

A. *Theoretical*

1. *The fusion strategies*

It will be convenient to consider the fusion strategies—usually termed 'sorting methods' —first. A number of alternatives have been suggested in different contexts, and the two selected for detailed study are described below. However, preliminary consideration must be given to two problems which may arise in all agglomerative systems, and which are related to the same cause—namely, that a hierarchical strategy involves irrevocable fusions; a 'bad' fusion early in the analysis is thus in principle capable of directing the subsequent fusions along an unprofitable path.

The first problem is that the data may be subject to appreciable error or bias in the sampling. Undesirable results can then be minimized by using a coefficient incorporating information from the population as a whole, such as the 'objectively weighted squared Euclidean distance' of Williams, Dale & McNaughton-Smith (1964); but in a complex situation the resulting dependence on overall population structure may destroy local concentrations of interest. A better solution would be to carry out a duplicated analysis and compare the results. In practice, it must be admitted that the usual procedure is simply to ignore this problem, but it is important that its existence be realized.

The second problem is simply that, unless the number of attributes is very large, ambiguities may be encountered. This can only be resolved by appeal to a different coefficient—which may not improve the situation—or, again, by importing information from the population as a whole or from outside the data. In our present study this problem, too, has been ignored; to avoid computational difficulties in the pilot programmes, the first of a set of ambiguities encountered has been used for action.

We proceed to outline our two chosen strategies:

(a) '*Nearest-neighbour*' *or* '*single-link*' *sorting*. This is the simplest of all agglomerative procedures. The process begins with the calculation of a similarity coefficient between all pairs of individuals; these coefficients, tagged with the numbers of the individuals concerned, are then sorted into a linear order with the most-similar pair at the beginning. All coefficients are then examined in turn and subjected to the following strategy:

(i) If neither of the members of the next most similar pair is already in a group (this is always true at the start of an analysis) designate them as forming a new group.
(ii) If one is in an existing group, add the other to the group.
(iii) If both are in different groups, add the groups.
(iv) If both are in the same group, discard.

The process continues until all individuals are fused into a single group. If there are n individuals, $\frac{1}{2}n(n-1)$ coefficients are calculated at the start, but there is no further calculation. The method is in theory lacking in power, since the structure of the groups as they form is not itself used in calculation; the analysis never rises above the information-level of a single individual.

(b) '*Centroid*' *sorting*. This process also begins with the calculation of all $2n(n-1)$ similarity coefficients, which are tagged but not in this case sorted into order. The strategy is then as follows:

(i) The most similar pair of individuals are *added together*, attribute by attribute, to form a new synthetic individual which is allotted the next available serial number.

(ii) The records of the individual members of the pair concerned are deleted, together with all coefficients involving either of them.

(iii) Coefficients are calculated between the new individual and all other remaining individuals; the process then returns to operation (i).

When all individuals have been fused into a single group, $(n-1)^2$ coefficients will have been calculated. The theoretical advantage of this method is that the groups grow in information content as the analysis proceeds, and become progressively less sensitive to errors and accidents in the data. More coefficients have to be calculated than for 'nearest-neighbour' sorting, so that the process is slower. More important than this is the fact that it requires more computer storage space, and is unsuitable for small computers. It is doubtless for this reason that the method, though long known in principle, has in the past been little used in classificatory studies.

2. *The coefficients*

We have used four coefficients, with two versions of one of them, making five in all; they are as follows:

(i) *Correlation coefficient.* This needs no definition. For qualitative data and 'nearest-neighbour' sorting it can be calculated from a 2×2 table as the Pearson ø-coefficient; for numerical data, and therefore for centroid sorting, the usual product-moment coefficient has been calculated. Special provision must be made for individuals lacking or possessing all attributes, since for such cases the coefficient is not defined. In the Canberra programme relationships with such individuals are allocated the impossible coefficient of $-2{\cdot}0$, which enables them to be segregated from the rest of the analysis.

(ii) *Squared Euclidean distance.* In a spatial model, let the co-ordinates of two individuals, or of the centroids of two groups, be $(x_{11}, x_{12}, \ldots, x_{1j}, \ldots, x_{1p})$ and $(x_{21}, x_{22}, \ldots, x_{2j}, \ldots, x_{2p})$; then the square of the distance between them is given by

$$\sum_{j=1}^{p} (x_{1j}-x_{2j})^2.$$

Qualitative data are accommodated by taking the jth co-ordinate for an individual as 1 if it possesses the attribute considered, and 0 if it lacks it; in the usual (a, b, c, d) symbolism of a 2×2 table, the squared distance between two qualitatively specified individuals then reduces to $(b+c)$. In centroid sorting, each attribute-entry is divided by the number of individuals in the group before calculating the distance; the values so obtained are the co-ordinates of the centroid of the group, and it is from this particular case that centroid methods derive their name.

(iii) *Standardized squared Euclidean distance.* This is merely a variant of (ii). It might plausibly be suggested that, in qualitative data, the joint presence of two rare attributes (or joint absence of two common ones) is more meaningful than the joint presence of two common attributes (or joint absence of two rare ones). To weight such joint occurrences appropriately, the attributes are standardized to zero mean and unit variance before the analysis begins.

(iv) *Non-metric coefficient.* We imply by this the coefficient, for qualitative data, $(b+c)/(2a+b+c)$. It is the complement of the familiar 'coefficient of floral community' which seems first to have been used by Czekanowski (1913), and which is monotonic with the coefficient $a/(a+b+c)$, probably first used by Jaccard (1908), and subsequently by Sneath in his early work (Sneath 1957) to avoid counting double-negative matches. Its quantitative form, in the symbols used in (ii) above, is $(\Sigma|x_{1j}-x_{2j}|)/\Sigma(x_{1j}+x_{2j})$, and as such (though in different symbols) is the familiar coefficient used by Curtis (1959) for ordination. For algebraic reasons which need not concern us here, the synthetic individuals of centroid sorting are again reduced by division to centroid co-ordinates before the calculation. The coefficient is undefined if both individuals being compared are everywhere zero. Since it is desirable that such individuals should be grouped together as identicals, the coefficient is put equal to zero if $(2a+b+c)$ or $\Sigma(x_{1j}+x_{2j})$ is zero.

Attention has been drawn elsewhere (Williams & Dale 1965) to some mathematical shortcomings of this coefficient; nevertheless, it has been used so often in early ecological work that we have felt it necessary to include it if only for its historical importance.

(v) *Information statistic.* The suggestion that statistics of this type should be used in classificatory problems is not novel—*vide*, e.g. Rescigno & Maccaccaro (1960). We deal with the derivation and relationships of the form we use in a parallel paper concerning computer problems (Lance & Williams 1966), and shall here content ourselves with a definition. Let a group of n individuals be specified by the presence or absence of p attributes, and let there be a_j individuals possessing the jth attribute. Then we define a statistic I, such that

$$I = pn \log n - \sum_{j=1}^{p} \left[a_j \log a_j + (n-a_j) \log (n-a_j) \right]$$

The statistic arises from the concept of entropy, and may be regarded as a measure of the disorder of the group; it becomes zero if all members of the group are identical. The most efficient route through the hierarchy is obtained by fusing those two individuals or groups which, on fusion, produce the smallest *increase* in I (I-gain or ΔI); but the absolute *value* of I for the resulting group is a consequent property of the group that may be of interest. For classificatory purposes the base of the logarithms is at arbitrary choice; we have utilized the tables of $n \log n$ to base e given in Kullback (1959).

The statistic is not defined for truly quantitative (i.e. continuously varying) data. In the qualitative case, if only two individuals are being fused, it reduces, in the symbolism of a 2×2 table, to $2(b+c) \log 2$; with 'nearest-neighbour' sorting, where all fusions are of this type, it therefore reduces to a constant multiple of squared Euclidean distance and the classification is identical in form with that produced by (ii).

3. *The problem of hierarchical levels*

The immediate outcome of any similarity analysis is a string of instructions for successive fusions of individuals, each fusion being associated with the value of the coefficient which brought it about. For assessment purposes, however, it is usually desirable not only to know the sequence of the fusions, but also to define in some way a set of levels which can be associated with these fusions.

To appreciate fully the issues involved, we must first consider the properties of the system defined by the ordered fusions, without regard to the coefficient values. If there are n dissimilar individuals, we have a dichotomously-branched system in which the n individuals are progressively fused into the entire population, passing through $(n-2)$ intermediate populations *en route*. This system, which we have called a hierarchy, and which Sokal & Sneath (1963) call a dendrogram, is topologically a *tree*, consisting of *nodes* joined by *line-segments*, with the individuals and the intermediate and final populations occupying the nodes. Any tree has certain important properties, viz.: (i) there are one fewer line-segments than there are nodes; (ii) the system is *connected*, in that there is a continuous route from any node (i.e. individual or population) to any other node; and (iii) if no line-segment is to be traversed more than once this route is unique and passes through a fixed string of nodes. These properties are invariant; the line-segments may be of any length, and a map of the system may be crumpled, twisted or stretched without losing these properties. A tree can always be represented in two dimensions. Since a hierarchical system is intrinsically directional, the only internodal routes of interest or meaning are those joining the population node to any of the ultimate nodes occupied by individuals; we shall refer to these as major routes.

We shall adopt the convention of disposing the individuals horizontally along a baseline with the population node above them and the intermediate nodes in the space between. It is now natural to regard the baseline as an abscissa, and a line perpendicular to this and passing through the population node as an ordinate. However, the height of any intermediate node above the baseline is meaningless without further definition; for the line-segments can be so adjusted that any intermediate node lies above any other.

Since each of the intermediate nodes represents a sub-population which may itself possess features of intrinsic interest, it is again natural to wish to import additional information so that the vertical distance of any such node above the baseline shall be associated with some meaningful property of the sub-population which occupies that node. This is the concept of *hierarchical levels.* A simple example would be to associate each node with the number of individual elements in the sub-population at that node; every node would now occupy a definite position along the ordinate (i.e. a definite distance above the baseline), and every major route would define a monotonic string of integers.

Universally, however, there has been an instinctive feeling that, since the tree itself is generated by a series of similarity coefficients, and since the formation of every population-node is associated with such a coefficient, it would be desirable to use these coefficients for the additional purpose of defining hierarchical levels. Our next problem, therefore, is to investigate the extent to which this is possible.

We consider two sub-populations (j) and (k) which fuse to give a third sub-population (i). Of the five coefficients here under consideration, the first four coefficients have this in common: they are (j, k) coefficients, in the sense that they provide a measure of the difference between (j) and (k), but they provide no measure of the heterogeneity of (i). However, it is reasonable to suppose that, the more dissimilar are (j) and (k), the more heterogeneous is (i), and it is thus reasonable to regard the (j, k) coefficient as a measure of the heterogeneity of (i) so far as a single division of fusion is concerned. If these measures could be accumulated over the hierarchy, a genuine measure of the overall heterogeneity of (i) could be obtained. Unfortunately, none of these four coefficients is additive in this sense; the convention in the past has therefore been to take the (j, k) coefficient, technically only the measure of a single fusion, as the best available measure of the heterogeneity of (i) taken over the whole of the hierarchy up to that point.

The situation is entirely different for the fifth coefficient, the information statistic; for this is an (i, jk) coefficient, defining the difference between (i) on the one hand and (j) and (k) jointly on the other, leaving the (j, k) measure undefined. Moreover, the coefficient is completely additive; if we write $I(i)$ for the total information content of (i), we have by definition:

$$I(j)+I(k)+\triangle I(i, jk) = I(i).$$

Individuals (or groups of identicals) have zero information content; so that if the I values are accumulated successively, a value of the information content (i.e. heterogeneity) is obtained which genuinely applies to the node attained. It is now possible to place the (i), (j) and (k) values on their appropriate point along the ordinate.

However, it is clearly, also possible, to use the convention which the other four coefficients impose on us for setting hierarchical levels; that is, to use the (i, jk) value for a single fusion—ΔI—as if it itself were a measure of the heterogeneity of (i). Only by so doing can we genuinely compare the hierarchies generated by all five coefficients, since the (i) values are at least then all subject to the same type of restriction. In our comparative assessment of results from the different methods, therefore, we shall use this convention throughout; the vertical scale will be that of the actual similarity coefficients (ΔI in the case of the information statistic), treated as properties of the group produced *after* the fusion they define.

4. *General criteria for comparison*

The use of a classificatory programme implies that the user has already decided that classification, and not ordination, is the aim; and the use of basically hierarchical system, rather than a clustering technique, likewise implies an interest in the actual path of fusion as well as in the groups. Given that the general requirement is to maximize the information specifically needed at the expense of other less relevant properties of the data, we can erect two basic criteria by which to compare the general effectiveness of the various methods under examination. These are as follows:

(a) The classification should be as clear-cut as possible, i.e. the hierarchy should consist of well-marked groups at well-separated levels as far as the data permit. First, this implies that the coefficient used should be somewhat sensitive to group size, so that the fusion of large groups is delayed as long as possible; the information statistic necessarily possesses this property, the correlation coefficient and Euclidean distance do not, and the properties of the non-metric coefficient are obscure. Secondly, the hierarchy should rise continuously to successively higher levels, i.e. the values used in constructing the hierarchy should be monotonic throughout; hierarchies from 'nearest neighbour' sorting are monotonic by definition, those from centroid sorting not necessarily so unless the coefficient is itself monotonic.

(b) The results must be profitable, i.e. they must suggest groupings which are ecologically meaningful when tested by appeal to other relevant information from outside.

Although on theoretical grounds alone, some of the methods under discussion appear to have certain intrinsic advantages over the others, the extent to which these operate in practice still requires empirical test. In the following section we compare the results obtained from parallel sets of analyses of two test-communities, using each of the two sorting strategies combined in turn with each of the five coefficients.* For ease of reference, we shall subsequently refer to analyses using 'nearest neighbour' sorting as the 'A' series, and centroid sorting as the 'B' series; the coefficients are numbered 1–5 in the order in which they have been discussed.

B. *Community analyses*

The test-communities chosen were two which have been used for earlier work in this series, namely 'Tumulus Heath' (20 sites/76 species) and 'Hoveton Great Broad' (56 sites/73

* Although for 'nearest neighbour' sorting the results from the information statistic are necessarily coincident with those from squared Euclidean distance (see p. 431), the two analyses are both included in all the comparisons for the sake of completeness.

species); their general ecological characteristics have already been described (Papers II, III and IV), and need not be repeated here. For all methods, each community was subjected to both a 'normal' and an 'inverse' analysis, i.e. to a classification of the sites in terms of the species present, and to a classification of the species in terms of the sites

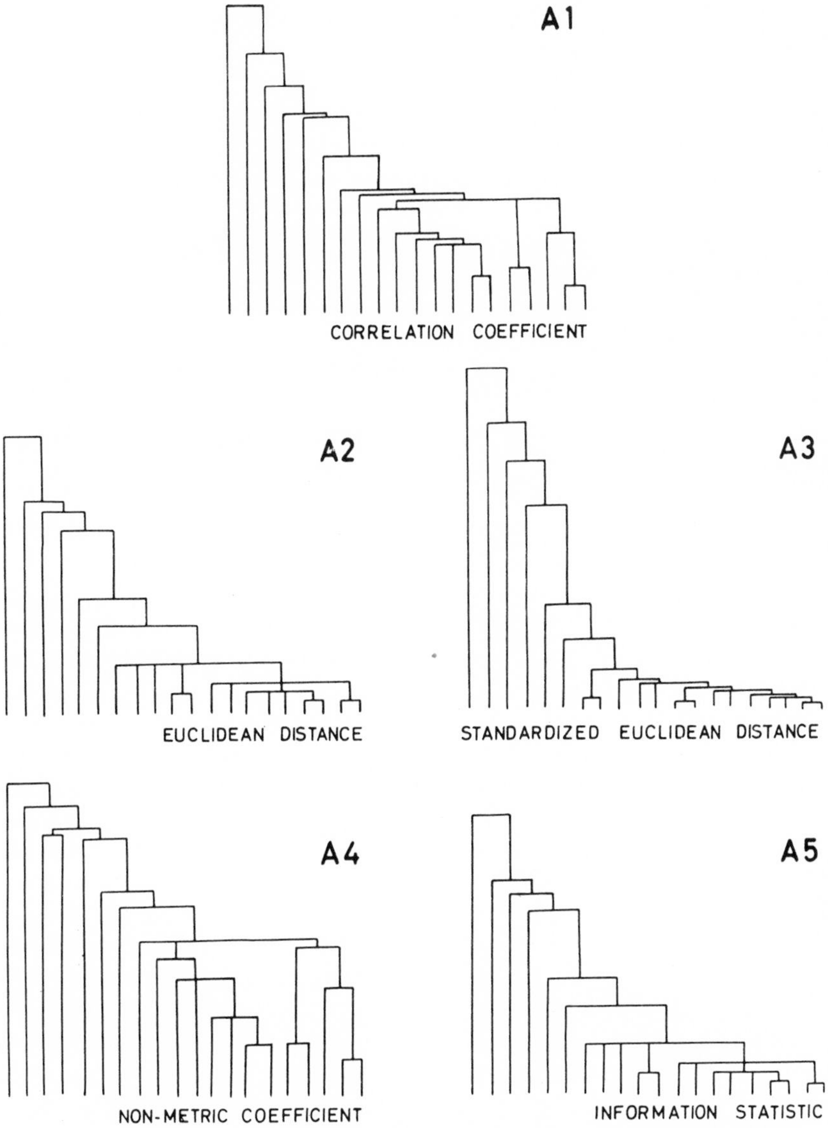

FIG. 1. Tumulus Heath: 'Similarity' analyses. Hierarchies from 'nearest neighbour' sorting methods.

in which they occur. In every case the hierarchy was plotted exactly as it emerged from the computer, with no subjective shuffling of the elements.

The hierarchies from the normal analysis of Tumulus Heath, reduced roughly to the same scale in terms of range of coefficient, are shown in Figs. 1 and 2. The other three sets (Tumulus Heath inverse; Hoveton normal and inverse), with larger and more complex hierarchies, are not depicted here for reasons of space; the results, however, are included in our general assessment.

It is obvious from the figures that the different analyses give very different results; we may now apply our basic criteria to them.

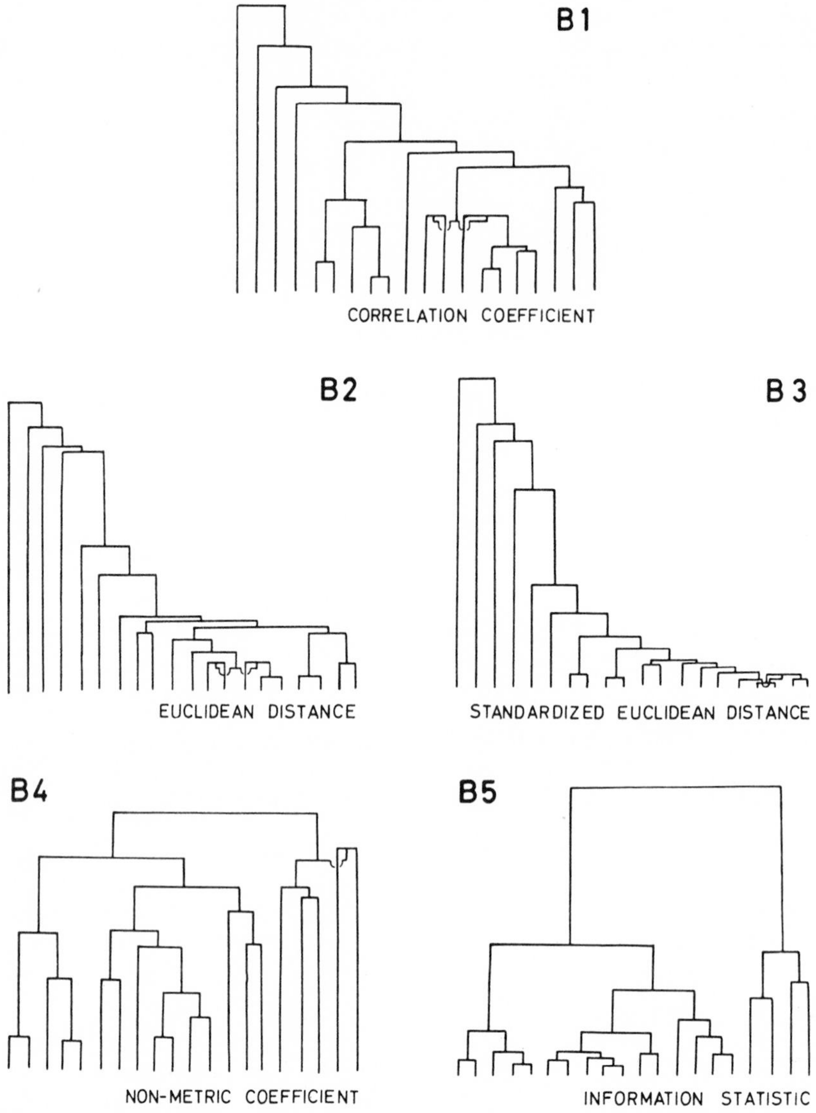

Fig. 2. Tumulus Heath: 'Similarity' analyses. Hierarchies from centroid sorting methods.

1. *General form of the hierarchy*

Here there are two separate features to be considered: (a) the degree of grouping, and (b) the distinctness of the groups depending on degree of what we shall loosely refer to as 'stratification'.

(a) *Grouping*. The essence of a useful classification is that the bulk of the individuals should be absorbed as quickly as possible into groups of higher order: we already know that the individuals (except identicals) are different from one another in some respect, and our main concern is to discover distinctive sets of individuals with properties in common. Furthermore, the groupings at a given level should preferably be of roughly comparable size as far as the data permit; otherwise extrinsic information needed for interpretation will be unequally distributed. In general, therefore, on *a priori* grounds a roughly symmetrical hierarchy is to be preferred to one with a high degree of 'chaining'.

By chaining, we mean the tendency for a given group to grow in size by the addition of single individuals or groups much smaller than itself, rather than by fusion with other groups of comparable size. Since chaining is due to inequality in the numbers in the two sub-populations concerned at each fusion in a hierarchy, a simple assessment of this tendency is possible. A dichotomous hierarchy defining n dissimilar individuals has $(n-1)$ junctions; let these be numbered in any order from 1 to $(n-1)$. Observe, at each junction, the *difference* between the numbers in the two sub-populations which fuse at that point, and let this difference at the ith node be denoted by δ_i. Then we define a coefficient of chaining, C, such that

$$C = \frac{2\sum_{i=1}^{n-1}\delta_i}{(n-1)\,(n-2)}$$

It is easily shown that this coefficient varies between zero for even divisions throughout (only attainable, of course, if n is a binary power) and unity for complete chaining.

Table 1. '*Chaining coefficient*' *values*

	'Nearest neighbour' sorting					Centroid sorting				
	A1	A2	A3	A4	(A5)	B1	B2	B3	B4	B5
TUMULUS HEATH										
Normal	0·77	0·88	0·86	0·67	(0·88)	0·55	0·75	0·79	0·18	0·18
Inverse	0·69	0·74	0·67	0·85	(0·74)	0·24	0·51	0·55	0·25	0·14
HOVETON										
Normal	0·37	0·35	0·68	0·38	(0·34)	0·26	0·37	0·62	0·21	0·07
Inverse	0·48	0·55	0·63	0·45	(0·55)	0·15	0·50	0·55	0·29	0·04
Mean	0·58	0·63	0·71	0·59	(0·63)	0·30	0·53	0·63	0·23	0·11

Table 1 gives the value of C, over both sorting strategies and all similarity coefficients, for the normal and inverse analyses of both test-communities. To summarize the major differences, the sums of squares of deviations from the grand mean were calculated as in preparation for an analysis of variance. First, this showed that the highest percentage of variation (31%) lies in the difference between the two sorting strategies, the 'A' analyses being in general more highly chained than the 'B'. This is perhaps only to be expected, since, with rather continuous data (as is frequent in ecology), the ability of an individual to link with *any member* of an existing group is likely to prejudice the formation of new groups. It is interesting in this connection that in Tumulus Heath, where the samples were all variants of heathland, not only are the chaining values generally higher than those for the more discontinuous Hoveton vegetation, but the distinction between the two sorting methods is somewhat more marked; in fact, although 18% of the total variation is caused by differences between the communities themselves, community–strategy interaction is still responsible for a further 4%.

Secondly, we must turn to the effect of the coefficients. Although the overall difference between them is appreciably less than that for the sorting strategies, this difference is nevertheless responsible for some 22% of the variation. However, first-order interactions between sorting strategy and coefficient account for a further 12%, due largely to the fact that the chaining values for the 'B' series are more variable than those for 'A'. Indeed, reference to Table 1 shows that the values for B2 and B3 approach the universally high 'A' values, and even surpass them in specific instances. The means for the first-order interactions between strategy and coefficients have been included in the table, from which it is clear that the overall difference between the methods lies mainly in the difference between the low values for B1, B4 and B5 and the high values for the remainder. These three thus fulfil our requirement for good grouping; and, of the three, B5 is clearly the best, though B4 and B1 are sufficiently close to merit further attention.

(b) *Stratification.* Though a tendency towards a symmetrical grouping is a valuable characteristic of a hierarchical method, this is not sufficient in itself. The groups must further be distinguished as sharply as possible from one another, and the levels at which they arise must be unequivocal: the picture is only confused if the criterion for ordering the groups runs counter in a given instance to that used in assessing their relative importance, i.e. the value of the coefficient. Moreover, since the essence of hierarchical classification is that the groupings towards the top have more intrinsic value than the rest, these need to be particularly discrete.

Table 2. *Number of 'reversals' in the value of the similarity coefficient*

	'Nearest neighbour' sorting					Centroid sorting				
	A1	A2	A3	A4	(A5)	B1	B2	B3	B4	B5
TUMULUS HEATH										
Normal	–	–	–	–	–	1	1	1	1	0
Inverse	–	–	–	–	–	2	3	4	2	0
HOVETON										
Normal	–	–	–	–	–	6	7	5	1	0
Inverse	–	–	–	–	–	10	14	9	5	0
Total	–	–	–	–	–	19	25	19	9	0

The 'nearest neighbour' methods cannot by definition give rise to reversals in the value of the coefficient for successive groups, but this is not true of the others. Reference to the figures for Tumulus Heath shows that, even in this simple community, B1–4 have one reversal each; in B4, the reversal is particularly serious in that it occurs in the upper part of the hierarchy.

The overall situation for the two test-communities is shown in Table 2, which gives the actual numbers of reversals produced by the different methods over all the analyses. B2, with a total of twenty-five, is particularly bad in this respect; B1 and B3, with nineteen each, are not much better; B4, with nine, still has too many for convenience; and it is only B5, with none, which meets this particular requirement completely.

Though the presence of reversals may confuse the stratification at certain points, the general picture is also affected by the proportion of the total range of the coefficient value occupied by successive fusions. Ideally, the most conspicuous changes should be towards the top, so that the major groupings stand out above the rest. Admittedly this can be adjusted by appropriate algebraic means; but since the solution is unique for a given population and cannot be used directly for another, such tricks are not to be recommended.

A rough guide to the general distribution of the groupings can be obtained by calculating the proportion of the total range of coefficient occupied by a given percentage of the fusions. In a complex population, it is usually possible to assess only the upper few, so that the bulk of the sub-populations can be ignored. For this reason, we have calculated, in Table 3, the proportion of the range occupied by the top 15% of fusions in all our

Table 3. *Proportion of total range of similarity coefficient value occupied by top 15% of fusions*

	'Nearest neighbour' sorting					Centroid sorting				
	A1	A2	A3	A4	(A5)	B1	B2	B3	B4	B5
TUMULUS HEATH										
Normal	0·39	0·36	0·42	0·19	(0·36)	0·36	0·18	0·37	0·18	0·69
Inverse	0·50	0·50	0·52	0·48	(0·50)	0·60	0·74	0·73	0·39	0·86
HOVETON										
Normal	0·44	0·18	0·29	0·69	(0·18)	0·69	0·17	0·41	0·56	0·91
Inverse	0·16	0·54	0·77	0·19	(0·54)	0·64	0·41	0·85	0·17	0·92
Mean	0·37	0·39	0·50	0·39	(0·39)	0·57	0·37	0·59	0·32	0·83

analyses. This again shows B5 to be far superior to the rest, and confirms us in our preference for this method on grounds of the evidence so far.

Finally, before we leave this section, we must look briefly at the B5 hierarchy in its genuine form, i.e. with its levels plotted according to *total* information content instead of mere information gain (see p. 433); we shall call this other version B5′. It is clear from Fig. 3, which shows the two forms of B5 side by side, that the desirable feature of marked

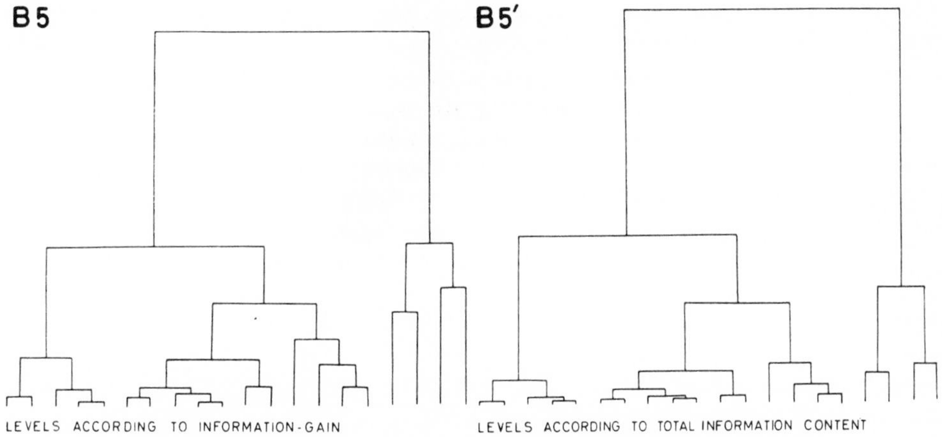

FIG. 3. Tumulus Heath: Comparison of hierarchial levels in the B5 method.

stratification has been accentuated by the new levels; moreover, there are slight changes in the relative levels of certain fusions which could be ecologically important.

2. *Ecological assessment*

Though a method may appear to have unassailable theoretical advantages, the acid test is whether the information it imparts is acceptable to the user—in this case, the ecologist. Frequently, a method developed *in vacuo* fails miserably when actually applied, either because of other unexpected and hitherto unexamined properties of the method itself, or because of special features of the data which make it inapplicable. We shall therefore complete this study by comparing in general terms the ecological information extracted from the two-test communities by the various methods.

(a) *Method of assessment.* The difficulty with a comparison involving the concept of ecological acceptability is to find objective criteria in an essentially subjective situation by which to differentiate between the analyses. Two courses are open to us in this respect: either we can examine the results from each analysis directly and assess the extent to which the groupings produced are readily interpretable in the light of other ecological experience; or we can erect a set of groupings *ab initio* from this experience and assess the extent to which these groupings are reflected by the analyses. Since ecological experience has many facets, some are likely to be more relevant to the one situation and some to the other; the two approaches will therefore not necessarily give exactly the same results. However, if only one is to be employed, the choice between the two is largely a matter of convenience. In present circumstances, we find it easier to adopt the second system.

With the given test-communities, the threshold for our acceptance of any of the hierarchical methods under study will be that the major groupings which arise shall not be fewer than, or markedly different from, those recognized intuitively as distinct ecological entities at the time the data were collected. The stability of these entities has already been partly confirmed by the fact that they reappeared (together with additional information) when these communities were subjected to association-analysis in our earlier work (Paper III, p. 723; Paper IV, pp. 792, 799); we therefore feel justified in setting a minimum standard based on them. For Tumulus Heath, we shall require a differentiation between wet heath, dry heath and grass-heath at least; and for Hoveton, a similar separation will be needed between open water, reedswamp, primary fen, mowing marsh, swamp carr and fen carr groupings. These we shall call our *standard ecological categories.*

As a second criterion, we may reasonably ask that the groupings in the different categories for either community shall all be differentiated at roughly the same hierarchial level; the very fact that these groupings are required to reflect the lowest level of intuitive recognition of distinctive vegetation types in the original subjective survey suggests that in some sense they should be equivalent in degree of ecological homogeneity.

Since it is clearly impracticable to give detailed consideration here to the separate results of forty different analyses, a crude allocation system for overall comparison was devised:

(i) Each quadrat and species was first subjectively allocated to one or other of the predetermined ecological categories on the basis of previous experience of the vegetation; individuals of uncertain ecological affinities, such as anomalous or ecotonal quadrats, or widely tolerant species, were left uncategorized. The categorized individuals were known as *standard units.*

(ii) Each grouping in each analysis was then examined for ecological homogeneity in terms of the standard units it contained. To qualify for recognition as a *standard grouping*, a group must have accumulated at least 50% of the total number of standard units in a given category without the addition of any unit or grouping of units of different type. To give as much latitude as possible, the uncategorized individuals were regarded as 'floaters', i.e. they could be attached to any group without destroying or adding to its ecological integrity.

(iii) The hierarchical level at which a standard grouping achieved 50% of its total possible membership was designated its *minimum level of organization*; the ultimate level which it reached before becoming contaminated by the addition of one or more extraneous units, or fused with another standard group, was designated its *level of maximum differentiation*. Where the picture was confused at a critical point by reversals in level (see p. 437), the sequence of fusions was allowed to override the relative values of the levels.

(b) *Results*. Table 4 gives the tally of standard groupings for all the hierarchies under examination, including the two versions of B5; a positive entry for a particular ecological category indicates the presence of a standard grouping of that type, while a negative entry means that the requisite degree of differentiation was not achieved. As a rough indication of the relative levels at which the various standard groupings became organized in any one analysis, the single lowest level over all the groupings at which a grouping was maximally differentiated was taken as a dividing line; those groupings which existed at this level are shown in bold-face type, while the others, whose minimum level or organization lay higher than this line, are shown in normal type.

It is clear from the table that even the very crude allocation system employed is sufficient to discriminate cleanly between the analyses. We shall first consider the emergence of the groupings themselves, and then the relative levels at which they arise.

The first point of interest lies in the general pattern of the table as a whole. For both Tumulus Heath and Hoveton, there are certain standard groupings which appear with all the methods. In Tumulus Heath, for instance, wet and dry heath are differentiated throughout in the normal analyses, while in Hoveton the open water and mowing marsh groupings appear with all methods on both normal and inverse sides; these entities are clearly sufficiently distinct for every method to extract them, whatever its sorting strategy or coefficient. Conversely, there are other groupings which appear infrequently, such as the 'normal' grass-heath grouping in Tumulus Heath, and the 'normal' swamp carr and 'inverse' fen carr in Hoveton; and it is the less well-circumscribed groupings like these which serve best to test the relative sensitivity of the different methods.

Secondly, there is some interest in comparing the relative efficiency of the extraction on the normal and inverse sides. For most of the analyses, the *number* of groupings produced is similar on both sides, though the *categories* may differ. However, for the two methods using the non-metric coefficient (A4 and B4), there is an overall tendency for the normal groupings to be better than the inverses. This is clearly a function of this particular coefficient, which is sensitive only to positive matches: since the species, being abstractions, are inherently more likely to be more variable than the quadrats, they will tend to have fewer occurrences in common to bind them together into discrete groups.

With regard to the overall number of groupings extracted by the individual methods, B5 is clearly best, with all eighteen possible groupings represented; B4, with fourteen groupings despite its failure on the inverse side, is next, followed by A1 and B1 with thirteen each. In general, the methods using the centroid strategy show a slightly better performance than their counterparts with 'nearest neighbour' sorting; but the fact that A3 and B3 show the lowest returns in the 'A' and 'B' series respectively, while those for A1 and B1 are relatively high, is a pointer to the importance also of the effect of the different coefficients.

We may now turn to our second criterion concerning the relative levels of differentiation of the standard groups extracted from a single analysis. Even with the extremely crude device we have used to identify the groupings which become organized only at high

Table 4. *Extraction of ecological groupings by the different methods*

	Tumulus Heath Normal: Wet heath	Tumulus Heath Normal: Dry heath	Tumulus Heath Normal: Grass-heath	Tumulus Heath Inverse: Wet heath	Tumulus Heath Inverse: Dry heath	Tumulus Heath Inverse: Grass-heath	Hoveton Normal: Open water	Hoveton Normal: Reedswamp	Hoveton Normal: Primary fen	Hoveton Normal: Mowing marsh	Hoveton Normal: Swamp carr	Hoveton Normal: Fen carr	Hoveton Inverse: Open water	Hoveton Inverse: Reedswamp	Hoveton Inverse: Primary fen	Hoveton Inverse: Mowing marsh	Hoveton Inverse: Swamp carr	Hoveton Inverse: Fen carr	Total standard groups (max. = 18)
A1	+	+	−	+	−	+	+	+	+	+	−	+	+	−	+	+	+	−	13
A2	+	+	−	+	−	+	+	−	+	+	−	+	+	−	−	+	+	−	11
A3	+	+	−	+	−	+	+	−	−	+	−	−	+	−	−	+	+	−	9
A4	+	+	−	−	−	−	+	+	+	+	−	+	+	−	−	+	+	−	10
(A5)	(+)	(+)	(−)	(+)	(−)	(+)	(+)	(−)	(+)	(+)	(−)	(+)	(+)	(−)	(−)	(+)	(+)	(−)	(11)
B1	+	+	−	+	−	−	+	+	+	+	−	+	+	+	+	+	+	−	13
B2	+	+	−	+	+	+	+	−	+	+	−	+	+	−	−	+	+	−	12
B3	+	+	−	+	+	+	+	−	−	+	−	−	+	−	−	+	+	−	10
B4	+	+	+	+	+	−	+	+	+	+	+	+	+	−	−	+	+	−	14
B5	+	+	+	+	+	+	+	+	+	+	+	+	+	+	+	+	+	+	18
B5′	+	+	+	+	+	+	+	+	+	+	+	+	+	+	+	+	+	+	18

levels, the general picture is clear. For the majority of methods, at least some of the groupings fail to form until at least one another grouping has already been fully differentiated and has lost its identity; in such a case, only a proportion of the total number of standard groupings extracted can exist together at any one hierarchical level, and the value of these groupings as ecological units for mapping or descriptive purposes (as in the original survey) is correspondingly reduced.

On this particular criterion, A4 and B4 are good, with only one failure each; but, because of its other advantages, the interest naturally centres on B5. Here, since we are now specifically concerned with hierarchical levels, we may legitimately extend our comparison to include both versions of the hierarchy which can be constructed from the B5 type of analysis (see p. 438); and we find that, whereas the form which is strictly comparable with the other hierarchies has two failures, the other—B5′—has none. The B5′ hierarchy is thus the only one which fulfils our initial requirements completely, both in the range of standard groupings extracted and the levels at which they are formed.

(c) *Discussion.* The immediate results from the crude tests applied for ecological assessment appear at first sight to eliminate all except B5 from further serious consideration. However, it could be argued that to concentrate exclusively on the ability of a method to extricate predetermined groupings at roughly predetermined levels is placing too much reliance on preconceived ideas as to the nature of the situation in the given test-communities. Ideally, a parallel appraisal should be made using the alternative method of assessment (see p. 439), in which the groupings themselves are examined directly in their own right, using an external measure of some sort to compare their ecological significance. Apart from the difficulty of erecting such a measure, the extra expenditure of time could hardly be justified here: the real difference between the methods lay not in the ecological *nature* of the groups, but whether substantial groupings were formed *at all*, i.e. those methods which failed to provide the requisite number of 'standard' groupings did not in general provide alternative groupings to be assessed. Some of the failures were certainly due to intermingling of individuals from allied ecological categories, such as reedswamp and open water, or the two types of carr; but these were failures due to lack of differentiation, rather than to the erection of genuinely new and unexpected groupings. In short, there was little evidence of migration of tagged individuals into new composite groups, such as would justify an independent assessment from other external evidence.

It is particularly interesting in this context that, in the few cases where substantial new-groupings did in fact emerge, these usually concerned the unallocated individuals; these groups were thus *additional* to the standard groups instead of replacing them. Since we have been primarily concerned with threshold criteria for the acceptance of a method we have so far ignored these supernumerary groupings in our general assessment. They were almost entirely restricted to B4 and B5 and, as we shall be examining the B5 hierarchies in more detail subsequently, we shall content ourselves for the moment with the comment that most of them seemed ecologically meaningful.

Before a final decision is made in favour of B5, B4 deserves a last consideration; though failing on some counts, such failures were usually rather trivial. For instance, detailed examination of the B4 hierarchies showed that failure to extract inverse standard groupings was due largely to the incorporation of an alien element at a critical stage, rather that to a genuine lack of grouping. It is, in fact, its emphasis on rather trivial features which operates particularly against B4 for ecological work: the inability of the coefficient to use information from negative matches means that chance records—such as entries for casual species—are given undue importance. Nevertheless, the overall performance of B4 is such that with more symmetrical data—as in certain types of taxonomic work—its particular disadvantages could easily be outweighed (see e.g. Watson, Williams & Lance 1966). For present purposes, however, B5 seems superior on all counts. We have therefore selected B5 alone for further consideration in the following paper; and we shall henceforth refer to B5—the method using centroid sorting and the information statistic as its coefficient—as *information-analysis*.

IV. GENERAL DISCUSSION

With interest increasing in the use of numerical methods, it is inevitable that techniques developed in one field of study should find their way into others. The history of taxonometrics is no exception to this: strategies and coefficients developed for vegetational work have been adopted for the classification of individual organisms, and *vice versa*. However, without some regard for differences in the nature of the material to be manipulated, such practices can easily lead to failure. For instance, a rather crude classificatory method can appear strikingly efficient if used on material already partly classified subjectively before analysis, as in much taxonomic work; but if the same method is then required to extra groupings from more continuous vegetational data, it may prove insufficiently sensitive for the purpose. The essence of this difference in material to be handled is concisely stated by Webb (1954): 'The majority of species are guaranteed some measure of objectivity, stability and discriminability by the genetic pattern .... No comparable factor is available to stabilize plant communities .... The fact is that the pattern of variation shown by the distribution of species among quadrats over the earth's surface chosen at random hovers in a tantalizing manner between the continuous and the discontinuous'.

With such material, it is only natural that plant ecologists have wavered between the relative merits of classification and ordination as a means of reflecting vegetational relationships. At first sight, therefore, a method which is sensitive only to fairly discrete groups and then 'chains' the other individuals might be regarded as a useful compromise between the two. However, although the chained part of a hierarchy may appear superficially like an ordination, the order in which the chained individuals appear is related only to their degree of similarity with the groupings previously formed; there is, in fact, no necessary immediate affinity between two juxtaposed chained individuals, and hierarchies showing excessive chaining have therefore little useful function to perform.

The very fact that different analyses of exactly the same set of data show every gradation between excessive chaining (as in A3) and fairly symmetrical grouping (as in B5), itself is an indication of the futility of arguments as to the 'real' nature of vegetation. All that such arguments mean, is that different observers are instinctively using different values to assess similarities and differences between one individual and another, or between an individual and a pre-erected group. Some ecologists may be most impressed in the field with likenesses between neighbouring vegetation samples, and are intuitively using 'nearest neighbour' sorting; others, with perhaps more power of integration, are using a mental process akin to centroid sorting; others, again, may be particularly struck by the presence of rare species and the absence of common ones, and are mentally using a measure like the standardized Euclidean coefficient; while, yet again, others may be especially susceptible to the size of the area covered by a given type of vegetation, and tend to make mental adjustments allied to the group-size sensitivity of the information statistic.

With such diverse conceptions as to what is 'important' in vegetational analysis, the only objective criterion we can use in selecting a 'best' method is to choose that method which will most efficiently perform the particular function which is asked of it. Our attitude here has been that, in a complex ecological situation, the clarity of exposition of the results is all-important. The great advantage of information-analysis in this respect is that, not only does it produce a clear-cut hierarchy, but the method itself is internally consistent so that different mathematical models are not confused: by minimizing the variables in the method of analysis itself, the variables in the situation under study are thus more clearly exposed.

ACKNOWLEDGMENTS

We are indebted to Mr P. Macnaughton-Smith, of the Home Office Research Unit, for his original suggestion of the possible application of the information statistic in the present context. The diagrams for this paper were drawn by Mr B. A. Lockyer.

SUMMARY

Agglomerative-polythetic methods (commonly known as 'similarity methods') of hierarchically classifying elements into sets can take a large number of different forms, according to: (a) the type of fusion strategy ('sorting method') employed; and (b) the coefficient used to measure similarity. Ten selected versions, using two different sorting methods combined in turn with five different coefficients, are tested empirically for their relative efficiency, using both theoretical and ecological criteria. The results from the comparative analyses of two test-communities show that, whereas 'centroid' sorting in general gives better results than 'nearest neighbour' sorting, there is also an interaction between sorting strategy and coefficient. The method combining centroid sorting with an information-statistic coefficient is shown to be greatly superior to the others in producing clear-cut and ecologically acceptable hierarchies; and this method, called *information analysis*, is selected for further test.

REFERENCES

Curtis, J. T. (1959). *The Vegetation of Wisconsin.* Wisconsin.

Czekanowski, J. (1913). *Zarys metod statystycznych* (*Die Grundzuge der statischen Metoden*). Warsaw. (*ex* Curtis, 1959).

Edwards, A. W. F. & Cavalli-Sforza, L. L. (1965). A method for cluster analysis. *Biometrics*, **21**, 362–75.

Jaccard, P. (1908). Nouvelles recherches sur la distribution florale. *Bull. Soc. vaud. Sci. nat.* **44**, 223–70. (*ex* Sokal & Sneath 1963).

Kulczynski, S. (1927). Zespoly roslin w Pieninach. (Die Pflanzen-associationen der Pieninen). *Bull. intern. Acad. pol. Sci. Lett., Cl. Sci. Math., et Nat.* Ser. B. Suppl. II, **57**, 203. (*ex* Curtis, 1959).

Kullback, S. (1959). *Information Theory and Statistics.* New York.

Lambert, J. M. & Dale, M. B. (1964). The use of statistics in phytosociology. *Adv. ecol. Res.* **2**, 59–99.

Lambert, J. M. & Williams, W. T. (1962). Multivariate methods in plant ecology. IV. Nodal analysis. *J. Ecol.* **50**, 775–802.

Lance, G. N. & Williams, W. T. (1966). Computer programs for hierarchical polythetic classification ('Similarity analyses'). *Brit. Comp. J.* (In press).

Macnaughton-Smith, P., Williams, W. T., Dale, N. B. & Mockett, L. G. (1964). Dissimilarity analysis: a new technique of hierarchical sub-division. *Nature, Lond.* **202**, 1034–5.

Rescigno, A. & Maccaccaro, W. B. (1960). The information content of biological classifications. In: *Symposium on Information Theory.* London.

Sneath, P. H. A. (1957). The application of computers to taxonomy. *J. gen. Microbiol.* **17**, 201–26.

Sokal, R. R. & Sneath, P. H. A. (1963). *Principles of Numerical Taxonomy.* San Francisco.

Watson, L., Williams, W. T. & Lance, G. N. (1966). Angiosperm taxonomy: a comparative study of some novel numerical techniques. *J. Linn. Soc.* (In press).

Webb, D. A. (1954). Is the classification of plant communities either possible or desirable? *Bot. Tidsskr.* **51**, 362–70.

Williams, W. T. & Dale, M. B. (1965). Fundamental problems in numerical taxonomy. *Adv. bot. Res.* **2**, 35–68.

Williams, W. T., Dale, M. B. & Macnaughton-Smith, P. (1964). An objective method of weighting in similarity analysis. *Nature, Lond.* **201**, 426.

Williams, W. T. & Lambert, J. M. (1959). Multivariate methods in plant ecology. I. Association-analysis in plant communities. *J. Ecol.* **47**, 83–101.

Williams, W. T. & Lambert, J. M. (1960). Multivariate methods in plant ecology. II. The use of an electronic digital computer for association-analysis. *J. Ecol.* **48**, 689–710.

Williams, W. T. & Lambert, J. M. (1961). Multivariate methods in plant ecology. III. Inverse association-analysis. *J. Ecol.* **49**, 717–29.

Williams, W. T. & Lance, G. N. (1965). Logic of computer-based intrinsic classifications. *Nature, Lond.* **201**, 159–61.

(*Received* 3 *November* 1965)

22

Reprinted with permission from *Syst. Zool.*, **21**(2), 174–186 (1972)

DISTANCE ANALYSIS IN BIOLOGY[1]

MAJOR M. GOODMAN

Abstract

Goodman, M. M. (Dept. of Statistics, North Carolina State University at Raleigh, North Carolina, 27607) 1972. Distance analysis in biology. Syst Zool., 21:174–186.—The mathematical relationships between generalized distances, Sokal's distances, distances based upon principal components, and Cavalli-Sforza and Edwards' chord distances are discussed, and the biological consequences of these relationships in the case of nontrivial correlation matrices are pointed out. While none of the techniques discussed are wholly satisfactory, a practical solution to the problems encountered in their deployment is suggested which combines two of the techniques (generalized distances and principal components). [Principal components; Sokal distance; generalized distance.]

Both numerical and non-numerical taxonomists as well as anthropologists and other quantitatively oriented scientists have employed several types of distance analyses based on morphological data (Sokal and Sneath, 1963). More recently similar techniques have also been used in the analyses of human blood group frequencies (Cavalli-Sforza and Edwards, 1967; Pollitizer, 1958, 1969; Pollitzer et al., 1967, 1970; Fitch and Neel, 1969). In addition, principal component analyses are frequently utilized in conjunction with, or as an alternative to, such techniques (Gower, 1966; Morrison, 1967). Certain of these techniques are mathematically related in simple ways which are both biologically important and generally unfamiliar.

FORMAT OF THE DATA

The data generally used are a set of vectors ($\bar{\mathbf{x}}_i$; i = 1, N), each of which contains n character means for one of the N taxa or populations being studied (such means may be frequencies) and an $n \times n$ covariance matrix, which may be derived from the above-mentioned vectors or from variation of individuals around the taxa means. For simplicity, it will be assumed that the data, $[\bar{\mathbf{x}}_i' = (x_{i1}, x_{i2}, \ldots, x_{in})]$, have been standardized by dividing the elements of the data vectors by the square roots of the corresponding diagonal elements of the $n \times n$ covariance matrix and that the covariance matrix has thereafter been converted to a correlation matrix, **R**.

GEOMETRIC INTERPRETATIONS

Sokal's Distances and Pearson's Coefficient of Racial Likeness

The N points represented by the N data vectors can generally be regarded as being scattered in n dimensional space. (If $N \leqslant n$, they can generally be represented in $[N-1]$ dimensional space. Linear dependencies among characters or among taxa may reduce the number of dimensions necessary for such a representation.) The Euclidean distance between two such points is

$$(1) \qquad D_{ij} = [\sum_{k=1}^{n} (x_{ik} - x_{jk})^2]^{\frac{1}{2}}$$

Two well-known examples of such distances are Sokal's (1961) distances and (when sample sizes are equal) Pearson's (1926) Coefficient of Racial Likeness. Such techniques ignore correlations among the characters studied and are thus most useful in cases where the characters are essentially independent. Analytical genetic comparisons of the most commonly used techniques of this type have been made elsewhere (Goodman, 1969). For correlated characters, several alternatives are available.

---

[1] Paper No. 3327 of the Journal Series of the North Carolina State University Agricultural Experiment Station, Raleigh, North Carolina. This investigation was supported by Public Health Service Grant GM 11546.

Principal Components

Principal component analysis is a simple rotation of the axes yielding a new set of coordinates for each point. These new coordinates $\{y_{ik}\}$ have correlation matrix **I** (the identity matrix) and variances λ_k, where the λ_k (where $\lambda_k \geqslant \lambda_{k+1} > 0$) are the non-zero characteristic roots of **R**.

$$(2) \qquad D_{ij} = [\sum_k (y_{ik} - y_{jk})^2]^{\frac{1}{2}}$$

remains the same, if the covariance matrix is non-singular, since the axes have only been rotated; the points themselves (or the distances between them) have not been shifted. In the case of a singular correlation matrix (e.g., when one character is exactly determined by a linear combination of the other characters), (1) will be larger than (2) due to the inclusion of redundant information in (1).

From (2) it is clear that the Euclidean distances based upon principal component analysis successively approach the Euclidean distances based on the original data as more and more of the principal components (the y_{ik}) are used [i.e., as k increases in (2)]. When the full set of principal components is used, the distances are the same as those calculated from the original data (unless the correlation matrix is singular), and no adjustments are made for correlations among characters. Such distances are simply Euclidean distances measured within a multidimensional correlation ellipse (a hyperellipsoid). Use of less than the full set of principal components in (2) (frequently only the first two or three are used) results in measuring distances within a correlation ellipsoid of fewer dimensions. Paraphrasing Edgar Anderson's (1949, p. 39) apt description of such a configuration, "If we think of a hypothetical individual having the lowest possible values of all of the characters being represented at one of the apices of a multidimensional cube and a second hypothetical individual with all the highest possible values of all of the characters being positioned at the opposite apex, then the individuals having varying combinations of intermediate values of the various characters form a narrow spindle through the center of the cube." Distances measured within such a "spindle" do not all have the same biological importance, as variation is more readily possible in certain directions than in others. To equalize variation, it is necessary to standardize the principal components. Such a standardization, which converts the multidimensional ellipse to a multidimensional sphere, will be discussed later together with generalized distances (Mahalanobis, 1936).

The fact that equal geometric distance may not indicate equal biological divergence is an all-too-familiar phenomenon to plant and animal breeders working with negatively correlated traits. Although such breeders do not generally attempt to develop the formal distance analyses relevant to their problems, they are well aware that genetic progress can more readily be achieved along the major axis of a correlation ellipse than along the minor axis. Thus cotton breeders have worked rather unsuccessfully for years to develop a reasonably productive, long fiber cotton with high strength, while high yielding, short but high strength cottons and long fibered, low strength cottons are much more readily obtained (Stephens, 1961). Similar results are well known for both ear length and diameter and also yield and lodging resistance in corn and for rate of gain and reproductive ability in poultry.

Generalized Distances

The generalized distance (Mahalanobis, 1936) between i and j is

$$(3) \qquad D_{ij}^m = [(\bar{\mathbf{x}}_i - \bar{\mathbf{x}}_j)'\mathbf{R}^{-1}(\bar{\mathbf{x}}_i - \bar{\mathbf{x}}_j)]^{\frac{1}{2}} = [\sum_k \{(y_{ik} - y_{jk})^2/\lambda_k\}]^{\frac{1}{2}}.$$

When **R** is singular, $\mathbf{R}^{-1}$ must be a generalized inverse of the correlation matrix, and if N (the number of taxa) is large, the D_{ij}^m can be calculated much more rapidly using the y_{ik} and the λ_k than with the $\bar{\mathbf{x}}_i$ and $\mathbf{R}^{-1}$ [for an alternative technique see Rightmire (1969)]. Equation (3) shows that the generalized distances are simply the corresponding distances obtained after adjusting Anderson's multidimensional ellipse to a multidimensional sphere (a hypersphere). Unfortunately, while the correlation ellipse disappears as a result of the transformation, all the problems associated with it do not. Interest in generalized distances is, in essence, confined to cases in which the correlation matrix, **R**, differs substantially from the identity matrix, **I**. (If this were not so, one would use simple Euclidean distance [equation (1)]). In such situations, certain characters are often essentially determined by the values of other characters, and, as a result, the standard errors of certain of the smaller λ_k may actually be larger than the λ_k themselves. Thus the corresponding values of $[(y_{ik}-y_{jk})^2/\lambda_k]$ in (3) can be greatly inflated in the computation of Mahalanobis' distances, even though they have little or no biological importance and even though they may partially or completely mask the effects of other terms of considerable biological significance.

INTERRELATIONSHIPS AND ALTERNATIVES

The relationship between generalized distances and principal components [equation (3)] apparently has not been widely recognized by biologists. From (3), it is immediately apparent that conclusions based upon the first few principal components (a procedure often followed) *must* differ from those based on the complete set of standardized principal components (the generalized distances). In addition, when the correlation matrix **R** is calculated from the $\bar{x}_i$ themselves $(i = 1, 2, \ldots, N)$, then there are certain restrictions on the values which the distances (the $D_{ij}{}^m$) may have (Gower, 1966). In general, for k characters the average value of all possible squared distances (excluding, of course, distances between a population and itself) is $2k$ for $k < (N-1)$; for $k \geqslant (N-1)$ the average value is $2(N-1)$. In fact, for $k \geqslant (N-1)$ all of the squared distances are equal to $2(N-1)$. From equation (3) it can be seen that as k increases $D_{ij}{}^m$ must increase also until $k = (N-1)$. Thus $2(N-1)$ is the upper limit for the square of $D_{ij}{}^m$ under such circumstances. (Note that exact linear dependencies among characters effectively lower the value of k, while exact linear dependencies among the populations effectively lower the value of N above). Clearly the utility of such distances decreases as more and more of the squared distances approach $2(N-1)$. The general equivalence between equations (1) and (2) strongly suggests that the use of nonstandardized principal components would not be justified in cases where the characters are so strongly correlated that Sokal's distances or Pearson's Coefficient of Racial Likeness would not be used. Thus in the case of correlated characters, only one procedure seems tenable: use of some, but not all, of the standardized principal components. One question remains: how many of the standardized principal components should be used?

No complete solution to this problem is known. Three extremes appear avoidable: a. ignoring correlations among characters and using (1), b. using only the first two or three non-standardized principal components [i.e., using k equal to 2 or 3 in (2)], and c. use of (3) which may include much "noise" due to near singularity of **R** as a result of high correlation among characters. A conservative approach would be to use only principal components for which $\lambda_k \geqslant K$:

$$(4) \qquad D_{ij}{}^* = [\sum_{k=1}^{q} \{(y_{ik}-y_{jk})^2/\lambda_k\}]^{\frac{1}{2}},$$

where $\lambda_k \geqslant K$ for $k \leqslant q$ and $\lambda_k < K$ for $k > q$. In (4), those principal components which contribute greatly to a decrease in the generalized variance (the determinant of the covariance matrix [in this case, the correlation matrix] which can be obtained as the product of the λ_k) are not used when K is approximately 1.0 or greater. For the full set of principal components, the generalized variance is often, for practical purposes, zero, and it may closely approximate zero (i.e., have a value of about 0.00001) well before the full set is reached. This procedure [equation (4) with $K \approx 1.0$], analogous to a procedure widely used in

TABLE 1.—THE CHARACTERS USED, THEIR MEANS, AND THEIR STANDARD DEVIATIONS (AMONG RACE MEANS).

Character No.	Character Name	Mean	Standard Deviation
1.	Ear pointing	2.01	.6608
2.	Kernel pointing	2.70	.5936
3.	Barren tips	.422	.4949
4.	Bulging butts	.187	.3907
5.	Pop endosperm frequency*	.205	.4890
6.	Flint endosperm frequency*	.576	.6103
7.	Floury endosperm frequency*	.496	.6150
8.	Dent endosperm frequency*	.260	.4715
9.	Sweet endosperm frequency*	.047	.2358
10.	White color frequency*	.620	.5799
11.	Yellow color frequency*	.621	.5879
12.	Other color frequency*	.378	.4893
13.	Altitude†	32.5	16.76
14.	Latitude	−7.19	16.63
15.	Longitude†	8.49	.7176
16.	Ear diameter	4.19	.7461
17.	Cob diameter	2.41	.5149
18.	Rachis diameter†	1.20	.1566
19.	Kernel length**	2.42	.1982
20.	Kernel width†	2.93	.3122
21.	Kernel thickness**	1.49	.1656
22.	Row number†	3.62	.3986
23.	Ear diameter/length index**	3.37	.3015
24.	Cob diameter/length index**	2.81	.2896
25.	Rachis diameter/length index**	2.31	.3372
26.	Kernel width/length index	77.0	14.16
27.	Kernel thickness/width index**	3.95	.2202
28.	(Kernel "volume")$^{1/3}$†	2.75	.2106
29.	Kernel length/ear length index**	2.07	.3356
30.	|Latitude|	15.7	9.035

* Transformed as $X = \mathrm{Sin}^{-1}\sqrt{Y}$
** Transformed as $X = \mathrm{Log}_e Y$
† Transformed as $X = \sqrt{Y}$

psychological applications of factor analysis (Rummel, 1970), is admittedly a practical compromise between techniques most commonly followed. While it does avoid certain of their more obvious deficiencies, it appears that an immediate solution to this problem will need to be obtained by practical experimentation rather than theoretical investigation. The distributions of the y_{ik} and λ_k obtained from sample correlation matrices are difficult, and they have not been subjected to thorough study.

Biologically this procedure (4) corresponds to transforming that portion of the correlation ellipsoid which seems most likely to be of biological significance to a multidimensional sphere, and then calculating the Euclidean distances between the transformed points. In such a space, the distance in any direction will hopefully have "the same significance in whatsoever direction and in whatsoever part of the new space it is measured." (Balakrishnan and Sanghvi, 1968, p. 862).

AN EXAMPLE

Table 1 lists thirty characters, their means, and their standard deviations calculated from the race means of 230 Latin American races of maize. The correlation matrix obtained from the 230 race means (not the within-race matrix) is available from the author. All possible distances among 27 well-known Mexican races of maize were calculated using the standard deviations of Table 1, the among 230 races correlation matrix, and the means of the Mexican races (see Table 2).

TABLE 2.—THE MEXICAN RACE MEANS FOR THE FIRST 15 CHARACTERS. THE REMAINING MEANS ARE AVAILABLE FROM WELLHAUSEN ET AL. (1952).

Race Number	Character Number														
	1	2	3	4	5	6	7	8	9	10	11	12	13	14	15
1	1	1	0	1	1.57	.00	.00	.00	.00	1.57	.00	.00	51.5	19.2	9.95
2	2	2	0	1	1.57	.00	.00	.00	.00	.00	1.57	.00	44.7	19.8	9.85
3	2	3	0	0	1.57	.00	.00	.00	.00	.00	.00	1.57	10.0	27.0	10.44
4	2	3	0	0	1.57	.00	.00	.00	.00	.00	.78	.78	10.0	20.8	9.38
5	1	3	0	1	.00	.00	1.57	.00	.00	1.57	.00	.00	51.5	19.2	9.95
6	3	3	0	0	.00	.00	1.57	.00	.00	1.57	.00	.00	10.0	25.0	10.34
7	2	3	0	0	.00	.67	.00	.87	.00	.00	.00	1.57	30.2	20.9	10.20
8	1	3	0	1	.00	1.57	.00	.00	.00	.88	.46	.46	49.0	17.0	9.59
9	2	3	1	0	.00	.00	.00	.00	1.57	.32	.32	1.10	38.7	21.1	10.10
10	1	2	0	0	.52	.41	.00	.86	.00	1.04	.46	.20	49.3	19.2	9.90
11	2	3	0	0	1.57	.00	.00	.00	.00	.78	.52	.52	6.8	22.1	10.25
12	2	3	0	0	.00	.56	.78	.47	.00	1.03	.22	.47	37.3	19.9	10.15
13	2	3	0	0	.00	1.57	.00	.00	.00	1.57	.00	.00	6.8	22.1	10.25
14	2	3	1	0	.00	.61	.00	.94	.00	.94	.61	.00	26.4	16.5	9.59
15	2	3	1	0	.00	.00	.00	1.57	.00	1.24	.00	.32	16.1	16.0	9.64
16	2	3	0	0	.00	.78	.78	.00	.00	.67	.87	.00	38.7	16.2	9.59
17	2	3	0	0	.00	.00	.00	1.57	.00	1.14	.41	.00	31.6	20.5	10.25
18	2	2	0	0	.00	.00	.00	1.57	.00	1.57	.00	.00	10.0	16.5	9.70
19	2	3	1	0	.00	.00	.00	1.57	.00	1.57	.00	.00	10.0	15.3	9.59
20	1	1	0	0	.00	.00	.00	1.57	.00	1.57	.00	.00	31.6	18.3	9.95
21	3	3	0	0	.00	.00	.45	1.09	.00	.78	.67	.30	26.5	16.6	9.59
22	3	3	0	0	.00	.00	.00	1.57	.00	1.23	.00	.30	22.4	19.8	9.85
23	2	3	0	0	.00	.00	.00	1.57	.00	1.17	.00	.38	16.9	16.1	9.70
24	1	2	0	1	.00	.52	.00	1.04	.00	1.57	.00	.00	46.2	19.7	9.95
25	3	3	0	0	.00	.46	.00	1.10	.00	1.57	.00	.00	42.1	20.4	10.05
26	1	2	0	1	.00	.50	.00	1.05	.00	1.21	.00	.33	43.6	20.9	10.05
27	2	3	0	0	.00	.67	.00	.87	.00	.98	.56	.00	38.3	17.4	9.85

Ear shape for each race was scored using grades 1 to 3 for conical to cylindrical ears. Barren tips were scored 0 (absent) or 1 (present), as were enlarged butts. Similarly, the presence (1) or absence (0) of each of the following kernel color characters was scored: yellow, white, or other. Altitude was measured in meters, and geographic source was recorded in two ways: latitude and longitude and distance from the equator (absolute value of latitude), measured in degrees. All measured botanical characters were recorded in centimeters, except for the kernel characters which were measured in millimeters. Simple transformations were used for certain characters in order to obtain more symmetric distributions for the 230 racial means. Detailed measurement procedures are found in Wellhausen et al. (1952) and Goodman and Paterniani (1969). The qualitative data on kernel and ear shape came from scores applied by the author to ears photographed in Wellhausen et al. (1952, 1957), Hatheway (1957), Roberts et al. (1957), Brieger et al. (1958), Brown (1960), Ramirez et al. (1961), Timothy et al. (1961, 1963), Grobman et al. (1961), and Grant et al. (1963). The geographic data, kernel color, and endosperm type came from the same sources, when data from typical collections were not available from the Committee on Preservation of Indigenous Strains of Maize (1954, 1955). Brazilian data from Paterniani (1967) and Goodman (1967, 1968, and unpublished) were also used.

Table 3 presents the characteristic roots of the correlation matrix in decreasing order, their cumulative percentages (they add to $n = 30$), and their cumulative products. The y_{ik} are the first k principal com-

TABLE 3. THE CHARACTERISTIC ROOTS OF THE CORRELATION MATRIX CALCULATED FROM 230 RACE MEANS, THE CUMULATIVE PERCENTAGES OF THE TOTAL VARIATION REPRESENTED BY THESE ROOTS, AND THEIR CUMULATIVE PRODUCTS.

Root No.	Root	Cum. %	Cum. Product	Root No.	Root	Cum. %	Cum. Product	Root No.	Root	Cum. %	Cum. Product
1.	5.6	19	5.6	11.	.80	87	1500.	21.	.12	99.6	.018
2.	4.8	35	27.	12.	.74	89	1100.	22.	.094	99.9	.0017
3.	3.8	47	100.	13.	.57	91	620.	23.	.0075	99.9	.00001
4.	2.4	55	250.	14.	.55	93	340.	24.	.0062	100.	.00000
5.	2.1	62	510.	15.	.53	95	180.	25.	.0046		
6.	1.7	68	860.	16.	.37	96	68.	26.	.0021		
7.	1.6	73	1300.	17.	.32	97	22.	27.	.0017		
8.	1.3	77	1700.	18.	.24	98	5.3	28.	.0010		
9.	1.1	81	1900.	19.	.18	99	.97	29.	.00078		
10.	0.97	84	1800.	20.	.16	99	.15	30.	.00056		

ponents for race i obtained by multiplying the first k characteristic vectors, $\mathbf{T}_k$, times the vector of means for race i, $\overline{\mathbf{X}}_i$ (Table 2; however, these data should be standardized by subtracting the overall mean vector $\overline{\boldsymbol{\mu}}$, [Table 1], and dividing by the standard deviations among races [also Table 1]):

$$y_{ik} = \mathbf{T}'_k \mathbf{D}^{-1}(\overline{\mathbf{X}}_i - \overline{\boldsymbol{\mu}}),$$

where $\mathbf{D}$ is a matrix whose diagonal elements are the standard deviations from Table 1 and whose other elements are zero. Figure 1 shows the relationships among the 27 Mexican races based upon the first two standardized principal components ($y_{i1}/\sqrt{\lambda_1}$ and $y_{i2}/\sqrt{\lambda_2}$). The first ten characteristic roots are essentially equal to or greater than one, so

$$D_{ij}^* = [\sum_{k=1}^{10} \{(y_{ik} - y_{jk})^2/\lambda_k\}]^{\frac{1}{2}}.$$

Sokal and Sneath's (1963) unweighted cluster analysis was used to obtain the dendrogram presented in Figure 2.

While neither of the Figures 1 and 2

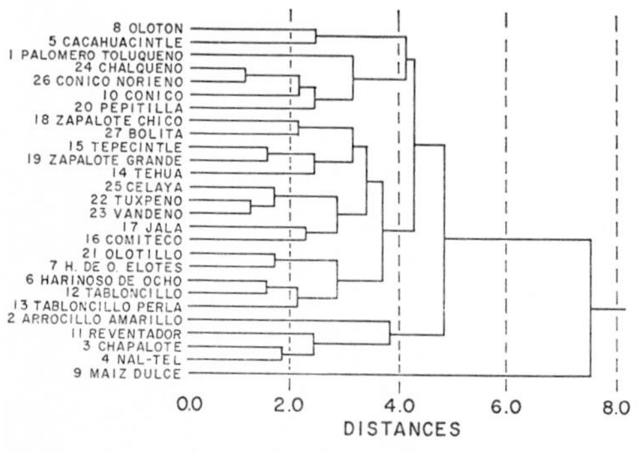

FIG. 2.—Dendrogram of the Mexican races of maize derived from distances calculated from the first ten standardized principal components using unweighted cluster analysis.

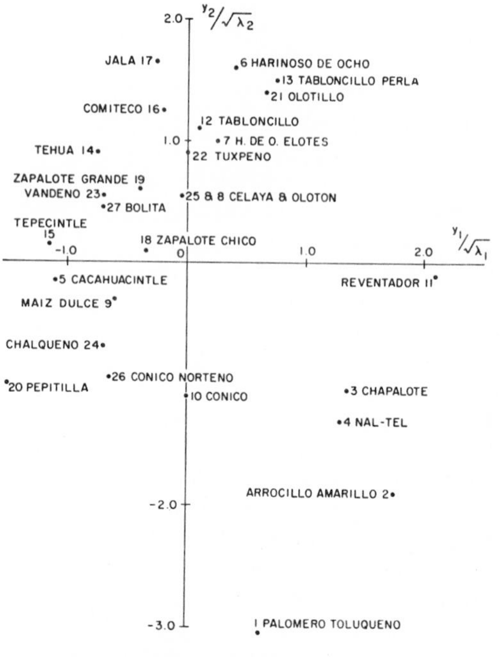

FIG. 1.—Relationships among Mexican races of maize on the basis of the first two standardized principal components.

is an unreasonable clustering of the Mexican races, clearly the arrangement in Figure 2, based on the D^*_{ij}, comes closest to agreeing with the interpretations of Wellhausen et al. (1952). From Figure 2 certain groupings are immediately apparent: (1) the Conico group consisting of Chalqueño and Conico Norteño (which do not seem to merit separate racial status), Conico, Pepitilla, and Palomero Tolqueño; (2) the group of relatively short, thick-eared dent races: Zapalote Chico, Bolita, Tepecintle, Zapalote Grande, and Tehua; (3) the group of large eared commercial dents and semi-dents: Celaya, Vandeño, Tuxpeño, Jala, and Comiteco; (4) the eight-rowed flour corns and their derivatives: Olotillo, Harinoso de Ocho Occidentales, Harinoso de Ocho, Tabloncillo, and Tabloncillo Perla; and (5) the Chapalote group: Reventador, Chapalote, and Nal-Tel. Groups (2) and (3), the thick-eared dents and the large eared commercial dents and semidents, are sufficiently similar that they could well be lumped into one cluster. Arrocillo Amarillo clusters very loosely with group (5), the Chapalote group, but its closest affinities are with group (1), the Conico group. A detailed study of the races of maize from Central America, which are not included in this study since complete data are lacking, will be needed before conclusions can be reached on the grouping of Oloton and Cacahuacintle. Wellhausen et al. (1952) suggest that these two races have been derived from Guatemalan races which were themselves derivatives of South American races. Cacahuacintle seems to have some affinity to the Conico group; it is interesting to note that the Conico group and Cacahuacintle share so many common plant characteristics that they are readily grouped on the basis of general field appearance. Comiteco is the only other race which appears to be closely related to Oloton. The latter is hypothesized to be a parent of the former by Wellhausen et al. (1952). Both Kelly and Anderson (1943) and Wellhausen et al. (1952) regard Maiz Dulce as being related to the South American sweet corns and not to the other races of maize in Mexico. This analysis can neither support nor refute a relationship with South American sweet corns, but it clearly suggests a lack of similarity with the other races of maize in Mexico.

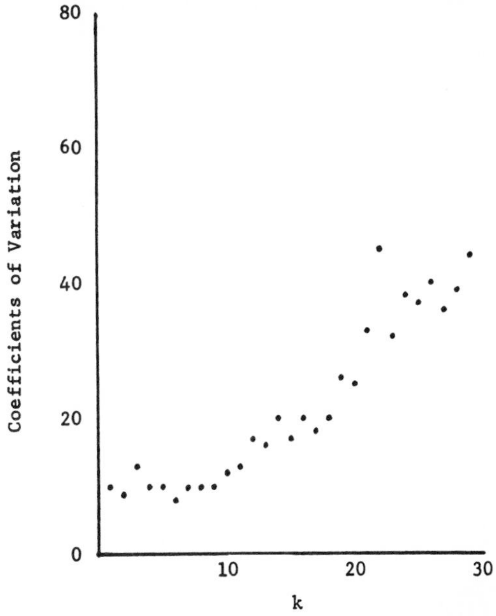

FIG. 3.—Actual coefficients of variation for the characteristic roots of a 30 × 30 matrix. The roots were obtained from 50 samples drawn with replacement of 40 mean vectors each from the population of 230 Latin American corn races.

The dendrogram based on generalized distances (not shown), calculated using the means from Table 2 and the among 230 races covariance matrix, is much less satisfactory in several respects. This should not be unexpected due to the near singularity of the correlation matrix. The cumulative product of the characteristic roots of the matrix clearly indicates this, becoming less than 1.0 after $k = 18$ and reaching .00001 by $k = 23$. Dendrograms based on the use of $k = 9$ and $k = 11$ principal components were less consistent with the generally accepted interpretations of Wellhausen et al. (1952) than was Figure 2.

To evaluate the sampling error of the λ_k used in this analysis, 50 random samples

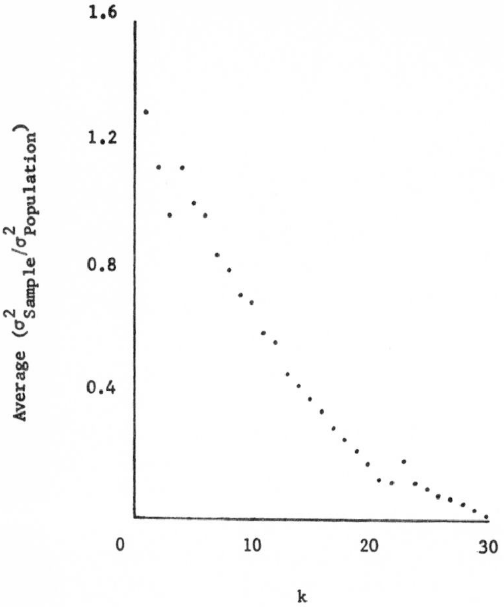

FIG. 4.—Mean values of the ratios of the variance of the sample to the variance of the population for each sample principal component for the samples of Figure 3.

of size 40 were drawn with replacement from the 230 races used in Table 1. For each sample a principal component analysis was carried out. Figure 3 shows the coefficients of variation are generally about 10% for λ_1 through λ_{10}, increasing gradually to about 20% for λ_{15} through λ_{19}, and increasing rapidly thereafter to 75% for λ_{30}. Since a particular λ_k does not necessarily correspond to the same axes in any two samples (the longest axis of the first sample may, for example, correspond most closely to the second longest of sample two), the relationship between the variance of the population and the variance of each of the samples was determined for each component of each sample. The mean values over the 50 samples for λ_k/σ^2_{pk}, where σ^2_{pk} and λ_k are the population variance for component k, and the sample variance, respectively, are shown in Figure 4. Clearly the λ_k correspond less and less to actual population variances for larger values of k.

While Figure 1 is less satisfactory than Figure 2, the first two principal components alone do a fair job of depicting the relationships among the races.

DISTANCES BASED ON FREQUENCY DATA

The geometric interpretations discussed previously are quite satisfactory for a wide range of applications. However, with frequency data and certain other types of data where the variance of a sample mean for a particular character of a specific population depends upon the mean itself, there may be cases where transformations to eliminate such relationships should be made before proceeding further. Illustrations of such transformations are found in most introductory statistical methods textbooks; they will not be discussed here. However, an approach which has been presented as a new and useful procedure for the analysis of blood group frequencies does merit examination. Cavalli-Sforza and Edwards (1967) and Edwards and Cavalli-Sforza (1964) have suggested transforming blood group frequencies (p_{lik}; $k = 1, m$; $\sum_{k=1}^{m} p_{lik} = 1$; $i = 1, N$) for a locus l with m alleles to m dimensional polar coordinates using the transformations:

$$\sin^2 \theta_{lik} = p_{lik}; \quad k = 1, m.$$

For each locus, l, the populations will lie on the circumference of a segment of an m dimensional sphere with radius 1.0 (since $\sum_{k=1}^{m} p_{lik} = \sum_{k=1}^{m} \sin^2 \theta_{lik} = 1.0$). They then suggest measuring the distance between populations i and j as either the length of the chord connecting the points representing the population means (in 1964) or as being directly proportional to such a distance (in 1967):

$$(5) \qquad \check{D}_{lij} = (2 - 2 \cos \varphi_{lij})^{\frac{1}{2}},$$

where φ_{lij} is the angle between the straight lines passing through the origin and the points representing the means of populations i and j for locus l. (The same technique is applicable to other frequencies [such as genotypic or phenotypic frequencies] as well as allelic frequencies). However,

$$\check{D}_{lij} = (2 - 2\cos\varphi_{lij})^{\frac{1}{2}}$$
$$= (2 - 2\sum_{k=1}^{m} \sin\theta_{lik}\sin\theta_{ljk})^{\frac{1}{2}}$$
$$= (2 - 2\sum_{k=1}^{m} \sqrt{p_{lik}}\sqrt{p_{ljk}})^{\frac{1}{2}} \tag{6}$$
$$= (\sum_{k=1}^{m} p_{lik} + \sum_{k=1}^{m} p_{ljk} - 2\sum_{k=1}^{m} \sqrt{p_{lik}}\sqrt{p_{ljk}})^{\frac{1}{2}}$$
$$= [\sum_{k=1}^{m} (\sqrt{p_{lik}} - \sqrt{p_{ljk}})^2]^{\frac{1}{2}}.$$

These equations demonstrate clearly the equivalence of Cavalli-Sforza and Edwards' technique to that of Sokal (1961) given in equation (1). The only differences are that the square roots of the allelic frequencies are used by Cavalli-Sforza and Edwards, rather than the frequencies themselves, and no attempt is made to standardize the frequencies. The allelic frequencies at a given locus *must* be correlated (since $\sum_{k=1}^{m} p_{lik} = 1.0$), and hence the square roots of the frequencies will generally be correlated as well. (For formulae for calculating approximate values of such correlations see Kendall and Stuart, 1963, p. 232). Cavalli-Sforza and Edwards are thus measuring distances within a correlation ellipsoid. Their distances will not be "isotropic with respect to evolutionary progress" (Edwards and Cavalli-Sforza, 1964, p. 72) since they will not have "the same significance in whatsoever direction and in whatsoever part of the new space" (Balakrishnan and Sanghvi, 1968, p. 862) that they are measured. Whereas a given distance along the major axis of the ellipsoid may be of little biological consequence, that same distance could actually be greater than the length of some of the shorter minor axes.

More importantly, Cavalli-Sforza and Edwards' technique ignores correlations among loci as well as correlations among allelic frequencies at a locus. They simply add distances over loci to obtain the total distance $\check{D}_{ij}$ between populations *i* and *j*

$$\check{D}_{ij} = \sum_{l} \check{D}_{ljk}. \tag{7}$$

This corresponds to ignoring the correlations between traits, a characteristic of Pearson's (1926) Coefficient of Racial Discrimination, and has essentially the same effect as measuring distances within an unadjusted correlational ellipse. The distinction is simply that

$$\sum_{i} d_i \neq (\sum_{i} d_i^2)^{\frac{1}{2}},$$

where the d_i are the distances calculated for each trait or locus. In fact, Fitch and Neel (1969) have suggested a modification of this type of distance analysis in which they actually use $(\sum d_i^2)^{\frac{1}{2}}$. They call this their *root model* in contrast to the *additive model* of Cavalli-Sforza and Edwards.

Balakrishnan and Sanghvi (1968) and Kurczynski (1970) have already made some of these points, but the latter authors have proposed using generalized distances analogous to those of Mahalanobis. Thus their distances will likewise contain a large amount of "noise" when some of the characteristic roots of their covariance matrices are very close to zero. In fact, one characteristic of distances obtained using a technique such as Sokal's distances, Pearson's C.R.L., or Cavalli-Sforza and Edwards' chord distance is that with reasonable care in selecting characters and obtaining data, the results obtained will be reasonable. They may imply more than the data merit due to mutual interdependence of the characters, but they are not subject to the "noise" problem encountered with generalized distances.

AN EXAMPLE

Approximate A_1, A_2, B, O, M, and N gene frequencies and Rh chromosome frequencies for 25 different human populations (Table 4) were taken from Mourant (1954) and Mourant et al. (1958). The choice of populations was arbitrary; the data are used as an example and should

TABLE 4.—APPROXIMATE A_1, A_2, B, O, M, AND N GENE FREQUENCIES AND RH CHROMOSOME FREQUENCIES FOR 25 DIFFERENT HUMAN POPULATIONS.

	Frequencies												
Population	p_1	p_2	q	r	CDE	CDe	Cde	cDE	cdE	cDe	cde	M	N
French	.21	.06	.06	.67	.00	.43	.01	.14	.01	.02	.39	.55	.45
Czech	.25	.04	.14	.57	.01	.42	.01	.15	.00	.01	.40	.53	.47
German	.22	.06	.08	.64	.02	.38	.03	.12	.01	.03	.41	.55	.45
Basque	.19	.04	.02	.75	.00	.38	.01	.07	.00	.01	.53	.54	.46
South Chinese	.18	.00	.15	.67	.00	.74	.00	.19	.00	.03	.04	.62	.38
Ainu	.23	.00	.28	.49	.00	.56	.00	.21	.19	.00	.04	.40	.60
Australian Aborigine	.22	.00	.00	.78	.03	.63	.04	.14	.00	.16	.00	.22	.78
New Guinea Native	.22	.00	.14	.64	.00	.95	.00	.04	.00	.01	.00	.21	.79
Maori	.34	.00	.00	.66	.00	.48	.01	.48	.00	.03	.00	.51	.49
Icelander	.13	.06	.06	.75	.00	.43	.00	.16	.02	.02	.37	.57	.43
Eskimo	.30	.00	.06	.64	.03	.59	.00	.33	.00	.05	.00	.83	.17
Br. Col. Indians	.11	.00	.00	.89	.02	.35	.02	.53	.00	.05	.03	.77	.23
Blood Indians	.58	.00	.01	.41	.04	.48	.00	.35	.03	.00	.10	.87	.13
Braz. Indians	.00	.00	.00	1.00	.03	.55	.00	.40	.00	.02	.00	.74	.26
Bantu	.10	.08	.11	.71	.00	.08	.05	.06	.00	.61	.20	.56	.44
N. African Arabs	.17	.06	.15	.62	.00	.43	.00	.14	.00	.09	.34	.54	.46
S. Asian Indians	.16	.03	.22	.59	.01	.57	.00	.14	.00	.04	.24	.69	.31
U. S. Negro	.10	.06	.13	.71	.00	.16	.02	.15	.00	.40	.27	.49	.51
Spanish	.27	.04	.06	.63	.02	.44	.01	.11	.00	.04	.38	.53	.47
Norwegian	.26	.06	.06	.62	.00	.43	.01	.13	.01	.02	.40	.55	.45
Mexican Indians	.02	.00	.01	.97	.00	.68	.00	.30	.00	.02	.00	.79	.21
Egyptian	.21	.05	.20	.54	.00	.46	.01	.14	.00	.24	.15	.52	.48
U. S. Chinese	.21	.00	.18	.61	.00	.71	.00	.20	.00	.02	.07	.58	.42
U. S. Japanese	.28	.00	.18	.54	.00	.63	.00	.32	.00	.00	.05	.56	.44
Navaho	.03	.00	.00	.97	.04	.38	.09	.31	.01	.17	.00	.91	.09

not be considered definitive of the populations. More recent and more accurate data are probably available elsewhere. The correlation matrix among populations was calculated after transforming the data by means of the arcsin transformation (which hopefully stabilizes the within population variances as well as eliminating functional dependence of such variances upon the mean frequencies). It should be noted that there were large correlations among alleles at different loci as well as among alleles at a locus. The characteristic roots of the matrix, their cumulative percentages, and their cumulative products are presented in Table 5. The distances were calculated as

$$D_{ij}^* = \{ \sum_{k=1}^{3} [(y_{ik} - y_{jk})^2 / \lambda_k] \}^{\frac{1}{2}}$$

There were 12 non-zero characteristic roots of which 4 were greater than one; however, cluster analyses based upon $k = 2$, 4, and 5 were less satisfactory than (although similar to) those based upon $k = 3$. Figure 5 presents the association among the 25 populations based upon the first two stan-

TABLE 5.—THE CHARACTERISTIC ROOTS OF THE CORRELATION MATRIX CALCULATED FROM THE TRANSFORMED FREQUENCIES OF TABLE 5, CUMULATIVE PERCENTAGES OF THE TOTAL VARIATION REPRESENTED BY THESE ROOTS, AND THEIR CUMULATIVE PRODUCTS.

Root No.	Root	Cum. %	Cum. Product	Root No.	Root	Cum. %	Cum. Product	Root No.	Root	Cum. %	Cum. Product
1.	4.2	32	4.2	5.	.94	87	28.	9.	.18	99.	.48
2.	3.1	56	13.	6.	.71	92	20.	10.	.057	99.9	.028
3.	1.7	69	22.	7.	.52	96	10.	11.	.015	100.0	.00041
4.	1.3	79	29.	8.	.27	98	2.7	12.	.0016		.00000

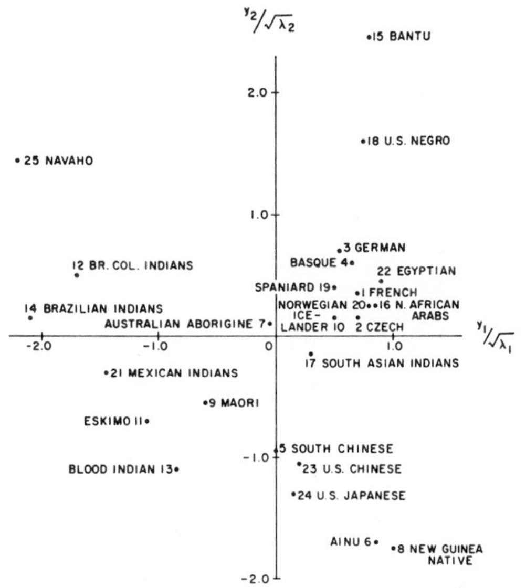

FIG. 5.—Associations among 25 human populations on the basis of the first two standardized principal components.

dardized principal components ($y_{i1}/\sqrt{\lambda_1}$ and $y_{i2}/\sqrt{\lambda_2}$), while Figure 6 presents the dendrogram based upon the unweighted cluster analysis of the D_{ij}^*.

Figure 5 shows a clustering of populations from Europe and northern Africa with other populations being scattered in reasonable proximity to our expectations. The position of the Australian aborigine population is closer to the European populations than one might expect, but this may

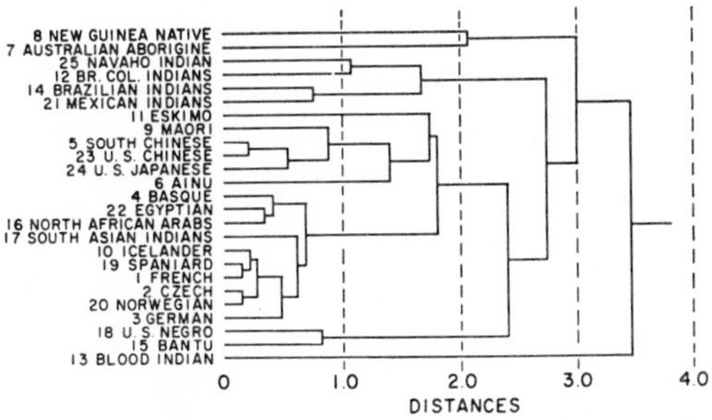

FIG. 6.—Dendrogram of 25 human populations derived from distances based upon the first three standardized principal components using unweighted cluster analysis.

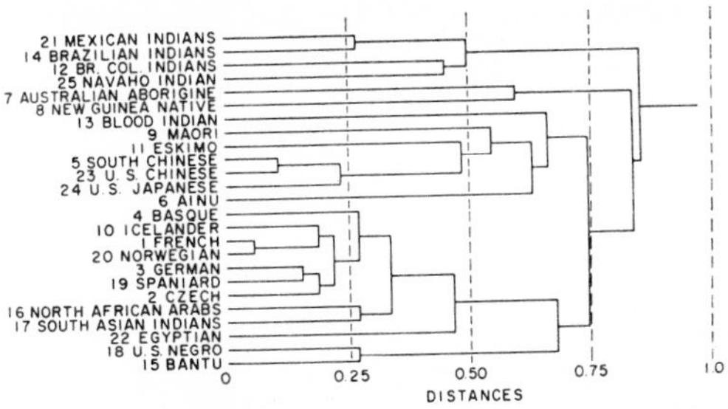

FIG. 7.—Dendrogram of 25 human populations derived from distances based upon Fitch and Neel's root model of Cavalli-Sforza and Edwards' chord distances using unweighted cluster analysis.

be in part due to limitations of a two dimensional drawing and possible Negroid contributions to the aboriginal Australian populations. Figure 6 first groups the European populations (with the exception of the Basques), then adds non-European populations peripheral to Europe. While the Eskimos, Ainu, and Maori are grouped with the Japanese and Chinese, the New Guinea natives and the Australian aborigines are not. Similarly the Blood Indians, atypical of most New World Indians, are not grouped with the other New World Indian populations. The close similarity between U. S. Negro and African Negro populations and between U. S. Chinese and the South Chinese populations is striking.

Figure 7 shows the dendrogram of the distances obtained by applying Fitch and Neel's (1969) root model of Cavalli-Sforza and Edwards' chord distances to the data from Table 4. The method of clustering was again Sokal and Sneath's (1963) unweighted procedure. The results are similar to those presented in Figure 6. The analogous dendrogram based upon generalized distances, calculated using the arcsin transformation of the means from Table 4 and the corresponding among populations covariance matrix, differed greatly in many respects. While perhaps not totally unrealistic, only three groups were well defined: (a) the Germans, Czechs and Spaniards, (b) the U. S. and South Chinese, and (c) the French and Norwegians. While the Bantu were not grouped with any other population, U. S. Negroes actually grouped very loosely with the British Colombian Indians.

REFERENCES

ANDERSON, E. 1949. Introgressive Hybridization. John Wiley and Sons, Inc., New York.

BALAKRISHNAN, V., AND L. D. SANGHVI. 1968. Distance between populations on the basis of attribute data. Biometrics 24:859–865.

BRIEGER, F. G., J. T. A. GURGEL, E. PATERNIANI, A. BLUMENSCHEIN, AND M. R. ALLEONI. 1958. Races of maize in Brazil and other eastern South American countries. Nat. Acad. Sci.-Nat. Research Council Publ. 593. Washington, D. C. 283 p.

BROWN, W. L. 1960. The races of maize in the West Indies. Nat. Acad. Sci. Nat. Research Council Publ. 792. Washington, D. C. 60 p.

CAVALLI-SFORZA, L. L., AND A. W. F. EDWARDS. 1967. Phylogenetic analysis: Models and estimation procedures. Evolution 21:550–570.

COMMITTEE ON PRESERVATION OF INDIGENOUS STRAINS OF MAIZE. 1954. Collections of Original Strains of Corn, I. Nat. Acad. Sci. Nat. Research Council. Washington, D. C. 300 p.

COMMITTEE ON PRESERVATION OF INDIGENOUS STRAINS OF MAIZE. 1955. Collections of Original Strains of Corn, II. Nat. Acad. Sci. Nat. Research Council. Washington, D. C. 298 p.

EDWARDS, A. W. F., AND L. L. CAVALLI-SFORZA. 1964. Reconstruction of evolutionary trees, p. 67–76. *In* V. H. Heywood and J. McNeill (eds.), Phenetic and phylogenetic classification. The Systematics Assoc. Publ. No. 6. London.

FITCH, W. M., AND J. V. NEEL. 1969. The phylogenetic relationships of some Indian tribes of Central and South America. Amer. J. Human Genetics 21:384–397.

GOODMAN, M. M. 1967. The races of maize: I. The use of Mahalanobis' generalized distances to measure morphological similarity. Fitotecnia Latinoamericano 4:1–22.

GOODMAN, M. M. 1968. The Races of Maize. II. Use of multivariate analysis of variance to measure morphological similarity. Crop Science 8:693–698.

GOODMAN, M. M. 1969. Measuring evolutionary divergence. Japanese J. Genetics 44(Suppl. 1): 310–316.

GOODMAN, M. M., AND E. PATERNIANI. 1969. The races of maize: III. Choices of appropriate characters for racial classification. Economic Botany 23:265–273.

GOWER, J. C. 1966. Some distance properties of latent roots and vector methods used in factor analysis. Biometrika 53:325–338.

GRANT, U. J., W. H. HATHEWAY, D. H. TIMOTHY, C. CASSALETT, AND L. M. ROBERTS. 1963. Races of maize in Venezuela. Nat. Acad. Sci. Nat. Research Council Publ. 1136. Washington, D. C. 92 p.

GROBMAN, A., W. SALHUANA, AND R. SEVILLA. 1961. Races of maize in Peru. Nat. Acad. Sci. Nat. Research Council Publ. 915. Washington, D. C. 374 p.

HATHEWAY, W. H. 1957. Races of maize in Cuba. Nat. Acad. Sci. Nat. Research Council Publ. 453. Washington, D. C. 75 p.

KELLY, I., AND E. ANDERSON. 1943. Sweet corn in Jalisco. Ann. Missouri Botan. Garden 30: 405–412.

KENDALL, M. G., AND A. STUART. 1963. The Advanced Theory of Statistics. Vol. 1, 2nd Ed., Charles Griffin and Company Ltd., London. 433 p.

KURCZYNSKI, T. W. 1970. Generalized distance and discrete variables. Biometrics 26:525–534.

MAHALANOBIS, P. C. 1936. On the generalized distance in statistics. Proc. Nat. Inst. Sci. India 2:49–55.

MORRISON, D. F. 1967. Multivariate Statistical Methods. McGraw-Hill Book Company, New York. 338 p.

MOURANT, A. E. 1954. The Distribution of the Human Blood Groups. Charles C Thomas, Springfield, Illinois. 438 p.

MOURANT, A. E., A. C. KOPEĆ, AND K. DOMANIEWSKA-SOBCZAK. 1958. The ABO Blood Groups. Charles C Thomas, Springfield, Illinois. 276 p.

PATERNIANI, E. 1967. Banco de germoplasma de milho. Racas de milho. *In* Relatorio Cientifico de 1967, p. 48, 59–70. Cadeira de Citologia e Genética, Escola Superior de Agricultura "Luiz de Queiroz," Piracicaba, S. P., Brazil. 117 p.

PEARSON, K. 1926. On the Coefficient of Racial Likeness. Biometrika 18:105–117.

POLLITZER, W. S. 1958. The Negroes of Charleston (S. C.); a study of hemoglobin types, serology, and morphology. Amer. J. Physical Anthropology 16:241–263.

POLLITZER, W. S. 1969. Ancestral traits, parental populations, and hybrids. Amer. J. Physical Anthropology 30:415–419.

POLLITZER, W. S., D. S. PHELPS, R. E. WAGGONER, AND W. C. LEYSHON. 1967. Catawba Indians: Morphology, genetics, and history. Amer. J. Physical Anthropology 26:5–14.

POLLITZER, W. S., D. RUCKNAGEL, R. TASHIAN, D. C. SHREFFER, W. C. LEYSHON, K. NAMBOODIRI, AND R. C. ELSTON. 1970. The Seminole Indians of Florida: Morphology and serology. Amer. J. Physical Anthropology 32:65–82.

RAMIREZ E., R., D. H. TIMOTHY, E. DÍAZ B., AND U. J. GRANT WITH G. E. NICHOLSON CALLE, E. ANDERSON, AND W. L. BROWN. 1960. Races of maize in Bolivia. Nat. Acad. Sci. Nat. Research Council Publ. 747. Washington, D. C. 159 p.

RIGHTMIRE, G. P. 1969. On the computation of Mahalanobis' generalized distance (D^2). Amer. J. Physical Anthropology 30:157–160.

ROBERTS, L. M., U. J. GRANT, R. RAMIREZ E., W. H. HATHEWAY, AND D. L. SMITH WITH P. C. MANGELSDORF. 1957. Races of maize in Colombia. Nat. Acad. Sci. Nat. Research Council. Publ. 510. Washington, D. C. 153 p.

RUMMEL, R. J. 1970. Applied Factor Analysis. Northwestern University Press. Evanston. 617 p.

SOKAL, R. R. 1961. Distance as a measure of taxonomic similarity. Syst. Zool. 10:70–79.

SOKAL, R. R., AND P. H. A. SNEATH. 1963. Principles of Numerical Taxonomy. W. H. Freeman and Co., San Francisco. 359 p.

STEPHENS, S. C. 1961. Species differentiation in relation to crop improvement. Crop Science 1:1–5.

TIMOTHY, D. H., W. H. HATHEWAY, U. J. GRANT, M. TORREGROZA C., D. SARRIA V., AND D. VARELA A. 1963. Races of maize in Ecuador. Nat. Acad. Sci. Nat. Research Council Publ. 975. Washington, D. C. 147 p.

TIMOTHY, D. H., B. PEÑA V., AND R. RAMÍREZ E. WITH W. L. BROWN AND E. ANDERSON. 1961. Races of maize in Chile. Nat. Acad. Sci. Nat. Research Council Publ. 847. Washington, D. C. 84 p.

WELLHAUSEN, E. J., A. FUENTES O., AND A. HERNÁNDEZ C. WITH P. C. MANGELSDORF. 1957. Races of maize in Central America. Nat. Acad. Sci. Nat. Research Council Publ. 511. Washington, D. C. 128 p.

WELLHAUSEN, E. J., L. M. ROBERTS, AND E. HERNANDEZ X. WITH P. C. MANGELSDORF. Races of Maize in Mexico. The Bussey Institution of Harvard University, Cambridge, Massachusetts. 223 p.

Received April 20, 1971.

ERRATUM

Page 183: Table 5 should read as follows:

Table 5 The characteristic roots of the correlation matrix calculated from the transformed frequencies of Table 4, cumulative percentages of the total variation represented by these roots, and their cumulative products

Root No.	*Root*	*Cum. %*	*Cum. Product*	*Root No.*	*Root*	*Cum. %*	*Cum. Product*	*Root No.*	*Root*	*Cum. %*	*Cum. Product*
1.	4.2	32	4.2	5.	.91	86	26.	9.	.18	99.	.51
2.	3.0	56	13.	6.	.77	92	20.	10.	.057	99.9	.029
3.	1.8	70	23.	7.	.56	96	11.	11.	.015	100.	.00043
4.	1.2	79	28.	8.	.25	98	2.9	12.	.0016		.00000

23

Reprinted from *Taxon,* **11**(2), 33–40 (1960)

THE COMPARISON OF DENDROGRAMS BY OBJECTIVE METHODS [1]

Robert R. Sokal and F. James Rohlf (Lawrence, Kansas) [2]

INTRODUCTION

The purpose of this paper is to present a technique for comparing dendrograms resulting from numerical taxonomic research with one another and with dendrograms produced by conventional methods. One of the most frequent ways of depicting the results of studies in numerical taxonomy (Sokal, 1960; Sneath and Sokal, 1962) is by so-called dendrograms or diagrams of relationships. These are tree-like schemes which indicate the affinity of taxa to their nearest relatives (on the basis of similarity or *phenetic* resemblance alone, without any necessary phylogenetic implications). These diagrams resemble the customary phylogenetic trees, but are preferred for classificatory purposes; first, because phylogenetic inferences are speculative, while similarities are factual; secondly, because they are quantitative evaluations of these similarities; and thirdly, because they lack some of the other meanings often implied in phylogenetic trees (Sneath and Sokal, 1962). Such dendrograms have been published in bacteriological work (Sneath and Cowan, 1958), in studies of bees (Michener and Sokal 1957; Sokal and Michener 1958), butterflies (Ehrlich, 1961), rice (Morishima and Oka, 1960), members of the nightshade genus *Solanum* (Soria and Heiser, 1961) and others.

With the increasing acceptance of the philosophy of numerical taxonomy an experimental phase in using various types of coefficients is beginning, which will involve the comparison ot the results of numerical taxonomic research based on these different coefficients. So far we have lacked a procedure for such comparisons. The cophenetic correlations which will be developed below provide an extremely simple and effective method for comparing dendrograms of various sorts.

Before proceeding to a detailed account of the technique, it will be useful to discuss briefly the four types of comparisons of dendrograms that we wish to make in numerical taxonomy and the reasons for them:

1. A major use of the methods proposed below will be to compare dendrograms arrived at on the basis of numerical taxonomic techniques with dendrograms prepared earlier on the basis of conventional taxonomic methods. In a crude manner such a comparison was undertaken by Michener and Sokal (1957). Such techniques will provide some estimate of the magnitude of the differences between numerical taxonomic classifications and the classifications produced on the basis of the currently used methods. While experience to date has shown that numerical taxonomic

---

[1]) For footnotes see end of article.

classifications are generally not too different from the conventional ones, it is desirable to quantify such comparisons.

2. Recent efforts in numerical taxonomy have resulted in the development of several methods of clustering of taxa from an original matrix of similarity values among taxa. These different methods include those of Sneath (1957), Sokal and Michener (1958) and Rogers and Tanimoto (1960) and are based on somewhat different approaches to cluster analysis. They will therefore necessarily result in somewhat different dendrograms. Within any method many variations of the technique can be employed. Thus the weighted variable group method of Sokal and Michener (1958) can be modified into a pair group method, both weighted and unweighted, and the method of averaging the relations can be modified from the Spearman sums of variables method to one of several possible alternatives. It is important to have a procedure for comparing the dendrograms resulting from these techniques to determine the amount of similarity among them.

3. As a useful by-product, not intended originally, the method developed below turned out to be a valuable test of the amount of distortion of the mutual relations among taxa introduced by a given clustering method. By comparing a dendrogram arrived at by a given technique with the actual similarity coefficients which exist between each pair of taxa within the study, one can obtain a measure of the amount of distortion which the clustering method leading to the dendrogram has imposed upon the system. This will be discussed in greater detail below by means of an example.

4. The occasion may arise when one may wish to compare two different conventional dendrograms of the same taxonomic group with each other. Previously there has not existed an objective and quantitative criterion for expressing the similarity of two dendrograms of the same taxa prepared by different authors. While not related directly to numerical taxonomy, such a comparison will be possible by means of the technique proposed below.

TECHNIQUE

We may start our considerations with a dendrogram as shown in Figure 1 (taken from Rohlf and Sokal, 1962). This represents the weighted pair group method of clustering, applied to 23 species of the bee subgenera *Chilosima* and *Ashmeadiella s. str.* from Sokal and Michener (1958), amended slightly by Rohlf and Sokal (1962). While in this example the units being clustered are species, other taxonomic units can be clustered by numerical taxonomy. The lowest ranked units possible would be individuals, which could be clustered to form groups of individuals phenetically resembling each other and representing infraspecific taxa, if all the individuals concerned were conspecific. Higher taxonomic entities, such as genera, tribes or families, could be used as units in a taxonomic study and they could be grouped to form still higher taxa. Sokal and Sneath (book, in preparation) have therefore called the lowest taxonomic unit being classified in any given study an *operational taxonomic unit* (OTU). These OTU's conventionally represent the tips of the dendrogram in any given study.

In a dendrogram the abscissa has no particular meaning, except for spacing out the original taxa employed in the study. The ordinate, on the other hand, represents similarity values, which may be on one of several conventionally used scales. Thus they may be association or resemblance coefficients (Sneath 1957; Sneath and Sokal, 1962), correlation coefficients (Sokal and Michener, 1958), as in Figure 1, or distances (Sokal, 1961). The upper end of the scale will represent maximum similarity, unity in the case of correlation and association and zero with distances, while the lower end of the scale will extend as far as is necessary in order to unite all stems in the particular dendrogram. The meaning of such a dendrogram is restricted to showing the level at which two or more stems join with each other to form a common stem. It should be clear that from such a diagram one can only obtain a rough idea of the relation of every OTU with every other one. Any given pair of OTU's may be more closely or more distantly related than is indicated by the level of the junction of the stems bearing them. This is so because the level of junction represents the average resemblance of the OTU's of one cluster with the OTU's of the second cluster.

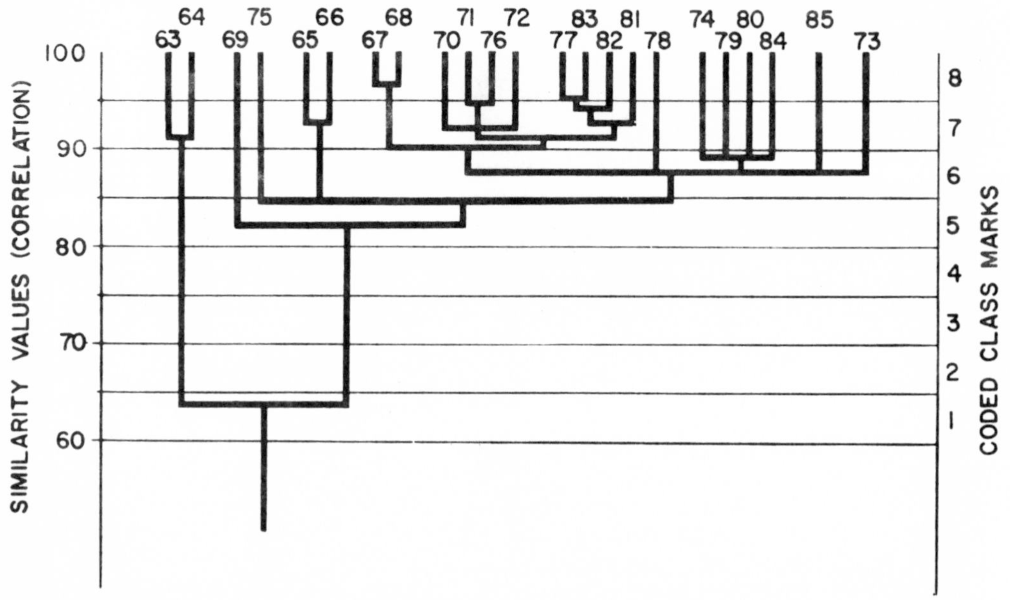

Fig. 1. Dendrogram or diagram of relationships among 23 species of the bee subgenera *Chilosima* and *Ashmeadiella s. str.*, taken from Rohlf and Sokal (1962) and based on data by Sokal and Michener (1958). The relationships were obtained by the weighted pair group method (WPGM). The ordinate is graduated in a Pearson product-moment correlation coefficient scale (coded by multiplication by 100). Numbers across the top of the figure are species code numbers which are identified in Rohlf and Sokal (1962) or Michener and Sokal (1957). Horizontal lines across the dendrogram are phenon lines defining taxa at the minimum level of similarity at which the phenon line cuts the ordinate. Class intervals along the similarity scale delimited by phenon lines have had their class marks coded (on the right hand side of the dendrogram).

In order to avoid misunderstandings it should be clearly stated that the word "related" in the context of this paper means only similar, and does not necessarily infer relationship by descent. According to the principles of numerical taxonomy (Sneath and Sokal, 1962), the process of classification is based only on observed differences and similarities, so-called phenetic evidence, and not on phylogenetic inferences and speculations, however important and interesting these may be from other viewpoints.

In order to avoid the historical and semantic connotations of the conventional system of higher categories (i.e. subgenera, genera, tribes families, etc.), Sneath and Sokal (1962) have introduced the idea of a *phenon* (distinct from the phenon concept of Camp and Gilly, 1943), which defines taxa by drawing horizontal lines across the dendrograms. All taxonomic units carried by single stems crossed by such a phenon line are called phenons with a numerical prefix indicating the vertical level of the phenon line. Thus, as is shown in Figure 1, there are five 85-phenons and two 75-phenons in our present study. In this manner we are able to define objectively the limits of any given taxon and avoid arguments as to what a genus, tribe or any other category is in a given group, since the level of cohesion of the group can be arbitrarily fixed by the investigator interested in defining the group. We are, however, prevented from drawing other than horizontal lines, since these would conflict with the aim of objectivity and repeatability of numerical taxonomy.

In devising methods of comparing dendrograms one of the initial schemes which occurred to us, but which we have so far not followed up, was to try to compare different arrangements of the same OTU's by counting the number of breaks and rearrangements necessary to convert one dendrogram into another. We have since adopted the simpler approach shown below.

We divide the range of similarity values along the ordinate into a suitable number of equal class intervals by drawing phenon lines as class limits across the dendrogram (see Figure 1, where the range of similarity values has been divided into eight classes). The number of classes into which the range of variation should be divided will depend upon the number of OTU's being classified. As a very rough guide, dendrograms involving less than ten OTU's need not be divided into more than four classes, while dendrograms involving as many as 100 OTU's should probably be divided into at least ten classes. A further consideration should be that the class intervals should be fine enough to reveal a reasonable amount of structural detail in the dendrogram to be analyzed. Persons planning to do such computations on a desk calculator should employ the minimum number of classes necessary. On the other hand, increasing the number of classes never does any harm from a statistical point of view. As a matter of fact the computer program which was developed by one of us (F.J.R.) divides the range of similarity values into 50 classes. Schemes could also be developed which would handle the actual similarity value at which two stems join. The probable statistical consequences of using actual juncture levels rather than cophenetic values are probably slight, based on the well known effect of grouping in frequency distributions.

Once the class intervals along the ordinate have been established, each class mark should be coded on a scale starting with unity at the lower end, i.e. the end having the lowest similarity value, and going up in unit steps. Thus with ten classes the highest class should be coded 10. These values will then be proportional to the similarity values, except in the case of distances, where they will be complementary and where coding in reverse, i.e. starting with unity at the highest level, might be advised. The coding is a computational convenience for desk calculator operations. Actual class marks can be used in digital computer programs.

We shall define the *cophenetic value* of two OTU's as the class mark of the class (between phenon lines) in which their stems are connected. For example, in Figure 1 we can see that species 63 and species 64 are connected in class interval 7. Hence their cophenetic value is 7. Similarly the cophenetic value of species 69 with species 71 is 5, since that is the level at which these OTU's are connected. The closer the relationship between OTU's, the higher will be their cophenetic values. It is convenient to record cophenetic values in matrix form, resembling a matrix of similarity values (see Table 1). The actual procedure of recording the cophenetic values for each pair of OTU's in a study, may be appreciably simplified, since the cophenetic values between all OTU's on any one stem with all other OTU's are identical. Thus, for example, in Figure 1 species 63 and 64 will have a cophenetic value of 1 with all of the remaining species.

TABLE 1.

	63	64	65	66	67	68	69	70	71	72	73	74	75	76	77	78	79	80	81	82	83	84	85
63	x																						
64	7	x																					
65	1	1	x																				
66	1	1	7	x																			
67	1	1	5	5	x																		
68	1	1	5	5	8	x																	
69	1	1	5	5	5	5	x																
70	1	1	5	5	7	7	5	x															
71	1	1	5	5	7	7	5	7	x														
72	1	1	5	5	7	7	5	7	7	x													
73	1	1	5	5	6	6	5	6	6	6	x												
74	1	1	5	5	6	6	5	6	6	6	6	x											
75	1	1	5	5	5	5	5	5	5	5	5	5	x										
76	1	1	5	5	7	7	5	7	7	7	6	6	5	x									
77	1	1	5	5	7	7	5	7	7	7	6	6	5	7	x								
78	1	1	5	5	6	6	5	6	6	6	6	6	5	6	6	x							
79	1	1	5	5	6	6	5	6	6	6	6	6	5	6	6	6	x						
80	1	1	5	5	6	6	5	6	6	6	6	6	5	6	6	6	6	x					
81	1	1	5	5	7	7	5	7	7	7	6	6	5	7	7	6	6	6	x				
82	1	1	5	5	7	7	5	7	7	7	6	6	5	7	7	6	6	6	7	x			
83	1	1	5	5	7	7	5	7	7	7	6	6	5	7	8	6	6	6	7	7	x		
84	1	1	5	5	6	6	5	6	6	6	6	6	5	6	6	6	6	6	6	6	6	x	
85	1	1	5	5	6	6	5	6	6	6	6	6	5	6	6	6	6	6	6	6	6	6	x

Matrix of cophenetic values for figure 1.

We are now able to proceed to the actual comparison of dendrograms. This is done quite simply by calculating an ordinary product-moment correlation coefficient between the corresponding elements of the two matrices of cophenetic values to be compared. We propose to call these coefficients *cophenetic correlations*. For these procedures the half-matrix can be imagined as strung out in single file, column by column. For n OTU's there will be $n\ (n-1)/2$ elements in a similarity coefficient matrix. When the number of classes is rather small a coefficient of association will have to be used between the cophenetic values to be compared. However, in studies of any magnitude, where we are likely to deal with six or more classes per range of similarity values (and thus a two-way frequency distribution involving 36 classes), ordinary product moment correlation is indicated. This can be done by desk calculation, but will preferably be done on a computer. Some transformation of the correlation coefficients may be necessary in order to insure homoscedasticity.

In case the two dendrograms to be compared contain OTU's which are not common to them, one should simply ignore the non-coincident OTU's and proceed as before.

In small studies, i.e. less than 10 OTU's, it is computationally simplest to perform the operations on a desk calculator. We have found that in larger studies it is more efficient to use a digital computer. An I.B.M. — 650 computer program (DENDRON I) developed by one of us (F.J.R.) may be obtained by writing to the authors. The program was designed to accept the output of our Weighted Pair Group Method (WPGM) programs (TAXON I and II), prepares dendrograms from these and computes cophenetic values if desired.

EXAMPLES

The correlation matrix of 23 species of the subgenera *Chilosima* and *Ashmeadiella s. str.* presented by Sokal and Michener (1958) was used as a test case for comparing the various dendrograms which resulted from several methods of clustering used by these authors.

The dendrogram resulting from the weighted pair group method is illustrated in Figure 1. It was compared with the weighted variable group method, the dendrogram of which can be found in figure 1 of Sokal and Michener (1958). Matrices of cophenetic values based on eight classes were prepared. Only eight classes were chosen to make the example simple to inspect and because the first trials of the method were performed on a desk calculator. Fifty classes were used in a repetition of some of the calculations on a computer, resulting in essentially identical cophenetic correlations. The matrix of cophenetic values for the weighted pair group dendrogram of Figure 1 can be seen in Table 1. A similar matrix was prepared for the weighted variable group dendrogram, but is not shown here. Figure 2 shows a two-way frequency distribution of the 253 pairs of species plotted with respect to the two sets of cophenetic values. From this figure we can see the generally high correlation between the two sets of cophenetic values, representing substantial agreement between the two dendrograms. When a Pearson product-moment correlation coefficient was computed for the two-way frequency distribution, a correlation of 0.95 was obtained. Cophenetic values based on eight classes were also computed for the dendrogram constructed by Sokal and Michener (1958) using the unweighted variable group method. A comparison of these cophenetic values with those discussed above is shown expressed as correlation coefficients in Table 2.

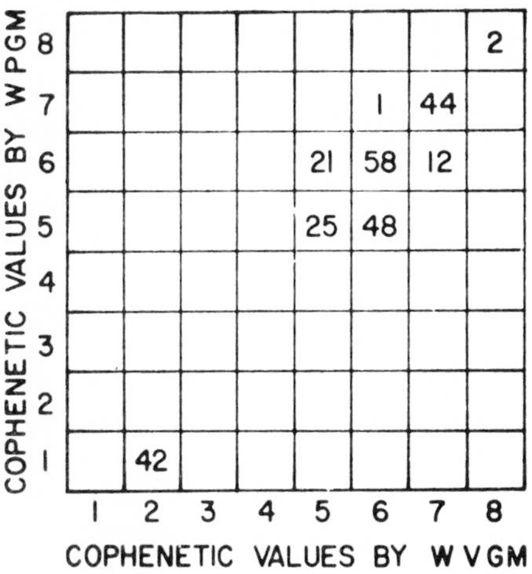

Fig. 2. Two-way frequency distribution of cophenetic values among the 23 species of bees for weighted pair group method (WPGM) and weighted variable group method (WVGM).

Also included in Table 2 is the weighted pair group method with the correlation between joining stems computed by an ordinary arithmetic average of the correlation coefficients, rather than by Spearman's sums of variables method. The last row and column of the correlation matrix in Table 2 represents correlations between the original correlation coefficients (between pairs of species) and the four sets of cophenetic values. These correlations give an indication of the amount of information lost by presenting the correlation matrix in the form of a dendrogram (i.e. the degree of the distortion between the original correlations and the cophenetic values).

TABLE 2.

	WPGM	WVGM	UVGM	WPGA	r
WPGM	x	.95	.84	.86	.80
WVGM	.95	x	.87	.89	.82
UVGM	.84	.87	x	.92	.83
WPGA	.86	.89	.92	x	.86
r	.80	.82	.83	.86	x

Matrix of correlation coefficients among the cophenetic values resulting from four dendrograms and the original correlation coefficients used to prepare the dendrograms. WPGM — weighted pair group method, WVGM — weighted variable group method, UVGM — unweighted variable group method, WPGA — weighted pair group method using arithmetic averages, and r — the original correlations.

Inspection of Table 2 shows that in general the correlation between the original correlation coefficients and the dendrograms resulting from them are high (in the 0.80's range). Of the various clustering methods that of averaging correlation coefficients, rather than Spearman's sums of variables method, seems to show a somewhat closer correlation with the original similarity values and hence the least amount of distortion of the original data. When we consider the various types of dendrograms we note that in general they are more correlated with each other than they are to the original correlations. It would appear that the closest relation between dendrograms is between the weighted pair group method and the weighted variable group method, as has been pointed out before. This validates the choice of the weighted pair group method for computational purposes as being essentially identical with the weighted variable group method, which has been chosen as more suitable for desk calculator operation.

We shall not pursue the implications of the results of this analysis in this paper. It should be obvious that by the use of cophenetic correlations we have a powerful method for evaluating the relative similarities of various dendrograms and the distortions due to different types of cluster analysis. The relative merits of different methods of clustering will be discussed in a later paper.

SUMMARY

The method developed in this paper makes quantitative comparisons among taxonomic dendrograms or diagrams of relationships. Dendrograms obtained by numerical taxonomy can be compared with each other, with dendrograms constructed by conventional taxonomic methods and with the original coefficients of similarity from which a given dendrogram has been derived. The method of comparison is based on the drawing of equidistant phenon lines horizontally across a dendrogram and on the computation of cophenetic values, which are the coded class marks of similarity classes in which any two taxa in a dendrogram are connected. Examples of the method are shown, using a portion of Michener and Sokal's bee data.

LITERATURE CITED

CAMP, W. H. and C. L. GILLY. 1943. The structure and origin of species. Brittonia 4: 323-385.

EHRLICH, P. R. 1961. Has the biological species concept outlived its usefulness? Systematic Zool. 10: 167-176.

MICHENER, C. D. and R. R. SOKAL. 1957. A quantitative approach to a problem in classification. Evolution 11: 130-160.

MORISHIMA, H. and H. OKA. 1960. The pattern of interspecific variation in the genus *Oryza:* its quantitative representation by statistical methods. Evolution 14: 153-165.

ROGERS, D. J., and T. T. TANIMOTO. 1960. A computer program for classifying plants. Science 132: 1115-1118.

ROHLF, F. J. and R. R. SOKAL. 1962. The description of taxonomic relationships by factor analysis. Systematic Zool. (in press).

SNEATH, P. H. A. 1957. The application of computers to taxonomy. J. Gen. Microbiol. 17: 201-226.

——. and S. T. COWAN. 1958. An electro-taxonomic survey of bacteria. J. Gen. Microbiol. 19: 551-565.

——. and R. R. SOKAL. 1962. Numerical taxonomy. (Nature, in press).

SOKAL, R. R. 1960. Die Grundlagen der numerischen Taxonomie. Proc. XI. Intern. Congr. Entomol. (in press).

——. 1961. Distance as a measure of taxonomic similarity. Systematic Zool. 10: 70-79.

——. and C. D. MICHENER. 1958. A statistical method for evaluating systematic relationships. Univ. of Kans. Sci. Bull. 38: 1409-1438.

Soria, J., and C. B. Heiser, Jr. 1961. A statistical study of relationships of certain species of the *Solanum nigrum* complex. Econ. Botany 15: 245-255.

FOOTNOTES

1. Contribution No. 1131 from the Department of Entomology, The University of Kansas, Lawrence, and paper No. 3 in a series entitled "Experiments in Numerical Taxonomy". Paper No. 1 in this series is Sokal (1961) and paper No. 2 is Rohlf and Sokal (1962).

2. This investigation was carried out during the tenure by F. J. Rohlf of a predoctoral fellowship from the Division of General Medical Sciences, United States Public Health Service. The authors are indebted to the University of Kansas General Research Fund for making computer time available at the University of Kansas Computation Center. Mrs. Ann Schlager prepared the drawings. Prof. C. D. Michener of the University of Kansas has read the manuscript and contributed useful suggestions.

24

Reprinted with permission from *Syst. Zool.*, **18**(3), 279–285 (1969)

ON THE COPHENETIC CORRELATION COEFFICIENT

JAMES S. FARRIS

Abstract

Some algebraic properties of the cophenetic correlation coefficient (CPCC) are derived. Conditions under which the CPCC is maximized for a phenogram are calculated, and a strategy for finding a phenogram with largest CPCC is described. The desirability of the CPCC as an optimality criterion for classifications is discussed.

Since its introduction by Sokal and Rohlf (1962), the cophenetic correlation coefficient (CPCC) has been widely used in numerical phenetic studies, both as a measure of degree of fit of a classification to a set of data and as a criterion for evaluating the efficiency of various clustering techniques. Paradoxically, little effort has been devoted to analysis of the properties of the CPCC. The present paper attempts partly to fill this gap in the theory of numerical taxonomy. I shall discuss the problems of producing a phenogram with optimal CPCC and the effect that optimizing the CPCC may have on the ability of the classification to retrieve information about operational taxonomic units (OTUs).

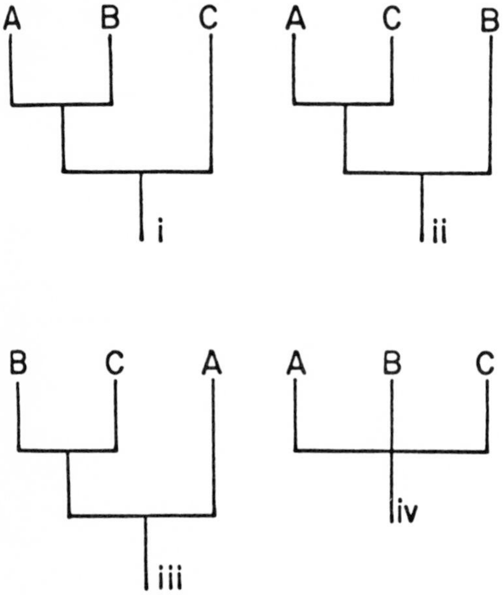

FIG. 1.—Four branching forms for three OTUs.

DEFINITIONS

Since the CPCC has been used largely in the context of finding optimal phenograms, I shall discuss the concept of fit to data primarily as it applies to phenograms and the cophenetic matrices of differences derived from them. I shall throughout use the term *branching form* to refer to the shape of a phenogram and the placement of the OTUs on the branches of the phenogram. Fig. 1 depicts the 4 different branching forms for 3 OTUs.

It may be taken as the defining property of phenograms, as that term is used here, that the cophenetic difference value derived from a particular linkage is constant over the OTUs affected by that linkage. For example, suppose that the phenetic difference matrix for the 3 OTUs of Fig. 1 is

OTU	A	B	C
A	–		
B	x	–	
C	y	z	– .

The matrix is assumed to be symmetric. For phenogram (i) of Fig. 1, the cophenetic difference matrix would be of the form

OTU	A	B	C
A	–		
B	f	–	
C	g	g	– .

C(A,C) and C(B,C) have the same value, g, because A and B link to C at the same level. The problem of optimizing the CPCC for branching form (i) of Fig. 1 is to find values f and g so that the correlation between the vectors (x,y,z) and (f,g,g) is maximized. Of course, it may be that branching form (ii) of Fig. 1 is a better one for optimizing the CPCC. It may be possible to find constants s and t such that the correlation between (x,y,z) and (s,t,s) is greater than the best possible value for the CPCC between (x,y,z) and (f,g,g). Since the CPCC depends on the branching form, the problem of finding optimal branching forms will also be treated.

Throughout this paper I shall refer to fits on matrices of phenetic *differences* with the understanding that similar comments apply to matrices of phenetic similarity.

BEST FIT FOR A GIVEN BRANCHING FORM

We consider first the problem of finding values for the matrix of cophenetic differences, C, so that the elements of C have maximum correlation with the elements of the phenetic difference matrix, P. If C and P have a certain CPCC, then so do P and matrix $A = C + a$, where a is any constant. I shall limit my treatment to the case in which C and P have the same average value of their elements.

We can maximize the CPCC by minimizing the residual variance about a regression line of the form

$$p' - \bar{p} = a + b(c - \bar{c}), \qquad (1)$$

where c is an element of the cophenetic difference matrix, $\bar{p}$ and $\bar{c}$ are the mean values of the elements of P and C, respectively, a and b are parameters of the regression, and p′ is the phenetic difference value predicted from c by the regression. Note that, while P is taken to be fixed data to which C is fitted, we are concerned with retrieving—that is, predicting—P from C. Hence the regression equations treat P as a dependent variable regressed on C.

The requirement that $\bar{c} = \bar{p}$ allows (1) to be reduced to

$$p' = bc. \qquad (2)$$

Now b, the slope of the regression line, depends only on the units of C (we assume P to be fixed). The value of b can be changed without altering the CPCC. For convenience, we choose the units of C so that $b = 1$. Then, (2) becomes

$$p' = c. \qquad (3)$$

The residual variance about the regression is proportional to

$$S^2 = \sum_{P} [p(c) - p'(c)]^2, \qquad (4)$$

the sum over the elements of P of the squared difference between the actual phenetic difference, p(c) associated with a given c value and p′(c), the predicted phenetic difference value. Since $p'(c) = c$ and $p[c(i,j)] = p(i,j)$, where i and j index rows and columns of the matrices P and C, (4) may be written

$$S^2 = \sum_{i,j<i} [p(i,j) - c(i,j)]^2. \qquad (5)$$

We sum only over the lower triangular submatrices of C and P, since C and P are symmetric.

It is apparent from (5) that the conventional restrictions we have placed on C imply that the C that maximizes the CPCC for a given P is the least-squares fit for P. This seems a reasonable strategy for establishing a criterion of efficiency of information-retrieval ability. If (5) is optimized, each c value is an unbiased predictor of its corresponding p values, and the c values selected minimize the average amount of error in predicting p(i,j) from c(i,j).

The CPCC is maximized by minimizing (5). The branching form specifies an array of sets, each set consisting of a number of values, p(i,j), for which the corresponding c(i,j) are all equal. As pointed out above, the elements $p(C,A) = x$ and $p(C,B) = z$ form such a set for branching form (i) of Fig. 1.

Let Q be a set of values, p(i,j), for which every corresponding c(i,j) equals a constant, d. Let Q have k elements, and let Q be indexed by a set of integers, L, with elements $1 \leqslant l \leqslant k$. The contribution to S^2 of Q can be written as

$$\sum_{L} [p(l) - d]^2, \qquad (6)$$

where p(l) is a value, p(i, j) in Q, indexed by l. We define

$$p^* = (1/k) \sum_{L} p(l), \qquad (7)$$

the average value of values p(i, j) in Q. We rewrite (6):

$$\sum_L [p(l) - d]^2 = \sum_L [(p(l) - p^*) - (d - p^*)]^2$$
$$= \sum_L [(p(l) - p^*)^2 - 2(p(l) - p^*) \times (d - p^*) + (d - p^*)^2]$$
$$= \sum_L (p(l) - p^*)^2 - 2(d - p^*) \sum_L [p(l) - p^*] + \sum_L (d - p^*)^2. \quad (8)$$

Since $\sum_L [p(l) - p^*] = 0$, (8) becomes

$$\sum_L [p(l) - d]^2 = \sum_L [p(l) - p^*]^2 + k(p^* - d)^2. \quad (9)$$

The first term of the right-hand side of (9) depends only on the p values, which are fixed for the branching form. The second term is minimized if $p^* = d$. Hence for a particular branching form, the CPCC is maximized if each c(i,j) value is equal to the arithmetic mean of all the values, p(i,j), to which it corresponds.

One existing method for finding phenograms, Unweighted Pair-Group Analysis (UPGA) (see Sokal and Sneath, 1963, and references therein), computes cophenetic difference values in just this way. We would predict that UPGA would yield higher CPCCs than other phenogram-producing methods, and this prediction is substantiated by the empirical findings of Sokal and Rohlf (1962).

OPTIMAL BRANCHING FORMS

Let us now consider some properties of branching forms that have maximum CPCC over the set of branching forms. Since C is so computed as to minimize the residual variance in P around the line described by (3), the least-squares slope of the linear regression of elements of P on elements of C is unity. Letting b again represent this slope, as in (2), and letting s(PC), $s^2(P)$, $s^2(C)$ denote the covariance of the elements of P with the elements of C, the variance of the elements of P, and the variance of the elements of C, we note

$$b = \frac{s(PC)}{s^2(C)} = 1, \quad (10)$$

by the definition of least-squares slope. Then

$$s(PC) = s^2(C), \quad (11)$$

and,

$$CPCC = \frac{s(PC)}{s(P)s(C)} = \frac{s^2(C)}{s(P)s(C)} = \frac{s(C)}{s(P)}, \quad (12)$$

or,

$$(CPCC)^2 = \frac{s^2(C)}{s^2(P)}, \quad (13)$$

by the definition of the linear correlation coefficient. Now define

$$E^2 = T[s^2(P) - s^2(C)], \quad (14)$$

where T is the degrees of freedom of $s^2(P)$. Then (13) can be rewritten

$$(CPCC)^2 = \frac{s^2(C)}{s^2(P)} = \frac{Ts^2(P) - E^2}{Ts^2(P)} = 1 - \frac{E^2}{Ts^2(P)}. \quad (15)$$

Since T and $s^2(P)$ are fixed, the CPCC is monotone decreasing on E^2. E^2 can be written as

$$E^2 = \sum_{i,j<i} [p(i, j)]^2 - \frac{[\sum_{i,j<i} p(i, j)]^2}{T + 1} - \sum_{i,j<i} [c(i, j)]^2 + \frac{[\sum_{i,j<i} c(i, j)]^2}{T + 1} \quad (16)$$

Since C and P have the same mean value of elements $\sum_{i,j<i} p(i, j) = \sum_{i,j<i} c(i, j)$, so that (16) becomes

$$E^2 = \sum_{i,j<i} [p(i, j)]^2 - \sum_{i,j<i} [c(i, j)]^2. \quad (17)$$

Now we partition the lower submatrix of P exhaustively into m disjoint sets of elements, Q(r), $1 \leqslant r \leqslant m$, each of the form of the set Q defined above. The fixed value of c(i,j) corresponding to any element p(m,n) in Q(r) will be denoted $p^*(r)$, and $p^*(r)$ is equal to the mean value of elements p(i,j) in Q(r).

Substituting in (17) we obtain

$$E^2 = \sum_{r}^{m} \{ \sum_{Q(r)} [p(i,j)]^2 - \sum_{Q(r)} [p^*(r)^2] \}$$

$$= \sum_{r}^{m} \sum_{Q(r)} [p(i,j) - p^*(r)]^2. \qquad (18)$$

Since each set Q(r) is induced by one linkage of a phenogram, (18) implies a strategy for finding best branching forms to maximize the CPCC. "Good" linkages—or clusters—induce sets Q(r) with small values of

$$e^2(r) = \sum_{Q(r)} [p(i,j) - p^*(r)]^2. \qquad (19)$$

"Bad" linkages induce high values of $e^2(r)$.

Current programs for forming phenograms typically form clusters containing OTUs with low average difference values. The fact that the CPCC can be maximized by minimizing E^2, however, suggests that a more efficient way to form phenograms with a high CPCC would be to cluster OTUs with similar columns in the P matrix. The two clustering strategies are not necessarily the same, as the following examples show:

Suppose that for OTUs A, B, C in Fig. 1, P is

	A	B	C
A	–	4	7
B	4	–	3
C	7	3	–

Since B and C are closest together, clustering according to similarity could lead to branching form (iii) of Fig. 1. Then C_{III} would be

	A	B	C
A	–	5.5	5.5
B	5.5	–	3
C	5.5	3	–

and $E^2_{III} = 4.5$. But p(A,B) and p(B,C) are nearly equal, so that branching form (ii) of Fig. 1, with C_{II},

	A	B	C
A	–		
B	3.5	–	
C	7	3.5	–

has $E^2_{II} = .5$.

Phenogram (ii) might be characterized as having an *inversion* in the last example, in that cluster A,C, with linkage difference value 7, is a subcluster of cluster A,B,C, with linkage difference value 3.5. Inversions are not the only way in which clustering to optimize the CPCC may differ from clustering by similarity. Suppose the 3 taxa of Fig. 1 have P:

	A	B	C
A	–		
B	9	–	
C	2	7	–

Clustering by similarity gives branching form (ii) of Fig. 1. If A, B, and C are OTUs, branching form (ii) also has the highest CPCC. But suppose that A and C are OTUs, while B is a large cluster of, say, 100 OTUs. The P matrix given above is then an abbreviation of a matrix with 102 rows and columns. Such a compacted matrix is employed in most of the stepwise clustering procedures for forming phenograms. If B is a cluster, the exact value of the CPCC for the data will depend on the internal structure of B. If we take the internal structure of B to be fixed, we can compute the order of the E^2 values for phenograms on A, B, and C, since the components, e^2, of E^2, are additive over disjoint subsets of the P matrix. Then we can compute that branching form (i) has the highest CPCC, branching form (iii) has the next highest, while branching form (ii)—previously optimal—has the lowest! Branching form (i) has an inversion for these data, but branching form (iii) does not. If B has a weight of 100, C_{III} is

	A	B	C
A	–		
B	8.95	–	
C	8.95	7	–

which is the best weighted least-squares fit to P when deviations in the B row and column are weighted 100 times as heavily as deviations elsewhere.

Consideration of E^2 shows that Pair Group Analysis can always do at least as well at maximizing the CPCC as can Variable Group Analysis (the reader is referred to

Sokal and Sneath, 1963, for a fuller discussion of these terms). For let $Q(r)$ be a set of elements of P as defined above and corresponding to a linkage on the phenogram at which more than two clusters are joined. Branching form (iv) of Fig. 1 shows such a linkage. The multiple linkage can be resolved into a sequence of two simpler linkages. Branching form (i), (ii), or (iii) of Fig. 1 could be interpreted as the result of resolving the multiple linkage of (iv). The Q-sets corresponding to the two new linkages replacing the multiple linkage will be designated $Q(a)$ and $Q(b)$. The union of $Q(a)$ and $Q(b)$ is $Q(r)$. Now the contribution of the multiple linkage to E^2 is

$$e^2(r) = \sum_{Q(r)} [p(i,j) - p^*(r)]^2, \qquad (20)$$

following (19) above. The contribution of the two simpler linkages is

$$e^2(a) + e^2(b) = \sum_{Q(a)} [p(i,j) - p^*(a)]^2 + \sum_{Q(b)} [p(i,j) - p^*(b)]^2. \qquad (21)$$

It is clear that $e^2(a) + e^2(b) \leq e^2(r)$. Then a Paired Group method is always sufficient to attain the maximum CPCC possible for a phenogram with a particular set of data. But this result should not be misinterpreted to mean that existing Paired Group methods will maximize the CPCC; or that any paired-group branching form has a higher CPCC than any variable-group branching form.

Now which branching forms will have highest CPCC? Uniting two OTUs or clusters at a particular level will result in both OTUs (or all members of both clusters) having the same cophenetic difference values at that level and at all later formed levels. Suppose, for example, the first step in clustering is to form a cluster of OTU A and OTU B. Then in every row of the cophenetic difference matrix, column A will have the same value as column B. The degree to which this situation will lower the CPCC is given by

$$V(A,B) = \sum_{i \neq A,B} (p(A,i) - M(i))^2 + (p(B,i) - M(i))^2, \qquad (22)$$

where $M(i) = (\frac{1}{2})\,(p(A,i) + p(B,i))$.

Then we would expect a "good" cluster of two OTUs to have a small V value, and a "bad" cluster of two OTUs to have a high V value.

Now we can construct a method for choosing a branching form step by step. We begin with the original P matrix with t rows and columns. We find a pair of columns, X and Y, such that $V(X,Y)$ is minimized and unite the corresponding OTUs (or clusters). Row and column X of P is replaced with the $M(i)$, $i \neq X,Y$, $p(X,X)$ retaining its original value. Row and column Y are deleted from the matrix. We repeat the process of joining columns that differ by small V values until only one element remains in the matrix. The order of joining of columns gives the branching order of a phenogram. At each stage, the weight of a row or column is equal to the number of OTUs in the cluster represented by that row or column. With weighted columns,

$$M(i) = \frac{k_X p(X,i) + k_Y p(Y,i)}{k_X + k_Y};$$

$$V(X,Y) = \left[\sum_{i \neq X,Y} k_i (p(X,i) - p(Y,i))^2\right] \times \left[\frac{k_X k_Y}{k_X + k_Y}\right], \qquad (23)$$

where k is the weight of a row or column. It is the dependence of V on the k values that produces the sensitivity, noted above, of the CPCC to cluster size.

A FORTRAN IV program to perform this clustering algorithm is available from the author.

DISCUSSION

Most existing methods for forming phenograms cluster together OTUs that are most highly similar. An exception is Hartigan's (1967) method, which seeks the best weighted least-squares fit of C to P by trial and error alterations of the branching form. All these procedures produce phenograms without inversions. The conventional clustering techniques such as the Unweighted

Pair Group Method do so because their clustering sequence is determined by similarity. Hartigan deliberately restricted his procedure to prevent inversions. Clustering to maximize the CPCC is accomplished by clustering together OTUs that have most highly similar columns of the P matrix; such clustering can produce inversions in the phenogram. The results of this paper could then be applied to constructing a phenogram-producing algorithm that would achieve higher values of the CPCC than do current methods. There is some doubt that such an algorithm should be used in taxonomy.

The objectives of phenetic taxonomy are largely concerned with constructing taxa of maximum information content. Just what a "taxon of maximum information content" means has never been precisely defined in the context of phenetics. If a taxon has high information content when all the OTUs in it have similar columns of the phenetic difference matrix, then the CPCC is a measure of information content for a classification. If, on the other hand, a taxon has high information content when the OTUs in it are themselves similar, then the CPCC does not *directly* measure the information content of the classification, and we should choose to optimize some criterion other than the CPCC.

A number of reasonable alternative criteria have already been introduced into the literature. Edwards and Cavalli-Sforza (1965) suggest choosing the branching form that minimizes the within-cluster sums of squared deviations of OTUs from cluster centroids. Orloci (1967) has devised a polythetic, agglomerative clustering technique that could be expected to be fairly efficient at optimizing the Edwards-Cavalli-Sforza criterion. Williams, Lambert, and Lance (1966) have devised a similar procedure that seeks to optimize the information content of clusters.

A significant feature of these approaches is that the form of the optimality criterion depends on the form of the measure of difference (or similarity) between OTUs, and not just on the relationship between P and C considered purely as strings of numbers. Choice of a "best" optimality criterion for classifications is thus in general dependent on the prior choice of a difference/similarity coefficient. That coefficient might be the usual information-statistic, the euclidean distance, the squared euclidean distance—or virtually any other function! In a sense, the choice of a similarity coefficient in this context is equivalent to a choice of the meaning of "information" in the phenetic premise that the best classification is the one with the highest "information" content.

The CPCC is a true measure of optimality of a classification only for a particular definition of taxonomic "information." Under the usual criterion that similar OTUs should be clustered together in a "good" classification, the CPCC is not a direct measure of optimality of classifications. Further, the problem of finding the most appropriate optimality criterion for classifications will have to be considered jointly with the question of what is the most appropriate measure of "similarity" between OTUs.

We might be tempted to conserve the widely employed CPCC as an optimality criterion by restricting phenograms to lack inversions. There is some precedent for doing so. As noted above, many clustering techniques effectively already use that restraint. Jardine, Jardine, and Sibson (1967) found it useful to treat phenograms as ultrametrics on OTUs. A phenogram that is an ultrametric cannot have any inversions. Still, it does not seem defensible to salvage the CPCC by restricting phenograms. To do so amounts to asserting that the CPCC is the right optimality criterion to use, provided we never really try to optimize it! At any rate, the no-inversion convention alone is not enough to cure all the curious properties of the CPCC. The sensitivity to cluster size remains. As seen in the second example above, disallowing inversions may simply permit the optimal phenogram to take on some other form that still does not properly describe the OTUs themselves. The cluster size sensitivity of the CPCC may have particularly unfavorable consequences for the stability of numerical classifications. One can conceive of cases in which different samples of OTUs from some group yield different CPCC-optimal classifications of the group due to random variations in the relative proportions of the major subgroups included in the study.

The cluster size sensitivity of the CPCC would seem to rule it out as a useful optimality criterion for numerical classifications. Related criteria with lower cluster size lability might profitably be investigated. An appropriate optimality criterion cannot be selected, however, until some decision is reached on the nature of an optimal classification: it will have to be specified whether the aim of taxonomy is to describe phenetic difference matrices or to describe OTUs.

SUMMARY

The CPCC (cophenetic correlation coefficient) is maximized when the fixed value of a set of elements in the cophenetic difference matrix is equal to the mean value of the corresponding elements in the phenetic difference matrix.

This fact implies that Unweighted Pair Group Analysis should be most successful among existing phenogram-producing clustering techniques at maximizing the CPCC. This prediction is in accord with the empirical findings of Sokal and Rohlf (1962).

For any set of data there exists a pair group phenogram with CPCC at least as great as the CPCC of any variable group phenogram.

The phenogram with greatest CPCC is one in which OTUs with most similar columns of the phenetic difference matrix are clustered together.

The CPCC is therefore not a direct measure of the degree to which the classification describes the distribution of character states.

If it is desired to term "optimal" those classifications in which most "similar" OTUs are clustered together, the CPCC should not be employed as an optimality criterion. The choice of an optimality criterion for classifications will have to be determined in conjunction with the choice of a measure of "similarity" between OTUs and an implied measure of "information" content of classifications.

REFERENCES

EDWARDS, A. W. F., AND L. L. CAVALLI-SFORZA. 1965. A method for cluster analysis. Biometrics, 21:363–375.

HARTIGAN, J. A. 1967. Representation of similarity matrices by trees. J. Amer. Stat. Assn., 62:1140–1158.

JARDINE, C. J., N. JARDINE, AND R. SIBSON. 1967. The structure and construction of taxonomic hierarchies. Math. Biosciences, 1:173–179.

ORLOCI, L. 1967. An agglomerative method for classification of plant communities. J. Ecol., 55:193–206.

SOKAL, R. R., AND F. J. ROHLF. 1962. The comparison of dendrograms by objective methods. Taxon, 11:33–40.

SOKAL, R. R., AND P. H. A. SNEATH. 1963. Principles of numerical taxonomy. Freeman, San Francisco. 359 p.

WILLIAMS, W. T., J. M. LAMBERT, AND G. N. LANCE. 1966. Multivariate methods in plant ecology. V. Similarity analysis and information-analysis. J. Ecol., 54:427–445.

Department of Biological Sciences, State University of New York, Stony Brook, New York, 11790.

25

Reprinted from *Computer J.*, **9**, 177–184 (1968)

The construction of hierarchic and non-hierarchic classifications

By N. Jardine and R. Sibson*

Many of the cluster methods that are used in the construction of classificatory systems operate on data in the form of a dissimilarity coefficient on a set of objects. In this paper we outline a theoretical framework within which the properties of such methods may be discussed. Certain conditions that a cluster method should satisfy are suggested, and a particular sequence of cluster methods which satisfies these conditions is described. The application of the sequence of methods is illustrated by a simple example.

(First received January 1968, and in revised form, February 1968)

Numerous methods for the derivation of classificatory systems from data in the form of a dissimilarity coefficient on a set of objects have been proposed. Some of these methods are described in: Sokal and Sneath (1963); Williams and Dale (1965); Lance and Williams (1967a, 1967b). Despite the very considerable number of papers describing and applying such methods, there have been relatively few attempts to construct a mathematical framework within which the properties of such methods may be investigated. Exceptions are the papers of: Bonner (1964); Watanabé (1965); Estabrook (1966); Johnson (1967); Jardine, Jardine, and Sibson (1967); and Jardine and Sibson (1968). Likewise, there have been relatively few reports of comparative studies in which a variety of methods were applied to the same data. Exceptions are the papers of: Boyce (1964); Minkoff (1965); Watson, Williams, and Lance (1966); and Sokal and Michener (1967).

If progress is to be made in the understanding of these potentially very useful methods it is important that analytical and empirical investigation of the properties of cluster methods should go hand in hand. Suppose, for example, that we wish to investigate the *stability* of the classification produced by a given cluster method as we increase the quantity of information used in calculating the dissimilarity coefficient on the set of objects to be classified. Before we can interpret the changes produced in the classification we must know whether the transformation of the dissimilarity coefficient by the cluster method is a continuous transformation. Similarly, we may wish to be able to compare the goodness-of-fit to given data of the classifications obtained using a variety of cluster methods. In this case we need some measure of the distortion imposed by a classification on a dissimilarity coefficient, and we must set up the analytical framework required to determine for each cluster method whether it minimises the distortion under certain conditions.

In Jardine, *et al.* (1967) and in Jardine and Sibson (1968) a formal model was established within which the mathematical properties of any cluster method operating on a dissimilarity coefficient may be investigated. Criteria of adequacy for a cluster method were suggested, and it was shown that the majority of the cluster methods currently in use fail to satisfy these criteria. A particular sequence of cluster methods which satisfy the criteria was described. In the following sections an informal account of the model is given, and a graph-theoretic description is used to illustrate the sequence of cluster methods and their application. First we consider cluster methods that may be used in obtaining a hierarchic classification, then we examine the more general problem of characterising non-hierarchic classifications, and describing cluster methods that may be used in their construction.

1. Hierarchic classificatory systems

The derivation of a hierarchic classificatory system from a dissimilarity coefficient is a two-stage process. The first stage is the derivation of a *dendrogram*. A dendrogram may be described informally as a hierarchy with numerical levels. The levels at which each pair of objects meet in a dendrogram, the *splitting-levels*, are determined by the dissimilarity coefficient from which the dendrogram is derived. The way in which the splitting-levels are determined depends upon the cluster method used. A hierarchic classification may be derived from a dendrogram by identifying the ordinal levels (ranks) of the hierarchy with numerical levels in the dendrogram. We shall call the sets of objects which are grouped at or below some numerical level in a dendrogram *clusters*. The sets of objects that are grouped at some rank (ordinal level) in a classificatory system we shall call *classes*. The classes of a given rank in a hierarchy consist of just those objects that cluster at or below the corresponding level in the dendrogram. Various rules may be devised for identifying the ranks of a hierarchy with the numerical levels in a dendrogram. For example we might make the identification correspond to some observed clumping of the splitting-levels in the dendrogram. Alternatively we may, as was suggested by Wirth, Estabrook, and Rogers (1966), use the dendrogram, together with suitable measures of the isolation and homogeneity of the clusters in the dendrogram, as a guide to the construction of a hierarchic classification.

* *King's College, Cambridge, England.*

It is the first stage, the passage from a dissimilarity coefficient to a dendrogram, that constitutes what is generally called cluster analysis (or a sorting-strategy). A hierarchic dendrogram may be given a numerical characterisation by indicating the splitting-level for each pair of objects (see Fig. 1). In general a hierarchic dendrogram may be characterised as a pair $S = (P, d)$, where P is a set and d is a function from pairs of elements of P to the non-negative real numbers, satisfying the conditions:

(1) $d(a, b) \geq 0$ for all $a, b \in P$;
(2*a*) $d(a, b) = 0$ if $a = b$;
(2*b*) $d(a, b) = 0$ only if $a = b$;
(2*c*) there exist $a, b \in P$ such that $d(a, b) \neq 0$ if $|P| > 1$;
(3) $d(a, b) = d(b, a)$ for all $a, b \in P$;
(4) $d(a, c) \leq \max\{d(a, b), d(b, c)\}$ for all $a, b, c \in P$.

In other words a hierarchic dendrogram is characterised by an *ultrametric*. A similar characterisation is given by Johnson (1967).

A *dissimilarity coefficient* will, in general, satisfy conditions (1), (2*a*), (2*c*) and (3); it will not necessarily satisfy condition (4) (the ultrametric inequality), or condition (2*b*). It may satisfy some weaker condition analogous to (4) such as the metric inequality:

(5) $d(a, c) \leq d(a, b) + d(b, c)$ for all $a, b, c \in P$.

A cluster method which transforms a dissimilarity coefficient into a hierarchic dendrogram may therefore be regarded as a method whereby the ultrametric inequality is 'imposed' on a dissimilarity coefficient. There are certain simple conditions which we may reasonably require any such transformation to satisfy.

(A) A unique result should be obtained from given data; that is, the transformation should be *well-defined*.

(B) Small changes in the data should produce small changes in the resultant dendrogram; that is, the transformation should be *continuous*.

(C) If the dissimilarity coefficient is already ultrametric it should be unchanged by the transformation.

(D) In some sense the result obtained should impose the minimum distortion upon the dissimilarity coefficient, subject to conditions (A)–(C) and (E)–(G).

A possible family of measures of the distortion imposed by a hierarchic dendrogram is given by:

$$\hat{\Delta}_\mu(d, D^*(d))$$
$$= [\Sigma|d(a, b) - D^*(d)(a, b)|^{1/\mu}]^\mu/[\Sigma d(a, b)^{1/\mu}]^\mu \quad (0 < \mu \leq 1)$$

$$\hat{\Delta}_0(d, D^*(d))$$
$$= [\max|d(a, b) - D^*(d)(a, b)|]/[\max d(a, b)]$$

where summation and maximisation are taken over all pairs $(a, b) \in P \times P$; d is the dissimilarity coefficient; and $D^*(d)$ is the ultrametric characterising the hierarchic dendrogram obtained by some transformation of the dissimilarity coefficient. Some of the $\hat{\Delta}_\mu$ are familiar in form: $\hat{\Delta}_0$ (obtained in the limit as $\mu \to 0$) is a normalised 'maximum modulus'; $\hat{\Delta}_1$ is a normalised 'mean modulus'; and $\hat{\Delta}_{1/2}$ is a normalised 'root mean square'. A similar suggestion was made by Ward (1962).

(E) The operation of the transformation should commute with multiplication of the dissimilarity coefficient by any strictly positive scalar; that is,

$$D^*(kd) = kD^*(d)$$

for any $k > 0$. The effect of this is to ensure that the transformation is independent of scale.

(F) The operation of the transformation should commute with any permutation of P: that is,

$$D^*d[(\rho \times \rho)] = [D^*(d)](\rho \times \rho)$$

where ρ is any permutation of P. The effect of this is to ensure that the transformation is independent of any preliminary labelling of the objects to be classified.

(G) If a cluster is excised and the transformation is applied to the restriction to it of the dissimilarity coefficient, the resultant dendrogram should be the restriction to that cluster of the original dendrogram; that is, (in view of condition (F)) if d has the form, as a matrix

$$\begin{bmatrix} d_1 & & e \\ & d_2 & \\ e^T & & d_l \end{bmatrix}$$

where the minimum entry outside the square diagonal blocks exceeds the maximum entry within them, then $D^*(d)$ has the form

$$\begin{bmatrix} D^*(d_1) & & e' \\ & D^*(d_2) & \\ e'^T & & D^*(d_l) \end{bmatrix}$$

where this matrix satisfies the same condition. This condition guarantees the consistency of the dendrograms obtained when the set P is extended or restricted in suitable ways.

It can be shown that the majority of the cluster methods currently in use fail to satisfy these conditions. For example, of the cluster methods described by Lance and Williams (1967a) all but one fail to satisfy the conditions; the *complete-link* (furthest-neighbour) method originally proposed by Sørensen (1948) fails by being ill-defined; the *centroid* method of Sokal and Michener (1958), the *median* method, the *group-average* method, and the *flexible* method suggested by Lance and Williams (1966b), all fail by being discontinuous. The flexible method may also fail to satisfy condition (C). Most known methods satisfy conditions (E)–(G).

The *single-link* (nearest-neighbour) method is the only method that we have seen that satisfies conditions (A)–(G). It can be given a simple graph-theoretic description which makes clear its defects as a method of classification (cf. Wirth *et al.*, 1966). Any dissimilarity coefficient on a set of objects can be characterised by a set of graphs, one for each value taken by the dissimilarity coefficient. The graph for a given value h of a dissimilarity coefficient on a set of objects has as vertices points representing the objects and edges joining just those pairs of points representing pairs of objects with dissimilarity $\leqslant h$. The single-link method produces clusters at each level that correspond to the components of the graph for that value of the dissimilarity coefficient (see Fig. 1). The defect of the single-link method is that it clusters together at a relatively low level objects linked by chains of intermediates. This defect is generally called *chaining*, but to call chaining a defect of the single-link method is rather misleading; the graph-theoretic description makes it clear that chaining is simply a description of what the method does.

The various average-link and centroid methods attempt to avoid this 'defect' by picking out clusters which are in some sense more homogeneous than those obtained by the single-link method, but in doing so they fall prey to the defect of discontinuity. In Jardine and Sibson (1968) it is suggested that the defects of the single-link method should be regarded as defects of hierarchic classification itself. It is further suggested that the best way to recover information of the kind that is concealed by chaining, for example information about the relative homogeneity of clusters, is to consider cluster methods which lead to overlapping (non-hierarchic) classificatory systems.

2. Non-hierarchic classificatory systems

The fact that there is a considerable body of consistent usage for hierarchic classificatory systems facilitates the construction of an appropriate logical model for such systems, for the terms which must be defined and the theorems which must be proved are known in advance. The situation is not as simple for non-hierarchic classificatory systems, for in this case there is no well-established consistent usage. Some of those who have discussed methods leading to systems of overlapping clusters are: Olson and Miller (1951); McQuitty (1956); Parker-Rhodes and Needham (1960); Needham (1961, 1965, 1967); Jancey (1966); and Lance and Williams (1967b). Intuitively we should expect that as the degree of overlap is allowed to increase so the accuracy of representation of the data should increase, although at the cost of increased complexity. In the limiting case where arbitrary overlap is allowed an exact representation of the original data should be obtained. These intuitions are precisely expressed in the generalised model covering both hierarchic and non-hierarchic classificatory systems given in Jardine and Sibson (1968).

A hierarchic classificatory system may be considered as a nested sequence of partitions of a set of objects. In Jardine and Sibson (1968) the notion of a partition is generalised by defining a *k-partition*. A k-partition allows a maximum of $k - 1$ objects in the overlaps between the classes that belong to it. A classificatory system may therefore be considered as a nested sequence of k-partitions; the system will be hierarchic in case $k = 1$ and overlapping in case $k > 1$. By a corresponding generalisation we may define the notion of a *k-dendrogram*. In a k-dendrogram clusters at a given level may overlap to the extent of $k - 1$ objects.

The single-link cluster method can be generalised and it can be shown that each of the sequence of cluster methods so defined satisfies conditions (A), (B), and (D)–(F) given on p. 178, and suitable generalisations of conditions (C) and (G). This sequence of methods we denote by (B_k). The first member of the sequence, B_1, is the *single-link* method leading to a hierarchic dendrogram (1-dendrogram). The second member of the sequence, B_2, may be called the *double-link* method, and leads to a dendrogram in which clusters may overlap to the extent of one object (a 2-dendrogram), and so on. If P is the set of objects, and $|P| = p$, where $|P|$ is the number of elements of P, then B_{p-1} gives an exact representation of the dissimilarity coefficient. It can be shown that the family of measures of distortion, $\hat{\Delta}_\mu(d, B_k(d))$, is monotone decreasing with increasing k, becoming zero in case $k = p - 1$. It can be shown also that the use of this sequence of methods enables us to recover information about the homogeneity of clusters that the single-link method fails to reveal.

The sequence of cluster methods (B_k) can be given a simple graph-theoretic description which generalises that given for the single-link method (see above). The clusters at level h in $B_k(d)$, the kth member of the sequence of k-dendrograms, are obtained as follows. A graph is drawn whose vertices represent the objects and whose edges join just those pairs of points which represent objects with dissimilarity $\leqslant h$. The maximal complete subgraphs (maximal subsets of the set of vertices in which all possible edges are present) are marked, and wherever the vertex sets of two such subgraphs intersect in at least k vertices further edges are drawn in to make the union of the two vertex sets into a complete subgraph. The process is repeated until there is no further alteration. If this process is carried out for each of the values taken by the dissimilarity coefficient the graph representation of $B_k(d)$ is obtained. This algorithm is not suitable for computation, but is useful since it illustrates the way in which the sequence of cluster methods (B_k) operates on a dissimilarity coefficient. The application of this algorithm is illustrated in **Fig. 1**. An algorithm suitable for computation is given on p. 180.

The maximal complete subgraphs given by each value of the dissimilarity coefficient can be regarded as the 'nuclei' of the clusters formed at each level in the sequence of k-dendrograms $(B_k(d))$. Several authors, notably Bonner (1964) and Needham (1961), have suggested that the recognition of maximal complete subgraphs should be the first stage in a cluster method. The

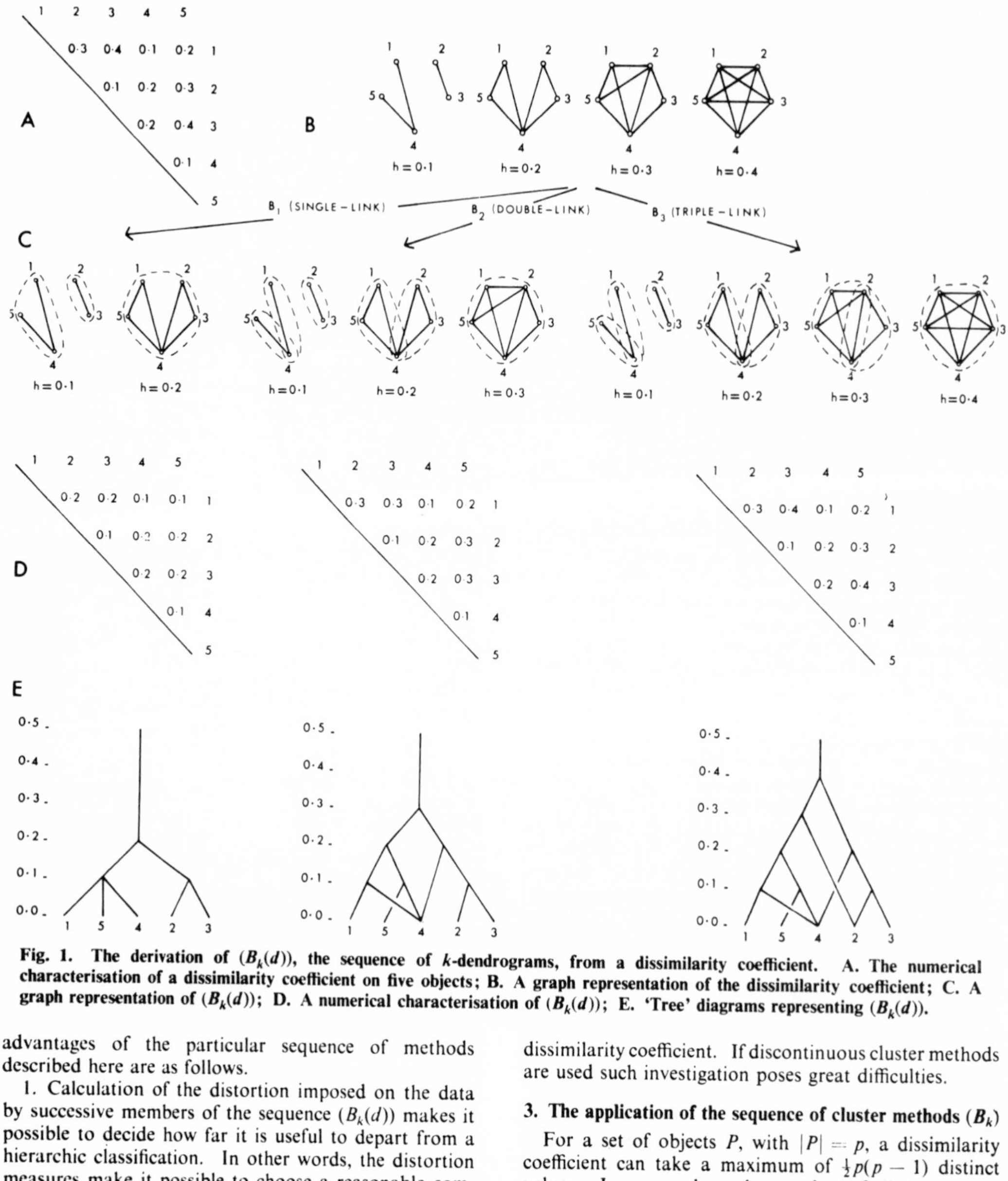

Fig. 1. The derivation of $(B_k(d))$, the sequence of k-dendrograms, from a dissimilarity coefficient. A. The numerical characterisation of a dissimilarity coefficient on five objects; B. A graph representation of the dissimilarity coefficient; C. A graph representation of $(B_k(d))$; D. A numerical characterisation of $(B_k(d))$; E. 'Tree' diagrams representing $(B_k(d))$.

advantages of the particular sequence of methods described here are as follows.

1. Calculation of the distortion imposed on the data by successive members of the sequence $(B_k(d))$ makes it possible to decide how far it is useful to depart from a hierarchic classification. In other words, the distortion measures make it possible to choose a reasonable compromise between complexity and accuracy of representation of the data.

2. Since each member of sequence of methods is a continuous transformation it is possible to investigate the stability of the classifications obtained when further information about the objects is used in computing the dissimilarity coefficient. If discontinuous cluster methods are used such investigation poses great difficulties.

3. The application of the sequence of cluster methods (B_k)

For a set of objects P, with $|P| = p$, a dissimilarity coefficient can take a maximum of $\frac{1}{2}p(p-1)$ distinct values. In cases where the number of distinct values approaches this maximum it becomes unfeasible to carry out the sequence of cluster methods (B_k) by hand, using the graph-theoretic algorithm, for more than about 10 objects. The following terminating algorithm is suitable for purposes of computing for up to about 25 objects.

1. List the subsets of P with exactly $k + 2$ elements in an arbitrary order.

2. Consider the value of d taken on each pair in the first subset. If d takes a unique maximum value on a single pair reduce this value to the next largest value taken by d on any pair from the subset. Otherwise leave d unchanged.

3. Repeat the process on the next subset starting with the modified d. Continue until all the subsets have been considered.

4. Repeat 2 and 3 until the list can be run through without further modification of d. The resultant dissimilarity coefficient is $B_k(d)$. This algorithm, together with the calculation of $\hat{\Delta}_{1/2}(d, B_k(d))$ for each member of the sequence, has been programmed for use on the Titan computer at the Cambridge University Mathematical Laboratory by Miss A. A. Houston.

Two practical points should be noted in using this algorithm. Firstly, the computation time increases rapidly with p (approximately as p^{k+2} for $k \ll p/2$). It is probable that a more economical algorithm could be devised. Secondly, certain difficulties arise in cases where condition (2*b*) on p. 178 is not satisfied by the dissimilarity coefficient: i.e. $d(a, b) = 0$, but $a = b$. Two options are available. One is to treat the objects as distinct and to apply the algorithm directly. Alternatively single-link clustering at level $h = 0$ may be applied to produce preliminary clusters. The dissimilarity between two such clusters is taken to be the minimum value of the dissimilarities between pairs, one member of the pair being from each cluster. The algorithm is then applied with those clusters as objects, and dissimilarities calculated as above. This process corresponds to the preliminary identification of objects not distinguished by the original dissimilarity coefficient. In general the first process leads to smaller distortion by the resultant k-dendrograms; if the dissimilarity coefficient satisfies the metric inequality (condition (5), p. 178) the two options are equivalent.

The representation of the sequence of k-dendrograms $(B_k(d))$ is itself a serious problem. The numerical characterisations are not very informative, and do not readily suggest useful classifications of the objects. The 'tree' representations, whilst useful for $k \leq 2$, become increasingly difficult to draw and interpret for larger values of k (see Fig. 1). The most useful way of presenting the information seems to be to indicate the clusters recognised at each level on graph diagrams (see Figs. 1 and 2). The difficulty in this method of presentation lies in finding a two-dimensional arrangement of points representing the objects which minimises the tangling between edges in the graphs for each level. A convenient way of obtaining such a representation is to apply non-metric multidimensional scaling to the dissimilarity coefficient (see Kruskal, 1964a, 1964b. The two-dimensional arrangement obtained by non-metric multidimensional scaling may then be modified to eliminate collinearity of points and to obtain a neater arrangement for purposes of display. The arrangement

Table 1

A dissimilarity coefficient (Mahalanobis' generalised distance) on nine populations of *Sagina apetala*

	1	2	3	4	5	6	7	8	9
1		9·28	6·53	9·73	1·79	6·41	6·98	13·04	3·24
2			4·65	5·21	10·21	7·30	3·66	4·72	10·09
3				6·01	7·55	2·98	2·63	9·30	6·21
4					11·20	7·73	8·37	4·48	10·15
5						4·62	6·90	12·45	4·68
6							3·68	11·21	3·80
7								8·38	7·13
8									12·75
9									

Table 2

The numerical characterisation of the first four members of the sequence of k-dendrograms, $(B_k(d))$

	1	2	3	4	5	6	7	8	9
1		3·80	3·80	4·72	1·79	3·80	3·80	4·72	3·24
2			3·66	4·72	3·80	3·66	3·66	4·72	3·80
3				4·72	3·80	2·98	2·63	4·72	3·80
4					4·72	4·72	4·72	4·48	4·72
5						3·80	3·80	4·72	3·24
6							2·98	4·72	3·80
7								4·72	3·80
8									4·72
9									

$k = 1$

$\hat{\Delta}_{1/2}(d, B_1(d)) = 0·528$

	1	2	3	4	5	6	7	8	9
1		6·21	6·21	6·21	1·79	4·68	6·21	6·21	3·24
2			4·65	5·21	6·21	4·65	3·66	4·72	6·21
3				6·01	6·21	2·98	2·63	6·01	6·21
4					6·21	6·01	6·01	4·48	6·21
5						4·62	6·21	6·21	4·68
6							3·68	6·01	3·80
7								6·01	6·21
8									6·21
9									

$k = 2$

$\hat{\Delta}_{1/2}(d, B_2(d)) = 0·367$

	1	2	3	4	5	6	7	8	9
1		7·30	6·53	7·73	1·79	6·41	6·90	8·38	3·24
2			4·65	5·21	7·30	7·30	3·66	4·72	7·30
3				6·01	6·53	2·98	2·63	8·38	6·21
4					7·73	7·73	7·73	4·48	7·73
5						4·62	6·90	8·38	4·68
6							3·68	8·38	3·80
7								8·38	6·90
8									8·38
9									

$k = 3$

$\hat{\Delta}_{1/2}(d, B_3(d)) = 0·227$

	1	2	3	4	5	6	7	8	9
1		9·28	6·53	9·28	1·79	6·41	6·98	9·30	3·24
2			4·65	5·21	9·28	7·30	3·66	4·72	9·28
3				6·01	7·13	2·98	2·63	9·30	6·21
4					9·28	7·73	8·37	4·48	9·28
5						4·62	6·90	9·30	4·68
6							3·68	9·30	3·80
7								8·38	7·13
8									9·30
9									

$k = 4$

$\hat{\Delta}_{1/2}(d, B_4(d)) = 0·146$

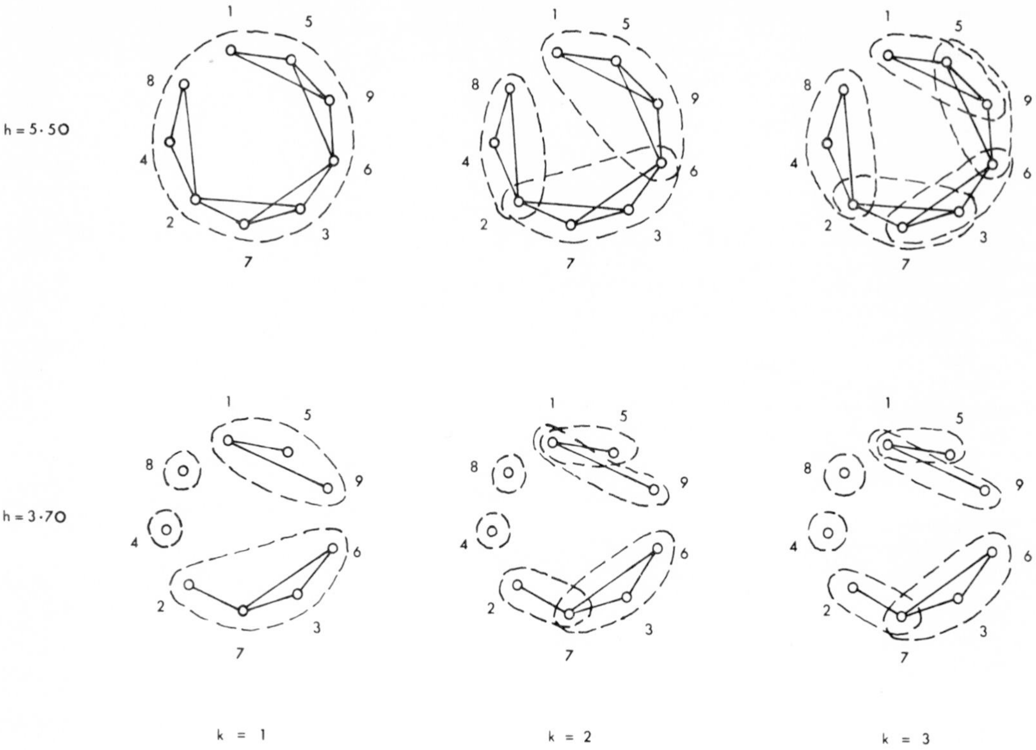

Fig. 2. The graph representations of $(B_k(d))$ for $k = 1$, 2 and 3 at levels $h = 3{\cdot}70$ and $5{\cdot}50$.

of points in Fig. 2 was produced by modification of the two-dimensional arrangement obtained using a program for non-metric multidimensional scaling written by J. B. Kruskal.

In deciding on the level or levels at which clusters obtained in a k-dendrogram should be recognised as classes in a classification of the set of objects, it is helpful to devise measures of the relative *isolation*, and of the relative *internal cohesion* (homogeneity) of the clusters. One possible measure of isolation is given by $(h_1 - h_0)$, where $h_1 = \min\{d(a, b) : a \in P-S, b \in S\}$, and h_0 is the lowest level at which the cluster S appears in the k-dendrogram; this measure generalises a measure suggested by Wirth *et al.* (1966). An appropriate measure of the internal cohesion of a cluster in a k-dendrogram at level h is given by

$$[\Sigma_+ (d(a, b) - h)]/\tfrac{1}{2}s(s-1)$$

where $s = |S|$, $d(a, b)$ is the value of the dissimilarity coefficient on the pair of objects (a, b) from S, and Σ_+ is summation over positive terms.

4. An illustrative example

In **Table 1** a dissimilarity coefficient on nine populations of the plant *Sagina apetala* is shown. The coefficient is the Generalised Distance computed from the values taken by twenty parameters on thirty individuals in each population; pooled variances and covariances were used (see Mahalanobis, 1936; Rao, 1952). The aim of the study was to find out if the data supports the recognition of three geographical subspecies within *Sagina apetala*.

In **Table 2** the numerical characterisations of the k-dendrograms obtained are shown for $k = 1 \ldots 4$. In each case $\hat{\Delta}_{1/2}(d, B_k(d))$ is given. In **Fig. 2** the clusters obtained for $k = 1 \ldots 3$ at levels $h = 3{\cdot}70$ and $5{\cdot}50$ are shown.

In the hierarchic dendrogram the distortion is very high and the clusters obtained shown low homogeneity. In $B_2(d)$ there is a marked drop in distortion and at $h = 5{\cdot}50$ three clusters having relatively high homogeneity are obtained. Subsequent members of the sequence reveal no useful clusters. The clusters obtained n $B_2(d)$ at $h = 5{\cdot}50$ assign the populations correctly to he subspecies recognised on independent grounds in Clapham and Jardine (1964). Populations 2 and 6 which occur in the overlaps between the clusters are ntermediates which occur where the geographical ranges of two of the subspecies overlap with that of the third. In this case, at least, the use of the sequence of cluster methods (B_k) is vindicated.

5. Conclusions

The theoretical model outlined in this paper is one in which classification is regarded as a two-stage process. The first stage is the derivation of a dissimilarity coefficient. The second stage is the transformation of a dissimilarity coefficient into a classificatory system, and it is this stage that we have considered in detail. The derivation of dissimilarity coefficients from discrete-state (attribute) data, quantitative data, and mixed data is the subject of a forthcoming paper by Jardine and Sibson.

An apparently very different kind of theoretical model was constructed by Watanabé (1965); his information-theoretic model covers a variety of methods that go directly from attribute data to a hierarchic classificatory system (for example the methods of Alexander and Mannheim, 1962; Alexander, 1963; Macnaughton-Smith, Williams, Dale, and Mockett, 1964; Lance and Williams, 1966a). The relation between such methods and methods of the kind that are covered by our model needs further investigation.

One obvious omission in our discussion of the sequence of methods described is the absence of any tests for the significance of the clusters obtained. A promising approach to devising such tests is to compare the clusters obtained by the sequence of cluster methods (B_k) on a given dissimiliarity coefficient, with those obtained by application of (B_k) to dissimilarity coefficients generated from some kind of random data.

We do not claim that the sequence of methods described here constitutes any kind of unique solution to the problems of cluster analysis. We do, however, suggest that the construction of mathematical models of the kind outlined here is essential if cluster methods are to be used in scientific investigations; for very little can be done to establish the empirical structure of data until the mathematical properties of the methods of analysis used are known.

Acknowledgements

We acknowledge with gratitude the advice and helpful criticism of Professor D. G. Kendall and Dr. R. M. Needham, and the assistance of Miss A. A. Houston, Mr. C. J. Jardine, Mrs. V. Cole, Mrs. H. M. Hunt, and Miss J. Champion. This work was carried out during the tenure of a grant from the King's College Research Fund by N. Jardine and of an S. R. C. Research Studentship by R. Sibson.

References

Alexander, C. (1963). HIDECS 3: four computer programs for the hierarchical decomposition of systems which have an associated linear graph, *M.I.T. Civil Engineering Systems Lab. Res. Rept.* *R* 63-28, Cambridge, Mass.

Alexander, C., and Mannheim, M. (1962). HIDECS 2: a computer program for the hierarchical decomposition of a set with an associated graph, *M.I.T. Civil Engineering Systems Publ. No.* 160, Cambridge, Mass.

Bonner, R. E. (1964). On some clustering techniques, *IBM Journ. Res. Devpt.*, Vol. 8, p. 22.

Boyce, A. J. (1964). The value of some methods of numerical taxonomy with reference to hominoid classification, in: *Phenetic and phylogenetic classification* (eds. Heywood and McNeill), Syst. Assoc. Publ. No. 6, p. 47.

Clapham, A. R., and Jardine, N. (1964). '*Sagina*' in: *Flora Europaea*, Vol. I (eds. Tutin *et al.*), Cambridge University Press, p. 148.

Estabrook, G. F. (1966). A mathematical model in graph theory for systematic biology, *Journ. Theor. Biol.*, Vol. 12, p. 297.

Jancey, R. C. (1966). Multidimensional group analysis, *Aust. J. Bot.*, Vol. 14, p. 127.

Jardine, C. J., Jardine, N., and Sibson, R. (1967). The structure and construction of taxonomic hierarchies, *Math. Biosc.*, Vol. 1, p. 173.

Jardine, N., and Sibson, R. (1968). A model for taxonomy, *Math. Biosc.* (in press).

Jardine, N., and Sibson, R. The measurement of dissimilarity, (forthcoming paper).

Johnson, S. C. (1967). Hierarchical clustering schemes, *Psychometrika*, Vol. 32, p. 241.

Kruskal, J. B. (1964a). Multidimensional scaling by optimising goodness-of-fit to a nonmetric hypothesis, *Psychometrika*, Vol. 29, p. 1.

Kruskal, J. B. (1964b). Nonmetric multidimensional scaling: a numerical method, *Psychometrika*, Vol. 29, p. 115.

Lance, G. N., and Williams, W. T. (1966a). Computer programs for monothetic classification ('association analysis'), *Computer Journal*, Vol. 8, p. 246.

Lance, G. N., and Williams, W. T. (1966b). A generalised sorting strategy for computer classifications, *Nature*, Vol. 212, p. 218.

Lance, G. N., and Williams, W. T. (1967a). A general theory of classificatory sorting strategies. I. Hierarchical systems, *Computer Journal*, Vol. 9, p. 373.

Lance, G. N., and Williams, W. T. (1967b). A general theory of classificatory sorting strategies. II. Clustering systems, *Computer Journal*, Vol. 10, p. 271.

Macnaughton-Smith, P., Williams, W. T., Dale, M. B., and Mockett, L. G. (1964). Dissimilarity analysis: a new technique of hierarchical subdivision, *Nature*, Vol. 202, p. 1034.

McQuitty, L. L. (1956). 'Agreement analysis', classifying persons by predominant pattern of response, *Brit. Journ. Stat. Psychol.*, Vol. 9, p. 5.

Mahalanobis, P. C. (1936). On the generalised distance in statistics, *Proc. Nat. Inst. Sci. India*, Vol. 2, p. 49.

MINKOFF, E. C. (1965). The effects on classification of slight alterations in numerical technique, *Syst. Zool.*, Vol. 14, p. 196.
NEEDHAM, R. M. (1961). The theory of clumps II, Cambridge Language Research Unit, M.L. 139, (mimeo).
NEEDHAM, R. M. (1965). Application of the theory of clumps, *Mechanical Translation*, Vol. 8, p. 13.
NEEDHAM, R. M. (1967). Automatic classification in linguistics, *Statistician*, Vol. 17, p. 45.
OLSON, E. L., and MILLER, R. L. (1951). A mathematical model applied to a study of the evolution of species, *Evolution*, Vol. 5, p. 256.
PARKER-RHODES, A. J., and NEEDHAM, R. M. (1960). The theory of clumps, Cambridge Language Research Unit, M.L. 126, (mimeo).
RAO, C. R. (1952). *Advanced statistical methods in biometric research*, John Wiley, New York.
SOKAL, R. R., and MICHENER, C. D. (1958). A statistical method for evaluating systematic relationships, *Univ. Kansas Sci. Bull.*, Vol. 38, p. 1409.
SOKAL, R. R., and MICHENER, C. D. (1967). The effects of different numerical techniques on the phenetic classification of bees of the *Hoplitis* complex (Megachilidae), *Proc. Linn. Soc. Lond.*, Vol. 178, p. 59.
SOKAL, R. R., and SNEATH, P. H. A. (1963). *Principles of numerical taxonomy*, W. H. Freeman & Co., San Francisco and London.
SØRENSEN, T. (1948). A method of establishing groups of equal amplitude in plant sociology . . ., *Vidensk. Selsk. Biol. Skr.*, Vol. 5, No. 4, p. 1.
WARD, J. H. (1963). Hierarchic grouping to optimise an objective function, *Journ. Amer. Stat. Assoc.*, Vol. 58, p. 236.
WATANABÉ, S. (1965). Une explication mathématique du classement des objets, in: *Information and prediction in science* (Eds. Dockx and Bernays), Academic Press, New York and London.
WATSON, L., WILLIAMS, W. T., and LANCE, G. N. (1966). Angiosperm taxonomy: a comparative study of some novel numerical techniques, *Journ. Linn. Soc. (Bot.)*, Vol. 59, p. 491.
WILLIAMS, W. T., and DALE, M. B. (1965). Fundamental problems in numerical taxonomy, *Adv. Bot. Res.*, Vol. 2, p. 35.
WIRTH, M., ESTABROOK, G. F., and ROGERS, D. J. (1966). A graph theory model for systematic biology, with an example for the Oncidiinae (Orchidaceae), *Syst. Zool.*, Vol. 15, p. 39.

Part IV

NONMETRIC SCALING

Editors' Comments on Papers 26 and 27

26 **KRUSKAL**
Multidimensional Scaling by Optimizing Goodness of Fit to a Nonmetric Hypothesis

27 **ROHLF**
An Empirical Comparison of Three Ordination Techniques in Numerical Taxonomy

Most of the techniques covered under the general rubric of multivariate data analysis have the basic assumption that relationships among variables or objects can be linearized with minimal loss of information. The emphasis of these methods is to represent derived variables as linear combinations of the original variables in such a way as to maximize some criterion such as explained variation or simple structure. On the other hand, one might simply seek the clearest geometric representation of the data so that interobject relationships may be assessed visually. With this goal in mind, one may relax the linearity assumptions and only require that the interobject relationships in the derived or reduced space be monotonically related to the original pairwise relationships: that is, if object 1 is nearer object 2 than object 3 in the original data, this relationship should also hold in the derived space. The objective of *nonmetric scaling,* as this procedure is known, is then to determine the configuration of objects in [Euclidean] space of minimal dimensions that best represents the original object distances. Thus, one seeks maximal compression of dimensionality while retaining fidelity of the original pairwise monotonic relationships among the objects.

For a given dimensionality, nonmetric methods use an iterative procedure to adjust the position of each object in turn so as to achieve the best possible correspondence with the original distances. A measure of this correspondence is the "stress" imposed upon the data by the configuration: the higher the stress, the greater the distortion in the geometric model. Ideally, one tries to represent the data in few dimensions; however, distortion usually increases with decreasing dimensionality so that the "optimal" solution is a compromise between stress and dimensionality. By plotting stress against dimensionality, the appropriate number of dimensions should be displayed as a more or less abrupt change in the slope of the curve, indicating that additional axes have little affect on reducing distortion. The special case of scaling in one dimension, or seriation, must be treated separately since usual procedures are not guaranteed to converge for this situation.

Nonmetric scaling has several advantages over other methods of ordination. Because less restrictive monotonic rather than linear fidelity is sought, some degree of nonlinearity can be accounted for in the geometric model, so fewer dimensions are usually required for the same degree of data explanation. Another feature is that interval relationships among objects are not required in the original distance matrix, and hence the method allows one to "metricize" nonmetric data. Finally, it tends to distribute distortions evenly across small and large distances, whereas factor analysis and cluster analysis tend to distort preferentially the smaller or larger distances, respectively (Rohlf, 1970).

As with most techniques of multivariate analysis, nonmetric scaling was developed outside biology, in this case principally in sociology and psychology. Although relevant background work can be traced to earlier papers (e.g., Coombs, 1958; Torgerson, 1958), Roger Shepard (1962) can be credited with introducing the present methodology of nonmetric scaling under the name *analysis of proximities*. Kruskal (Paper 26), however, presented more precise formulations of the technique and provided necessary algorithms for computations. Several more recent papers on the subject include Beals et al. (1968), Green et al. (1968), Guttman (1968), and Klahr (1969). For a more comprehensive history and general discussion of the methodology, the reader is encouraged to consult two recent treatises on the subject by Green and Carmone (1970) and Shepard et al. (1972).

Nonmetric techniques are a rather recent development and as yet they have been little utilized in biology. Holloway and Jardine (1968) and Rohlf (1970; Paper 27) have been the first to apply the method to biological data. Recently, Atchley (1974) has used this technique in the solution of a taxonomic problem. Rohlf (Paper 27) compared several methods of *Q*-mode ordination, including principal components analysis, and concluded that Kruskal's technique provided a less distorted geometric picture of object interrelationships than did the other methods. Because of the profusion of nominal and ordinal data in systematics and ecology, nonmetric scaling methods will undoubtedly command greater attention in the future. In addition, even though no applications are currently available, the individual-differences scaling method of Carroll (1969; Carroll and Chang, 1969), which not only scales the objects but also scales the way different individuals view these objects, may have considerable utility in certain areas of biology.

BIBLIOGRAPHY

Atchley, W. R. 1974. A quantitative taxonomic analysis of *Leptoconops torrens* and *L. carteri* (Diptera: Ceratopogonidae). *J. Med. Entomol. 11:*467–470.

Beals, R. W., D. H. Krantz, and A. Tversky. 1968. Foundations of multidimensional scaling. *Psychol. Rev. 75:*127–142.

Carroll, J. D. 1969. *Individual Differences and Multidimensional Scaling.* Bell Telephone Laboratories, Murray Hill, N.J.

———, and J. J. Chang. 1969. *A New Method for Dealing with Individual Differences in Multidimensional Scaling.* Bell Telephone Laboratories, Murray Hill, N.J.

Coombs, C. H. 1958. An application of a nonmetric model for multidimensional analysis of similarities. *Psychol. Rept. 4:*511–518.

Green, P. E., and F. J. Carmone. 1970. *Multidimensional Scaling and Related Techniques in Marketing Analysis.* Allyn and Bacon, Inc., Boston. 203 pp.

———, F. J. Caromone, and P. J. Robinson. 1968. Nonmetric scaling: an exposition and overview. *Wharton Quart. 2:*27–41.

Guttman, L. 1968. A general nonmetric technique for finding the smallest coordinate space for a configuration of points. *Psychometrika 33:*469–506.

Holloway, J. D., and N. Jardine. 1968. Two approaches to zoogeography: A study based on the distribution of butterflies, birds, and bats in the Indo-Australian area. *Proc. Linn. Soc. London 179:*153–188.

Klahr, D. 1969. A Monte Carlo investigation of the statistical significance of Kruskal's nonmetric scaling procedure. *Psychometrika 34:*319–333.

Kruskal, J. B. 1964. Nonmetric multidimensional scaling: a numerical method. *Psychometrika 29:*115–129.

Rohlf, F. J. 1970. Adaptive hierarchical clustering schemes. *Syst. Zool. 19:*58–82.

Shepard, R. N. 1962. The analysis of proximities: multidimensional scaling with an unknown distance function: I. *Psychometrika 27:*125–140; II. *Psychometrika 27:*219–245.

———, A. K. Romney, and S. B. Nerlove, eds. 1972. *Multidimensional scaling: Vol. I, Theory; Vol. II, Applications.* Seminar Press, New York.

Torgerson, W. S. 1958. *Theory and Methods of Scaling.* John Wiley & Sons, Inc., New York. 460 pp.

26

Reprinted from *Psychometrika,* **29**(1), 1–27 (1964)

MULTIDIMENSIONAL SCALING BY OPTIMIZING GOODNESS OF FIT TO A NONMETRIC HYPOTHESIS

J. B. KRUSKAL

BELL TELEPHONE LABORATORIES
MURRAY HILL, N. J.

Multidimensional scaling is the problem of representing n objects geometrically by n points, so that the interpoint distances correspond in some sense to experimental dissimilarities between objects. In just what sense distances and dissimilarities should correspond has been left rather vague in most approaches, thus leaving these approaches logically incomplete. Our fundamental hypothesis is that dissimilarities and distances are monotonically related. We define a quantitative, intuitively satisfying measure of goodness of fit to this hypothesis. Our technique of multidimensional scaling is to compute that configuration of points which optimizes the goodness of fit. A practical computer program for doing the calculations is described in a companion paper.

The problem of multidimensional scaling, broadly stated, is to find n points whose interpoint distances match in some sense the experimental dissimilarities of n objects. Instead of dissimilarities the experimental measurements may be similarities, confusion probabilities, interaction rates between groups, correlation coefficients, or other measures of proximity or dissociation of the most diverse kind. Whether a large value implies closeness or its opposite is a detail and has no essential significance. What is essential is that we desire a monotone relationship, either ascending or descending, between the experimental measurements and distances in the configuration.

We shall refer only to dissimilarities and similarities, but we explicitly include in these terms all the varied kinds of measurement indicated above. We also note that similarities can always be replaced by dissimilarities (for example, replace s_{ij} by $k - s_{ij}$). Since our procedure uses only the rank ordering of the measurements, such a replacement does no violence to the data.

According to Torgerson ([17], p. 250), the methods in use up to the time of his book follow the general two-stage procedure of first using a one-dimensional scaling technique to convert the dissimilarities or similarities into distances, and then finding points whose interpoint distances have approximately these values. The statistical question of goodness of fit is treated separately, not as an integral part of the procedure. Despite the success these methods have had, their rationale is not fully satisfactory. Due to the nature of the one-dimensional scaling techniques available, these methods either accept the averaged dissimilarities or some fixed transformation of them as distances or else use the variability of the data as a critical element in forming the distances.

A quite different approach to multidimensional scaling may be found in Coombs [5]. However, its rationale is also subject to certain criticisms.

A major advance was made by Roger Shepard [15a, b], who introduced two major innovations. First, he introduced as the central feature the goal of obtaining a monotone relationship between the experimental dissimilarities or similarities and the distances in the configuration. He clearly indicates that the satisfactoriness of a proposed solution should be judged by the degree to which this condition is approached. Monotonicity as a goal was proposed earlier [for example, see Shepard ([14], pp.333–334) and Coombs ([5], p. 513)], but never so strongly. Second, he showed that simply by requiring a high degree of satisfactoriness in this sense and without making use of variability in any way, one obtains very tightly constrained solutions and recovers simultaneously the form of the assumed but unspecified monotone relationship. In other words, he showed that the rank order of the dissimilarities is itself enough to determine the solution. (In a later section we state a theorem which further clarifies this situation.) Thus his technique avoids all the strong distributional assumptions which are necessary in variability-dependent techniques, and also avoids the assumption made by other techniques that dissimilarities and distances are related by some fixed formula. In addition, it should be pointed out that Shepard described and used a practical iterative procedure for finding his solutions with the aid of an automatic computer.

However, Shepard's technique still lacks a solid logical foundation. Most notably, and in common with most other authors, he does not give a mathematically explicit definition of what constitutes a solution. He places the monotone relationship as the central feature, but points out ([15a], p. 128) that a low-dimensional solution cannot be expected to satisfy this criterion perfectly. He introduces a measure of departure δ from this condition [15a, pp. 136–137] but gives it only secondary importance as a criterion for deciding when to terminate his iterative process. His iterative process itself implies still another way of measuring the departure from monotonicity.

In this paper we present a technique for multidimensional scaling, similar to Shepard's, which arose from attempts to improve and perfect his ideas. Our technique is at the same statistical level as least-squares regression analysis. We view multidimensional scaling as a problem of statistical fitting—the dissimilarities are given, and we wish to find the configuration whose distances fit them best.

"To fit them best" implies both a goal and a way of measuring how close we are to that goal. Like Shepard, we adopt a monotone relationship between dissimilarity and distance as our central goal. However, we go further and give a natural quantitative measure of nonmonotonicity. Briefly, for any given configuration we perform a monotone regression of distance upon dissimilarity, and use the residual variance, suitably normalized, as our quantitative measure. We call this the *stress*. (A complete explanation is given in the next section.) Thus for any given configuration the stress measures how well that configuration matches the data.

Once the stress has been defined and the definition justified, the rest of the theory follows without further difficulty. The solution is defined to be the best-fitting configuration of points, that is, the configuration of minimum stress.

There still remains the problem of computing the best-fitting configuration. However, this is strictly a problem of numerical analysis, with no psychological implications. (The literature reflects considerable confusion between the main problem of definition and the subsidiary problem of computation.) In a companion paper [12] we present a practical method of computation, so that our technique should be usable on many automatic computers. (A program which should be usable at many large computer installations is available on request.)

In our two papers we extend both theory and the computational technique to handle missing data and certain non-Euclidean distances, including the city-block metric. It would not be difficult to extend the technique further so as to reflect unequal measurement errors.

We wish to express our gratitude to Roger Shepard for his valuable discussions and for the free use of his extensive and valuable collection of data, obtained from many sources. All the data used in this paper come from that collection.

The Stress

In this section we develop the definition of stress. We remark in advance that since it will turn out to be a "residual sum of squares," it is positive, and the smaller the better. It will also turn out to be a dimensionless number, and can conveniently be expressed as a percentage. Our experience with experimental and synthetic data suggests the following verbal evaluation.

Stress	Goodness of fit
20%	poor
10%	fair
5%	good
2½%	excellent
0%	"perfect"

By "perfect" we mean only that there is a perfect monotone relationship between dissimilarities and the distances.

Let us denote the experimentally obtained dissimilarity between objects i and j by δ_{ij}. We suppose that the experimental procedure is inherently symmetrical, so that $\delta_{ij} = \delta_{ji}$. We also ignore the self-dissimilarities δ_{ii}.

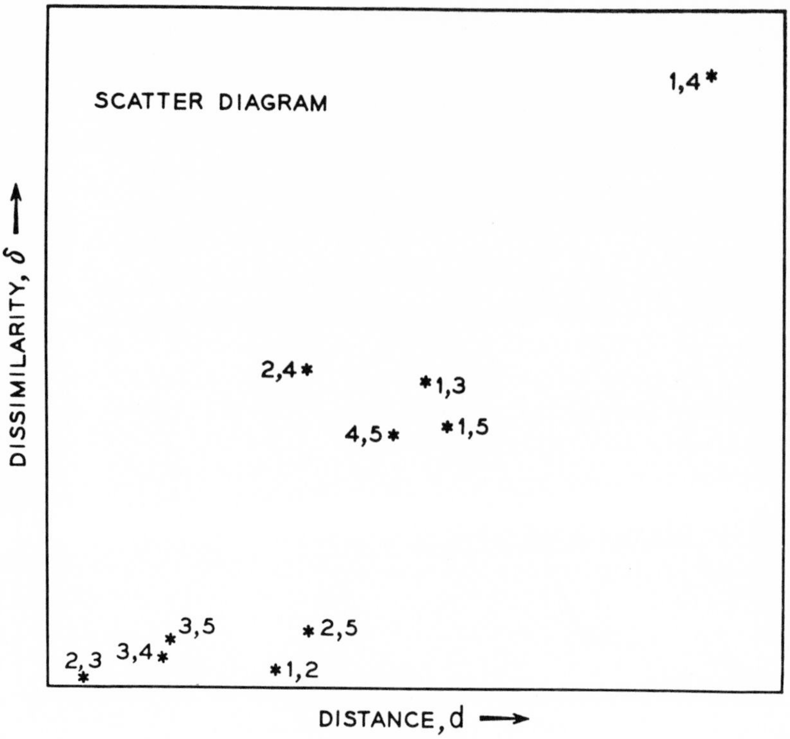

FIGURE 1

Thus with n objects, there are only $n(n - 1)/2$ numbers, namely δ_{ij} for $i < j; i = 1, \cdots, n - 1; j = 2, \cdots, n$. We ignore the possibility of ties; that is, we assume that no two of these $n(n - 1)/2$ numbers are equal. Later in the paper we will be able to abandon every one of the assumptions given above, but for the present they make the discussion much simpler. Since we assume no ties, it is possible to rank the dissimilarities in strictly ascending order:

$$\delta_{i_1 j_1} < \delta_{i_2 j_2} < \delta_{i_3 j_3} < \cdots < \delta_{i_M j_M} .$$

Here $M = n(n - 1)/2$.

We wish to represent the n objects by n points in t-dimensional space. Let us call these points $x_1, \cdots, x_n$. We shall suppose for the present that we know what value of t we should use. Later we discuss the question of determining the appropriate value of t. (Formally and mathematically, it is possible to use any number of dimensions. The appropriate value of t is a matter of scientific judgment.)

Let us suppose we have n points in t-dimensional space. We call this a *configuration*. Our first problem is to evaluate how well this configuration represents the data. Later on we shall want to find the configuration which

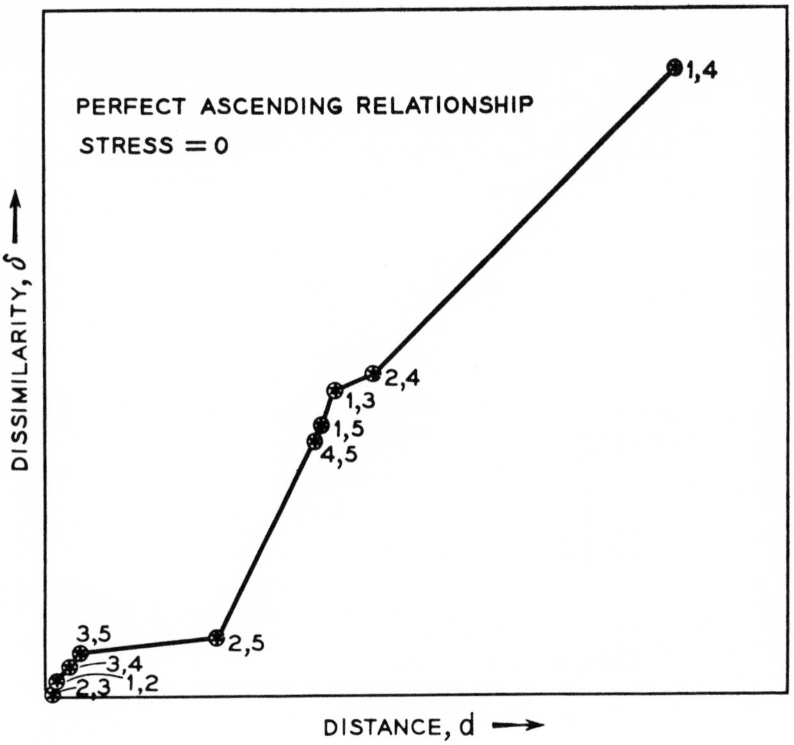

FIGURE 2

represents the data best. At the moment, however, we are only concerned with constructing the criterion by which to judge configurations. To do so, let d_{ij} denote the distance from x_i to x_j. If x_i is expressed in orthogonal coordinates by

$$x_i = (x_{i1}, \cdots, x_{is}, \cdots, x_{it}),$$

then we have

$$d_{ij} = \sqrt{\sum_{s=1}^{t} (x_{is} - x_{js})^2}.$$

In order to see how well the distances match the dissimilarities, large with large and small with small, let us make a scatter diagram (Fig. 1). There are M stars in the diagram. Each star corresponds to a pair of points, as shown. Star (i, j) has abscissa d_{ij} and ordinate δ_{ij}. This diagram is fundamental to our entire discussion. We shall call it simply the *scatter diagram.*

Let us first ask "What should perfect match mean?" Surely it should mean that whenever one dissimilarity is smaller than another, then the

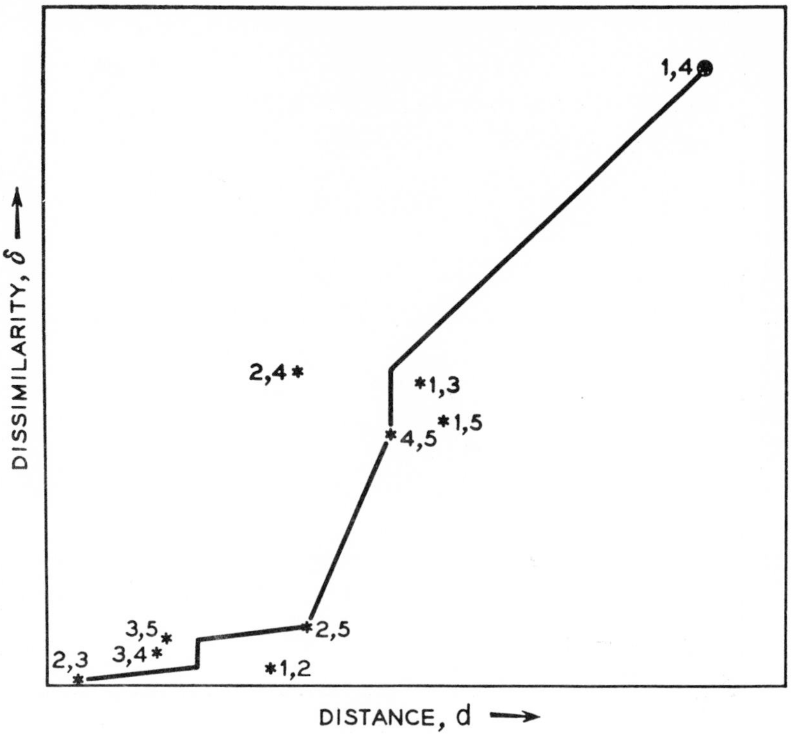

FIGURE 3

corresponding distances satisfy the same relationship. In other words, perfect match should mean that if we lay out the distances d_{ij} in an array

$$d_{i_1j_1}, d_{i_2j_2}, d_{i_3j_3}, \cdots, d_{i_Mj_M}$$

corresponding to the array of dissimilarities given above, then the smallest distance comes first, and the other distances follow in ascending order. In terms of the scatter diagram, this means that as we trace out the stars one by one from bottom to top, we always move to the right, never to the left. This fails in Fig. 1, but holds in Fig. 2.

To measure how far a scatter diagram such as Fig. 1 departs from the ideal of perfect fit, it is natural to fit an ascending curve to the stars as in Fig. 3 and then to measure the deviation from the stars to the curve. This is precisely what we do. However, the details are of critical importance.

Should we measure deviations between the curve and stars along the distance axis or along the dissimilarity axis? The answer is "along the distance axis." For if we measure them along the dissimilarity axis, we shall find ourselves doing arithmetic with dissimilarities. This we must not do, because

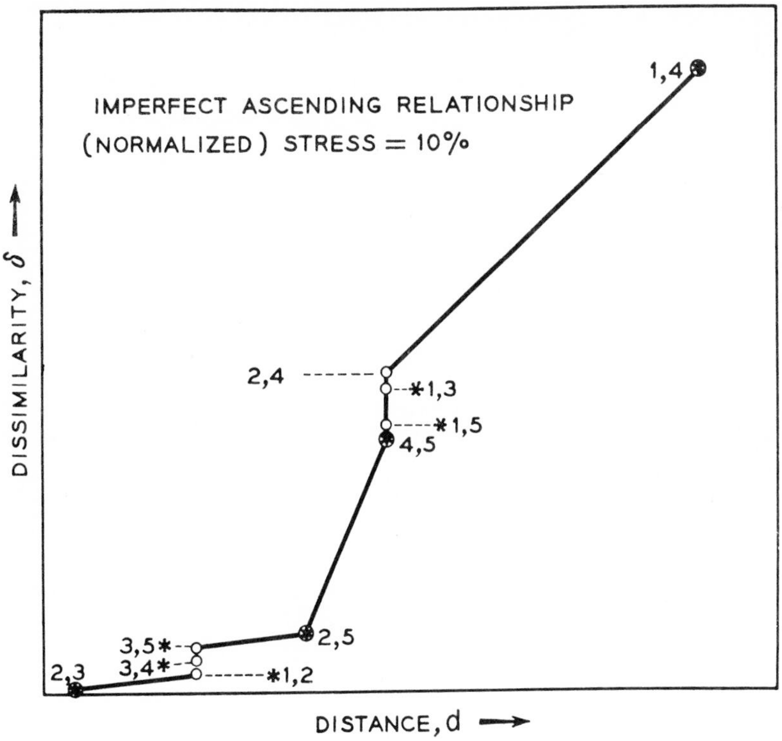

FIGURE 4

we are committed to using only the rank ordering of the dissimilarities. To say the same thing in a different way, we wish to measure goodness of fit in such a way that monotone distortion of the dissimilarity axis will not have any effect. This clearly prevents us from measuring deviations along the dissimilarity axis.

Having decided to measure the deviations along the distance axis, we next see that we do not actually need the whole curve, but only M points on it, as shown in Fig. 4. The rest of the curve does not enter into the calculation of deviations. We may continue to talk of fitting a curve, but all we mean is fitting the points.

Each point we fit shares the value of δ with the corresponding star, but has its own value of d. If a star is located at (d_{ij}, δ_{ij}), then we denote the corresponding point by $(\hat{d}_{ij}, \delta_{ij})$. Thus fitting the curve means no more than fitting the values of $\hat{d}_{ij}$.

We realize of course that the numbers $\hat{d}_{ij}$ are not distances. There is no configuration whose interpoint distances are $\hat{d}_{ij}$. The $\hat{d}_{ij}$ are merely a monotone sequence of numbers, chosen as "nearly equal" to the d_{ij} as possible, which we use as a reference to measure the nonmonotonicity of the numbers

d_{ij} . To simplify the discussion, we delay the precise definition of $\hat{d}_{ij}$ for a little while.

The fitted curve was of course intended to be ascending. Phrased in terms of the M points $(\hat{d}_{ij}, \delta_{ij})$ which in effect constitute the curve, this means that as we trace out these points from bottom to top, we never move to the left but only to the right. Phrased in terms of the numbers $\hat{d}_{ij}$, it means that when they are arranged in the standard order

$$\hat{d}_{i_1 j_1}, \hat{d}_{i_2 j_2}, \hat{d}_{i_3 j_3}, \cdots, \hat{d}_{i_M j_M},$$

then each $\hat{d}_{ij}$ is greater than or equal to the one before it, namely

$$\hat{d}_{i_1 j_1} \leqq \hat{d}_{i_2 j_2} \leqq \hat{d}_{i_3 j_3} \leqq \cdots \leqq \hat{d}_{i_M j_M} \qquad \text{(Mon)}.$$

Whenever any numbers satisfy these inequalities, we shall say that they are *monotonically related* to the d_{ij} .

Now suppose we have the fitted values $\hat{d}_{ij}$, which satisfy (Mon) of course. Then the horizontal deviations are $d_{ij} - \hat{d}_{ij}$. How shall we combine these many individual deviations into a single overall deviation? Following a time-honored tradition of statistics, we square each deviation and add the results:

$$\text{raw stress} = S^* = \sum_{i<j} (d_{ij} - \hat{d}_{ij})^2.$$

Except for normalization, this is our measure of goodness of fit. It measures how well the given configuration represents the data. And very prosaic looking it is too—nothing more than the old familiar "residual sum of squares" associated with so many fitting techniques. It is special in only two ways: first, in the use of distance axis deviations; second, because of the fact that the fitted curve is chosen not from a "parametric" family of curves, such as polynomials or trigonometric series, but from a "nonparametric" family of curves, namely, all monotone ascending curves.

The raw stress still lacks certain desirable properties. Most notably, while it is clearly invariant under rigid motions of the configuration (rotation, translations, and reflections), it is not invariant under uniform stretching and shrinking of the configuration. In other words, if we stretch the configuration $x_1, \cdots, x_n$ by the factor k to the configuration $kx_1, \cdots, kx_n$, that is, replace each point $(x_{i1}, \cdots, x_{it})$ by $(kx_{i1}, \cdots, kx_{it})$, then the raw stress changes. In fact, it changes from S^* to k^2S^* because the numbers $\hat{d}_{ij}$ also change by the factor k. Surely sheer enlargement of a configuration should not change how well it fits the data, for the relationships between the distances do not change. An obvious way to cure this defect in the raw stress is to divide it by a scaling factor, that is, a quantity which has the same quadratic dependence on the scale of the configuration that raw stress does. Such a scaling factor is easily found. We use

$$T^* = \sum_{i<j} d_{ij}^2 .$$

Thus

$$\frac{S^*}{T^*} = \frac{\sum_{i<j} (d_{ij} - \hat{d}_{ij})^2}{\sum_{i<j} d_{ij}^2}$$

is a measure of goodness of fit which has all the desirable properties of S^*, and in addition is invariant under change of scale, that is, uniform stretching or shrinking. This is the normalization. (Another plausible scaling factor is the variance of the numbers d_{ij}. We plan to compare these scaling factors elsewhere.)

Finally, it is desirable to use the square root of this expression, which is analogous to choosing the standard deviation in place of the variance. Thus our definition of the normalized stress is

$$\text{stress} = S = \sqrt{\frac{S^*}{T^*}} = \sqrt{\frac{\sum_{i<j} (d_{ij} - \hat{d}_{ij})^2}{\sum_{i<j} d_{ij}^2}}.$$

Again we emphasize that this measures how well the given configuration represents the data. Smaller stress means better fit. Zero stress means "perfect" fit in our special sense.

Now it is easy to define the $\hat{d}_{ij}$. They are the numbers which minimize S (or equivalently, S^*) subject to the constraint (Mon). Thus we may condense our entire definition of stress into the following formula.

$$\begin{aligned} S(x_1, \cdots, x_n) &= \text{stress of the fixed configuration } x_1, \cdots, x_n \\ &= \min_{\substack{\text{numbers } \hat{d}_{ij} \\ \text{satisfying (Mon)}}} \sqrt{\frac{\sum (d_{ij} - \hat{d}_{ij})^2}{\sum d_{ij}^2}}. \end{aligned}$$

We point out that this minimization is accomplished not by varying a trial set of values for the $\hat{d}_{ij}$, but rather by a rapid, efficient algorithm which is described in detail in the companion paper [12].

Now that we have defined the stress, we have a quantitative way of evaluating any configuration. Clearly the configuration we want is the configuration whose stress is a minimum, for this is the configuration which best fits the data. Thus we define

$$\text{stress in } t \text{ dimensions} = \min_{\substack{\text{all } t\text{-dimensional} \\ \text{configurations}}} S(x_1, \cdots, x_n),$$

and we define the best-fitting configuration to be the one which achieves this minimum stress.

How do we find the minimum-stress configuration? We may answer this question at three levels. At the intuitive level, we may describe the procedure as one of successive approximation. We start with an arbitrary configuration, move all the points a little so as to improve it a bit, and then repeat this procedure until we reach the configuration from which no improvement is possible. Typically, anywhere from 15 to 100 such steps are necessary to reach the final configuration. Roughly speaking, we move points x_i and x_j closer together if $\hat{d}_{ij} < d_{ij}$, and apart in the opposite case, so as to make d_{ij} more like $\hat{d}_{ij}$. Of course, each point x_i is subject to many such motions at once, and usually these will be in partial conflict.

At the theoretical level, we see that our problem is to minimize a function of many variables, namely $S(x_1, \cdots, x_n)$. Actually the stress S is a function of nt variables, as each vector x_i has t coordinates. The problem of minimizing a function of many variables is a standard problem in numerical analysis, and to solve it we adopt a widely used iterative technique known as the "method of gradients" or the "method of steepest descent."

Finally, at the practical level, we give in a companion paper [12] all the important details necessary to perform this iterative technique successfully.

An Example

To illustrate these ideas, we use synthetic data based on a 15-point configuration in the plane. Our configuration is shown by the + signs in Fig. 11. It was used by Shepard ([15b], p. 221) and taken by him from Coombs and Kao ([6], p. 222). To create the 105 dissimilarities we applied a monotone distortion to the interpoint distances, and then added independent random normal deviates to the distorted distances. Specifically,

$$\delta_{ij} = -(0.9) \exp [-(1.8)d_{ij}] - 0.1 + \eta_{ij} ,$$

where η_{ij} is normal with mean 0 and standard deviation 0.01.

We analyze these synthetic data in two dimensions ($t = 2$). The arbitrary starting configuration is shown by numbered circles in Fig. 5. (This and many later figures were created automatically by the computer with the aid of the General Dynamics Electronics Model SC-4020 Highspeed Microfilm Printer.) The lines show the motion of the first iteration to the next, slightly better configuration. The stress of the first configuration is 47.3%. After one iteration it is down to 44.3%. After ten iterations the configuration has become that in Fig. 6, with stress 2.92%. (For most practical purposes the calculation could stop here, as the configuration hardly changes after this.) After fifty iterations the minimum-stress configuration shown in Fig. 7 is reached; its stress is 2.48%. The scatter diagrams of these three configura-

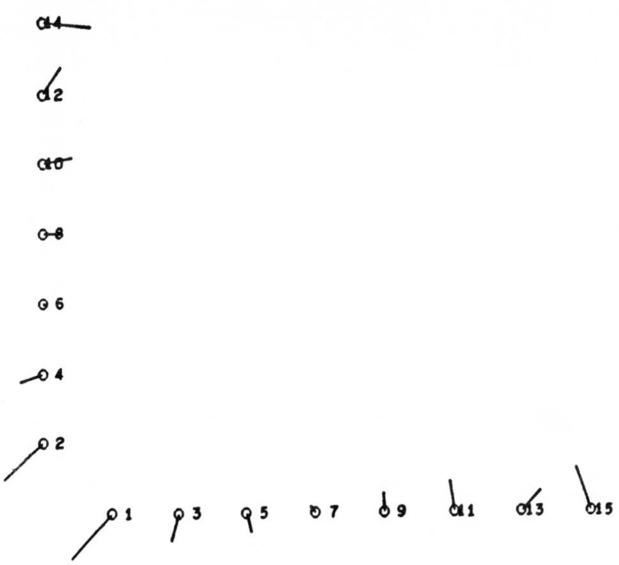

FIGURE 5
Initial Configuration (Coombs and Kao Data)

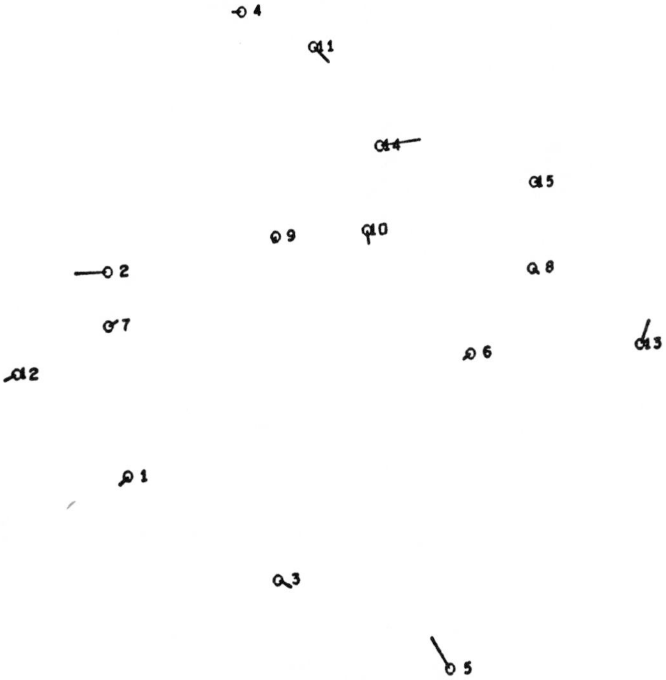

FIGURE 6
Configuration After 10 Iterations (Coombs and Kao Data)

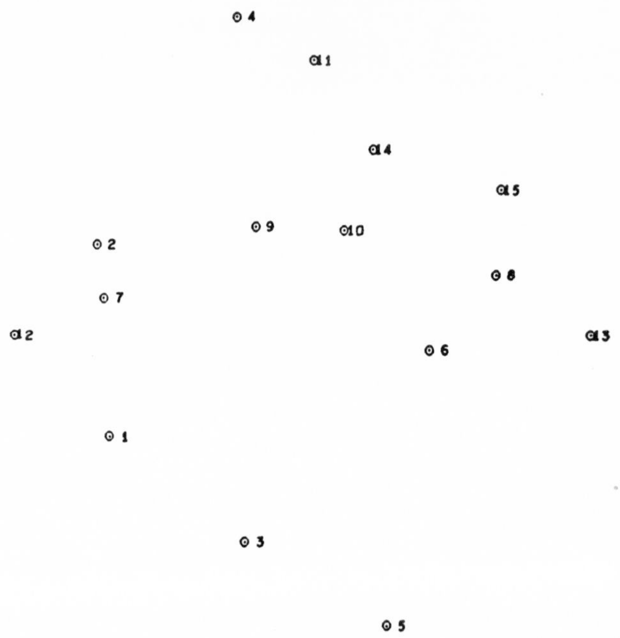

FIGURE 7
Configuration After 50 Iterations (Coombs and Kao Data)

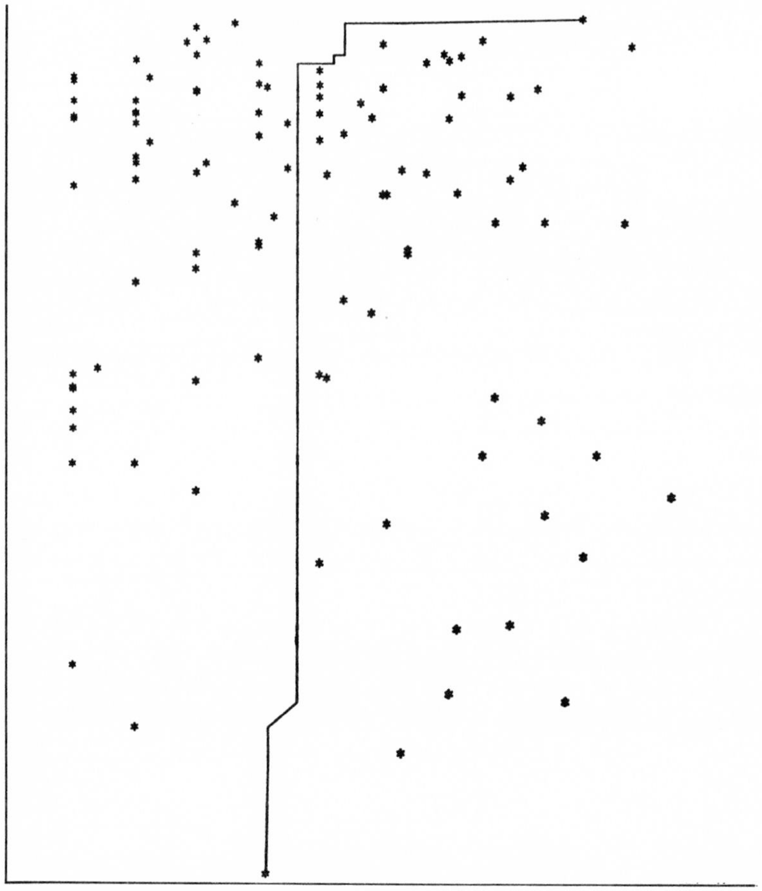

FIGURE 8
Initial Scatter Diagram (Coombs and Kao Data)

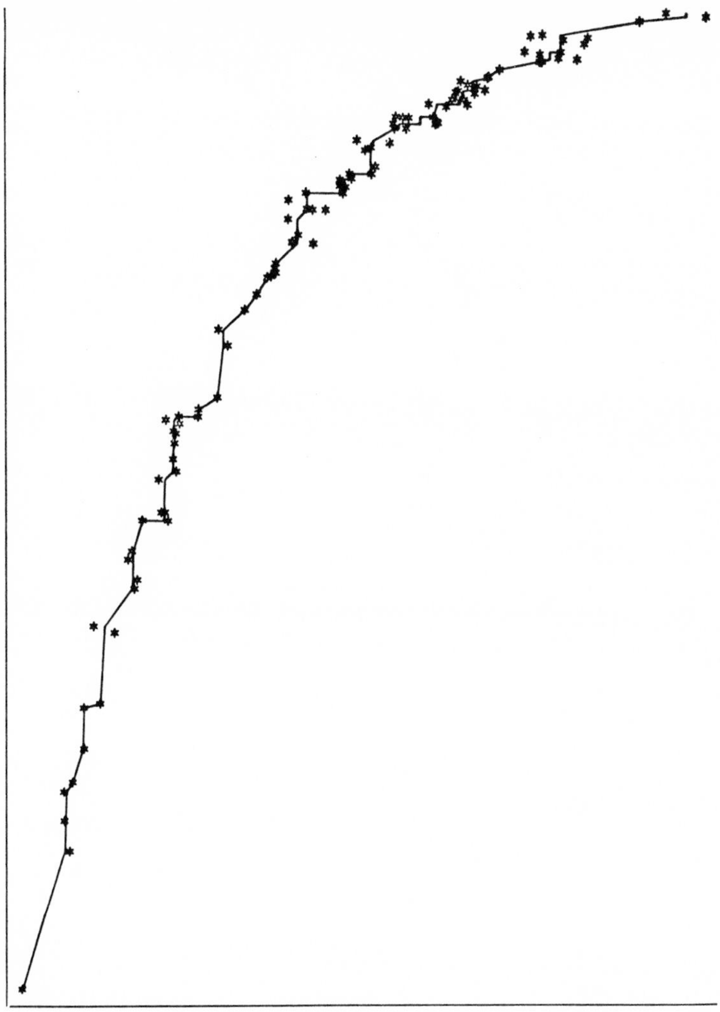

FIGURE 9

Scatter Diagram After 10 Iterations (Coombs and Kao Data)

tions are shown in Figs. 8, 9, and 10. The monotone distorting function has been accurately recovered, and is displayed in the last of these scatter diagrams.

To show how accurately the original configuration has been recovered, we display in Fig. 11 the recovered configuration together with the original configuration of Coombs and Kao. The recovered configuration has been reflected and rotated by eye into best apparent agreement with the original

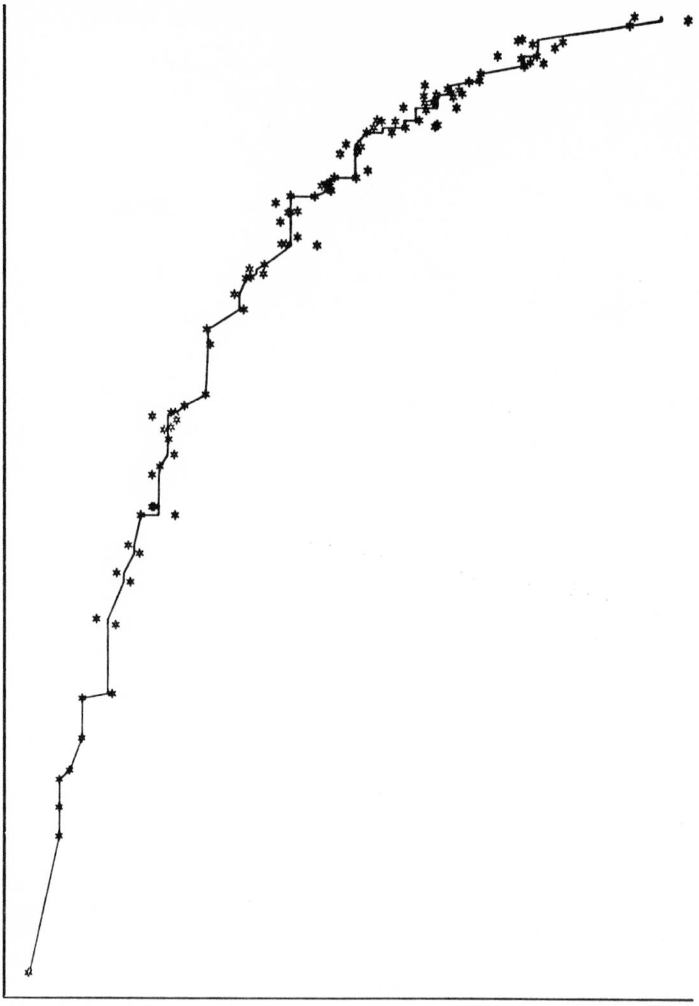

FIGURE 10

Scatter Diagram After 50 Iterations (Coombs and Kao Data)

configuration for this purpose. Since the angular position of the recovered configuration is quite arbitrary, this is entirely legitimate.

Another obvious way of measuring how nearly alike the two configurations are is to compare the distances $d_{ij}^{(1)}$ within one configuration with the distances $d_{ij}^{(2)}$ within the other. Corresponding distances differ typically by 3.16%. More precisely, the expression

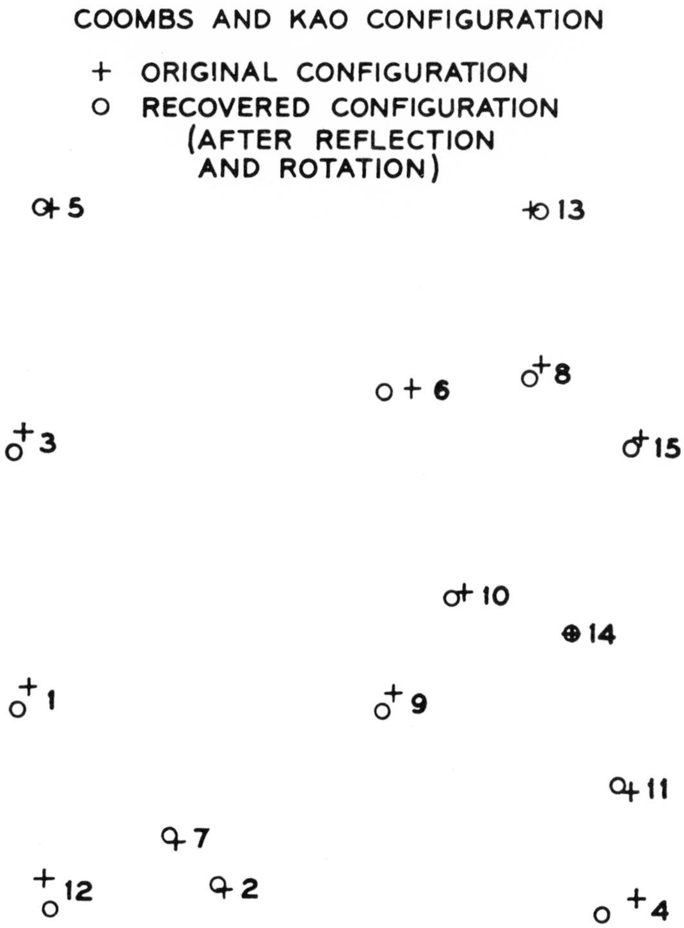

FIGURE 11

$$\sqrt{\frac{\sum_{i<j} (d_{ij}^{(1)} - d_{ij}^{(2)})^2}{\sum_{i<j} \left(\frac{d_{ij}^{(1)} + d_{ij}^{(2)}}{2}\right)^2}}$$

has the value 0.0316.

How Many Dimensions?

So far we have assumed that the number of dimensions to be used is fixed and known. In practice, this is seldom the case. The final determination of how many coordinates to recover from the data rests ultimately with the scientific judgment of the experimenter. However, we can suggest certain aids.

The analysis should be done in several dimensions, and a graph plotted to show the dependence of minimum stress on dimension. Of course, as t increases, minimum stress decreases. For $t \geqq n - 1$, the minimum stress is always 0. (Perfect match can always be managed with n points in $n - 1$ dimensions.) It is reasonable to choose a value of t which makes the stress acceptably small, and for which further increase in t does not significantly reduce stress. Good data sometimes exhibit a noticeable elbow in the curve, thus pointing to the appropriate value of t.

A second criterion lies in the interpretability of the coordinates. If the t-dimensional solution provides a satisfying interpretation, but the $(t + 1)$-dimensional solution reveals no further structure, it may be well to use only the t-dimensional solution. A third criterion can be used if there is an independent estimate of the statistical error of the data. The more accurate the data, the more dimensions one is entitled to extract.

To study the question of dimensionality, we first use synthetic data. Separate sets of ten, fifteen, and twenty random points in six dimensions were chosen. The actual distances were used as dissimilarities δ_{ij} . Fig. 12 shows how stress varies with dimension for these three sets of data. A perfect match is obtained in six dimensions. The ten-point curve displays a distinct elbow, which strongly suggests the use of three dimensions. Of course, with error-free synthetic data, further coordinates may be successfully extracted, but even with excellent experimental data this curve would make the use of more than three dimensions quite dubious. (Examination of the original configuration of ten points shows that by chance it lies very nearly in a three-dimensional subspace.) The fifteen- and twenty-point curves are much less clear. If we obtained curves similar to these but without perfect fit in six dimensions from real data, then three dimensions would seem advisable, four would also seem reasonable, and five might be justified by other considerations, such as good interpretability or independent indications of very low variability in the data.

Let us illustrate these ideas with data from Indow and Uchizono [9]. (The dissimilarities themselves did not appear in the paper. We thank Professor Indow for providing them.) They obtained direct judged dissimilarities between 21 colors of constant brightness, using an ingenious technique. It may seem obvious that the analysis should be done in two dimensions. However, there is the possibility that colors of constant brightness may be best described as lying on a *curved* two-dimensional surface. If this should be the case, we would want $t = 3$. In any case, it is instructive to see what happens. Fig. 13 shows the dependence of stress on dimension. The elbow in the curve at dimension 2 confirms our natural expectation that two dimensions are appropriate, but does not completely rule out the possibility that three dimensions might become appropriate with more comprehensive data of the same sort. Figs. 14 and 15 show the configuration and the scatter diagram

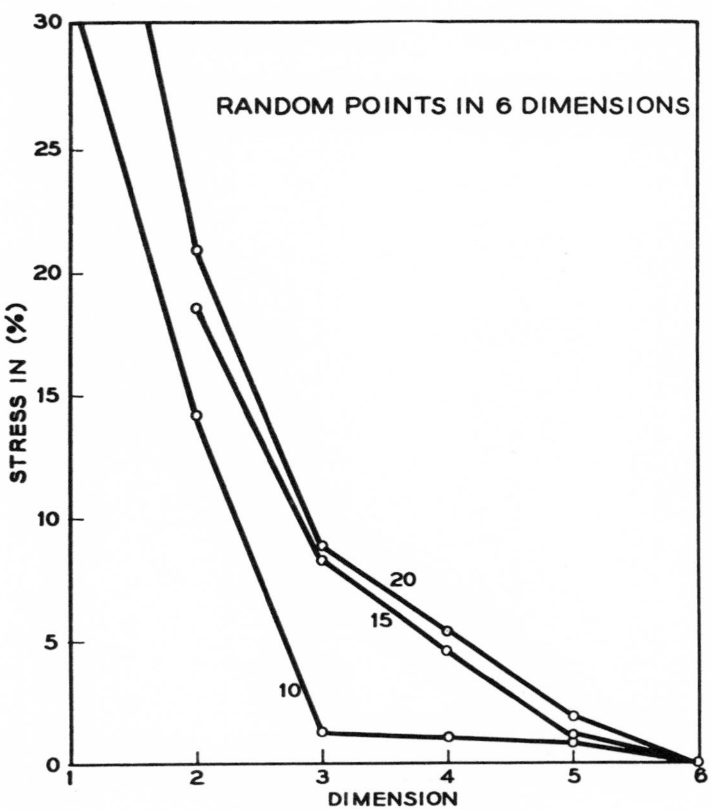

FIGURE 12

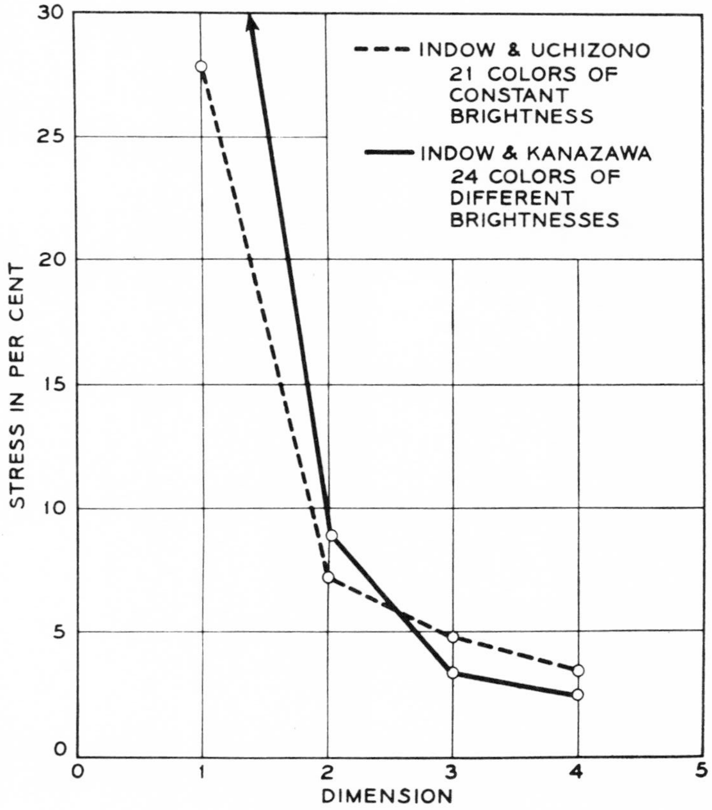

FIGURE 13

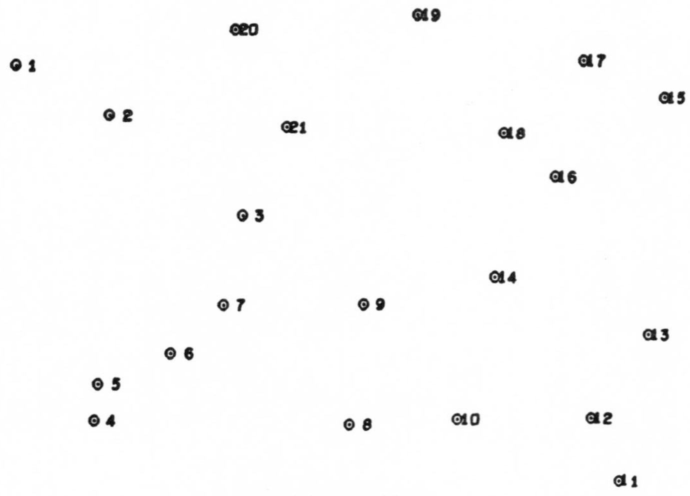

FIGURE 14
Configuration for 21 Colors of Constant Brightness
(Indow and Uchizono)

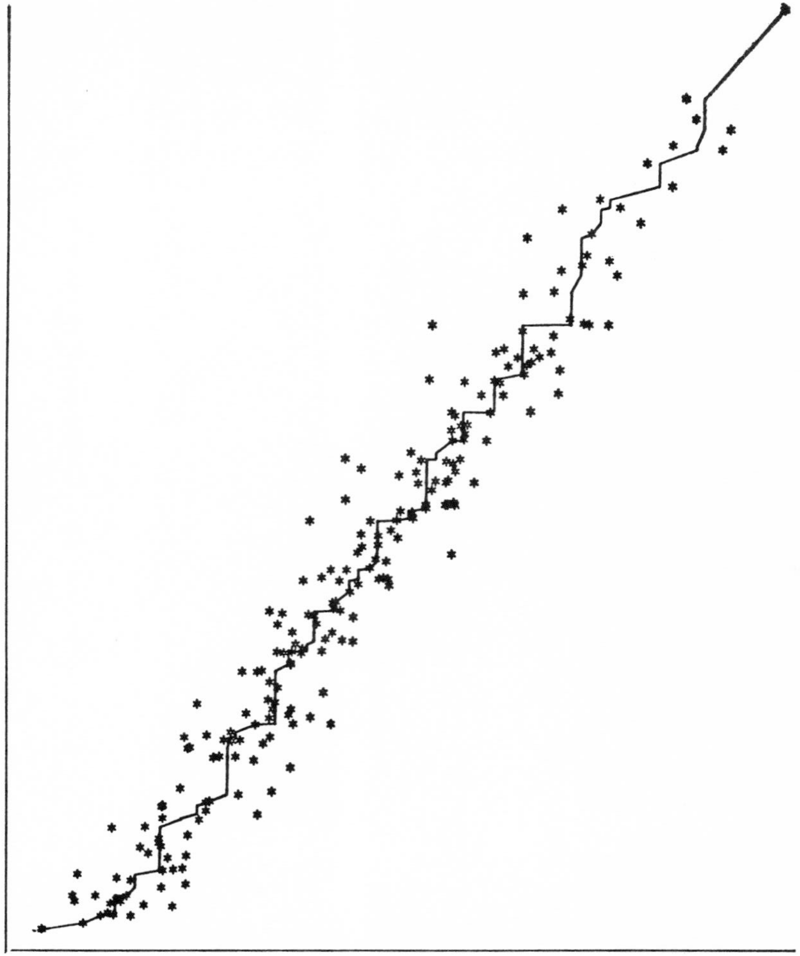

FIGURE 15
Scatter Diagram for 21 Colors of Constant Brightness
(Indow and Uchizono Data)

when the dimension is two. The configuration, which resembles the one given by Indow and Uchizono, corresponds roughly to the Munsell diagram for the 21 colors, but with considerable stretching and shrinking in various places. The scatter diagram, with a stress of 7.27%, would be classified as fair-to-good.

A very similar experiment by Indow and Kanazawa [10] supplies a

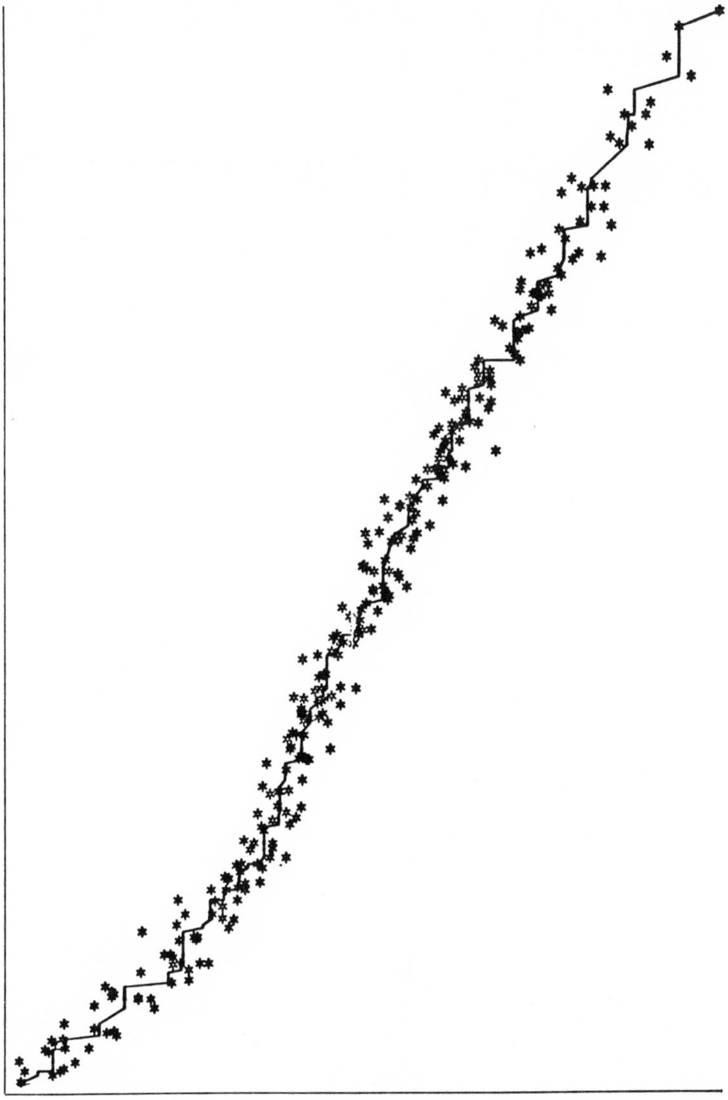

FIGURE 16
Scatter Diagram for 24 Colors of Varying Brightness (Indow and Kanazawa Data)

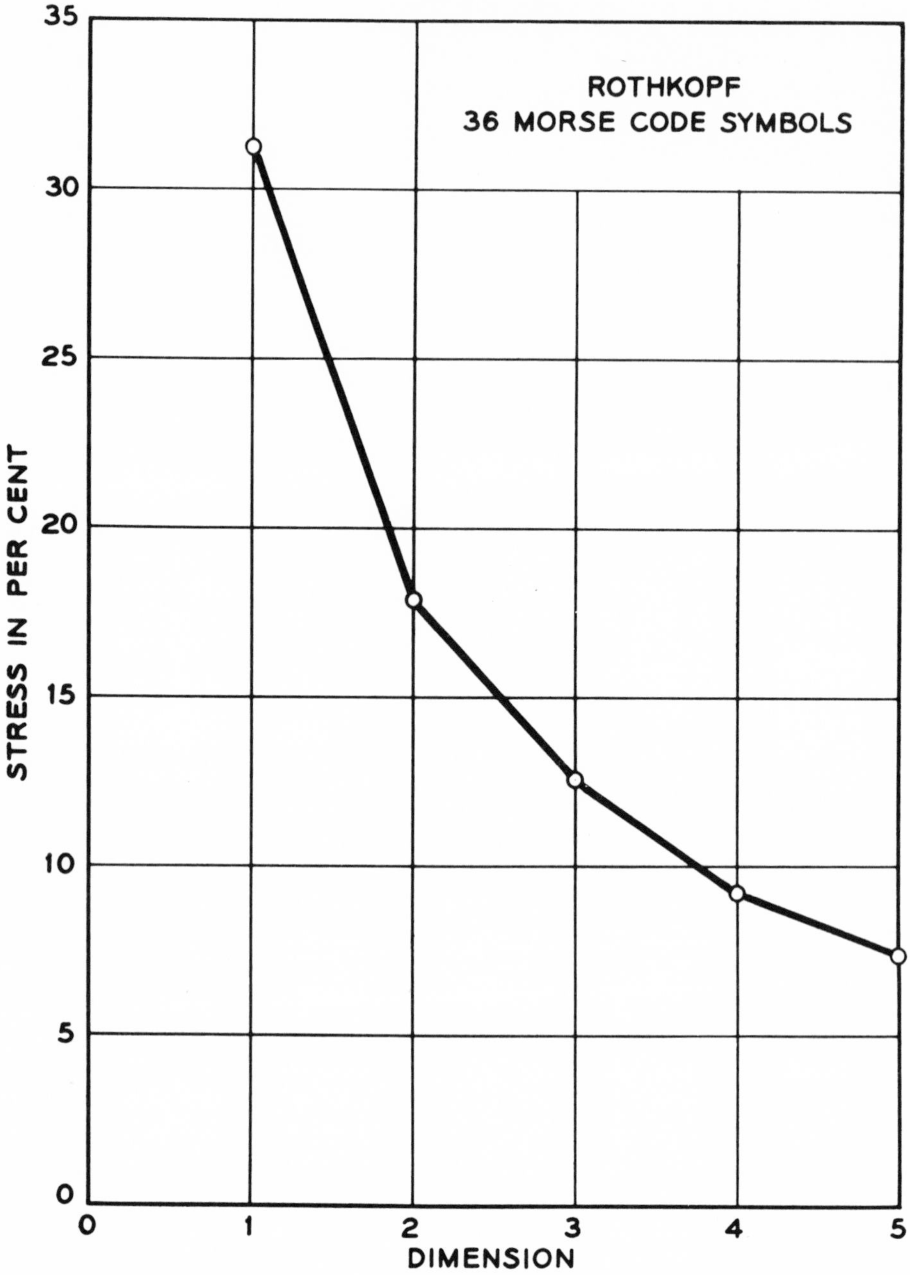
35
30
25
20
15
10
5
0
STRESS IN PER CENT
ROTHKOPF
36 MORSE CODE SYMBOLS
0
1
2
3
4
5
DIMENSION

FIGURE 17

second illustration. In this experiment 24 colors of differing brightness were used. Fig. 13 fits well with our expectation that three dimensions are appropriate. The reason that the stress is fairly small in two dimensions is that after rotation to principal axes the third recovered coordinate varies over only half the range of the first two coordinates. This third coordinate corresponds approximately to brightness. The scatter diagram in three dimensions (Fig. 16) has a stress of 3.67%, and would be classified as fair-to-excellent. Our configuration in three dimensions resembles that obtained by Indow and Kanazawa.

Our third illustration is based on the confusions between 36 Morse Code symbols from E. Rothkopf [13]. An analysis of these and other data, using our technique and our computer program, appears in Shepard [16]. We have calculated the stress of the best-fitting configuration in one, two, three, four, and five dimensions (Fig. 17). The figure does not clearly show the number of dimensions needed, but suggests that two is the minimum and four the maximum. However, Shepard [16] found a very lucid and convincing interpretation for the two-dimensional solution, while he could extract no further structure from the three-dimensional solution. Thus he successfully extracted two coordinates, but expressed some doubt about the value of extracting a third.

Missing Data, Nonsymmetry, and Ties

Suppose some of the dissimilarities are missing, either by error or by design. (When n is large, say $n = 50$ or 60, there are a great many dissimilarities. It may be adequate and economical to obtain data covering only some of them.) How shall we measure stress? It seems natural to generalize the definition given before by simply omitting, both in the numerator S^* and the denominator T^*, the terms which correspond to the missing dissimilarities. We accept this generalization, and incorporate it throughout the rest of the paper.

This idea may be considered simply as a special case of weights being attached to the various dissimilarities to reflect varying uncertainties of measurement. However, we shall not in this paper further pursue this notion of weights, nor certain still more general weighting schemes which come easily to mind.

Suppose that the measurement procedure is not inherently symmetrical, so that $\delta_{ij} \neq \delta_{ji}$. If we are willing to assume that δ_{ij} and δ_{ji} are measurements of the same underlying quantity, and differ only because of statistical fluctuation, then two natural procedures are open to us. One is to form symmetrical measurements by averaging δ_{ij} and δ_{ji}. A more interesting procedure is to generalize the definition of stress by letting the summations for S^* and T^* extend over all $i \neq j$ (rather than $i < j$). Also in some situations the self-dissimilarities δ_{ii} may be meaningful, and one may wish to let the summations include the cases $i = j$.

Suppose there are ties, that is, dissimilarities which by chance are precisely equal to one another. The reader will recall that the numbers $\hat{d}_{ij}$, used in our formula for the stress, were defined as those numbers which minimize S^* subject to the constraint that they are monotonely related to the dissimilarities δ_{ij}. How shall we interpret this constraint in the presence of ties?

There are two approaches. One, which we call the primary approach because it seems preferable, is to say that when $\delta_{ij} = \delta_{kl}$ we do not care which of d_{ij} and d_{kl} is larger nor whether they are equal or not. Consequently we do not wish to downgrade the configuration if $d_{ij} \neq d_{kl}$, and hence do not wish the stress to reflect the inequality. The way we accomplish this is by not constraining $\hat{d}_{ij}$ and $\hat{d}_{kl}$. Consequently the terms $(d_{ij} - \hat{d}_{ij})^2$ and $(d_{kl} - \hat{d}_{kl})^2$ are permitted to be zero, except as prevented by other constraints. Thus in case of the primary approach our only constraints on the $\hat{d}_{ij}$ are these, which are equivalent to (Mon).

$$\text{(I)} \qquad \text{Whenever} \quad \delta_{ij} < \delta_{kl}, \quad \text{then} \quad \hat{d}_{ij} \leqq \hat{d}_{kl}.$$

The secondary approach is to say that $\delta_{ij} = \delta_{kl}$ is evidence that d_{ij} ought to equal d_{kl}, and to downgrade a configuration if this is not so. Consequently the stress ought to reflect this inequality. The way we accomplish this is by imposing the constraint $\hat{d}_{ij} = \hat{d}_{kl}$. Then if $d_{ij} \neq d_{kl}$, the terms $(d_{ij} - \hat{d}_{ij})^2$ and $(d_{kl} - \hat{d}_{kl})^2$ cannot be zero and reflect our displeasure at the inequality of d_{ij} and d_{kl}. Thus in the secondary approach to ties, the constraints on the $\hat{d}_{ij}$ are as follows.

$$\text{(II)} \qquad \begin{cases} \text{Whenever} \quad \delta_{ij} < \delta_{kl}, & \text{then} \quad \hat{d}_{ij} \leqq \hat{d}_{kl}. \\ \text{Whenever} \quad \delta_{ij} = \delta_{kl}, & \text{then} \quad \hat{d}_{ij} = \hat{d}_{kl}. \end{cases}$$

The place in which the difference between these two approaches actually takes effect is deep inside the algorithm for finding the $\hat{d}_{ij}$. Details are given in the companion paper [12]. We remark that it is very simple to build optional use of both approaches into a computer program, and we have done this.

Non-Euclidean Distance

We plan to discuss elsewhere the full degree to which our procedure may be generalized. In principle, there appears to be no reason why the definition of stress could not be used with almost any kind of distance function at all. However, computing the minimum-stress configuration with more general distance functions may offer difficulties.

For a certain class of non-Euclidean distance functions our procedure is quite practical, and has been fully implemented in our computer program. The numerical techniques we describe below fully cover this generalization.

We refer to distance functions generally known in mathematics as the L_p-norms or l_p-norms, but occasionally referred to as Minkowski r-metrics. For any $r > 1$, define the r-distance between points $x = (x_1, \cdots, x_t)$ and $y = (y_1, \cdots, y_t)$ to be

$$d_r(x, y) = \left[\sum_{s=1}^{t} | x_s - y_s |^r \right]^{1/r}.$$

This is just like the ordinary Euclidean formula except that rth power and rth root replace squaring and square root. Then d_r is a genuine distance. In particular, it satisfies the triangle inequality, namely

$$d_r(x, z) \leqq d_r(x, y) + d_r(y, z).$$

[For proof of this fact, see for example Kolmogorov and Fomin ([11], pp. 19–22) or Hardy, Littlewood, and Polya ([8], pp. 30–33).] If $r = 2$, then d_r is ordinary Euclidean distance. If $r = 1$, then d_r is the so-called "city block" or "Manhattan metric" distance.

The Minkowski r-metrics share several properties with ordinary Euclidean distance. In particular, if we displace two points x and y by the same vector z, then the distance between them does not change. In symbols,

$$d_r(x, y) = d_r(x + z, y + z).$$

If we stretch vectors x and y by a scalar factor k, then the distance stretches by a factor k. In symbols,

$$d_r(kx, ky) = kd_r(x, y).$$

However, the Minkowski r-metrics differ sharply from Euclidean distance when rotations are involved. Any rigid rotation leaves Euclidean distances unchanged. The only rigid rotations which leave d_r unchanged in general are those rotations which transform coordinate axes into coordinate axes.

The numerical significance of these properties is brought out in another section. However, we point out here that while a configuration may be freely rotated when Euclidean distances are being used, it may not be when more general distances are used. We do not need to worry explicitly about finding the preferred angular orientation of the configuration, since the iterative minimization process automatically does this for us. However, we must be aware that the coordinate axes have a significance for d_r that they do not have for Euclidean distance.

As an illustration we use experimental data by Ekman [7]. He obtained direct judged similarities of 14 pure spectral colors. We have analyzed his data for several values of r. In every case we obtain the familiar color circle, very similar to the configuration obtained by Shepard [15a], though the precise shape, spacing, and angular orientation varies with r. Fig. 18 shows the stress of the best-fitting configuration as a function of r. We see that a

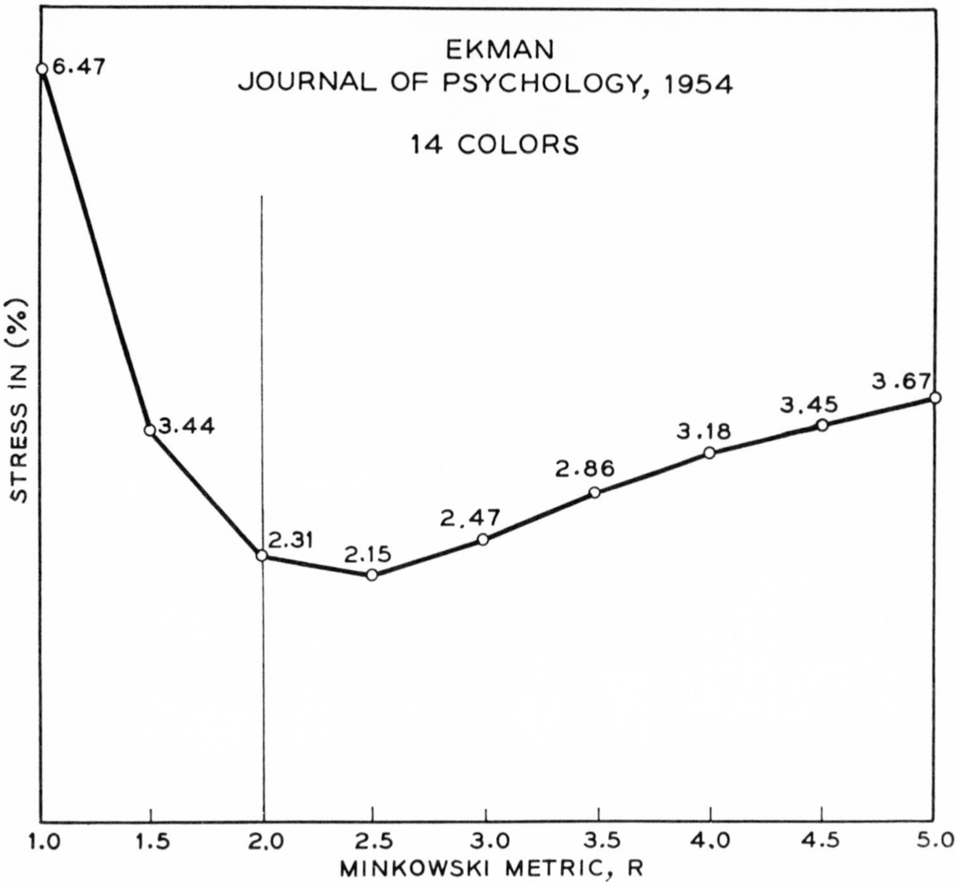

FIGURE 18

value of 2.5 for r gives the best fit. We do not feel that this demonstrates any significant fact about color vision, though there is the hint that subjective distance between colors may be slightly non-Euclidean. However, it illustrates an approach to non-Euclidean distance that could be of significance in various situations.

Miscellaneous Remarks

The idea of recovering metric information from nonmetric information is not new. A quite different application of this idea, as well as a theoretical discussion, can be found in two papers by Aumann and Kruskal [2, 3]. (See particularly pp. 118–120 in the earlier paper.) Though the situation is not presented there as a psychological one, it does not differ from psychological situations in any essential way. The "subjects," called there "The Board" and consisting of Naval officers, are assumed to make certain comparisons, e.g., which of two simple logistic allocations is superior, as a result of some hypothetical quantitative process of which they are not aware. By using a fairly small number of such comparisons, the experimenter determines with limited uncertainty the numerical values which enter into this quantitative process.

Another very interesting discussion of converting nonmetric information into metric information may be found in Abelson and Tukey [1].

In this paper we assume that there is a true underlying configuration of points in Euclidean t-dimensional space, that we can ascertain only the linear ordering of the interpoint distances, and that we wish from this nonmetric information to recover the configuration. Of course, perfect recovery can at best mean construction of a configuration which differs from the original by rigid motions and uniform expansions, for such transformations leave the linear ordering of distances unchanged. Such transformations are called "similarities," and by a known geometrical theorem any transformation in which every distance is multiplied by a fixed constant is a similarity. Thus perfect recovery means construction of a configuration which is geometrically similar to the original.

If the configuration has only a finite number of points, then of course perfect reconstruction is not possible. However, if the number of points is large compared to the number of dimensions, then usually the reconstructed configuration must closely resemble the original. (We note that Shepard was the first to give a practical demonstration that in several dimensions a reasonable number of points are usually tightly constrained.) If the configuration is infinite, perfect recovery may very well be possible. In particular it is possible to prove that if A and B are subsets of Euclidean t-dimensional space (that is, configurations), and if f is a 1-to-1 mapping from A to B which preserves both strict inequality and equality of distances, then f must be a similarity if only A is big enough. A is big enough if it is all of t-space, or if it is a truly t-dimensional convex subset, or even if it is merely a dense subset of the latter.

It is interesting to compare our technique with Shepard's. His iterative procedure closely resembles ours. Indeed, this whole paper is the outcome of the author's attempt to rationalize Shepard's successful iterative procedure. It is possible to describe his procedure in our terms thus. If d_{ij} is the mth largest distance, define $\hat{\delta}_{ij}$ to be the mth largest dissimilarity; instead of making the influence of x_j on x_i proportional to $d_{ij} - \hat{d}_{ij}$ as we do, he makes it proportional to $\hat{\delta}_{ij} - \delta_{ij}$. It does not appear possible to describe his procedure as one which minimizes some particular measurement of nonmonotonicity.

As far as results go, both procedures yield very similar configurations. Shepard's technique yields smoother-looking curves for dissimilarity versus distance. As actually programmed our procedure is substantially faster than Shepard's, but this probably reflects programming improvements rather than anything more fundamental.

It is interesting to read Bartholomew [4], who is concerned with testing whether parameters are equal, subject to the assumption that they are linearly ordered. (See especially p. 37.) His maximum-likelihood estimate of these parameters bears essentially the same relationship to the observations that our $\hat{d}_{ij}$ bear to d_{ij}. Furthermore, his expression U_k, which plays an important role in his paper and in the likelihood ratio, is essentially the same as our raw stress S^*. In fact it might be possible to interpret our minimum-stress configuration as being a maximum-likelihood estimate in some natural sense.

Summary

To give multidimensional scaling a firm theoretical foundation, we have defined a natural goodness of fit measurement which we call the stress. The stress measures how well any given configuration fits the data. The desired configuration is the one with smallest stress, which we find by methods of numerical analysis. The stress of this best-fitting configuration is a measure of goodness of fit.

Shepard first brought out clearly that what we *should* be looking for in multidimensional scaling is a monotone relation between the experimental data and the distances in the configuration. The stress is no more than a quantitative measurement of how well this holds.

REFERENCES

[1] Abelson, R. P. and Tukey, J. W. Efficient conversion of nonmetric information into metric information. *Proc. Amer. statist. Ass. Meetings, Social statist. Section*, 1959, 226–230.

[2] Aumann, R. J. and Kruskal, J. B. The coefficients in an allocation problem. *Naval Res. Logistics Quart.*, 1958, **5**, 111–123.

[3] Aumann, R. J. and Kruskal, J. B. Assigning quantitative values to qualitative factors in the Naval electronics problem. *Naval Res. Logistics Quart.*, 1959, **6**, 1–16.

[4] Bartholomew, D. J. A test of homogeneity for ordered alternatives. *Biometrika*, 1959, **46**, 36–48.

[5] Coombs, C. H. An application of a nonmetric model for multidimensional analysis of similarities. *Psychol. Rep.*, 1958, **4**, 511–518.

[6] Coombs, C. H. and Kao, R. C. On a connection between factor analysis and multidimensional unfolding. *Psychometrika*, 1960, **25**, 219–231.

[7] Ekman, G. Dimensions of color vision. *J. Psychol.*, 1954, **38**, 467–474.

[8] Hardy, G. H., Littlewood, J. E., and Polya, G. *Inequalities*. (2nd ed.) Cambridge, Eng.: Cambridge Univ. Press, 1952.

[9] Indow, T. and Uchizono, T. Multidimensional mapping of Munsell colors varying in hue and chroma. *J. exp. Psychol.*, 1960, **59**, 321–329.

[10] Indow, T. and Kanazawa, K. Multidimensional mapping of colors varying in hue, chroma and value. *J. exp. Psychol.*, 1960, **59**, 330–336.

[11] Kolmogorov, A. N. and Fomin, S. V. *Elements of the theory of functions and functional analysis*. Vol. 1. *Metric and normed spaces*. Translated from the first (1954) Russian edition by Leo F. Boron. Rochester, N. Y.: Graylock Press, 1957.

[12] Kruskal, J. B. Nonmetric multidimensional scaling: A numerical method. *Psychometrika*, (accepted for publication, June, 1964).
[13] Rothkopf, E. Z. A measure of stimulus similarity and errors in some paired-associate learning tasks. *J. exp. Psychol.*, 1957, **53**, 94–101.
[14] Shepard, R. N. Stimulus and response generalization: A stochastic model relating generalization to distance in psychological space. *Psychometrika*, 1957, **32**, 325–345.
[15] Shepard, R. N. The analysis of proximities: Multidimensional scaling with an unknown distance function. *Psychometrika*, 1962, **27**, 125–139, 219–246.
[16] Shepard, R. N. Analysis of proximities as a technique for the study of information processing in man. *Human Factors*, 1963, **5**, 19–34.
[17] Torgerson, W. S. *Theory and methods of scaling.* New York: Wiley, 1958.

Manuscript received 4/11/63
Revised manuscript received 7/16/63

27

Reprinted with permission from *Syst. Zool.*, **21**, 271–280 (1972)

AN EMPIRICAL COMPARISON OF THREE ORDINATION TECHNIQUES IN NUMERICAL TAXONOMY

F. JAMES ROHLF

Abstract

Rohlf, F. James (Ecology and Evolution, State University of New York, Stony Brook, New York 11790) 1972. An empirical comparison of three ordination techniques in numerical taxonomy. Syst. Zool., 21:271–280.—This study reports on comparisons of Kruskal's nonmetric multidimensional scaling analysis, principal components analysis, and Gower's principal coordinates analysis. Nine different sets of data were used. It was found that nonmetric scaling gave better results (as measured by the correlation between the distances in the k-dimensional configuration and the original distances). It was also found that principal coordinates analysis gave a better fit than principal components analysis when missing values were present. [Nonmetric scaling; principal components; principal coordinates.]

INTRODUCTION

This study is an empirical comparison of Kruskal's nonmetric multidimensional scaling analysis (MDSCALE), principal components analysis (PCA), and Gower's principal coordinates analysis (PCRDA). Earlier studies by Holloway and Jardine (1968) and Rohlf (1970) indicated preference for the results obtained by MDSCALE over those obtained with PCA. Comparisons in this study were made over nine data matrices of different sizes, phenetic structure, and taxonomic rank. To make a meaningful comparison of these techniques possible it was necessary to investigate various coefficients which could be used to measure the effectiveness of an ordination technique in summarizing the patterns of phenetic structure found in a similarity or dissimilarity matrix.

MATERIAL AND METHODS

Table 1 describes the battery of nine data sets used in the present study. Due to storage requirements for the multidimensional scaling computer program the data sets employed could not have more than about 70 OTU's. Taxonomically the various data sets range from random samples of individuals within a supposedly homogeneous population (PIGEON data) to species selected to represent variation in the order Lepidoptera (BUTTERFLY data). The data sets also differ considerably in phenetic structure. The PIGEON data show very little structure whereas the RAT data exhibit distinct clusters corresponding to different age groups of individuals.

Three types of ordination analysis were employed:

1. A description of Kruskal's nonmetric multidimensional scaling (MDSCALE) and the algorithms employed can be found in Kruskal (1964a and b) and in the text by Green and Carmone (1970). Examples of its use in biology are given in Holloway and Jardine (1968) and Rohlf (1970). Briefly, MDSCALE places t points in a space of k dimensions so as to minimize stress, S, between the configuration of points and the original distance matrix. Average taxonomic distance, d, (Sokal, 1961) was used as input to MDSCALE and PCRDA in order for the results to be consistent with those obtained by PCA. The equation for stress is

$$S = \sqrt{\frac{\Sigma(d^*_{ij} - \hat{d}_{ij})^2}{\Sigma(d^*_{ij} - \bar{d}^*_{ij})^2}}$$

where d^*_{ij} is the distance between OTU's i and j in the reduced space of k dimensions, $\bar{d}^*$ is the mean of these distances, and $\hat{d}_{ij}$ is the value of d^*_{ij} expected assuming that there is a monotone relationship between the d^*'s and the original d's. MDSCALE starts with either a random configuration of points or a configuration obtained from a previous solution (of dimension $k-1$) and moves all points slightly so as to decrease stress. The process is repeated until either the maximum number of iterations allowed has been reached or stress no longer decreases (as in all cases analyzed in the present study). The presence of numerous NC's (*No Comparisons*, i.e., missing values) in the data causes no particular computational problems. In fact solutions are possible when there are even many missing values in the distance matrix itself as long as the number of defined elements is "large" relative to $k \times t$ (Green and Carmone, 1970).

2. Principal components analysis of the matrix of correlations among the characters with the projections of the OTU's onto the principal axes (PCA) has often been used as an ordination technique in numerical taxonomy (e.g., Basford et al., 1968; Rohlf, 1968, 1970; Seal, 1964).

The computational steps involved are as follows:

a. Compute an $n \times n$ matrix of correlations, **R**, among the n characters.

b. Extract k eigenvectors, **F**, from **R**.

c. Normalize **F** so that the length of each vector is unity.

d. Postmultiply the $k \times n$ matrix **F** by the $n \times t$ standardized data matrix yieldin the $k \times t$ matrix of projections, **P**, of the t OTU's onto the k principal axes.

If NC's are present step a must be modified. Each correlation coefficient must then be computed independently using only those OTU's for which both characters have been recorded. With many NC's one or more correlations may become undefined (as for example in the PUPAE data). Such correlations have been arbitrarily set equal to 0. Data with NC's often results in eigenvectors with negative eigenvalues in step b (fortunately they are usually not among the largest eigenvalues in the matrix). Step d must also be modified; the only practical procedure when there are numerous NC's is to replace each NC by the mean for the character in which it is found (zero for a data matrix standardized by characters).

3. Principal coordinate analysis (PCRDA) is described in Gower (1966) and Green and Carmone (1970). An example of its use is Sokal and Rohlf (1970). In this technique a matrix of similarities or dissimilarities between OTU's is transformed so that the elements of the first k eigenvectors extracted from this transformed matrix can be interpreted as coordinates of the OTU's in a space of k dimensions. If $\mathbf{D} = [d_{ij}]$ is a matrix of Euclidean distances (such as the average taxonomic distance, $d_{ij} = [(1/n)\Sigma(X_{ki} - X_{kj})^2]^{\frac{1}{2}}$ then define $\mathbf{A} = [a_{ij}] = [-\frac{1}{2} d_{ij}^2]$. Then the transformed matrix is $\mathbf{G} = [g_{ij}] = [a_{ij} - \bar{a}_i - \bar{a}_j + \bar{\bar{a}}]$ where $\bar{a}_i$, $\bar{a}_j$, and $\bar{\bar{a}}$ are means of the elements of **A** for variables i, j, and the entire matrix, respectively.

If there are no NC's the matrix of the first k eigenvectors of **G** is proportional to the matrix **P** obtained by PCA. However, even if NC's are present (d_{ij} is then computed using only characters with no NC's for either OTU i or OTU j), the matrix can still be interpreted as giving the coordinates of the t OTU's with respect to k orthogonal axes. This technique is computationally more efficient than PCA when $t < n$. Since some of the matrices analyzed contained numerous NC's (see Table 1) it was of interest to compare these two techniques in these data sets.

The best method for comparing the results of these analyses is not clear. Usually stress is used to measure the efficiency of MDSCALE but the percentage of variation explained, PE, by the first k dimensions is used for PCRDA and PCA. While a percentage of 100% implies a stress of zero, the converse is not true since the two

TABLE 1. DESCRIPTION OF DATA SETS USED IN STUDY.

Code name used in present study		Source	Taxonomic level	Organisms	Char.	OTU's	NC's
A	PUPAE	Rohlf (1967)	family	Mosquitoes	74	45	Yes
B	BUTTERFLY	Ehrlich and Ehrlich (1967)	order	Butterflies	185	33	Yes
C	FROG	Olsen and Miller (1958) Table 77	single population	*Rana pipiens*	50	20	No
D	PIGEON	Olsen and Miller (1958) Tables 79, 80, and 81	single population sampled at different times	Domestic pigeons	26	60	No
E	CORAL	Powers and Rohlf (1971)	order	Hawaiian reef corals of the order Scleractinia	54	54	Yes
F	CAMIN	Rohlf and Sokal (1967)		hypothetical	84	29	Yes
G	RAT	Olsen and Miller (1958), Tables 107, 108 and 109	single population, individuals measured at different ages	*Rattus norvegicus*	33	60	No
H	BEE	Eickwort (1969)	subfamily	genera of augochlorine bees	97	31	Yes
I	NEMATODE	Ferris (unpublished)	—	various nematodes	22	71	No
J	SNAIL	Murray (unpublished)	population sample	*Partula* sp.	54	62	Yes

coefficients measure fundamentally different aspects of goodness of fit. The product moment correlation, r_{dd^*} between d_{ij} and d^*_{ij} for $i < j$ is similar to stress in that it measures the extent to which the d^*'s can be used to summarize the relative magnitudes of the d's. However, r_{dd^*} presumes a linear relationship between d and d^*. Gower (1972) has proposed a new coefficient R^2 (which he now calls M^2 to avoid confusion with the multiple correlation coefficient) which also could be used. It is defined as the sum of the t squared distances between the positions of each OTU as determined by one analysis and its position in a second analysis (he shows rotation, reflection, and translation of one configuration so as to get the maximal fit). An advantage of this technique is that the t squared distances being summed are independent whereas the $t(t-1)/2$ quantities used to compute both r_{dd^*} and S clearly are not. However, Gower (personal communication) has found that minimizing M^2 is equivalent to maximizing PE. Therefore M^2 was not included in the present study. For each data set S and r_{dd^*} were computed for MDSCALE. The percentage of variation explained and r_{dd^*} were computed for PCRDA and PCA based upon $k = 1, 2, 3, \ldots, 10$ dimensions.

All computations were performed using the programs which are a part of the NT-SYS system of numerical taxonomy programs (version III for the IBM 360) developed by the present author with the assistance of John Kishpaugh and David Kirk. The MDSCALE program was based on the program MDSCAL-4M written by Dr. F. J. Carmone, Jr.

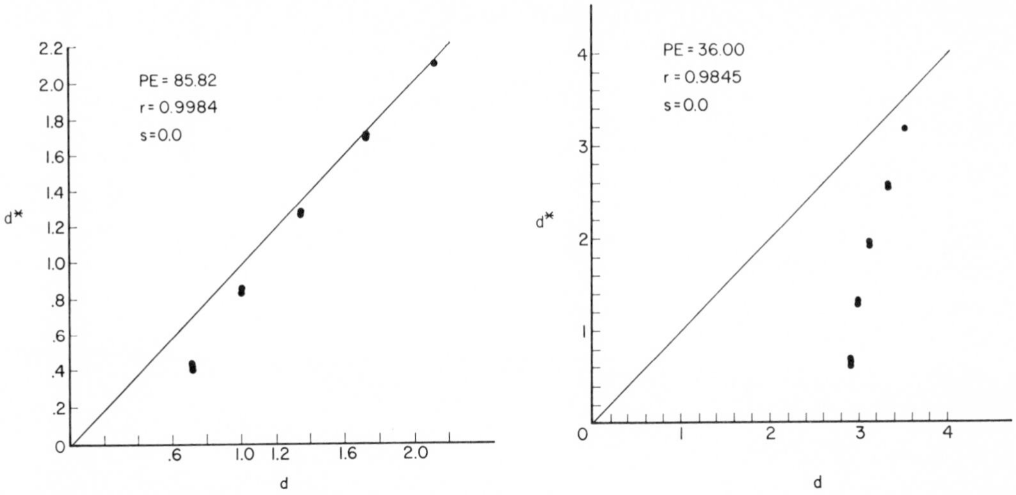

FIG. 1.—Bivariate scatter diagram of d based upon the data in Table 2 and d^* based upon a $k = 1$ dimensional solution from PCRDA. A. data of Table 2a and B. data of Table 2b. Straight line represents the relationship to be expected if $d = d^*$.

RESULTS

In the interpretation of the results the first problem encountered was that of deciding which coefficient to use for comparing the different analyses of the various data sets. The simple example shown in Table 2 and Figure 1 was designed to provide an intuitive understanding of some of the properties of r_{dd^*}, S, and PE. Table 2 contains two sets of artificial data

TABLE 2. TWO SETS OF ARTIFICIAL DATA.

a.		OTU's 1	2	3	4	5
	1	0	1	3	4	5
	2	1	0	0	0	0
Characters	3	0	1	0	0	0
	4	0	0	1	0	0
	5	0	0	0	1	0
	6	0	0	0	0	1

b.		OTU's 1	2	3	4	5
	1	0	1	3	4	5
	2	5	0	0	0	0
Characters	3	0	5	0	0	0
	4	0	0	5	0	0
	5	0	0	0	5	0
	6	0	0	0	0	5

in which the basic pattern of similarities is one dimensional. Characters 2 to 6 simply serve to indicate the extent to which each OTU has its own unique dimensions of variability. In the second data set the OTU's have a much higher level of uniqueness. If the distances between the OTU's for $k = 1$ from either PCRDA or PCA are plotted against distances based directly upon the data, we obtain the result shown in Figure 1a and 1b, respectively. Note how the percentage explained, PE, is strongly affected by the amount of uniqueness (in the sense of factor analysis) present. Stress is zero for both data sets since there is a perfect (but nonlinear) relation between d and d^*. For this reason the traditional use of PE would seem to be of relatively minor value in numerical taxonomy since one usually interprets the distances as giving some kind of a ranking. There seems little interest in their absolute values. Choice between S and r_{dd^*} depends upon the meaningfulness of the possible nonlinear relationships between d and d^*, a question which demands further study. All three coefficients were employed in the present study but S and r_{dd^*} are emphasized.

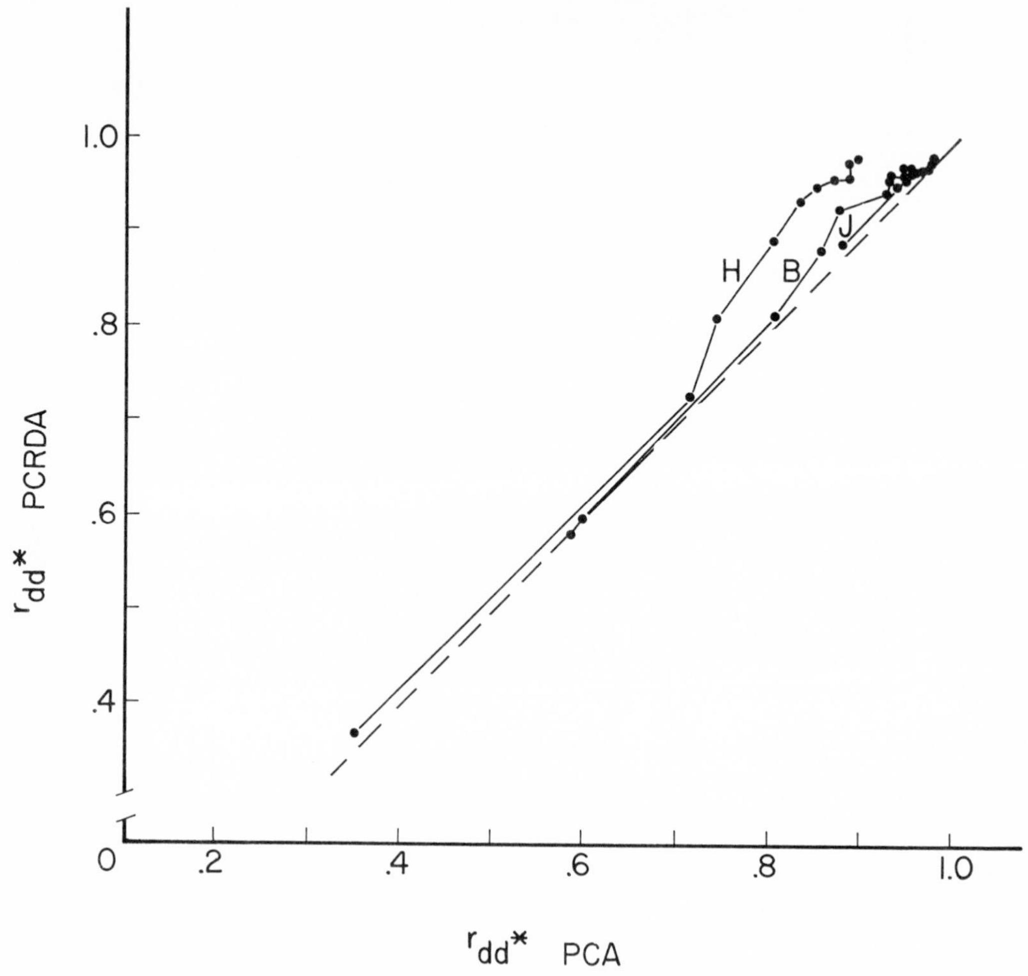

FIG. 2.—Plot of r_{dd^*} (ordinate) based on PCRDA versus r_{dd^*} (abscissa) from PCA for $k = 1, 2, 3, \ldots, 10$ for three data sets (B, H, J) in which NC's are present. Line segments connect results obtained on the same data set for $k = 1, 2, \ldots, 10$. The dashed line represents the relationship which would be obtained if there were no NC's. See Table 1 for a key to the code letters and text for further explanation for this and succeeding figures.

A second problem was that of assessing the relative efficiency of PCRDA and PCA when missing values are present in the data. Figure 2 shows a comparison of r_{dd^*} based upon PCRDA and PCA for three data sets (H, B & J) in which missing values are present. In this figure and succeeding ones, the results obtained for $k =$ 1 to 10 dimensions for a given data set are connected. Due to space limitations the details are not too clear but the overall trends are, I believe, sufficiently apparent. As can be seen PCRDA yielded higher correlations than did PCA (the only exception being for the BUTTERFLY data (B) with $k = 1$. This result implies that the treatment of missing values is more satisfactory in PCRDA. This is perhaps not too surprising due to the rather ad hoc technique used in PCA of simply replacing each missing value by the mean for the corresponding character when computing the projection matrix **P**. Thus one might expect (and it has been noted in practice,

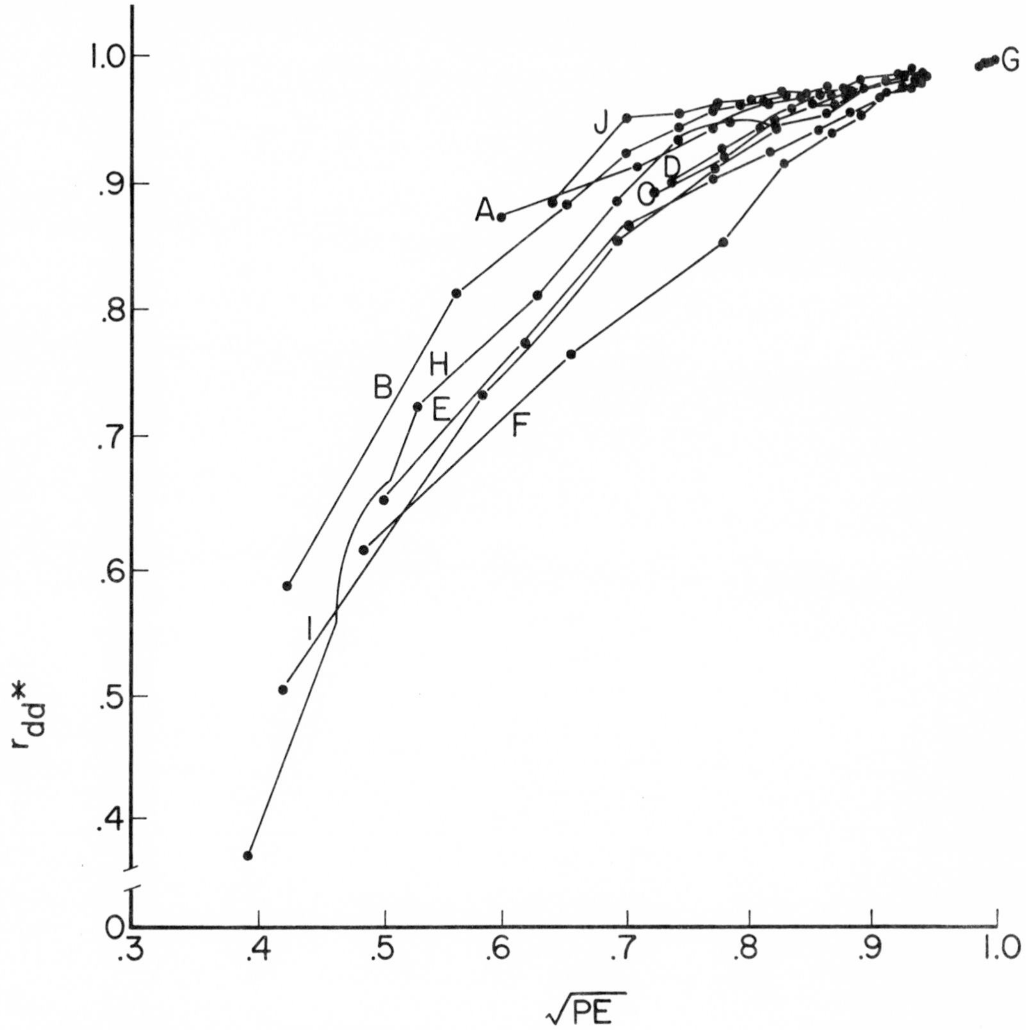

FIG. 3.—Plot of r_{dd*} versus the square root of percentage of variation explained $\sqrt{PE}$, for $k = 1, 2, 3, \ldots, 10$ dimensional solutions for PCRDA various data sets.

e.g., Schnell, 1970) that OTU's with large numbers of missing values are shown to be closer to the centroid by PCA than they are by PCRDA. For these reasons PCRDA can be recommended not only when there are fewer OTU's than characters but in all cases where there is an appreciable number of missing values. Only PCRDA is used in the following analyses for comparison with MDSCALE.

To present the results of a final analysis in which PCRDA and MDSCALE are compared using r_{dd^*} it is necessary to investigate the relationship between r_{dd^*} and $(PE)^{\frac{1}{2}}$ and between r_{dd^*} and S for each of these analyses.

Figure 3 shows a plot of r_{dd^*} versus $(PE)^{\frac{1}{2}}$ for PCRDA for each of the test data sets. The square root of PE was used since it was found to make the relationships more linear and because a squared correlation can be interpreted as a proportion of variation explained (see, for example, Sokal and Rohlf, 1969). Thus it seemed logical

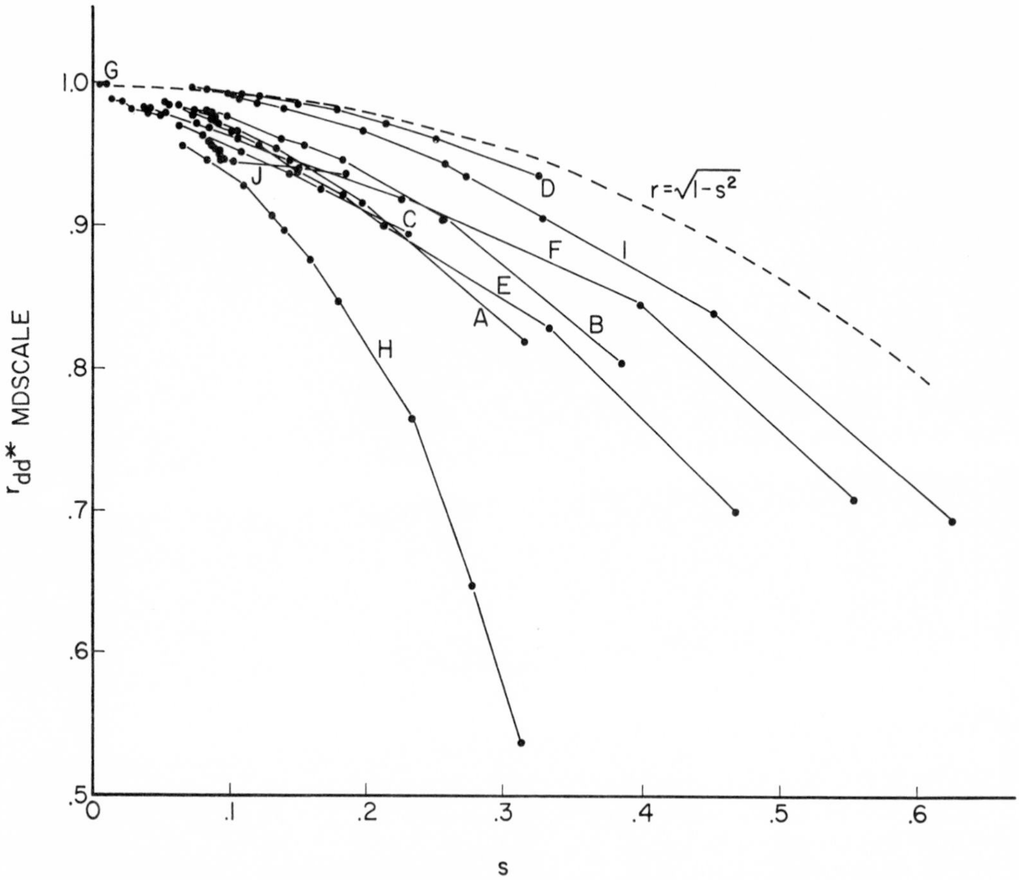

FIG. 4.—Plot of r_{dd*} (ordinate) versus stress (abscissa) for $k = 1, 2, 3, \ldots, 10$ dimensional solutions for MDSCALE for various data sets. The dashed line represents the relation $r = (1 - S^2)^{1/2}$.

to plot either r^2 versus PE, or r versus $(\mathrm{PE})^{\frac{1}{2}}$. As can be observed in Figure 3 there is a monotonic relationship between the two variables within each of the data sets but there is considerable variation among the various data sets. The SNAIL data set (J) has a much higher curve (larger values of r_{dd^*} relative to similar values of $(\mathrm{PE})^{\frac{1}{2}}$ and the CAMIN data set (F) a much lower curve than that obtained for the other data sets. This perhaps implies that the SNAIL data set has a much larger contribution due to unique variation in each species, thus causing $(\mathrm{PE})^{\frac{1}{2}}$ to be much lower than one might expect based upon the value of r_{dd^*} obtained. The RAT data set (G) is distinctive in that the first dimension accounts for almost all of the variation present.

Figure 4 shows a plot of r_{dd^*} versus S for MDSCALE for each of the test data sets. Here the PIGEON data set (D) yielded the highest curve (and also the closest to the $r = (1 - S^2)^{\frac{1}{2}}$ curve representing the relationship expected if the relationship between d and d^* were linear). The BEE data set (H) has the lowest curve, reflecting the fact that the relationship between d and d^* was very nonlinear for these data (the smaller distances were magnified and the larger distances compressed relative to the intermediate distance values in the k dimensional representation). It should also be noted that

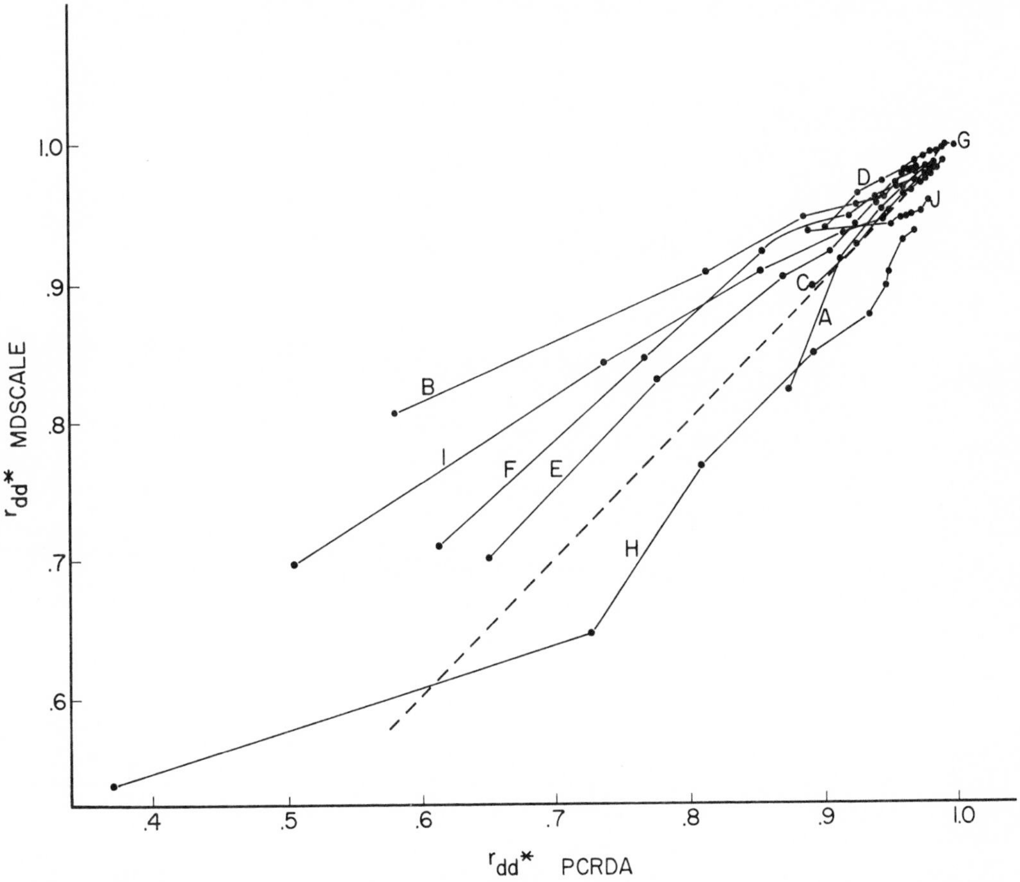

FIG. 5.—Plot of r_{dd^*} values obtained by MDSCALE (ordinate) plotted against those obtained by PCRDA (abscissa). Points above the dashed line represent solutions in which MDSCALE yielded higher correlations than that obtained for PCRDA for the same data set.

the curve for the SNAIL data set (J) (inexplicably) runs counter to the trend shown by the others.

Finally, Figure 5 compares r_{dd^*} obtained for MDSCALE and those obtained for PCRDA for each of the data sets. In most cases MDSCALE yielded higher correlations than did PCRDA. The BEE (H) and SNAIL (J) data are consistent exceptions except for $k=1$. PCRDA yielded a higher correlation for the PUPAE (A) data for $k=1$ only. MDSCALE gave higher correlations except for $k=1$. These exceptional cases for $k = 1$ may be due to the fact that MDSCALE has a tendency to become trapped in local minima in the one-dimensional case in certain data sets (Jardine, 1971). If an optimal one dimensional solution is required then the procedure proposed by Sibson (1971) could be used. MDSCALE was particularly effective in representing the structure in the BUTTERFLY (B) data.

DISCUSSION

The main problem in making adequate comparisons among these methods is to select a proper measure of goodness of fit ("optimality criterion") which can be used on both PCRDA and MDSCALE and which also measures that aspect of fit which one considers most important. The correlation coefficient, r_{dd^*}, has the advantage of simplicity of computation, applicability to results from both sets of analyses, and the fact that it measures relative agreement rather than an absolute agreement between d and d^*. Stress, S, has an advantage in that it takes into account nonlinear (but monotone) relations between d and d^*. But, all things being equal, I believe that one would prefer the relations to be linear unless a good fit is not possible for small values of k.

An examination of the data sets (H, J) which yielded higher r_{dd^*} values for PCRDA than for MDSCALE (Fig. 5) revealed that the relationship between d and d^* obtained by MDSCALE was distinctly nonlinear. In the SNAIL data (J) this was probably due to the fact that there were extremely distinct clusters present in the data. A frequency distribution of the distance coefficients shows two nonoverlapping modes. The MDSCALE configuration greatly emphasized the distinctness of these two modes (see also Anderson, 1971). Figure 4 also indicates that something is aberrant in the way MDSCALE treated these data. In the BEE data (H) there is simply a slight, nonlinear relationship between d and d^*, with no obvious reason for it. In the PUPAE data (A) there is a distinct nonlinear relationship only for $k=1$.

In a sense the comparisons made in the present paper are unfair. PCA and PCRDA are designed to maximize percentage of variation explained whereas MDSCALE minimizes stress. Thus neither can be expected to maximize r_{dd^*}. But this is one of the aims in the present paper—to demonstrate that the currently used criteria are not necessarily appropriate.

The actual configurations of points usually obtained are quite similar for these three techniques. The differences between the results of PCA and PCRDA are in the relative placement of the OTU's with the larger numbers of NC's (PCA places them closer to the centroid). The main differences between the results of MDSCALE and both PCA and PCRDA are: 1) differences between close OTU's are in general shown much more accurately by MDSCALE and 2) the shorter and the longer distances between OTU's are not necessarily shown to the same scale by MDSCALE. A compression of the shorter distances and a stretching of the longer distances was noted by Rohlf (1970, Figure 9d).

Of course other considerations also enter into the question of deciding the optimal technique to be used in practice. The currently available computer programs for MDSCALE require considerable amounts of core storage in order to avoid an exorbitant amount of time spent in input and output of intermediate results. A program for PCRDA or PCA needs only enough space to hold the lower one half of the symmetric matrix to be factored and the factor matrix itself.

The following general recommendations are suggested as a result of the present study: the MDSCALE solution is to be preferred unless there is a large number of OTU's. If there are NC's and/or fewer OTU's than characters then PCRDA should be considered next. Only if there are no NC's and/or many more OTU's than characters should PCA be employed. But as we have seen above there are sometimes exceptional cases so that one should try both MDSCALE and PCRDA in a critical study. A further consideration is the fact that PCA implies the use of "pythagorean distances" only, whereas both MDSCALE and PCRDA work on any Euclidean distance and can often be employed on non-Euclidean distances as well.

ACKNOWLEDGMENTS

This work was supported in part by grant (GB-20496) from the National Science Foundations. This paper is contribution No. 49 of the program in Ecology and Evolution at the State University of New York at Stony Brook. Drs. R. R. Sokal and J. C. Gower read the manuscript and made many helpful suggestions. Michael

Sullivan assisted in the setting up and the processing of the data through the various computer programs on the University's IBM 360/65 computer. Their help is gratefully acknowledged.

REFERENCES

ANDERSON, A. J. B. 1971. Numerical examination of multivariate soil samples. J. Math. Geol. 3:1–14.

BASFORD, N. L., J. E. BUTLER, C. A. LEONE, AND F. J. ROHLF. 1968. Immunologic comparisons of selected Coleoptera with analyses of relationships using numerical taxonomic methods. Syst. Zool. 17:388–406.

EHRLICH, P. R. AND A. H. EHRLICH. 1967. The phenetic relationships of the butterflies I. Adult taxonomy and the nonspecificity hypothesis. Syst. Zool. 16:301–317.

EICKWORT, G. C. 1969. A comparative morphological study and generic revision of the Augochlorine bees (Hymenoptera: Halictidae). Univ. Kansas Sci. Bull. 48:325–524.

GOWER, J. C. 1966. Some distance properties of latent root and vector methods used in multivariate analysis. Biometrika 53:325–338.

GOWER, J. C. 1972. Statistical methods of comparing different multivariate analyses of the same data. *In* Mathematics in the Archaeological and Historical Sciences. Eds. F. R. Hodson, D. G. Kendall, and P. Tantu, Edinburgh: Edinburgh Univ. Press. pp. 138–149.

GREEN, P. E. AND F. J. CARMONE. 1970. Multidimensional scaling and related techniques in marketing analysis. Allyn and Bacon, Boston XV + 203 pp.

HOLLOWAY, J. D. AND N. JARDINE. 1968. Two approaches to zoogeography: A study based on the distribution of butterflies, birds and bats in the Indo-Australian area. Proc. Linn. Soc. London 179:153–188.

JARDINE, N. 1971. Patterns of differentiation between human local populations. Phil. Trans. Roy. Soc. London 263:1–33.

KRUSKAL, J. B. 1964. Multidimensional scaling by optimizing goodness of fit to a nonmetric hypothesis. Psychometrika 29:1–27.

OLSON, E. C. AND R. L. MILLER. 1958. Morphological integration. University of Chicago: Chicago XV + 317 pp.

ROHLF, F. J. 1968. Stereograms in numerical taxonomy. Syst. Zool. 17:246–255.

ROHLF, F. J. 1970. Adaptive hierarchical clustering schemes. Syst. Zool. 19:58–82.

ROHLF, F. J. AND R. R. SOKAL. 1967. Taxonomic structure from randomly and systematically scanned biological images. Syst. Zool. 16: 246–260.

SCHNELL, G. D. 1970. A phenetic study of the suborder Lari (Aves) I. Method and results of principal components analyses. Syst. Zool. 19: 35–57.

SEAL, H. L. 1964. Multivariate statistical analysis for biologists. Wiley: New York XI + 207 pp.

SIBSON, R. 1971. Some thoughts on sequencing methods. Proc. Anglo-Romanian Conf. on Math. in the Historical and Archaeological Sciences, Mamaia, Romania, 1970.

SOKAL, R. R. 1961. Distance as a measure of taxonomic similarity. Syst. Zool. 10:70–79.

SOKAL, R. R. AND F. J. ROHLF. 1969. Biometry. Freeman: San Francisco, XXI + 776 pp.

SOKAL, R. R. AND F. J. ROHLF. 1970. The intelligent ignoramus, an experiment in numerical taxonomy. Taxon 19:305–319.

(Manuscript received January 25, 1972)

AUTHOR CITATION INDEX

SUBJECT INDEX

QH
323.5
M838 Multivariate statistical methods among-groups covariation.